SCHAUM'S OUTLINE OF

THEORY AND PROBLEMS

OF

THERMODYNAMICS
Second Edition

•

MICHAEL M. ABBOTT, Ph.D.
HENDRICK C. VAN NESS, D.Eng.
The Howard P. Isermann Department of Chemical Engineering
Rensselaer Polytechnic Institute

•

SCHAUM'S OUTLINE SERIES

McGRAW-HILL, INC.
New York   St. Louis   San Francisco   Auckland   Bogotá   Caracas   Lisbon
London   Madrid   Mexico City   Milan   Montreal   New Delhi
San Juan   Singapore   Sydney   Tokyo   Toronto

MICHAEL M. ABBOTT is Professor of Chemical Engineering at Rensselaer Polytechnic Institute, from which he received the B.Ch.E. in 1961 and the Ph.D. in 1965. From 1965 to 1969, he worked with Exxon Research and Engineering Company. He is author or coauthor of several books and numerous articles on engineering thermodynamics.

HENDRICK C. VAN NESS is Institute Professor Emeritus of Chemical Engineering at Rensselaer Polytechnic Institute, where he has taught since 1956. He received his B.S. and M.S. degrees from the University of Rochester, and the D.Eng. from Yale University. He has been Visiting Professor at several universities, and is the 1988 recipient of the Warren K. Lewis Award of the American Institute of Chemical Engineers.

 This book is printed on recycled paper containing a minimum of 50% total recycled fiber with 10% postconsumer de-inked fiber. Soybean based inks are used on the cover and text.

Schaum's Outline of Theory and Problems of
THERMODYNAMICS

6 7 8 9 10 11 12 13 14 15 SH SH 9 8 7 6 5 4 3

ISBN 0-07-000042-5

Sponsoring Editor, David Beckwith
Production Supervisor, Denise Puryear
Editing Supervisor, Meg Tobin

Cover design by Amy E. Becker.

Library of Congress Cataloging-in-Publication Data

Abbott, Michael M.
   Schaum's outline of theory and problems of thermodynamics /
Michael M. Abbott, Hendrick C. Van Ness. -- 2d ed.
      p.   cm. -- (Schaum's outline series)
   Includes index.
   ISBN 0-07-000042-5
   1. Thermodynamics.   I. Van Ness, H. C. (Hendrick C.)   II. Title.
III. Title: Theory and problems of thermodynamics.
TJ265.A19   1989
536'.7--dc20
                                          89-2576
                                          CIP

Preface

This outline presents the fundamental principles of classical thermodynamics, and illustrates by numerous worked examples and problems many of their applications in science and engineering. As a supplement, or as a primary textbook, it should prove useful at the undergraduate and first-year graduate level.

Chapters 1 through 5 form the core of the outline, and are appropriate for students in all areas of science and technology. The first and second chapters deal with basic principles and present the two fundamental laws of thermodynamics. Chapter 3 develops the mathematical framework of the subject, and is included at this point largely for subsequent reference. Thus it need not be given detailed study in sequence. Chapters 4 and 5 treat the behavior of PVT systems.

The remaining chapters are somewhat more specialized. Of particular interest to engineers, Chapters 6 and 8 are devoted to the treatment of flow processes. Chapter 7 constitutes an introduction to chemical thermodynamics, and should be especially useful to chemists, chemical engineers, biologists, and materials scientists and engineers.

We acknowledge with thanks the contributions of Professor Charles Muckenfuss, of Debra L. Saucke, and of Eugene N. Dorsey, whose efforts produced computer programs for calculation of the thermodynamic properties of steam and ultimately the Steam Tables of Appendices D and E. These are based on "The 1976 IFC (International Formulation Committee) Formulation for Industrial Use: A Formulation of the Thermodynamic Properties of Ordinary Water Substance," as published in the "ASME Steam Tables," 4th ed., App. I, pp. 11–29, The Am. Soc. Mech. Engrs., New York, 1979.

MICHAEL M. ABBOTT
HENDRICK C. VAN NESS

Contents

CONTENTS

Chapter 1

Fundamental Concepts and First Principles

1.1 BASIC CONCEPTS

Energy

Thermodynamics is concerned with energy and its transformations. The laws of thermodynamics are general restrictions which nature imposes on all such transformations. These laws cannot be derived from anything more basic; they are primitive. Moreover, the expression of these laws requires the use of words that are themselves primitive in that they have no precise definitions and no synonyms. *Energy* is such a word, and it was used in the very first sentence of this paragraph. Energy is a mathematical abstraction that has no existence apart from its functional relationship to variables or coordinates that do have a physical interpretation and that can be measured. For example, the kinetic energy of a given mass of material is a function of its velocity, and it has no other reality.

The first law of thermodynamics is merely a formal statement asserting that energy is conserved. Thus it represents a primitive statement about a primitive concept. Moreover, energy and the first law are coupled: The first law depends on the concept of energy, but it is equally true that energy is an *essential* thermodynamic function precisely because it allows formulation of the first law.

System and Surroundings

Any application of the first law to a discrete portion of the universe requires the definition of a *system* and its *surroundings*. A system can be any object, any quantity of matter, any region of space, etc., selected for study and set apart (mentally) from everything else, which then becomes the surroundings. The systems of interest in thermodynamics are finite, and the point of view taken is *macroscopic* rather than *microscopic*. That is to say, no account is taken of the detailed structure of matter, and only the coarse characteristics of the system, such as its temperature and pressure, are regarded as thermodynamic coordinates. These are advantageously dealt with because they have a direct relation to our sense perceptions and are measurable.

The imaginary envelope which encloses a system and separates it from its surroundings is called the *boundary* of the system, and it may be imagined to have special properties which serve either (*a*) to *isolate* the system from its surroundings or (*b*) to provide for *interaction* in specific ways between system and surroundings. An *isolated* system can exchange neither matter nor energy with its surroundings. If a system is not isolated, its boundaries may permit either matter or energy or both to be exchanged with its surroundings. If the exchange of matter is allowed, the system is said to be *open*; if only energy and not matter may be exchanged, the system is *closed* (but not isolated), and its mass is constant. The exchange of energy can occur by two modes, *heat* and *work*.

Potential energy and kinetic energy are considered in both mechanics and thermodynamics. These forms of energy result from the position and motion of a system as a whole and are regarded as the *external* energy of the system. The special province of thermodynamics is the energy *interior* to matter, energy associated with the internal state of a system, and called *internal energy*. When a sufficient number of thermodynamic coordinates, such as temperature and pressure, are specified, the internal state of a system is determined and its internal energy is fixed.

State

When a system is isolated, it is not affected by its surroundings. Nevertheless, changes may occur within the system that can be detected with measuring devices such as thermometers and pressure gages. However, such changes are observed to cease after a period of time, and the system is said to have reached a condition of *internal equilibrium* such that it has no further tendency to change.

For a closed system which may exchange energy with its surroundings, a final static condition

1

may also eventually be reached such that the system is not only internally at equilibrium but also in *external equilibrium* with its surroundings.

An *equilibrium state* represents a particularly simple condition of a system, and is subject to precise mathematical description because in such a state the system exhibits a set of identifiable, reproducible properties. Indeed, the word *state* represents the totality of macroscopic properties associated with a system. Certain properties are readily detected with instruments, such as thermometers and pressure gages. The existence of other properties, such as internal energy, is recognized only indirectly. An equilibrium state of a system is established (that is, all properties of the system are fixed) when a certain number of properties are arbitrarily set at fixed values. This number depends on the system, and is generally small; it represents the number of properties that may be treated as independent variables or as the thermodynamic coordinates of the system.

Any system exhibiting a set of identifiable properties has a thermodynamic state, whether or not the system is at equilibrium. Moreover, the laws of thermodynamics have general validity, and their application is not limited to equilibrium states. The importance of equilibrium states in thermodynamics derives from the fact that a system at equilibrium exhibits a set of *fixed* properties which are independent of time and which may therefore be measured with precision. Moreover, such states are readily reproduced from time to time and from place to place.

Process

When a closed system is displaced from equilibrium, it undergoes a *process*, during which its properties change until a new equilibrium state is attained. During such a process the system may be caused to interact with its surroundings so as to exchange heat and/or work in a way that produces in the system or surroundings changes considered desirable.

Dimensions and Units

The fundamental and primitive concepts which underlie all physical measurements, and all properties, are taken to be time, distance, mass, absolute temperature, electric current, amount of substance, and luminous intensity. For these primary dimensions, there must be set up scales of measure, each divided into specific units of size. The internationally accepted basic units for the seven quantities are, respectively, the *second* (s), the *meter* (m), the *kilogram* (kg), the *kelvin* (K), the *ampere* (A), the *mole* (mol), and the *candela* (cd). They form the basis for the SI (Système International) or International System of Units.

The Kelvin scale of absolute temperature is related to the common Celsius scale, which takes 0 °C as the freezing point and 100 °C as the normal boiling point of water, by the equation

$$T/K = t/°C + 273.15$$

When $T = 0$ K, $t = -273.15$ °C; thus the absolute zero of temperature is 273.15 °C below the freezing point of water. Absolute temperatures are required in most thermodynamic calculations.

The amount of substance contained in a system is given as a number of moles. A mole is defined as the amount of substance made up of as many elementary entities (e.g., molecules) as there are atoms in 0.012 kg (12 grams) of carbon-12. This number of atoms is $6.022\,52 \times 10^{23}$, *Avogadro's number*. For example, the mass of oxygen containing this number of molecules is 0.032 kg (32 grams), and this is the *molar mass* of oxygen, formerly called its molecular weight. An amount of oxygen with a mass of 32 kg makes up a kilomole (kmol).

Molar properties are widely used. Thus the molar volume of a substance is the volume occupied by 1 mol, and is expressed in basic SI units as $m^3 \cdot mol^{-1}$. Molar density is the reciprocal of molar volume.

Certain derived quantities are important in thermodynamics; two examples are force and pressure. Force is determined through Newton's second law of motion as the product of mass m and acceleration a:

$$F = ma$$

Thus force is measured in the composite unit $kg \cdot m \cdot s^{-2}$, which for convenience is called the *newton* (N). Pressure is defined as force per unit area, and is measured in $N \cdot m^{-2}$, which for convenience is called the *pascal* (Pa). Thus $1 \text{ Pa} = 1 \text{ N} \cdot m^{-2} = 1 \text{ kg} \cdot m^{-1} \cdot s^{-2}$. An additional pressure unit, widely used in industrial practice, is the *bar*, defined as 100 000 Pa. Its popularity derives from the fact that it is approximately equal to the *standard atmosphere*, defined as 101 325 Pa, the approximate average pressure exerted by the earth's atmosphere at sea level.

The use of multiplier prefixes, such as k for kilo, denoting one thousand times the prefixed unit, is very common. Table 1-1 lists names and symbols for the decimal multiples and submultiples that most frequently appear in applications. Thus we have, for example, that $1 \text{ mm} = 10^{-3} \text{ m}$ and $1 \text{ MPa} = 10^3 \text{ kPa} = 10^6 \text{ Pa} = 10 \text{ bar}$.

Table 1-1 Prefixes for Multiples of SI Units

10^{-12}	pico	p	10^1	deca	da
10^{-9}	nano	n	10^2	hecto	h
10^{-6}	micro	μ	10^3	kilo	k
10^{-3}	milli	m	10^6	mega	M
10^{-2}	centi	c	10^9	giga	G
10^{-1}	deci	d	10^{12}	tera	T

EXAMPLE 1.1 The acceleration of gravity on the surface of Mars is $3.74 \text{ m} \cdot s^{-2}$. The *mass* of a man as determined on earth is 76.2 kg. His *weight* on Mars is the *force* exerted on him by Martian gravity, and is given by

$$F = mg_{\text{Mars}} = (76.2 \text{ kg})(3.74 \text{ m} \cdot s^{-2}) = 285 \text{ kg} \cdot m \cdot s^{-2} = 285 \text{ N}$$

The English engineering system uses units that are related to SI units by fixed conversion factors. Thus the *foot* (ft) is 0.3048 m [i.e., $12 \times (2.54 \text{ cm})$] and the *pound mass* ($lb_m$) is 0.453 592 37 kg. The *pound force* (lb_f) is defined as that force which accelerates 1 pound mass 32.174 feet per second per second. Newton's law must here include a dimensional proportionality constant if it is to be reconciled with this definition. Thus, we write

$$F = \frac{1}{g_c} ma$$

whence

$$1 \text{ lb}_f = \frac{1}{g_c} (1 \text{ lb}_m)(32.174 \text{ ft} \cdot s^{-2})$$

or $g_c = 32.174 \text{ lb} \cdot \text{ft} \cdot \text{lb}_f^{-1} \cdot s^{-2}$. The pound force is equivalent to 4.448 221 6 N.

Since force and mass are different concepts, a pound force and a pound mass are different quantities, and their units cannot be canceled against one another. When an equation contains both lb_f and lb_m, the dimensional constant g_c must also appear in the equation to make it dimensionally correct.

In English engineering units, the Martian acceleration of gravity is $12.27 \text{ ft} \cdot s^{-2}$ and the mass of the man is 168 lb_m. His weight on Mars is therefore

$$F = \frac{1}{g_c} mg_{\text{Mars}} = \left(\frac{1}{32.174 \text{ lb}_m \cdot \text{ft} \cdot \text{lb}_f^{-1} \cdot s^{-2}} \right)(168 \text{ lb}_m)(12.27 \text{ ft} \cdot s^{-2}) = 64.1 \text{ lb}_f$$

Since g, the acceleration of gravity on the earth's surface is approximately equal numerically to g_c, the man's weight on earth, in lb_f, is approximately equal to his mass, in lb_m. However, his weight on Mars, in lb_f, is numerically different from his mass, which is still 168 lb_m.

PVT Systems

The simplest thermodynamic system consists of a fixed mass of a pure isotropic fluid uninfluenced by chemical reactions or external fields. Such systems are characterized by the three measurable coordinates pressure P, volume V, and temperature T, and are called *PVT systems*. However, experiment shows that these three coordinates are not all independent; fixing any two of them determines the third. Thus there must be an *equation of state* that interrelates these three coordinates for equilibrium states. This equation may be expressed in functional form as

$$f(P, V, T) = 0$$

In principle the equation of state may be solved for one of the coordinates in terms of the others; for example, $V = V(P, T)$.

The simplest equation of state is that for an ideal gas:

$$PV = RT$$

where V is molar volume, R is the *universal gas constant*, and T is absolute temperature. Dimensionally, R is clearly (pressure) × (molar volume)/(temperature). A commonly used value is $R = 8.314 \text{ m}^3 \cdot \text{Pa} \cdot \text{mol}^{-1} \cdot \text{K}^{-1}$; equivalents are listed in Appendix B.

1.2 MECHANICAL WORK

Work in thermodynamics always represents an exchange of energy between a system and its surroundings. Mechanical work occurs when a force acting on the system moves through a distance. As in mechanics this work is defined by the integral

$$W = \int F \, dl$$

where F is the component of the force acting in the direction of the displacement dl. In differential form this equation is written

$$\delta W = F \, dl \qquad (1.1)$$

where δW represents a differential quantity of work.

It is not necessary that the force F actually *cause* the displacement dl, but it must be an external force. We adopt the usual sign convention that the value of δW is *negative* when work is done *on* the system and *positive* when work is done *by* the system.

In thermodynamics one often finds work done by force distributed over an area, i.e., by a pressure P acting through a volume V, as in the case of a fluid pressure exerted on a piston. In this event, (1.1) is more conveniently expressed as

$$\delta W = P \, dV \qquad (1.2)$$

where P is an external pressure exerted on the system.

The unit of work, and hence the unit of energy, comes from the product of force and distance, or of pressure and volume. The SI unit of work and energy is therefore the newton-meter, which is called the *joule* (J). This is the one and only internationally recognized unit of energy. Power is the time rate of doing work, and the SI unit is the *watt* (W), which is defined as work done at the rate of $1 \text{ J} \cdot \text{s}^{-1}$.

In the English engineering system the unit of work and energy is often the *foot–pound force*. Another common unit is the Btu. The kilowatt and the *horsepower* are commonly used units of power. Conversion factors are listed in Appendix A.

EXAMPLE 1.2 A gas is confined initially to a volume V_i in a horizontal cylinder by a frictionless piston held in place by latches. When the piston is released, it is forced outward by the internal pressure of the gas acting on the interior piston face. A constant external pressure P acts on the external piston face and resists the motion of the piston. We calculate the work of the system—taken as the piston, cylinder, and the gas—as the gas expands to a final volume V_f such that the piston is in an equilibrium position with the gas pressure equal to the external pressure P.

From (1.2) we have for a constant external pressure P

$$W = P \, \Delta V = P(V_f - V_i)$$

If $P = 140$ kPa and $V_f - V_i = 0.01$ m^3,

$$W = (140)(0.01) = 1.40 \text{ kPa} \cdot \text{m}^3 = 1.40 \text{ kJ}$$

Note that we have made no use of the pressure of the gas in the cylinder. Even if we had been given the initial gas pressure, we could not have used it, because we must deal with forces *external* to the system. Had we taken the gas alone as our system, we would have needed the pressure exerted by the interior face of the piston on the gas and we would have needed to know this pressure as a function of V.

1.3 OTHER MODES OF THERMODYNAMIC WORK

In Section 1.2 we considered work for PVT systems. Different kinds of systems, described by other coordinates, are also important, and they are subject to work done by forces other than pressure. Thus we have electrical work, work of magnetization, work of changing surface area, etc. The total work done on a system can be expressed as

$$\delta W = \sum_i Y_i \, dX_i$$

where Y_i represents a generalized force and dX_i represents a generalized displacement. The summation allows superposition of different work modes.

A major difficulty in application of thermodynamics to new types of systems is in proper identification of the forces and displacements. In general this can be accomplished only by experiment; thermodynamics provides no a priori recipe for identification of the Y_i and X_i. However, as illustrated in the following example, *mechanical* work can be handled with little difficulty by appealing to (*1.1*).

EXAMPLE 1.3 Consider the deformation of a bar of length l subjected to an axial tensile load F. As the force F is applied, the bar elongates, and by (*1.1*) we have

$$\delta W = -F \, dl$$

The minus sign is required to satisfy our sign convention that work done *on* a system be negative. (Both F and dl are considered positive. For a compressive load both F and dl would be negative, and the minus sign would still be required.) Note that the volume of the bar may also change and that work may be done by the hydrostatic pressure exerted by the surrounding atmosphere. However, both the volume change and the pressure are small, and the associated work is negligible.

One commonly expresses the force acting on a bar as a stress, or force per unit area acting within the deformed material:

$$\text{Stress} \equiv \sigma = \frac{F}{A}$$

where A is the cross-sectional area of the bar. The practical unit of stress in the SI is the kPa or the MPa. Also, elongations are referred to a characteristic length. Thus the *natural strain* ε is related to l by

$$d\varepsilon = \frac{dl}{l}$$

The work is now written as $\delta W = -\sigma A l \, d\varepsilon$; since Al is the volume of the bar, $\delta W = -V\sigma \, d\varepsilon$.

For a finite process,

$$W = -\int_{\varepsilon_1}^{\varepsilon_2} V\sigma \, d\varepsilon$$

and a relationship between σ and ε (such as Hooke's law) is required for evaluation of the integral. In many applications V is nearly constant, and may be taken outside the integral.

1.4 HEAT

Heat, like work, is regarded in thermodynamics as energy in transit across the boundary separating a system from its surroundings. However, quite unlike work, heat transfer results from a temperature difference between system and surroundings, and simple contact is the only requirement for heat to be transferred by conduction. Heat is not regarded as being stored in a system. When energy in the form of heat is added to a system, it is stored as kinetic and potential energy of the microscopic particles that make up the system. The units of heat are those of work and energy.

The sign convention used for a quantity of heat Q is opposite to that used for work. Heat *added* to a system is given by a *positive* number, whereas heat *extracted* from a system is given by a *negative* number.

1.5 REVERSIBILITY

Because attention is focused on the system rather than on the surroundings in any application of thermodynamics, the equations of thermodynamics must pertain to the properties of the system. The difficulty here is suggested by the discussion of mechanical work in Section 1.2. Thus, the expression for the work, given by (*1.2*), involves the external pressure P.

However, for a special kind of process, it *is* always possible to base calculations on the properties of the system. This kind of process is termed *reversible*, and it plays a central role in the formulation of thermodynamics. A process is said to be reversible *if its direction can be reversed at any point by an* infinitesimal *change in external conditions*. The implications of this definition can most easily be explained through specific examples.

Discussion of an Adiabatic Process

Figure 1-1 represents a piston/cylinder apparatus which may be operated so as to bring about the expansion or compression of the gas trapped in the cylinder. We imagine the piston and cylinder to be perfect heat insulators so that no heat flows to or from the gas, which is taken as the system. Any process carried out in the absence of such heat transfer is said to be *adiabatic*. Thus we consider here processes driven by mechanical forces only.

The piston is shown initially held in place by a set of small weights, each of mass m, which may be slid from the piston as indicated. For a real apparatus the removal of a single small weight may or may not cause the piston to rise. It depends on the static friction between the piston and cylinder; the piston may well stick. If it does so, we presume that the removal of a number of small weights will allow the piston to break free and to rise rapidly. It then oscillates up and down, with decreasing amplitude, ultimately coming to rest a distance Δl above its initial position. Work is accomplished in this process by virtue of the elevation of the piston and remaining weights against the force of gravity. The oscillations of the piston are damped out by friction between the piston and cylinder and by the gradual conversion both in the gas and in the surrounding air of gross directed motion into chaotic molecular motion. These are *dissipative effects*, and they reduce the ultimate height attainable by the piston; thus the process produces less than the maximum possible work. Clearly, such a process is *irreversible*; an *infinitesimal* change in external conditions is hardly adequate to reverse the direction of a piston moving with finite velocity. All real processes are accompanied by dissipative effects, and all are therefore irreversible.

We can, however, *imagine* processes that are free from dissipative effects. For the process shown in Fig. 1-1, we imagine that the apparatus exists in an evacuated space and that the piston slides without friction in the cylinder. The initial state of the system is now one of equilibrium, with the upward force exerted by the gas on the piston face exactly equal to the downward force of gravity on the piston and weights. If we ignore the very small gravity-induced pressure gradient in the gas, then the pressure in the system is uniform and is identical with the external pressure exerted by the piston face. The sudden removal of a single weight from a system so perfectly in balance causes the piston to rise; moreover, it surely oscillates, only gradually settling down to a new equilibrium position as the result of the damping effect of the viscous gas. This dissipative effect is reduced, but not eliminated, when smaller mass increments are removed from the piston: even the removal of an infinitesimal mass leads to piston oscillations of infinitesimal amplitude and to a consequent

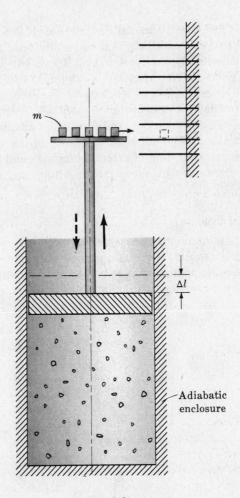

Fig. 1-1

dissipative effect. However, we may imagine a process wherein small mass increments are removed one after the other at a rate such that the piston's rise is continuous, with oscillation only at the end of the process.

The limiting case (removal of a succession of infinitesimal masses from the piston) is approximated when the masses m in Fig. 1-1 are replaced by a pile of powder, blown in a very fine stream from the piston. During this process the piston rises at a uniform but very slow rate, and the powder collects in storage at ever higher levels. The system is never more than differentially displaced from internal equilibrium or from equilibrium with its surroundings. If the removal of powder from the piston is stopped and the direction of transfer of powder is reversed, the process reverses direction and proceeds backward along its original path. Both the system and its surroundings are ultimately restored to their initial conditions in this reverse compression process. The original expansion process is *reversible*; so also is the reverse compression process.

Mechanical Reversibility

An important feature of the reversible process is that the system is never more than infinitesimally displaced from internal equilibrium. This implies that the system is always in an identifiable state of uniform temperature and pressure and that an equation of state is always applicable to the system. In addition the system is also never more than infinitesimally displaced from mechanical equilibrium with its surroundings, and this means that the internal pressure is always in virtual balance with the forces external to the system. Because of these circumstances, for a reversible process the expression for work (1.2), $\delta W = P\,dV$, may be evaluated through use of the *system* pressure for P. This is so, first, because the system pressure is well defined and is uniform, and, second, because the system pressure almost exactly balances the external forces. These two conditions are necessary but not sufficient for a process to be reversible. They are, however, both necessary and sufficient for the use of the system pressure in the calculation of work by (1.2). Processes that occur in such a way that these two conditions are fulfilled are said to be *mechanically reversible*.

Because the processes just considered were taken to be adiabatic, we have had no occasion to consider the role of heat transfer. The adiabatic enclosure of the system imposes a restraint on the system that makes irrelevant any question of external thermal equilibrium between the system and its surroundings. However, uniformity of temperature within the system implies internal thermal equilibrium.

Discussion of an Isothermal Process

Figure 1-2 illustrates an apparatus for carrying out expansion and compression processes isothermally rather than adiabatically. It shows a piston/cylinder assembly which operates exactly as before, except that it is placed in contact with a *heat reservoir* at temperature T. A heat reservoir is a body capable of absorbing or giving off unlimited quantities of heat without any change in

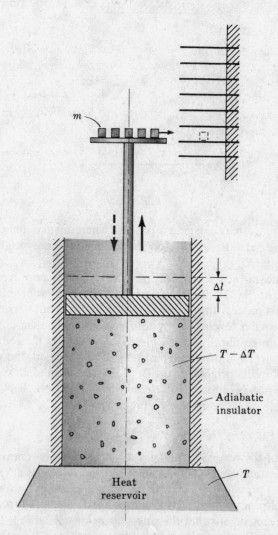

Fig. 1-2

temperature. The atmosphere and the oceans approximate heat reservoirs, usually exploited as heat-sink reservoirs. A continuously operating furnace and a nuclear reactor are equivalent to heat-source reservoirs.

The additional consideration that enters with respect to Fig. 1-2 is that heat may flow from the heat reservoir (part of the surroundings) to the gas (the system) or the reverse. When a system comes to equilibrium with its surroundings by virtue of long contact, it is observed that both have the same temperature. This condition is called external thermal equilibrium. When their temperatures are different, we observe changes, and attribute the changes to a flow of heat from the warmer to the cooler region. Thus we regard a temperature difference as the driving force for heat transfer. When there is no temperature difference, there is no driving force, and no heat transfer. Irreversible processes occur as the result of finite driving forces; reversible processes as the result of infinitesimal driving forces. Thus heat transfer is irreversible when the temperature difference is finite, and becomes reversible only when the temperature difference is infinitesimal.

With respect to Fig. 1-2, the removal of a finite weight m from a frictionless piston causes finite changes in the system, and these include a temperature drop in the gas from its initial equilibrium value T to $T - \Delta T$. Thus heat flows at a finite rate from the heat reservoir to the gas. The process is irreversible in all respects, including the heat transfer. When the piston is unloaded at an infinitesimal rate, the temperature drop is infinitesimal, and heat is transferred reversibly.

Summary

Irreversibilities always lower the efficiencies of processes. Their effect in this respect is identical with that of friction, which is in fact a prime cause of irreversibility. Conversely, no process more efficient than a reversible process can even be imagined. The reversible process is an abstraction, an idealization, which is never achieved in practice. It is, however, of enormous utility because it allows calculation of work from knowledge of the system properties alone. In addition it represents a standard of perfection that cannot be exceeded because:

(1) It places an upper limit on the work that may be *obtained* for a given work-producing process.

(2) It places a lower limit on the work *input* for a given work-requiring process.

EXAMPLE 1.4 We wish to calculate the work done when an ideal gas expands isothermally and reversibly in a piston-and-cylinder assembly.

Since the process is mechanically reversible, P in (1.2) is the gas pressure, and it may be eliminated by the ideal-gas law, $P = RT/V$, where V is the molar volume. Thus $\delta W = RT\, dV/V$, and integration at constant T from the initial to the final state gives the work per mole of gas:

$$W = RT \ln (V_f/V_i)$$

Since $P_i V_i = P_f V_f$, this may also be written

$$W = RT \ln (P_i/P_f)$$

For expansion of a gas at 300 K from an initial pressure 10 bar to a final pressure 1 bar, we have

$$W = (8.314)(300)(\ln 10) = 5740 \text{ J} \cdot \text{mol}^{-1} = 5.74 \text{ kJ} \cdot \text{mol}^{-1}$$

1.6 FIRST LAW OF THERMODYNAMICS

For a closed (constant-mass) system the first law of thermodynamics is expressed mathematically by

$$\Delta E = Q - W \qquad (1.3)$$

where ΔE is the *total* energy change of the system, Q is heat added to the system, and W is work done by the system. The first law of thermodynamics merely gives quantitative expression to the principle of energy conservation. In words, it says that the total energy change of a closed system is equal to the heat transferred *to* the system minus the work done *by* the system.

The total energy change ΔE can be split into several terms, each representing the change in energy of a particular form:

$$\Delta E = \Delta E_K + \Delta E_P + \Delta U$$

where ΔE_K is the change in kinetic energy, ΔE_P is the change in gravitational potential energy, and ΔU is the change in internal energy. Kinetic energy and potential energy are defined by

$$E_K \equiv \tfrac{1}{2}mu^2 \qquad E_P \equiv mgz$$

where u is the velocity, z is the elevation above a datum level, and g is the local acceleration of gravity. These energy functions are common to both mechanics and thermodynamics.

The internal energy function U, however, is peculiar to thermodynamics. It represents the kinetic and potential energies of the molecules, atoms, and subatomic particles that constitute the system on a microscopic scale. There is no known way to determine absolute values of U. Fortunately, only changes ΔU are needed and these can be found from experiment. It is also found by experiment that U is fixed whenever the state of the system is fixed. If ΔE is expanded in the first-law expression, we get the equation

$$\Delta E_K + \Delta E_P + \Delta U = Q - W$$

In the frequent case where the kinetic and potential energies of the system do not change, this equation becomes

$$\boxed{\Delta U = Q - W} \qquad\qquad (1.4)$$

or, in differential form,

$$dU = \delta Q - \delta W \qquad\qquad (1.5)$$

and all energy exchange with the surroundings serves to change just the internal energy. If in addition the process is adiabatic, then $Q = 0$ and (1.4) becomes

$$\Delta U = -W \qquad \text{(adiabatic)}$$

This last equation shows that for a system changed adiabatically from one equilibrium state to another work should be *independent of path*, for ΔU should depend only on the end states. Experiment shows it to be so, and this is the primary evidence that U is indeed a state function.

With respect to (1.5), note that the differential signs on Q and W are written δ, whereas we write dU. The *quantities* Q and W differ in a fundamental way from the *property* U. A property like U always has a value, dependent only on the state of the system. A process which changes the state changes U. Thus dU represents an infinitesimal *change* in U, and integration gives a difference between two values of the property:

$$\int_{U_1}^{U_2} dU = U_2 - U_1 = \Delta U$$

On the other hand, Q and W are not properties of the system and depend on the *path* of the process; thus δ is used to denote an infinitesimal *quantity*. Integration gives not a difference between two values but a finite quantity:

$$\int \delta Q = Q \qquad \text{and} \qquad \int \delta W = W$$

Thus integration of (1.5) yields (1.4).

EXAMPLE 1.5 (a) What velocity must be attained by a mass of 1 kg in order that it have a kinetic energy of 1 kJ? (b) To what elevation must a mass of 1 kg be raised in order that it have a potential energy of 1 kJ?

(a) By definition of kinetic energy, $E_K = \frac{1}{2}mu^2$. Since the required energy is 1 kJ = 1000 N·m, we have $1000 = \frac{1}{2}(1)u^2$, whence $u = 141.4$ m·s^{-1}.

(b) By the definition of gravitational potential energy, $E_P = mgz$. For an energy of 1 kJ, or 1000 N·m, we have $1000 = (1)(9.81)(z)$, from which $z = 101.9$ m.

EXAMPLE 1.6 Steel wool is contained in a cylinder in an atmosphere of pure oxygen. The cylinder is fitted with a frictionless piston which maintains the oxygen pressure at 1 bar. The iron in the steel wool reacts very slowly with the oxygen to form Fe_2O_3. Heat is removed from the apparatus during the process so as to keep the temperature constant at 25 °C. For the reaction of 2 mol of iron,

$$2Fe + 1.5O_2 \rightarrow Fe_2O_3$$

831.08 kJ of heat is removed. Taking the system as the contents of the cylinder, we wish to calculate Q, W, and ΔU for the process.

Both Fe and Fe_2O_3 are solids which occupy a negligible volume compared with the gaseous oxygen. We may therefore take the total volume of the system, V^t, as the volume of oxygen, which we assume is an ideal gas. Thus $V^t = nRT/P$, where n is the number of moles of gaseous oxygen present. During the process T and P are constant, and only n changes; whence the total volume change is $\Delta V^t = (RT/P)\Delta n$.

Since pressure is constant, the work of the process is $W = P\Delta V^t = RT\Delta n$. According to the reaction equation, 1.5 mol of oxygen is consumed, so that $\Delta n = -1.5$ mol; thus

$$W = (8.314)(298.15)(-1.5) = -3720 \text{ J} = -3.72 \text{ kJ}$$

The minus sign shows that work is done on the system by the piston. We are given that $Q = -831.08$ kJ; thus

$$\Delta U = Q - W = -831.08 + 3.72 = -827.36 \text{ kJ}$$

This decrease in internal energy reflects the changes in bond energies caused by the chemical reaction.

1.7 ENTHALPY

Special thermodynamic functions are defined as a matter of convenience. The simplest such function is the *enthalpy H*, explicitly defined for any system by the mathematical expression

$$\boxed{H \equiv U + PV} \tag{1.6}$$

Since the internal energy U and the PV product both have units of energy, H also has units of energy. Moreover, since U, P, and V are all properties of a system, H must be a property too.

Whenever a differential change occurs in a system, its properties change. From (1.6) the change in H is related to other property changes by

$$dH = dU + d(PV) \tag{1.7}$$

EXAMPLE 1.7 Show that the heat added to a closed PVT system undergoing a mechanically reversible, constant-pressure process is equal to ΔH.

By (1.7) at constant pressure, $dH = dU + P\,dV$. But for a mechanically reversible process, $\delta W = P\,dV$; and by (1.5), $dU = \delta Q - \delta W$. Therefore, $dH = \delta Q$.

For a finite process, $\Delta H = Q$.

EXAMPLE 1.8 Liquid carbon dioxide at 235 K has vapor pressure 1075 kPa and specific volume 0.9011×10^{-3} m^3·kg^{-1}. At these conditions it is a *saturated liquid*, i.e., a liquid at its boiling point. We arbitrarily assign the value 0.00 to its (specific) enthalpy, and this becomes the basis for all enthalpy and internal-energy values.

For instance, the internal energy of the saturated liquid is calculated from (1.6):

$$U^l = H^l - PV^l = -PV^l = -(1075 \times 10^3)(0.9011 \times 10^{-3}) = -968.7 \text{ J·kg}^{-1} = -0.9687 \text{ kJ·kg}^{-1}$$

The latent heat of vaporization of carbon dioxide at 235 K and 1075 kPa is 317.4 $kJ \cdot kg^{-1}$. The specific volume of the *saturated vapor* produced by vaporization at these conditions is 0.0357 $m^3 \cdot kg^{-1}$. From Example 1.7 we have that $\Delta H = Q$ for a constant-pressure, reversible process in a closed system. Since the latent heat is defined for this kind of process,

$$\Delta H^{lv} \equiv H^v - H^l = 317.4 \text{ kJ} \cdot kg^{-1}$$

However, by assignment, $H^l = 0$, whence $H^v = 317.4 \text{ kJ} \cdot kg^{-1}$. Correspondingly, the internal energy of the saturated vapor is

$$U^v = H^v - PV^v = 317.4 - (1075)(0.0357) = 279.0 \text{ kJ} \cdot kg^{-1}$$

1.8 NOTATION

We have not yet adopted a notation that makes explicit the amount of material in the system to which our symbols refer. For example, the definition $H \equiv U + PV$ can be written for any amount of material. The properties H, U, and V are *extensive*; that is, they are directly proportional to the mass of the system considered. Temperature T and pressure P are *intensive*, independent of the extent of the system. With respect to extensive properties, we shall let the plain symbols such as U, H, and V refer to a unit amount of material, either to a unit mass or to a mole. With a unit mass as the basis the properties are often called *specific* properties; e.g., specific volume. With a mole as the basis they are called molar properties; e.g., molar enthalpy. To represent a total property of a system we may multiply the specific or molar property by the mass or number of moles of the system; e.g., $V^t = mV$, $H^t = nH$.

1.9 HEAT CAPACITY

The amount of heat which must be added to a closed PVT system in order to accomplish a given change of state depends on how the process is carried out. Only for a reversible process for which the path is fully specified is it possible to relate the heat to a property of the system. On this basis we define *heat capacity* in general by

$$C_X = \left(\frac{\delta Q}{dT}\right)_X$$

where X indicates a reversible process of specified path. We could define a number of heat capacities according to this prescription; however, for PVT systems only two are in common use. These are C_V, *heat capacity at constant volume*, and C_P, *heat capacity at constant pressure*. Both definitions presume the system closed and of constant composition.

The property

$$C_V \equiv \left(\frac{\delta Q}{dT}\right)_V$$

is a measure of the amount of heat required to increase the temperature by dT when the system is heated in a reversible process at constant volume. That C_V is a property of the system follows from (1.5), which for a constant-volume, reversible process becomes $dU = \delta Q$, because no work can be done under the imposed restrictions. Thus

$$\boxed{C_V \equiv \left(\frac{\partial U}{\partial T}\right)_V} \qquad (1.8)$$

is an alternative definition of C_V, and since U, T, and V are all properties of the system, C_V must also be a property.

Similarly,

$$C_P \equiv \left(\frac{\delta Q}{dT}\right)_P$$

is a measure of the amount of heat required to increase the temperature by dT when the system is heated in a reversible process at constant pressure. Example 1.7 showed that $\delta Q = dH$ for a reversible, constant-pressure process; thus

$$C_P \equiv \left(\frac{\partial H}{\partial T}\right)_P \qquad (1.9)$$

is an alternative definition of C_P which shows C_P to be a property of the system.

We may also write (1.8) and (1.9) as

$$dU = C_V \, dT \qquad \text{(const. } V) \qquad\qquad (1.10)$$
$$dH = C_P \, dT \qquad \text{(const. } P) \qquad\qquad (1.11)$$

As these equations relate properties only and do not depend on the process causing changes in the system, no restriction with respect to reversibility is necessary. However, the change must be between equilibrium states.

Heat Capacity of an Ideal Gas

Equation (1.10) shows that in general the internal energy of a closed PVT system may be considered a function of T and V, $U = U(T, V)$. Then

$$dU = \left(\frac{\partial U}{\partial T}\right)_V dT + \left(\frac{\partial U}{\partial V}\right)_T dV$$

In view of (1.8) this becomes

$$dU = C_V \, dT + \left(\frac{\partial U}{\partial V}\right)_T dV \qquad\qquad (1.12)$$

Application of this equation requires values for $(\partial U/\partial V)_T$, and these must in general be determined by experiment. There is, however, one special and important case—that when $(\partial U/\partial V)_T = 0$ or, equivalently, $U = U(T)$—and this is part of the definition of an ideal gas. Thus the complete definition of an ideal gas requires that at all temperatures and pressures:

$$PV = RT \qquad\qquad U = U(T)$$

The ideal gas is, of course, an idealization, and no real gas exactly satisfies these equations over any finite range of temperature and pressure. However, real gases approach ideal behavior at low pressures, and in the limit as $P \rightarrow 0$ do in fact meet the above requirements. Thus the equations for an ideal gas provide good approximations to real-gas behavior at low pressures, and have the virtue of simplicity. For example, (1.12) becomes

$$dU = C_V \, dT \qquad \text{(ideal gas)} \qquad\qquad (1.13)$$

an equation always valid for an ideal gas regardless of the process.

EXAMPLE 1.9 We wish to show that for an ideal gas H is a function of temperature only and to find a relationship between C_P and C_V for an ideal gas.

By definition, $H = U + PV$. But $PV = RT$, so that $H = U + RT$. Since U is a function of T only, this must also be the case for H.

Differentiating (1.6) with respect to T,

$$\frac{dH}{dT} = \frac{dU}{dT} + \frac{d(PV)}{dT}$$

where we have used total derivatives because H, U, and PV are functions of T only. This being the case we can also write

$$\left(\frac{\partial H}{\partial T}\right)_P = \frac{dH}{dT} = C_P(T) \qquad \text{and} \qquad \left(\frac{\partial U}{\partial T}\right)_V = \frac{dU}{dT} = C_V(T)$$

Also $d(PV)/dT = R$. The required relationship is then

$$C_P = C_V + R \tag{1.14}$$

Since $H = U + RT$ for an ideal gas, one can use (1.13) and (1.14) to show that

$$dH = C_P\, dT \qquad \text{(ideal gas)} \tag{1.15}$$

an equation also valid for an ideal gas regardless of the process.

The ratio of heat capacities is often denoted by

$$\gamma \equiv \frac{C_P}{C_V}$$

and is a useful quantity in calculations for ideal gases.

As mentioned above, real gases approach ideality only in the zero-pressure limit. If a real gas at a pressure approaching zero is imagined to remain an ideal gas as it is compressed to a finite pressure, then the resulting state is known as an *ideal-gas state*. Real gases at zero pressure have properties which reflect their individuality, and this continues to be true for gases in the hypothetical ideal-gas state. Thus we often use ideal-gas heat capacities, these being different for different gases and functions of temperature only. The temperature dependence is often taken as quadratic:

$$C_P^{\text{ig}} = a + bT + cT^2$$

where a, b, and c are constants and C_P^{ig} indicates heat capacity for the ideal-gas state.

The effect of temperature is negligible for monatomic gases such as helium and argon, and for these gases the molar heat capacities are given by

$$C_V^{\text{ig}} = \frac{3}{2} R \qquad C_P^{\text{ig}} = \frac{5}{2} R \qquad \gamma = 1.67$$

For diatomic gases such as O_2, N_2, and H_2 the heat capacities change rather slowly with temperature, and near room temperature have the approximate values

$$C_V^{\text{ig}} = \frac{5}{2} R \qquad C_P^{\text{ig}} = \frac{7}{2} R \qquad \gamma = 1.40$$

For polyatomic gases such as CO_2, NH_3, CH_4, etc., the heat capacities vary appreciably with temperature and differ from gas to gas. So no values can be given that are generally valid approximations. Values of γ usually are less than 1.3.

EXAMPLE 1.10 We wish to show that for an ideal gas with constant heat capacities, undergoing a reversible, adiabatic compression or expansion,

$$\frac{T_2}{T_1} = \left(\frac{V_1}{V_2}\right)^{\gamma-1} \tag{1}$$

Since the process is adiabatic, (1.5) becomes $dU = -\delta W = -P\, dV$. But $dU = C_V\, dT$ and $P = RT/V$; thus

$$C_V\, dT = -RT\, \frac{dV}{V} \qquad \text{or} \qquad \frac{dT}{T} = -\frac{R}{C_V}\, \frac{dV}{V}$$

By (1.14), $R/C_V = \gamma - 1$; therefore

$$\frac{dT}{T} = -(\gamma - 1)\, \frac{dV}{V}$$

Integrating from (T_1, V_1) to (T_2, V_2) with γ constant,

$$\ln \frac{T_2}{T_1} = (\gamma - 1) \ln \frac{V_1}{V_2} \quad \text{or} \quad \frac{T_2}{T_1} = \left(\frac{V_1}{V_2}\right)^{\gamma - 1}$$

EXAMPLE 1.11 If an ideal gas for which $C_V^{ig} = (3/2)R$ and $C_P^{ig} = (5/2)R$ expands reversibly and adiabatically from an initial state $T_1 = 450$ K and $V_1 = 3$ L to a final volume $V_2 = 5$ L, the final temperature is given by Example 1.10 as

$$T_2 = T_1\left(\frac{V_1}{V_2}\right)^{\gamma - 1} = (450)\left(\frac{3}{5}\right)^{0.67} = 320 \text{ K}$$

$(1 \text{ L} = 10^{-3} \text{ m}^3$; but the computation does not depend on this.)

The work done during this process is

$$W = -\Delta U = -C_V^{ig}\, \Delta T = -(3/2)(8.314)(320 - 450) = 1620 \text{ J} \cdot \text{mol}^{-1}$$

The positive value of W correctly indicates that in the expansion process work is done by the system.

The enthalpy change of the gas can also be calculated. By (1.15),

$$\Delta H = C_P^{ig}\, \Delta T = (5/2)(8.314)(-130) = -2700 \text{ J} \cdot \text{mol}^{-1}$$

Solved Problems

BASIC CONCEPTS (Section 1.1)

1.1 Pressure measurements are often made with manometers, in which the height of a fluid column indicates the pressure. What is the relation between pressure and fluid height?

The weight of a column of fluid of height h and cross-sectional area A is given by Newton's second law as $F = mg = \rho A h g$, where ρ is the mass density of the fluid. The hydrostatic pressure exerted by the fluid is

$$P = \frac{F}{A} = \rho g h$$

Thus P is directly proportional to h.

For mercury at $0\,°C$, $\rho = 13.5951 \times 10^3 \text{ kg} \cdot \text{m}^{-3}$, and for a standard gravitational field, $g = 9.806\,65 \text{ m} \cdot \text{s}^{-2}$. Thus, for a column of mercury 1 mm in height in a standard gravitational field at $0\,°C$,

$$P = (13.5951 \times 10^3 \text{ kg} \cdot \text{m}^{-3})(9.806\,65 \text{ m} \cdot \text{s}^{-2})(1 \times 10^{-3} \text{ m}) = 133.322 \text{ kg} \cdot \text{m}^{-1} \cdot \text{s}^{-2} = 133.322 \text{ Pa}$$

This is a unit of pressure called the *torr*.

1.2 In the English engineering system of units absolute temperatures are measured on the Rankine scale, where $T\,(\text{R}) = 1.8[T\,(\text{K})]$; conventional temperatures are measured on the Fahrenheit scale, where $t\,(°\text{F}) = T\,(\text{R}) - 459.67$. (*a*) Does the rankine represent a larger or smaller temperature unit or interval than the kelvin? (*b*) What is the relationship between the kelvin and the Celsius degree? (*c*) What is the relationship between the rankine and the Fahrenheit degree? (*d*) Is the Fahrenheit degree a larger or smaller unit than the Celsius degree?

(*a*) A temperature in rankines is always larger than the temperature in kelvins by the factor 1.8. Thus the rankine must be a smaller unit of temperature or a smaller temperature interval than the kelvin, and is in fact smaller by the factor 1.8.

(*b*) The Celsius temperature scale is obtained from the Kelvin scale merely by a shift of the zero point. Thus the Celsius degree and the kelvin represent temperature units or intervals of exactly the same size.

(c) The Fahrenheit temperature scale and the Rankine scale differ only in respect to the zero point of the scale, and the Fahrenheit degree and the rankine are units of exactly the same size.

(d) The Fahrenheit degree is smaller than the Celsius degree by the factor 1.8; they are related exactly as the rankine and the kelvin.

All these relationships are easily seen in Fig. 1-3.

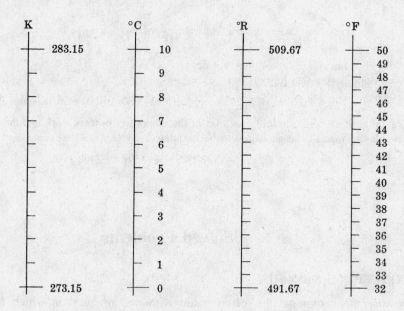

Fig. 1-3

1.3   Find the volume occupied by 3 mol of an ideal gas at 2 bar and 350 K.

From Appendix B, $R = 83.14 \, \text{cm}^3 \cdot \text{bar} \cdot \text{mol}^{-1} \cdot \text{K}^{-1}$; hence

$$V = \frac{nRT}{P} = \frac{(3)(83.14)(350)}{2} = 43\,650 \, \text{cm}^3$$

1.4   A deceleration of 25 g's, where $1 \, \text{g} = 9.8066 \, \text{m} \cdot \text{s}^{-2}$, is fatal to humans in most automobile accidents. What force acts on a 70-kg man subjected to this deceleration?

By Newton's second law, $F = ma = -(70)(25)(9.8066) = -17\,161 \, \text{N}$ (or about $4000 \, \text{lb}_\text{f}$). The minus sign originates with the negative acceleration and ultimately indicates that the force applied to the man is in a direction opposite to his displacement.

WORK   (Sections 1.2–1.5)

1.5   Show that (1.2) is consistent with the basic defining equation for work, (1.1).

The work done by a fluid pressure P acting over a piston of area A and moving through a volume change $dV = A \, dl$ (see Fig. 1-4) is given by (1.2) as

$$\delta W = P \, dV = (F/A)(A \, dl) = F \, dl$$

which is (1.1).

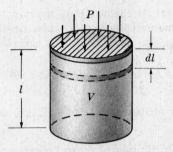

Fig. 1-4

1.6 A cylinder fitted with a sliding piston contains a volume of gas V which exerts a pressure P on the piston. The gas expands slowly, pushing the piston outward. The following data are taken:

P/bar	V/m^3
15 (initial)	0.0300 (initial)
12	0.0361
9	0.0459
6	0.0644
4	0.0903
2 (final)	0.1608 (final)

Calculate the work done by the gas on the piston.

The work is given by (1.2) in integral form,

$$W = \int_{V_i}^{V_f} P \, dV$$

The data provide the relationship of P to V that is required for evaluation of the integral. This is

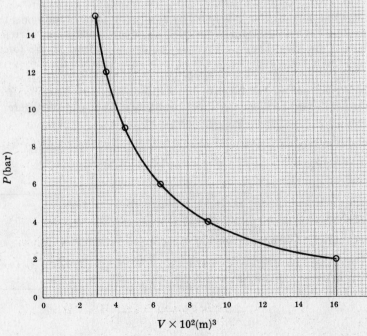

Fig. 1-5

conveniently accomplished graphically by means of a plot of P versus V (Fig. 1-5). The area below the curve between the initial and final values of V represents the work, and this area can be estimated by counting squares. Each large square represents 0.04 bar $\cdot$ m^3; and counting by rows, from the bottom up, we obtain a total of

$$6.6 + 4.5 + 2.2 + 1.3 + 0.8 + 0.4 + 0.2 + 0.0 = 16.0$$

large squares. Thus,

$$W = (16.0)(0.04 \text{ bar} \cdot \text{m}^3)(100 \text{ kJ} \cdot \text{bar}^{-1} \cdot \text{m}^{-3}) = 64.0 \text{ kJ}$$

1.7   The numerical P-V data of Problem 1.6 can be very closely fit by the equation $PV^{1.2} = 0.2232$. Rework Problem 1.6, making use of this expression for the evaluation of the work integral.

The P-V relation given is of the form $PV^{\delta} = k$. Substitution for P in the work integral gives

$$W = k \int_{V_i}^{V_f} \frac{dV}{V^{\delta}} = \frac{k}{1 - \delta}(V_f^{1-\delta} - V_i^{1-\delta}) = \frac{kV_f^{1-\delta} - kV_i^{1-\delta}}{1 - \delta}$$

But $k = P_i V_i^{\delta} = P_f V_f^{\delta}$, and appropriate substitution for k provides

$$W = \frac{P_f V_f^{\delta} V_f^{1-\delta} - P_i V_i^{\delta} V_i^{1-\delta}}{1 - \delta} = \frac{P_f V_f - P_i V_i}{1 - \delta}$$

Substitution of numerical values gives

$$W = \frac{(2)(0.1608) - (15)(0.0300)}{1 - 1.2} = 0.6418 \text{ bar} \cdot \text{m}^3 = 64.18 \text{ kJ}$$

Note that the expression $PV^{\delta} = k$ is purely empirical; its use implies no assumption as to ideality of the gas in the cylinder. The use of an expression of this form to represent the P-V relation for a real gas in compression and expansion processes is fairly common.

1.8   Show that the reversible work of changing the area of a liquid surface is given by $\delta W = -\gamma \, dA$, where γ is *surface tension*.

Consider a film of liquid (such as a soap solution) held on a wire framework as represented in Fig. 1-6. The film is made up of two surfaces and a thin layer of liquid in between. The force F required to hold the movable wire in position is found by experiment to be proportional to l, *but independent of x*. This means that F is not influenced by the thickness of the film and must therefore result from stresses in the surfaces of the film. Thus we have *surface tension* defined as the force per unit length acting perpendicular to any line in the surface or at its boundary. When the force F is moved to the right, the film is extended, and liquid moves from the bulk region between the surfaces to the surfaces. Thus new surface is formed, not by stretching the original surface but by generation from the bulk of the film. Since there are two surfaces associated with the liquid film shown in the figure, the relation between the force F and the surface tension γ at equilibrium is

$$\gamma = \frac{F}{2l}$$

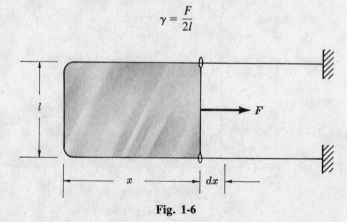

Fig. 1-6

The work of moving F through the distance dx is $\delta W = -F\,dx = -2l\gamma\,dx$, where the minus sign is inserted so as to conform with the sign convention. However, $dA = 2l\,dx$, and therefore

$$\delta W = -\gamma\,dA$$

This is the work required to form the new area dA. For pure materials γ is a function of temperature only, and the surface is considered a thermodynamic system for which the coordinates are γ, A, and T. Such a system is necessarily an open system, because it is not possible to have a surface without also having a bulk liquid phase.

1.9 What is the minimum work required to form the surface associated with a water mist of particle radius 5 μm? Assume that 0.001 m^3 of water at 50 °C is turned into mist also at 50 °C. The surface tension of water at 50 °C is 0.063 N $\cdot$ m^{-1}.

The surface-to-volume ratio of a spherical drop is

$$\frac{A}{V} = \frac{4\pi r^2}{(4/3)\pi r^3} = \frac{3}{r}$$

where r is the radius of the sphere. For a volume of 0.001 m^3 and a radius of 5×10^{-6} m,

$$A = \frac{3V}{r} = \frac{3(0.001)}{5 \times 10^{-6}} = 600 \text{ m}^2$$

By Problem 1.8,

$$W = -\int_0^A \gamma\,dA = -\gamma A = -(0.063 \text{ N} \cdot \text{m}^{-1})(600 \text{ m}^2) = -37.8 \text{ J}$$

Thus, 37.8 J of work must be performed on the water.

1.10 Show that the universal gas constant may have units of (molar energy)/(temperature).

The SI units for R are

$$(\text{N} \cdot \text{m}^{-2})(\text{m}^3 \cdot \text{mol}^{-1})(\text{K}^{-1}) = \text{N} \cdot \text{m} \cdot \text{mol}^{-1} \cdot \text{K}^{-1}$$

Since N $\cdot$ m is a joule, we have from Appendix B that $R = 8.314$ J $\cdot$ mol^{-1} $\cdot$ K^{-1}.

FIRST LAW OF THERMODYNAMICS (Sections 1.6–1.9)

1.11 By applying Newton's second law of motion to the translation of a rigid body, illustrate the origin of the kinetic- and potential-energy terms in the energy equation for a purely mechanical system.

Newton's second law for this system is

$$F = ma = m\,\frac{du}{dt} \tag{1}$$

where u is the velocity of the body and F is the total external force acting on the body parallel to its displacement dl. The total work done by the body against the force is then

$$W = -\int F\,dl \tag{2}$$

Substituting (1) into (2), we obtain

$$W = -\int m\,\frac{du}{dt}\,dl = -\int m\,\frac{dl}{dt}\,du = -\int mu\,du = -\int d\left(\frac{mu^2}{2}\right)$$

or

$$W = -\Delta\left(\frac{mu^2}{2}\right) \equiv -\Delta E_K \tag{3}$$

Equation (3) is a perfectly general expression for the total mechanical work done by a rigid body in translation, and the equation is not based on any assumptions as to the nature of the force F. However, F is conveniently considered the sum of two types of forces, *body* forces F_B and *surface* forces F_S:

$$F = F_B + F_S \tag{4}$$

Body forces are so called because they act throughout the *volume* of a system; surface forces act on an *area* of the bounding surface of a system. From (2) and (4), then, the total work can be considered the sum of two work terms:

$$W = W_B + W_S \tag{5}$$

where

$$W_B = -\int F_B \, dl \tag{6}$$

$$W_S = -\int F_S \, dl \tag{7}$$

Body forces are *conservative* forces. This means they can be derived from a *potential function* $\Phi(l)$, which depends only on the location of the system, by differentiation with respect to the position coordinate. Thus for the case at hand

$$F_B = -\frac{d\Phi(l)}{dl} \tag{8}$$

Substituting (8) in (6), we obtain

$$W_B = -\int \left[-\frac{d\Phi(l)}{dl} \right] dl = \int d\Phi = \Delta\Phi \tag{9}$$

Because the difference $\Delta\Phi$ depends only on the initial and final positions of the system, and not on the path followed between these positions, the work done against body forces is *independent of path*. Defining the potential energy E_P as $E_P \equiv \Phi$, we can write (9) as

$$W_B = \Delta E_P \tag{10}$$

Surface forces (e.g., friction) are in general *nonconservative*, and we usually write expressions like (7) for the work done against such forces. Combination of (5) and (10) gives

$$W = \Delta E_P + W_S \tag{11}$$

Equation (11) is an alternative to (3), and the two equations may be applied to the same process. Doing this, and rearranging, we obtain the energy equation

$$-W_S = \Delta E_K + \Delta E_P \tag{12}$$

When there are no surface forces (12) reduces to

$$\Delta E_K + \Delta E_P = 0 \qquad \text{or} \qquad E_K + E_P = \text{constant}$$

which is the familiar "principle of conservation of energy" of classical mechanics.

The work term in the first law of thermodynamics usually represents work done by surface forces, and the usual thermodynamic system is, of course, not a rigid body.

1.12 The gravitational force of attraction between homogeneous, spherical bodies A and B is given by

$$F_{\text{grav}} = -\frac{Gm_A m_B}{r^2} \tag{1}$$

where G is a constant ($=6.670 \times 10^{-11} \text{ N} \cdot \text{m}^2 \cdot \text{kg}^{-2}$), m_A and m_B are the masses of the two bodies, and r is the distance between their centers. Find an expression for the gravitational potential function Φ_{grav}, and show that the gravitational potential energy of a small body at a height z above the earth's surface is approximately

$$E_P = mgz \tag{2}$$

where $g = \text{constant} = GM/R^2$, $m = $ mass of the body, $M = $ mass of the earth, $R = $ radius of the earth.

The gravitational force is a body force and is conservative. From (8) of Problem 1.11, Φ_{grav} is related to F_{grav} by

$$F_{grav} = -\frac{d\Phi_{grav}}{dr} \tag{3}$$

Combination of (1) and (3) gives

$$\frac{d\Phi_{grav}}{dr} = \frac{Gm_A m_B}{r^2}$$

which yields on integration an *exact* expression for Φ_{grav}:

$$\Phi_{grav} = -\frac{Gm_A m_B}{r} + C \tag{4}$$

where C is a constant of integration. The gravitational potential energy E_P is by definition equal to the gravitational potential function; hence (4) can be written for an earth/body system as

$$E_P = -\frac{GMm}{R+z} + C$$

or

$$E_P = -\frac{GMm}{R\left(1 + \dfrac{z}{R}\right)} + C \tag{5}$$

where $r = R + z$ from the statement of the problem. Now, for small elevations, $z/R \ll 1$, and the first two terms of the binomial expansion give

$$\left(1 + \frac{z}{R}\right)^{-1} \approx 1 - \frac{z}{R}$$

Thus (5) becomes

$$E_P \approx -\frac{GMm}{R}\left(1 - \frac{z}{R}\right) + C = \frac{GMm}{R^2}\,z + C' \tag{6}$$

where C' is a new constant for a given m. Since only *differences* in E_P are of importance, C' may be taken as zero: this is equivalent to choosing a datum of zero potential energy at the surface of the earth, and leads to the desired result (2).

1.13   An elevator of mass 2268 kg rests at a level 7.62 m above the base of the elevator shaft. It is raised to 76.2 m above the base of the shaft, where the cable holding it breaks. The elevator falls freely to the base of the shaft where it is brought to rest by a strong spring. The spring assembly is designed to hold the elevator at the position of maximum spring compression by means of a ratchet mechanism. Assuming the entire process to be frictionless and taking g as the standard acceleration of gravity, $9.8066 \ \text{m} \cdot \text{s}^{-2}$, calculate:

(a)   The gravitational potential energy of the elevator in its initial position, relative to the base of the shaft.

(b)   The potential energy of the elevator in its highest position, relative to the base of the shaft.

(c)   The work done in raising the elevator.

(d)   The kinetic energy and velocity of the elevator just before it strikes the spring.

(e)   The potential energy of the fully compressed spring.

(f)   The energy of the system made up of the elevator and spring (1) at the start of the process, (2) when the elevator reaches its maximum height, (3) just before the elevator strikes the spring, and (4) after the elevator has come to rest. Use the numbers just given as subscripts on symbols to designate the stages of the process to which the symbols apply.

(a)   The gravitational potential energy of the elevator at stage 1 is given by

$$E_{P1} = mgz_1 = (2268)(9.8066)(7.62) = 169\,480 \text{ J} = 169.48 \text{ kJ}$$

(b)   The gravitational potential energy at stage 2 is ten times that at stage 1: $E_{P2} = 1694.8$ kJ.

(c)   The first law of thermodynamics applied between stage 1 and stage 2 reduces to

$$\Delta E_P = -W$$

because ΔE_K, ΔU, and Q are all presumed negligible for this process. Thus $-W = E_{P2} - E_{P1}$, or $W = E_{P1} - E_{P2} = -1525.3$ kJ. (The minus sign merely means that work was done on the system, rather than by the system.)

(d)   During the process from stage 2 to stage 3, the elevator falls freely, subject to no surface forces, and therefore $W = 0$. It is also assumed that $Q = 0$. Thus the first law becomes

$$\Delta E_K + \Delta E_P = 0 \qquad \text{or} \qquad E_{K3} - E_{K2} + E_{P3} - E_{P2} = 0$$

But E_{K2} and E_{P3} are zero. Therefore $E_{K3} = E_{P2} = 1694.8$ kJ and

$$u_3 = \sqrt{\frac{2E_{K3}}{m}} = \sqrt{\frac{(2)(1694.8 \times 10^3)}{2268}} = 38.66 \text{ m} \cdot \text{s}^{-1}$$

(e)   Since the elevator comes to rest at the bottom of the shaft at stage 4, it ends up with zero potential and kinetic energy. All of its former energy is stored in the spring as elastic potential energy; the amount of this stored energy is 1694.8 kJ.

(f)   If the elevator and the spring together are taken as the system, the initial energy of the system is the potential energy of the elevator, or 169.5 kJ. The total energy of the system can change only if energy is transferred between it and the surroundings. As the elevator is raised, work is done on the system by the surroundings in the amount of 1525.3 kJ. Thus the energy of the system when the elevator reaches its maximum height is $169.5 + 1525.3 = 1694.8$ kJ. Subsequent changes occur entirely within the system, with no energy transferred between the system and the surroundings. Hence the total energy of the system must remain constant at 1694.8 kJ. It merely changes form from potential energy of position (elevation) of the elevator to kinetic energy of the elevator to potential energy of configuration of the spring.

1.14   Water flows over a waterfall 100 m in height. Consider 1 kg of the water, and assume that no energy is exchanged between this 1 kg and its surroundings. (a) What is the potential energy of the water at the top of the falls with respect to the base of the falls? (b) What is the kinetic energy of the water just before it strikes the bottom? (c) After the 1 kg of water enters the river below the falls, what change has occurred in its state?

Taking the 1 kg of water as the system, and noting that it exchanges no energy with its surroundings, we may set Q and W equal to zero and write the first-law energy equation as

$$\Delta E_K + \Delta E_P + \Delta U = 0$$

This equation applies to any part of the process.

(a) $$E_P = mgz = (1 \text{ kg})(9.8066 \text{ m} \cdot \text{s}^{-2})(100 \text{ m}) = 980.66 \text{ J}$$

(b)   During the free fall of the water no mechanism exists for conversion of potential or kinetic energy into internal energy. Thus ΔU is zero, and

$$\Delta E_K + \Delta E_P = E_{K2} - E_{K1} + E_{P2} - E_{P1} = 0$$

We may take $E_{K1} \approx 0$ and $E_{P2} = 0$; then $E_{K2} = E_{P1} = 980.66$ J.

(c) As the 1 kg of water strikes bottom and joins with other masses of water to re-form a river, there is much turbulence, which has the effect of converting kinetic energy into internal energy. During this process ΔE_P is essentially zero, and therefore

$$\Delta E_K + \Delta U = 0 \quad \text{or} \quad \Delta U = E_{K2} - E_{K3}$$

However, as the downstream river velocity is assumed to be small, E_{K3} is negligible, and

$$\Delta U = E_{K2} = 980.66 \text{ J}$$

The overall result of the process is the conversion of potential energy of the water into internal energy of the water. This change in internal energy is manifested by a temperature rise. Since energy in the amount of $4184 \text{ J} \cdot \text{kg}^{-1}$ is required for a temperature rise of 1 °C in water, the temperature increase is $980.66/4184 = 0.234$ °C.

1.15 The driver of a 1350-kg car, coasting down a hill, sees a red light at the bottom, for which he must stop. His speed at the time the brakes are applied is $28 \text{ m} \cdot \text{s}^{-1}$ and he is 30 m vertically above the bottom of the hill. How much energy as heat must be dissipated by the brakes if wind and other frictional effects are neglected?

Equation (1.3), $\Delta E = Q - W$, is applicable to the automobile as the system. The total energy change ΔE is made up of two parts,

$$\Delta E_K = \frac{m \, \Delta u^2}{2} \quad \text{and} \quad \Delta E_P = mg \, \Delta z$$

Since there are no moving surface forces applied to the car, W is zero, and our energy equation becomes

$$Q = \frac{m \, \Delta u^2}{2} + mg \, \Delta z = \frac{(1350)(0^2 - 28^2)}{2} + (1350)(9.8066)(0 - 30) = -926\,400 \text{ J}$$

or -926.4 kJ. The minus sign indicates that heat must pass from the system to the surroundings.

1.16 A rigid tank which acts as a perfect heat insulator and which has a negligible heat capacity is divided into two unequal parts, A and B, by a partition. Different amounts of the same ideal gas are contained in the two parts of the tank. The initial conditions for both parts of the tank are as shown in Fig. 1-7.

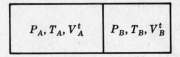

$$P_A, T_A, V_A^t \quad \mid \quad P_B, T_B, V_B^t$$

Fig. 1-7

Find expressions for the equilibrium temperature T and pressure P reached after removal of the partition. Assume that C_V, the molar heat capacity of the gas, is constant and that the process is adiabatic.

Since the tank is rigid, no external force acting on its contents (taken as the system) moves, and consequently $W = 0$. In addition, the process is adiabatic, and therefore $Q = 0$. Thus by the first law the total internal energy of the gas in the tank is constant. This may be expressed by the equation

$$\Delta U_A^t + \Delta U_B^t = 0$$

where the two terms refer to the quantities of gas initially in the two parts of the tank. For an ideal gas with constant heat capacities, $\Delta U^t = nC_V \, \Delta T$; thus the energy equation may be written

$$n_A C_V (T - T_A) + n_B C_V (T - T_B) = 0 \qquad (1)$$

or $\qquad\qquad\qquad\qquad n_A (T - T_A) + n_B (T - T_B) = 0 \qquad (2)$

where n_A and n_B are the numbers of moles of gas initially in parts A and B of the tank. By the ideal-gas equation, applied to the two initial compartments,

$$n_A = \frac{P_A V_A^t}{R T_A} \quad \text{and} \quad n_B = \frac{P_B V_B^t}{R T_B} \tag{3}$$

Combination of (2) and (3) and solution for T leads to

$$T = T_A T_B \left(\frac{P_A V_A^t + P_B V_B^t}{P_A V_A^t T_B + P_B V_B^t T_A} \right) \tag{4}$$

Application of the ideal-gas equation to the final state of the system gives the pressure as

$$P = \frac{(n_A + n_B) R T}{V_A^t + V_B^t}$$

Substitution for n_A and n_B by (3) and for T by (4) leads upon reduction to

$$P = \frac{P_A V_A^t + P_B V_B^t}{V_A^t + V_B^t} \tag{5}$$

The result (5) can be obtained in quite a different way. Multiplication of (1) by C_P/C_V gives

$$n_A C_P (T - T_A) + n_B C_P (T - T_B) = 0 \tag{6}$$

For an ideal gas with constant heat capacities, $\Delta H^t = n C_P \Delta T$. Therefore (6) may be written

$$\Delta H_A^t + \Delta H_B^t = 0$$

Since our original energy equation was $\Delta U_A^t + \Delta U_B^t = 0$, we may subtract to get

$$\Delta H_A^t - \Delta U_A^t + \Delta H_B^t - \Delta U_B^t = 0$$

However, by the definition of enthalpy, $\Delta H^t - \Delta U^t = \Delta(PV^t)$. Therefore

$$\Delta(PV^t)_A + \Delta(PV^t)_B = 0 \quad \text{or} \quad PV_A^t - P_A V_A^t + PV_B^t - P_B V_B^t = 0$$

which yields (5).

1.17   In Example 1.10 an equation was derived to relate T and V for an ideal gas with constant heat capacities that undergoes a reversible, adiabatic compression or expansion. Develop the analogous expression that relates T and P.

The ideal-gas equation gives

$$\frac{P_2}{P_1} \frac{T_1}{T_2} = \frac{V_1}{V_2} \quad \text{or} \quad \left(\frac{P_2}{P_1} \right)^{\gamma - 1} \left(\frac{T_1}{T_2} \right)^{\gamma - 1} = \left(\frac{V_1}{V_2} \right)^{\gamma - 1} = \frac{T_2}{T_1}$$

where the last equality follows from (1) of Example 1.10. Hence,

$$\left(\frac{P_2}{P_1} \right)^{\gamma - 1} = \left(\frac{T_2}{T_1} \right)^{\gamma} \quad \text{or} \quad \frac{T_2}{T_1} = \left(\frac{P_2}{P_1} \right)^{(\gamma - 1)/\gamma}$$

1.18   A rigid cylinder contains a "floating" piston, free to move within the cylinder without friction. Initially, it divides the cylinder in half, and on each side of the piston the cylinder holds 1 mol of the same ideal gas at 5 °C and 1 bar. An electrical resistance heater is installed on side A of the cylinder as shown in Fig. 1-8, and it is energized so as to cause the temperature in side A to rise slowly to 170 °C. If the tank and the piston are perfect heat insulators and are of negligible heat capacity, calculate the amount of heat added to the system by the resistor. The molar heat capacities of the gas are constant and have the values $C_V = (3/2)R$ and $C_P = (5/2)R$.

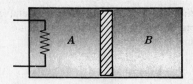

Fig. 1-8

We may write the first law to apply to the entire contents of the cylinder taken as the system:

$$\Delta U^t = Q - W$$

Since no force external to the system moves, $W = 0$. Hence

$$Q = n_A \, \Delta U_A + n_B \, \Delta U_B = C_V(\Delta T_A) + C_V(\Delta T_B) \qquad (1)$$

it being given that $n_A = n_B = 1$. In this equation ΔT_A is known from the data, but ΔT_B is unknown and must be found. As heat is added to A, the temperature and pressure of the gas in A rise, and the gas expands, pushing the piston to the right. This compresses the gas in B, and since the piston and cylinder are nonconducting and of negligible heat capacity, the process in B is adiabatic. The assumption of reversibility is also reasonable, because the piston is frictionless and the process is slow. For a reversible, adiabatic compression of an ideal gas with constant heat capacities, we have from Problem 1.17

$$\frac{T_{B2}}{T_1} = \left(\frac{P_2}{P_1}\right)^{(\gamma-1)/\gamma} \qquad (2)$$

Note that no letter subscripts are needed on P, because the pressure is always uniform throughout the cylinder. The total volume of the cylinder V^t is constant and is given by the ideal-gas equation applied both for the initial conditions and for the final conditions:

$$V^t = (n_A + n_B)RT_1/P_1 \qquad \text{(initial conditions)}$$
$$V^t = (n_A RT_{A2} + n_B RT_{B2})/P_2 \qquad \text{(final conditions)}$$

Equating these expressions and setting $n_A = n_B = 1$, we get

$$2\frac{P_2}{P_1} = \frac{T_{A2}}{T_1} + \frac{T_{B2}}{T_1} \qquad (3)$$

By combination of (2) and (3),

$$2\frac{P_2}{P_1} = \frac{T_{A2}}{T_1} + \left(\frac{P_2}{P_1}\right)^{(\gamma-1)/\gamma}$$

The ratio P_2/P_1 can now be determined by trial. We have

$$T_{A2} = 170 + 273.15 = 443.15 \text{ K} \qquad T_1 = 5 + 273.15 = 278.15 \text{ K}$$

and $(\gamma - 1)/\gamma = (\frac{5}{3} - 1)/\frac{5}{3} = 0.4$. Thus,

$$2\left(\frac{P_2}{P_1}\right) = \frac{443.15}{278.15} + \left(\frac{P_2}{P_1}\right)^{0.4} \qquad \text{whence} \qquad \frac{P_2}{P_1} = 1.3624$$

We now determine T_{B2} by (2):

$$T_{B2} = (278.15)(1.3624)^{0.4} = 314.76 \text{ K}$$

or 41.61 °C. Thus, by (1),

$$Q = \tfrac{3}{2}(8.314)[(170 - 5) + (41.61 - 5)] = 2514 \text{ J}$$

1.19 The differential equation relating hydrostatic pressure to fluid depth is (see Problem 1.1) $dP = -\rho g \, dz$, where ρ is the local mass density, g is the local acceleration of gravity, and z is elevation above a datum level in the fluid. Apply this equation to develop an expression for atmospheric pressure as a function of elevation above the surface of the earth.

For the temperatures and pressures involved, air behaves essentially as an ideal gas, for which $PV = RT$. The relation between ρ and V, the molar volume of air, is $\rho = M/V$, where M is the molar mass of air. Substitution in the hydrostatic formula gives

$$dP = -\frac{Mg}{V}\, dz$$

and elimination of V by the ideal-gas equation provides

$$\frac{dP}{P} = -\frac{Mg}{RT}\, dz \tag{1}$$

In order to integrate this equation we need to know how the atmospheric temperature T varies with elevation.

The simplest assumption is that T is constant. In this case (1) can be integrated immediately:

$$\int_{P_0}^{P} \frac{dP}{P} = -\frac{Mg}{RT_0} \int_0^z dz$$

whence
$$\ln \frac{P}{P_0} = -\frac{Mgz}{RT_0} \quad \text{or} \quad P = P_0 \exp\left(-\frac{Mgz}{RT_0}\right) \tag{2}$$

where P_0 is the pressure at the datum level, for which z is zero. Note that g has been assumed independent of z. Equation (2) is known as the *barometric equation*, and is strictly applicable where T is independent of z. This condition obtains in the region of the atmosphere known as the *stratosphere*, at elevations between 11 km and 25 km.

Below 11 km there is continual movement of great masses of air upward and downward. These air masses expand and cool as they rise, and compress and warm as they descend. If we assume this process to be reversible and adiabatic, then the relation $T = \text{const} \times P^{(\gamma-1)/\gamma}$ (Problem 1.17) implies

$$\frac{dT}{T} = \left(\frac{\gamma-1}{\gamma}\right) \frac{dP}{P}$$

If in this equation we eliminate dP/P by (1) and solve for dT/dz, we get

$$\frac{dT}{dz} = -\left(\frac{\gamma-1}{\gamma}\right)\left(\frac{Mg}{R}\right) \equiv K \tag{3}$$

where the quantity K is a constant, provided that γ and g are taken as independent of T and z. This equation shows that the rate of change of temperature with elevation is constant, and measurements show that this is essentially so below 11 km. Integration of (3) provides the linear equation $T = T_0 + Kz$, where T_0 is the temperature at $z = 0$. Substitution of this equation into (1) gives

$$\frac{dP}{P} = \left(\frac{\gamma}{\gamma-1}\right) K \frac{dz}{T_0 + Kz}$$

Therefore,

$$\int_{P_0}^{P} \frac{dP}{P} = \left(\frac{\gamma}{\gamma-1}\right) K \int_0^z \frac{dz}{T_0 + Kz}$$

whence
$$P = P_0\left(1 + \frac{Kz}{T_0}\right)^{\gamma/(\gamma-1)} \tag{4}$$

We have found two expressions, (2) and (4), which are based on the two extremes of possible temperature variation in the atmosphere. A more general empirical equation relating T and P is

$$\frac{T}{T_0} = \left(\frac{P}{P_0}\right)^{(\delta-1)/\delta}$$

where δ is a number between 1 and γ. When $\delta = \gamma$ we have the adiabatic case, which yields (4), and when $\delta = 1$ we have the isothermal case, which yields (2). The actual situation is in between. Measurements show that $\delta \approx 1.24$, and the equation for atmospheric pressure as a function of elevation is exactly like (4), but with γ replaced by δ:

$$P = P_0\left(1 + \frac{Kz}{T_0}\right)^{\delta/(\delta-1)} \tag{5}$$

where
$$K = \frac{dT}{dz} = -\left(\frac{\delta - 1}{\delta}\right)\left(\frac{Mg}{R}\right)$$

Table 1-2 gives results for various elevations as calculated by the three equations (2), (4), and (5); in each case, $T_0 = 300$ K and $P_0 = 1$ bar.

Table 1-2

	P/bar		
z/km	(2)	(4) $\gamma = 1.4$	(5) $\delta = 1.24$
0.5	0.9447	0.9441	0.9443
1	0.8922	0.8905	0.8911
2	0.7961	0.7899	0.7920
5	0.5655	0.5367	0.5466
10	0.3198	0.2517	0.2757
20	0.1022	0.0250	0.0494

Supplementary Problems

BASIC CONCEPTS   (Section 1.1)

1.20   (a) Show that the difference between internal and external pressures for a spherical liquid drop is given by

$$P_i - P_e = \frac{2\gamma}{r}$$

where P_i is the internal pressure, P_e is the external pressure, γ is the surface tension of the liquid, and r is the radius of the drop. (b) Explain why this pressure difference for an air-filled bubble is $4\gamma/r$ rather than $2\gamma/r$. (c) For water at 50 °C, $\gamma = 0.063$ N $\cdot$ m^{-1}. What is the pressure difference $P_i - P_e$ for a drop of radius 5 μm?   Ans.   (c) 25.2 kPa

1.21   (a) What pressure in torr (see Problem 1.1) is equivalent to standard atmospheric pressure, 101.325 kPa? (b) What height of water at 25 °C in a standard gravitational field is equivalent to standard atmospheric pressure? The density of water at 25 °C is 0.9971×10^{-3} m$^3 \cdot$ kg^{-1}.
Ans.   (a) 760 torr   (b) 10.36 m

WORK   (Sections 1.2–1.5)

1.22   (a) A certain gas obeys the molar equation of state $P(V - b) = RT$, where R is the universal gas constant, and b is a constant such that $0 < b < V$ for all V. Derive an expression for the work obtained from a reversible, isothermal expansion of one mole of this gas from an initial volume V_i to a final volume V_f. (b) If the gas in (a) were an ideal gas, would the same process produce more or less work?

Ans.   (a) $W = RT \ln\left(\dfrac{V_f - b}{V_i - b}\right)$ (b) less

1.23   Figure 1-9 shows the relation of pressure to volume for a closed PVT system during a reversible process. Calculate the work done by the system for each of the three steps 12, 23, and 31, and for the entire process 1231.   Ans. $W_{12} = 105$ kJ $W_{23} = -60$ kJ $W_{31} = 0$ $W_{1231} = 45$ kJ

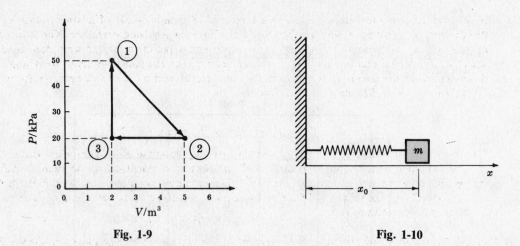

Fig. 1-9   Fig. 1-10

1.24 In Example 1.3 the *natural strain* ε was defined in terms of the length l of an axially stressed bar by the differential equation $d\varepsilon = dl/l$. Similarly, the *engineering strain* ε' can be defined by the equation $d\varepsilon' = dl/l_0$, where l_0 is the initial (unstrained) length of the bar. Show that ε and ε' are related by the equation $\varepsilon = \ln(1 + \varepsilon')$ and that the two measures of strain differ by less than 1% for ε' less than about 0.02.

1.25 (a) For a bar with no permanent strain, the stress σ is proportional to ε for small strains at constant temperature. Thus $\sigma = E\varepsilon$, where E is a constant property of the material. Show that the work done when such a bar is strained reversibly by an axial tensile load is given by $W = -VE\varepsilon^2/2$, where the volume V has been assumed constant. (b) A cylindrical brass rod of diameter 10 mm and initial length 100 mm is reversibly stretched by 0.1 mm. Calculate the work done if the volume remains constant. For brass, $E = 9 \times 10^{10}$ Pa.   *Ans.*   (b) $W = -0.35$ J

FIRST LAW OF THERMODYNAMICS   (Sections 1.6–1.9)

1.26 In some books the first law of thermodynamics for a closed system, equation (*1.3*), is written $\Delta E = Q + W$. Explain.   *Ans.* W is work done *on* the system *by* the surroundings.

1.27 (a) In the notation of Problem 1.12, give an exact expression for the acceleration of gravity g as a function of height z. (b) At what height does g become 1% of the value at the surface of the earth?   *Ans.*   (a) $g \equiv -F_{\text{grav}}/m = GM/(R + z)^2$ (b) $z = 9R$

1.28 A mass m is attached to a spring as shown in Fig. 1-10. When left by itself, the system assumes the position shown, where x_0 is the equilibrium position of the center of mass. If m is displaced from x_0, an elastic restoring force

$$F_e = -k\xi \tag{1}$$

acts on the mass, where k is the spring constant and ξ is the displacement ($\xi = x - x_0$). The work done by m against F_e can be incorporated in the mechanical-energy equation by treating F_e as a body force and proceeding as in Problems 1.11 and 1.12. (a) What are the *basic* SI units for k? (b) Show that the sign convention for force implied by (*1*) is consistent with the adopted coordinate system. (c) Define the elastic potential function Φ_e by $F_e = -d\Phi_e/d\xi$. If the zero of elastic potential energy is taken at the equilibrium position, show that $E_P = k\xi^2/2$. (d) It follows from part (c) and Problem 1.11 that the work W_B done by the system m against the body force F_e as m moves away from its equilibrium position is always positive. Why? (e) Write the mechanical-energy equation for the rectilinear translation of m if no forces other than F_e act on the mass. Assume that the *total* energy E_T of the system is known, and make use of the relationship $u = dx/dt = d\xi/dt \equiv \dot{\xi}$.
Ans.   (a) $\text{kg} \cdot \text{s}^{-2}$ (d) Work is done *on* the spring by the mass both in compressing and extending it.
      (e) $m\dot{\xi}^2 + k\xi^2 = 2E_T$

1.29   A scientist proposes to determine the heat capacities of liquids by use of a *Joule calorimeter*. In this device work is done by a paddle wheel on a liquid in an insulated container. The heat capacity is calculated from the measured temperature rise of the liquid and the measured work done by the paddle wheel. It is assumed that there is no heat exchanged between the liquid and its surroundings. To check this assumption, the scientist performs a preliminary experiment on 10 mol of benzene, for which C_P is $133.1 \ \text{J} \cdot \text{mol}^{-1} \cdot {}^{\circ}\text{C}^{-1}$. His data are as follows:

$$\text{Work done by the paddle wheel} = 6256 \ \text{J}$$
$$\text{Temperature rise of the liquid} = 4 \ {}^{\circ}\text{C}$$

If both C_P and the pressure on the liquid remain constant during the experiment, show that these results are not consistent with the stated assumptions, and offer an explanation for the inconsistency.

Ans.   The data do not satisfy the first law. The disparity could be due to a heat leak to the surroundings, of magnitude 932 J, or to an error of measurement; for example, an error of 0.7 °C in the temperature rise.

1.30   Figure 1-11 depicts two reversible processes undergone by one mole of an ideal gas. Curves T_a and T_b are isotherms, paths 23 and 56 are isobars, and paths 31 and 64 are isochores (paths of constant volume). Show that W and Q are the same for processes 1231 and 4564.

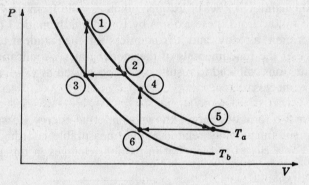

Fig. 1-11

1.31   An automobile tire is inflated to 270 kPa at the beginning of a trip. After three hours of high-speed driving the pressure is 300 kPa. What is the internal-energy change of the air in the tire between pressure measurements? Assume that air is an ideal gas with a constant heat capacity $C_V = \frac{5}{2}R$ and that the internal volume of the tire remains constant at $0.057 \ \text{m}^3$.   *Ans.* $\Delta U^t = 4.275 \ \text{kJ}$

1.32   A perfectly insulated rigid container of total volume V^t is divided into two parts by a partition of negligible volume. One side of the partition contains n moles of an ideal gas with constant heat capacities at a temperature T_i, and the other side is evacuated. If the partition is broken, calculate Q, W, and ΔU for the ensuing process, and calculate the final temperature and pressure, T_f and P_f, of the gas.
Ans. $Q = 0$ $W = 0$ $\Delta U = 0$ $T_f = T_i$ $P_f = nRT_i/V^t$

Chapter 2

The Second Law of Thermodynamics

2.1 AXIOMATIC STATEMENTS OF THE FIRST AND SECOND LAWS. ENTROPY

We pointed out in Chapter 1 that the concept of energy is primitive, and that the first law of thermodynamics, expressing the requirement that energy be conserved, is also primitive. The conservation principle originated in mechanics, where its application to rigid bodies in the absence of friction provides a relation between the external forms of energy and mechanical work. This principle of conservation of mechanical energy is readily tested by experiment and found valid. It is also directly related to Newton's second law of motion. Although historically slow, the transition from a limited conservation principle in mechanics to an all-inclusive conservation law in thermodynamics does not today seem difficult. The key step is the recognition that heat is a form of energy and that the quantity called internal energy is an intrinsic property of matter. The test is experiment and experience; now that all this is firmly established, the simplest procedure is to formalize the basic principles of thermodynamics in a set of axioms, taken as valid from the beginning. The multitude of consequences which follow from these axioms by formal mathematical deduction have been amply compared with experiment already, and it is pointless for each student to retrace the historical path that led to discovery of the fundamentals of thermodynamics. Application of the axioms presented below to familiar situations will lead to results easily recognized as valid, and no other justification of the axioms should be necessary.

Axiom 1: There exists a form of energy, known as internal energy U, which is an intrinsic property of a system, functionally related to the measurable coordinates which characterize the system. For a closed system, not in motion, changes in this property are given by

$$dU = \delta Q - \delta W \qquad (1.5)$$

Axiom 2 (First Law of Thermodynamics): The *total* energy of any system and its surroundings, considered together, is conserved.

The first axiom asserts the existence of a function called internal energy and provides a relationship connecting it with measurable quantities. This equation in no way explicitly defines internal energy: *there is no definition*. What is provided is a means for calculation of changes in this function: absolute values are unknown. The second axiom depends upon the first, and is regarded as one of the fundamental laws of science.

Nature places a second restriction on all processes, and this is given formal statement by two additional axioms:

Axiom 3: There exists a property called *entropy S*, which is an intrinsic property of a system, functionally related to the measurable coordinates which characterize the system. For a reversible process, changes in this property are given by

$$dS = \frac{\delta Q}{T} \qquad (2.1)$$

Axiom 4 (Second Law of Thermodynamics): The entropy change of any system and its surroundings, considered together, is positive and approaches zero for any process which approaches reversibility.

The third axiom does for the entropy function what the first axiom does for the internal energy, asserting its existence and providing a relationship which connects it with measurable quantities. Again there is no explicit definition; entropy is another primitive concept. Equation (2.1) allows calculation of changes in this function, but does not permit determination of absolute values. The fourth axiom clearly depends on the third and is, of course, its motivation. The second law of thermodynamics is a conservation law only for reversible processes, which are unknown in nature. All natural processes result in an increase in total entropy. The mathematical expression of the second law is simply:

$$\boxed{\Delta S_{\text{total}} \geqq 0} \tag{2.2}$$

where the label "total" indicates that both the system and its surroundings are included. The equality applies only to the limiting case of a reversible process.

EXAMPLE 2.1 Show that any flow of heat between two heat reservoirs at temperatures T_H and T_C, where $T_H > T_C$, must be from the hotter to the cooler reservoir.

A heat reservoir is by definition a body with an infinite heat capacity, and its entropy change is given by (2.1) regardless of the source or sink of the heat. The reason is that no irreversibilities occur *within* the reservoir, and changes in the reservoir depend only on the quantity of heat transferred and not on where it comes from or goes to. When a finite quantity of heat is added to or extracted from a heat reservoir, the reservoir undergoes a finite entropy change *at constant temperature*, and (2.1) becomes

$$\Delta S = \frac{Q}{T}$$

Let the quantity of heat Q pass from one reservoir to the other. The magnitude of Q is the same for both reservoirs, but Q_H and Q_C are of opposite sign, for heat added to one reservoir (considered positive) is heat extracted from the other (considered negative). Therefore, $Q_H = -Q_C$. By (2.1),

$$\Delta S_H = \frac{Q_H}{T_H} = \frac{-Q_C}{T_H} \qquad \text{and} \qquad \Delta S_C = \frac{Q_C}{T_C}$$

Thus

$$\Delta S_{\text{total}} = \Delta S_H + \Delta S_C = \frac{-Q_C}{T_H} + \frac{Q_C}{T_C} = Q_C\left(\frac{T_H - T_C}{T_H T_C}\right)$$

According to the second law, (2.2), ΔS_{total} must be positive; therefore

$$Q_C(T_H - T_C) > 0$$

Since by the problem statement $T_H > T_C$, Q_C must clearly be positive, and must therefore represent heat *added to* the reservoir at T_C. Thus heat must flow from the higher-temperature reservoir at T_H to the lower-temperature reservoir at T_C, a result which is entirely in accord with experience.

Example 2.1 deals with pure heat transfer, for which the temperature difference $T_H - T_C$ is the driving force. For such a process our results show that ΔS_{total} becomes zero only for $T_H = T_C$, which is the condition of thermal equilibrium between the two heat reservoirs. Reversible heat transfer occurs when the two reservoirs have temperatures that differ only infinitesimally.

EXAMPLE 2.2 What restrictions do the laws of thermodynamics impose on the production of work by a device that exchanges heat with the two reservoirs of Example 2.1, but which itself remains unchanged? Such a device is known as a *heat engine*.

Let us suppose that the device exchanges heat Q_H with the reservoir at T_H and Q_C with the reservoir at T_C. The symbols Q_H and Q_C refer to the heat reservoirs, and the signs on their numerical values are therefore determined by whether heat is added to or extracted from the heat reservoirs. Thus the entropy changes of the heat reservoirs are given by

$$\Delta S_H = \frac{Q_H}{T_H} \quad \text{and} \quad \Delta S_C = \frac{Q_C}{T_C}$$

These same heat quantities apply to the engine, but with opposite signs. Thus we let Q_H' and Q_C' be symbols for the heat exchanges taken with respect to the engine, where

$$Q_H = -Q_H' \quad \text{and} \quad Q_C = -Q_C'$$

The total entropy change resulting from any process involving just the engine and the heat reservoirs is

$$\Delta S_{\text{total}} = \Delta S_H + \Delta S_C + \Delta S_{\text{engine}}$$

Since the engine remains unchanged, the last term is zero, and we have

$$\Delta S_{\text{total}} = \frac{Q_H}{T_H} + \frac{Q_C}{T_C} \tag{1}$$

The first law as written for the *engine* has the form (1.4):

$$\Delta U = Q - W$$

where ΔU is the internal-energy change of the engine, W is the work done by the engine, and Q represents all heat transfer *with respect to the engine*. Thus

$$\Delta U_{\text{engine}} = Q_H' + Q_C' - W$$

Again, since the engine remains unchanged, $\Delta U_{\text{engine}} = 0$ and therefore

$$W = Q_H' + Q_C' = -Q_H - Q_C \tag{2}$$

Combination of (1) and (2) to eliminate Q_H gives

$$W = -T_H \Delta S_{\text{total}} + Q_C\left(\frac{T_H}{T_C} - 1\right) \tag{3}$$

As applied to a heat engine, this equation is valid within two limits. First, we are dealing with a work-producing device, for which W must be positive, with a limiting value of zero. At this limit the engine is completely ineffective: the process reduces to simple heat transfer between the reservoirs, and (3) reduces to the result of Example 2.1.

The second limit on (3) is represented by the reversible process, for which ΔS_{total} becomes zero by the second law, and W attains its maximum value for given values of T_H and T_C. In this case (3) reduces to

$$W = Q_C\left(\frac{T_H}{T_C} - 1\right) \tag{4}$$

From this result it is clear that for W to have a positive finite value, Q_C must also be positive and finite. This means that even in the limiting case of reversible operation it is necessary that heat Q_C be *rejected from* the engine and *absorbed by* the cooler reservoir at T_C.

Combination of (2) and (4), first, to eliminate W and, second, to eliminate Q_C, provides the following two equations:

$$\frac{Q_C}{T_C} = \frac{-Q_H}{T_H} \tag{5}$$

and

$$\frac{W}{-Q_H} = 1 - \frac{T_C}{T_H} \tag{6}$$

These are known as *Carnot's equations*, and apply to all reversible heat engines operating between fixed temperature levels, i.e., to all *Carnot engines*.

The various quantities considered in this example are shown schematically in Fig. 2-1.

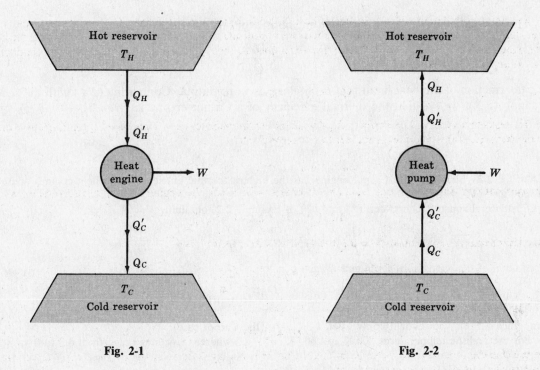

Fig. 2-1   Fig. 2-2

2.2 HEAT ENGINES AND HEAT PUMPS

In Carnot's equations, (5) and (6) of Example 2.2, the Q's refer to the heat reservoirs, and since Q_H represents a negative number (heat out of the reservoir), the minus sign in front of Q_H makes the term positive. Carnot's equations are commonly written without the minus signs; in this event the sign convention for Q is disregarded, and all Q's are taken as positive numbers. This is perhaps better done with the aid of absolute-value signs:

$$\left| \frac{Q_C}{Q_H} \right| = \frac{T_C}{T_H} \quad \text{and} \quad \left| \frac{W}{Q_H} \right| = 1 - \frac{T_C}{T_H} \qquad (2.3)$$

The first equation (2.3) shows that the ratio of heat discarded by a Carnot heat engine at T_C to the heat taken in at T_H is equal to the ratio of the two absolute temperatures. Thus the only way to make $|Q_C|$ zero is to discard heat to a reservoir at absolute zero of temperature. Since no reservoir at a temperature even approaching absolute zero is naturally available on earth, there is no practical means by which we can operate a heat engine so that it discards no heat.

This observation with respect to heat engines is so basic that its formal statement is often regarded as an alternative expression of the second law of thermodynamics: *It is impossible to construct an engine that, operating in a cycle, will produce no effect other than the extraction of heat from a reservoir and the performance of an equivalent amount of work.* This is the *Kelvin/Planck statement* of the second law. All heat engines must discard part of the heat they take in, and the natural heat reservoirs available to absorb this discarded heat are the atmosphere, lakes and rivers, and the oceans. The temperatures of these are of the order of 300 K.

The practical heat reservoirs at T_H are objects such as furnaces maintained at high temperature by the combustion of fossil fuels and nuclear reactors maintained at high temperature by fission of radioactive elements. The common components of all stationary power plants that generate our electricity are a high-temperature source of heat; a heat engine, which may be very complex; and a sink for the discharge of waste heat, namely our environment. This discharge of waste heat to the environment, or *thermal pollution*, is an inevitable consequence of the second law of thermodynamics.

The *thermal efficiency* η of any heat engine is arbitrarily defined as

$$\eta \equiv \frac{|W|}{|Q_H|} \tag{2.4}$$

i.e., the fraction of the heat input that is obtained as work output. Comparing (2.4) with the second equation (2.3), we see that the thermal efficiency of a Carnot engine is given by

$$\eta^* = 1 - \frac{T_C}{T_H} \tag{2.4*}$$

EXAMPLE 2.3 Show that the Carnot engine has the highest thermal efficiency of all heat engines operating between given T_C and T_H.

First, we eliminate Q_C between (1) and (2) of Example 2.2, obtaining

$$\Delta S_{\text{total}} = -\frac{Q_H}{T_C}\left(1 - \frac{T_C}{T_H} + \frac{W}{Q_H}\right) = \frac{|Q_H|}{T_C}\left(1 - \frac{T_C}{T_H} - \frac{|W|}{|Q_H|}\right)$$

Substitution of the expressions for η and η^* gives

$$\Delta S_{\text{total}} = \frac{|Q_H|}{T_C}(\eta^* - \eta)$$

from which $\eta \leqq \eta^*$, with equality only when $\Delta S_{\text{total}} = 0$ (the Carnot case).

For the realistic temperatures 300 K and 600 K, $\eta^* = 0.5$ whereas η-values range from 0.3 to 0.4.

A reversible heat engine may be reversed to run as a *heat pump* or *refrigerator*, as shown schematically in Fig. 2-2. Equations (2.3) are equally valid for a Carnot heat pump, because all quantities are the same as for a Carnot heat engine operating between the same two temperature levels. The only difference is that the directions of heat transfer are reversed, and work is required rather than performed. The work is used to "pump" heat from the heat reservoir at the lower temperature T_C to the reservoir at the higher temperature T_H. Refrigerators work on this principle. The reservoir at T_C is the "cold box" and the reservoir at T_H is the environment to which heat is discarded.

The fact that work is required in order to produce a refrigeration effect is again a basic observation which may be formalized by another negative statement that gives expression to the second law of thermodynamics: *It is impossible to construct a device that, operating in a cycle, will produce no effect other than the transfer of heat from a cooler to a hotter body.* This is known as the *Clausius statement* of the second law.

The important parameter for a heat pump or refrigerator is the ratio of heat removed at low temperature to the work input:

$$\omega \equiv \frac{|Q_C|}{|W|} \tag{2.5}$$

This quantity is called the coefficient of performance, and is obtained for a Carnot heat pump by dividing the first of (2.3) by the second:

$$\omega^* = \frac{T_C}{T_H - T_C} \tag{2.5*}$$

EXAMPLE 2.4 To maintain a freezer box at $-40\,°C$ on a summer day when the ambient temperature is $27\,°C$, heat is removed at the rate of 1.25 kW. What is the maximum possible coefficient of performance of the freezer, and what is the minimum power that must be supplied to the freezer?

The maximum coefficient of performance is obtained when the freezer operates reversibly (i.e., as a Carnot refrigerator) between the temperature levels

$$T_C = -40 + 273.15 = 233.15 \text{ K} \qquad \text{and} \qquad T_H = 27 + 273.15 = 300.15 \text{ K}$$

in which case

$$\omega^* = \frac{233.15}{300.15 - 233.15} = 3.48$$

Since $|\dot{Q}_C| = 1.25$ kW, the minimum power requirement is

$$|\dot{W}| = \frac{|\dot{Q}_C|}{\omega^*} = \frac{1.25 \text{ kW}}{3.48} = 359 \text{ W}$$

2.3 ENTROPY OF AN IDEAL GAS

The first law for a closed PVT system is given in differential form by (1.5): $dU = \delta Q - \delta W$. Now, for a reversible process, we may write (2.1) as $\delta Q = T\,dS$; in addition, (1.2) gives $\delta W = P\,dV$. Combining these three equations, we have

$$\boxed{dU = T\,dS - P\,dV} \tag{2.6}$$

This is a *general* equation relating the properties of a closed PVT system. We have simply taken an especially simple process for the purpose of derivation; since the resulting equation involves properties only, it must be independent of the process. It is in fact the *fundamental property relation* for a closed PVT system, and all other property relations are derived from it, as will be shown in Chapter 3.

For the special case of an ideal gas, we have from (1.13) that $dU = C_V\,dT$, and (2.6) yields

$$dS = C_V \frac{dT}{T} + \frac{P}{T}\,dV$$

However, for an ideal gas $P/T = R/V$, whence

$$dS = C_V \frac{dT}{T} + R \frac{dV}{V} \qquad \text{(ideal gas)} \tag{2.7}$$

This equation shows that the entropy of an ideal gas is a function of T and V; i.e., $S = S(T, V)$. If we integrate (2.7) between the limits T_1 and T_2, the result is

$$\Delta S = \int_{T_1}^{T_2} C_V \frac{dT}{T} + R \ln \frac{V_2}{V_1} \tag{2.8}$$

Alternatively, we can consider the entropy a function of T and P, i.e., $S = S(T, P)$. Differentiating $PV = RT$ and then dividing by PV gives

$$\frac{dV}{V} + \frac{dP}{P} = \frac{dT}{T}$$

Eliminating dV/V between this and (2.7), and substituting $C_V = C_P - R$, we find:

$$dS = C_P \frac{dT}{T} - R \frac{dP}{P} \qquad \text{(ideal gas)} \tag{2.9}$$

$$\Delta S = \int_{T_1}^{T_2} C_P \frac{dT}{T} - R \ln \frac{P_2}{P_1} \tag{2.10}$$

EXAMPLE 2.5 For an ideal gas with constant heat capacities, (2.7) and (2.9) integrate to give

$$S = C_V \ln T + R \ln V + S_0 \qquad \text{and} \qquad S = C_P \ln T - R \ln P + S_0'$$

where S_0 and S_0' are constants of integration. Show that a third equation of this kind is

$$S = C_V \ln P + C_P \ln V + S_0''$$

Of the many ways to derive the required equation one of the simplest is to start with the first equation above and to eliminate T by the ideal-gas equation. This gives

$$S = C_V \ln P + C_V \ln V - C_V \ln R + R \ln V + S_0$$

But

$$C_V \ln V + R \ln V = (C_V + R) \ln V = C_P \ln V \quad \text{and} \quad S_0 - C_V \ln R = S_0''$$

Combination of the last three equations yields the desired result. Note that V is *molar* volume.

EXAMPLE 2.6 Derive equations relating T and V, T and P, and P and V for the reversible, adiabatic compression or expansion of an ideal gas with constant heat capacities.

For a reversible process $dS = \delta Q / T$. If in addition the process is adiabatic, then $\delta Q = 0$, and so $dS = 0$. In other words, a reversible, adiabatic process is *isentropic*.

If dS is set equal to zero in (2.7), the result is

$$\frac{dT}{T} = -\frac{R}{C_V} \frac{dV}{V}$$

For an ideal gas $R = C_P - C_V$, and therefore

$$\frac{dT}{T} = -\left(\frac{C_P - C_V}{C_V}\right) \frac{dV}{V} = -(\gamma - 1) \frac{dV}{V}$$

where $\gamma = C_P / C_V$. Integration with γ constant yields

$$\ln \frac{T_2}{T_1} = \ln \left(\frac{V_1}{V_2}\right)^{\gamma - 1} \quad \text{or} \quad \frac{T_2}{T_1} = \left(\frac{V_1}{V_2}\right)^{\gamma - 1}$$

This same result was obtained in Example 1.10 by a different method. An equivalent statement is

$$TV^{\gamma - 1} = \text{constant}$$

If we treat (2.9) in an analogous way, we get

$$\frac{T_2}{T_1} = \left(\frac{P_2}{P_1}\right)^{(\gamma - 1)/\gamma} \quad \text{or} \quad \frac{T}{P^{(\gamma - 1)/\gamma}} = \text{constant}$$

Equating the two expressions just obtained for T_2/T_1 gives

$$\left(\frac{V_1}{V_2}\right)^{\gamma} = \frac{P_2}{P_1} \quad \text{or} \quad PV^{\gamma} = \text{constant}$$

EXAMPLE 2.7 Show that for an ideal gas with constant heat capacities the slope of a PV curve for an isentropic process is negative and that it has a larger absolute value than the slope of a PV curve for an isotherm at the same values of P and V.

It was shown in Example 2.6 that for an isentropic process, $PV^{\gamma} = k$, or $P = kV^{-\gamma}$, where k is a constant. Differentiation of this equation gives

$$\frac{dP}{dV} = k(-\gamma)V^{-\gamma - 1} = -\gamma(kV^{-\gamma})(V^{-1})$$

or

$$\frac{dP}{dV} = -\gamma \frac{P}{V} \quad \text{(isentropic)}$$

Since γ, P, and V are all positive, dP/dV is negative.

For an isotherm: $PV = RT$, or $P = RTV^{-1}$. Differentiation gives

$$\frac{dP}{dV} = RT(-1)(V^{-2}) = -(RTV^{-1})V^{-1}$$

or

$$\frac{dP}{dV} = -\frac{P}{V} \quad \text{(isothermal)}$$

Again the slope dP/dV is negative; but since $\gamma > 1$, the isentropic curve has a steeper negative slope than the isothermal curve passing through the same point. This is illustrated in Fig. 2-3.

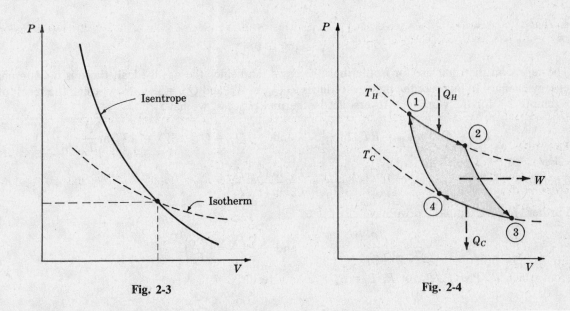

Fig. 2-3   Fig. 2-4

2.4 CARNOT CYCLE FOR AN IDEAL GAS

Our object now is to determine the shape of the cycle on a PV diagram that would be traced out by a Carnot heat engine operating between heat reservoirs at T_H and T_C. The engine considered is a piston-and-cylinder apparatus with an ideal gas as the working medium. Since the engine must be capable of performing continuously and must itself not undergo any permanent change, it must return periodically to its initial state; i.e., it must operate in a cycle. As the ideal gas in the engine changes state, it goes through a sequence of values of P and V, and these represent a closed cycle on a PV diagram.

If the engine is to operate reversibly, it must absorb heat Q_H from the hot reservoir at temperature T_H, and *only* at T_H; and it must discard heat Q_C to the cold reservoir at temperature T_C, and *only* at T_C. This means that steps of the cycle during which heat is transferred must be isothermal, either at T_H or at T_C, and all other steps must represent adiabatic, reversible processes between these two temperatures. The only possible combination of such steps that represents a work-producing cycle is shown in Fig. 2-4, which displays a *Carnot cycle* for an ideal gas.

At point 1 the apparatus is in thermal equilibrium with the hot reservoir at T_H, and the piston is at the inner limit of its stroke. The gas is now allowed to expand reversibly and isothermally along path $1 \rightarrow 2$. During this step, heat Q_H flows into the gas to compensate for the work done. At point 2 the apparatus is insulated from both heat reservoirs and the gas expands reversibly and adiabatically along path $2 \rightarrow 3$. As shown in Example 2.7, path $2 \rightarrow 3$ is steeper than path $1 \rightarrow 2$, and it leads away from the isotherm T_H and terminates at isotherm T_C. At point 3 equilibrium is established with the cold reservoir at T_C, and a reversible isothermal compression along path $3 \rightarrow 4$ carries the piston inward. The work done on the gas during this step is compensated for by removal of heat Q_C. This step terminates at a point such that reversible, adiabatic compression will complete the cycle along path $4 \rightarrow 1$.

The enclosed area in Fig. 2-4 represents $\oint P\,dV$ taken around the entire cycle, which is the net work of the engine for one complete cycle. This net work is the sum of the work terms for the various steps:

$$W = W_{12} + W_{23} + W_{34} + W_{41}$$

Processes $2 \rightarrow 3$ and $4 \rightarrow 1$ are adiabatic, and for them the first law and the relation $U = U(T)$ for an ideal gas give

$$W_{23} + W_{41} = [U(T_H) - U(T_C)] + [U(T_C) - U(T_H)] = 0$$

so that

$$W = W_{12} + W_{34}$$

The terms on the right are for isothermal processes, and since the gas is ideal, there is no internal energy change. In this case the first law reduces to $Q_{12} = W_{12}$ and $Q_{34} = W_{34}$. Now, using the result of Example 1.4 for the work of a reversible, isothermal process, we have

$$Q_H = Q_{12} = W_{12} = RT_H \ln \frac{P_1}{P_2} \qquad \text{and} \qquad Q_C = Q_{34} = W_{34} = RT_C \ln \frac{P_3}{P_4}$$

Therefore

$$\frac{Q_C}{Q_H} = \frac{T_C}{T_H} \frac{\ln (P_3/P_4)}{\ln (P_1/P_2)}$$

Further, from Example 2.6 we have

$$\frac{T_H}{T_C} = \left(\frac{P_2}{P_3}\right)^{(\gamma-1)/\gamma} \qquad \text{and} \qquad \frac{T_H}{T_C} = \left(\frac{P_1}{P_4}\right)^{(\gamma-1)/\gamma}$$

from which $P_2/P_3 = P_1/P_4$, or $P_1/P_2 = P_4/P_3$. As a result,

$$\frac{Q_C}{Q_H} = -\frac{T_C}{T_H} \qquad \text{or} \qquad \frac{|Q_C|}{|Q_H|} = \frac{T_C}{T_H}$$

which is the first Carnot equation (2.3). Combination with the first-law equation for the cycle, $|W| = |Q_H| - |Q_C|$ leads immediately to

$$\frac{|W|}{|Q_H|} = 1 - \frac{T_C}{T_H}$$

which is the second Carnot equation (2.3).

The fact that we regenerate the Carnot equations by consideration of the properties of an ideal gas as it serves as the working medium of a Carnot heat engine, means that the temperature used in the ideal-gas equation is entirely consistent with the absolute temperature used in (2.1), the basic relation connecting entropy with measurable quantities. This has been implicitly assumed all along, but now we have established a formal connection.

Many cyclic processes are possible besides the Carnot cycle; examples are the Otto cycle of gasoline engines and the Diesel cycle. Any reversible cyclic process is susceptible to the same kind of thermodynamic analysis as the Carnot cycle.

EXAMPLE 2.8 Develop an expression for the thermal efficiency of a reversible heat engine operating on the Otto cycle with an ideal gas of constant heat capacity as the working medium.

The Otto cycle consists of two constant-volume steps during which heat is transferred, connected by reversible adiabatic steps (isentropes), as shown in Fig. 2-5. For the various steps of the cycle, we have

$$
\begin{array}{llll}
\textbf{Step } 1 \rightarrow 2 & Q_{12} = 0 & W_{12} = -\Delta U_{12} = -C_V(T_2 - T_1) \\
\textbf{Step } 2 \rightarrow 3 & W_{23} = 0 & Q_{23} = \Delta U_{23} = C_V(T_3 - T_2) \\
\textbf{Step } 3 \rightarrow 4 & Q_{34} = 0 & W_{34} = -\Delta U_{34} = -C_V(T_4 - T_3) \\
\textbf{Step } 4 \rightarrow 1 & W_{41} = 0 & Q_{41} = \Delta U_{41} = C_V(T_1 - T_4)
\end{array}
$$

By (2.4), the thermal efficiency of the cycle is

$$\eta = \frac{W_{\text{cycle}}}{Q_{\text{in}}} = \frac{W_{12} + W_{34}}{Q_{23}} = \frac{-C_V(T_2 - T_1) - C_V(T_4 - T_3)}{C_V(T_3 - T_2)} = 1 - \frac{T_4 - T_1}{T_3 - T_2}$$

Since steps $1 \rightarrow 2$ and $3 \rightarrow 4$ represent isentropic processes, we may write (see Example 2.6)

$$\frac{T_1}{T_2} = \left(\frac{V_2}{V_1}\right)^{\gamma-1} \qquad \text{and} \qquad \frac{T_4}{T_3} = \left(\frac{V_3}{V_4}\right)^{\gamma-1}$$

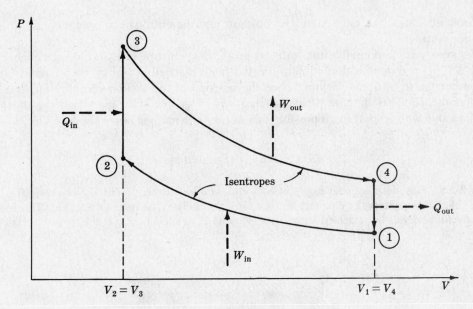

Fig. 2-5

However, $V_2 = V_3$ and $V_1 = V_4$, and therefore the right sides of the above equations are equal. As a result

$$\frac{T_1}{T_2} = \frac{T_4}{T_3} \qquad \text{or} \qquad T_4 = \frac{T_1 T_3}{T_2}$$

Elimination of T_4 in the equation for η gives

$$\eta = 1 - \frac{T_1}{T_2} = 1 - \left(\frac{V_2}{V_1}\right)^{\gamma-1}$$

The quantity V_1/V_2 is known as the *compression ratio r*, and this provides yet another expression for thermal efficiency:

$$\eta = 1 - \left(\frac{1}{r}\right)^{\gamma-1}$$

Clearly η gets larger as r increases, but η becomes unity only as $r \to \infty$.

2.5 ENTROPY AND EQUILIBRIUM

In the mathematical statement of the second law, $\Delta S_{\text{total}} \geqq 0$, the subscript "total" indicates that both the system and its surroundings must be taken into account. An alternative expression makes this explicit:

$$\Delta S_{\text{system}} + \Delta S_{\text{surroundings}} \geqq 0$$

The first law may be expressed similarly:

$$\Delta E_{\text{system}} + \Delta E_{\text{surroundings}} = 0$$

where E represents energy in general.

An *isolated* system is a system completely cut off from its surroundings, and changes in such a system have no effect on the surroundings. In this case we need consider the system only, and our equations become

$$\Delta S_{\text{system}} \geqq 0$$
$$\Delta E_{\text{system}} = 0$$

For an isolated system the total energy is constant, and the entropy can only increase or in the limit remain constant.

If the system is in an equilibrium state, its properties, entropy included, do not change. Thus the equality $\Delta S_{system} = 0$ implies that equilibrium has been reached, whereas the inequality $\Delta S_{system} > 0$ implies a change toward equilibrium. Since the entropy of the system can only *increase*, this must mean that the equilibrium state of an isolated system is that state for which the entropy has its maximum value with respect to all possible variations. The mathematical condition for this maximum is

$$dS_{system} = 0 \qquad \text{(isolated system)}$$

EXAMPLE 2.9 Consider a cylinder closed at both ends and containing gas divided into two parts by a piston, as in Fig. 2-6. Let the cylinder be a perfect heat insulator that isolates the system it contains. The piston will be a heat conductor and will be imagined frictionless. How is P_1 related to P_2, and how is T_1 related to T_2, at equilibrium?

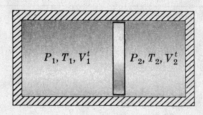

Fig. 2-6

Since the system is isolated, we may apply the equilibrium criterion $dS_{system} = 0$. To do so we need expressions for dS_1^t and dS_2^t, and these come from the property relation (2.6) solved for dS. Thus

$$dS_1^t = \frac{dU_1^t}{T_1} + \frac{P_1}{T_1} dV_1^t \qquad dS_2^t = \frac{dU_2^t}{T_2} + \frac{P_2}{T_2} dV_2^t$$

The sum of these gives dS_{system}:

$$dS_{system} = \frac{dU_1^t}{T_1} + \frac{dU_2^t}{T_2} + \frac{P_1}{T_1} dV_1^t + \frac{P_2}{T_2} dV_2^t$$

Since the system is isolated, its internal energy is constant, as is its volume. Thus

$$dU_2^t = -dU_1^t \qquad \text{and} \qquad dV_2^t = -dV_1^t$$

and we can write for equilibrium

$$dS_{system} = \left(\frac{1}{T_1} - \frac{1}{T_2} \right) dU_1^t + \left(\frac{P_1}{T_1} - \frac{P_2}{T_2} \right) dV_1^t = 0$$

For this equation to be valid for all imaginable variations in the independent variables U_1^t and V_1^t, the coefficients of dU_1^t and dV_1^t must be zero. Thus

$$\frac{1}{T_1} - \frac{1}{T_2} = 0 \qquad \text{and} \qquad \frac{P_1}{T_1} - \frac{P_2}{T_2} = 0$$

From these it is obvious that for equilibrium

$$T_1 = T_2 \qquad \text{and} \qquad P_1 = P_2$$

This answer was evident from the start. The important point is that our abstract criterion for equilibrium leads to results known to be correct. It can be applied to nontrivial problems as well.

Solved Problems

HEAT ENGINES AND HEAT PUMPS   (Sections 2.1–2.2)

2.1   (a) A reversible heat engine operates between a heat reservoir at $T_1 = 670$ K and another heat reservoir at $T_2 = 270$ K. If 100 kJ of heat is transferred to the heat engine from the reservoir at T_1, how much work is done by the engine? (b) The engine would produce more work if the temperature of the cold reservoir could be lowered to 220 K. Someone suggests that a finite "cold box" be held at 220 K by refrigeration, and proposes to remove heat from the "cold box" by a reversible heat pump which discharges heat to the original reservoir at $T_2 = 270$ K. The scheme is represented in Fig. 2-7. If, as before, $Q_1 = 100$ kJ, calculate the *net* work W of the process, and compare the result with the answer to part (a).

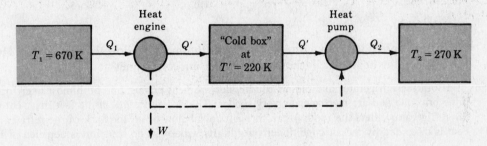

Fig. 2-7

(a)   By the second equation (2.3),

$$\frac{|W|}{|Q_1|} = 1 - \frac{T_2}{T_1} = 1 - \frac{270}{670} = 0.597$$

whence $|W| = 0.597\,|Q_1| = (0.597)(100) = 59.7$ kJ.

(b)   The result must be the same as in part (a), because the process represented is just a more complicated, but entirely equivalent, heat engine operating reversibly between the same two temperatures as the engine in part (a). However, if one needs proof, it is given as follows. The net work W is the difference between the total work of the heat engine and the work delivered to the heat pump:

$$W = |W|_{\text{engine}} - |W|_{\text{pump}} \tag{1}$$

The second equation (2.3) gives, for the engine,

$$|W|_{\text{engine}} = |Q_1|\left(1 - \frac{T'}{T_1}\right) = \frac{|Q_1|}{T_1}(T_1 - T') \tag{2}$$

and, for the heat pump, (2.5) and (2.5*) give

$$|W|_{\text{pump}} = \frac{|Q'|}{T'}(T_2 - T') \tag{3}$$

The first (2.3), applied to the engine, yields

$$|Q'| = |Q_1|\frac{T'}{T_1} \tag{4}$$

Combining (1) through (4) gives

$$W = \frac{|Q_1|}{T_1}(T_1 - T') - \frac{|Q_1|}{T_1}(T_2 - T') = |Q_1|\left(\frac{T_1 - T_2}{T_1}\right)$$

the formula used in part (a).

2.2 A thermoelectric device cools a small refrigerator and discards heat to the surroundings at 25 °C. The maximum electric power for which the device is designed is 100 W. The heat load on the refrigerator (that is, the heat leaking through the walls which must be removed by the thermoelectric device) is 350 W. What is the minimum temperature that can be maintained in the refrigerator?

The thermoelectric device is a heat pump, subject to the same thermodynamic limitations as any other heat pump. The minimum temperature is attained when the device operates reversibly at maximum power. For a reversible heat pump, (2.5) and (2.5*) give

$$\left|\frac{W}{Q_C}\right| = \frac{T_H}{T_C} - 1 \quad \text{or} \quad T_C = \frac{T_H}{|W/Q_C| + 1}$$

Substitution of $T_H = 25 + 273.15 = 298.15$ K; $W = 100$ W; and $Q_C = 350$ W yields

$$T_C = \frac{298.15}{(100/350) + 1} = 231.89 \text{ K} \quad \text{or} \quad -41.26 \text{ °C}$$

2.3 A kitchen is provided with a standard refrigerator and a window air conditioner. How is it that the refrigerator heats the kitchen whereas the air conditioner cools it?

Both the refrigerator and the air conditioner are heat pumps, and both most likely operate on the same principle and are mechanically very similar. The difference lies in the fact that heat discarded by the refrigerator enters the room (often through coils below or on the back of the refrigerator) whereas heat is discarded by the air conditioner outside the window. The situation is depicted in Fig. 2-8.

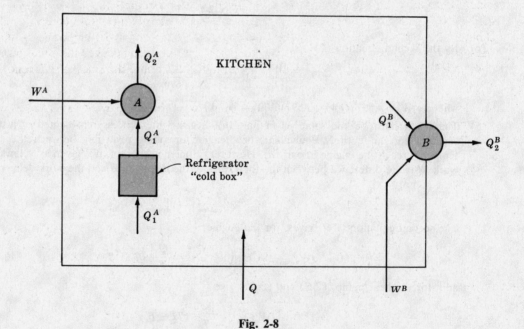

Fig. 2-8

The heat pumps are indicated by circles; the one designated A is in the refrigerator, and B is in the air conditioner. The work to drive both heat pumps is presumed to be supplied electrically, and this energy comes from outside the kitchen as shown. The refrigerator heat pump A removes heat $|Q_1^A|$ from the refrigerator "cold box" and discards heat $|Q_2^A|$ into the kitchen. An energy balance on heat pump A shows that $|Q_2^A| = |Q_1^A| + |W^A|$. Now if the "cold box" of the refrigerator stays at constant temperature, then the heat pump removes just the amount of heat that passes into the cold box, through its walls, from the kitchen. Thus the net effect of the refrigerator on the kitchen is to add energy as heat in an amount equal to the work $|W^A|$ of running the refrigerator.

The air conditioner holds the kitchen temperature constant by pumping heat $|Q_2^B|$ outdoors. The first law, applied to the entire kitchen, gives

$$|Q_2^B| = |Q| + |W^A| + |W^B|$$

where Q is the heat leakage into the kitchen.

2.4   An inventor claims to have devised a cyclic engine which exchanges heat with reservoirs at 25 °C and 250 °C, and which can produce 0.45 kJ of work per kJ of heat extracted from the hot reservoir. Is his claim feasible?

The efficiency of the proposed device cannot be higher than the Carnot efficiency; see Example 2.3. By (2.4),

$$\eta = \frac{0.45}{1.00} = 0.45$$

while the Carnot efficiency is given by (2.4*) as

$$\eta^* = 1 - \frac{298.15}{523.15} = 0.43$$

The claim is impossible.

2.5   (a) With reference to Example 2.8 and Fig. 2-5, take the working medium of an engine operating on the Otto cycle to be an ideal gas for which $C_V = (5/2)R$ and $C_P = (7/2)R$. The conditions at point 1 are $P_1 = 1$ bar and $T_1 = 60$ °C. The compression ratio of the engine is such as to make $P_2 = 15$ bar; heat is added in step $2 \to 3$ sufficient to make $T_3 = 2170$ °C. Determine the values of Q and W for the four steps of the cycle, the net work of the cycle, and the thermal efficiency of the engine. (b) If the engine accomplishes the *same changes of state as in* (a), but if each step is now 75% efficient, recompute the thermal efficiency of the engine.

(a)   In order to compute the required quantities by means of the equations of Example 2.8, we need first to determine the temperatures at the various points of the cycle. Given are

$$T_1 = 60 + 273.15 = 333.15 \text{ K} \qquad T_3 = 2170 + 273.15 = 2443.15 \text{ K}$$

Since step $1 \to 2$ is reversible and adiabatic, we have

$$T_2 = T_1 \left(\frac{P_2}{P_1}\right)^{(\gamma-1)/\gamma}$$

with $\gamma = C_P/C_V = 7/5$, $(\gamma-1)/\gamma = 2/7 = 0.286$. Thus

$$T_2 = (333.15)(15)^{0.286} = 722.77 \text{ K}$$

Example 2.8 provides the formula

$$T_4 = \frac{T_1 T_3}{T_2} = \frac{(333.15)(2443.15)}{722.77} = 1126.13 \text{ K}$$

With $C_V = (5/2)R = (5/2)(8.314)$ J·mol^{-1}·K^{-1}, the equations for Q and W of Example 2.8 yield:

Step $1 \to 2$ $Q_{12} = 0$ $W_{12} = -\Delta U_{12} = -C_V(T_2 - T_1) = -8098.25$ J·mol^{-1}
Step $2 \to 3$ $W_{23} = 0$ $Q_{23} = \Delta U_{23} = C_V(T_3 - T_2) = 35\,758.10$ J·mol^{-1}
Step $3 \to 4$ $Q_{34} = 0$ $W_{34} = -\Delta U_{34} = -C_V(T_4 - T_3) = 27\,374.26$ J·mol^{-1}
Step $4 \to 1$ $W_{41} = 0$ $Q_{41} = \Delta U_{41} = C_V(T_1 - T_4) = -16\,482.09$ J·mol^{-1}

The net work of the cycle is $W_{\text{cycle}} = W_{12} + W_{34} = 19\,276.01$ J·mol^{-1}; and, as derived in Example

2.8, the thermal efficiency of the cycle is

$$\eta = 1 - \frac{T_1}{T_2} = 1 - \frac{333.15}{722.77} = 0.539$$

(b) For an engine that operates irreversibly, but goes through the *same changes of state* as the reversible engine of part (a), we may calculate the work of each step by assigning an efficiency to the process. For steps that produce work the efficiency is defined as the ratio of the actual work to the reversible work:

$$\text{Efficiency} \equiv \frac{W}{W_{rev}} \quad \text{(work produced)}$$

This quantity is always less than unity, because it reflects the extent to which irreversibilities reduce the work output in comparison with the reversible work. For steps that require work, the reversible work is the minimum work, for irreversibilities always increase the work requirement. So in this case the efficiency is defined by

$$\text{Efficiency} \equiv \frac{W_{rev}}{W} \quad \text{(work required)}$$

This efficiency, too, is always less than unity.

For the present problem, step $1 \rightarrow 2$ requires work, and for an efficiency of 75% we have

$$W_{12} = \frac{W_{rev}}{0.75} = \frac{-8098.25}{0.75} = -10\,797.67 \text{ J} \cdot \text{mol}^{-1}$$

From the first law we have $\Delta U_{12} = Q_{12} - W_{12}$; but ΔU_{12} is the same as before, namely 8098.25 J $\cdot$ mol^{-1}. Thus

$$Q_{12} = \Delta U_{12} + W_{12} = 8098.25 - 10\,797.67 = -2699.42 \text{ J} \cdot \text{mol}^{-1}$$

and we see that in the irreversible engine this step cannot be adiabatic. Step $3 \rightarrow 4$ produces work, and for an efficiency of 75% the amount is given by

$$W_{34} = (0.75)W_{rev} = (0.75)(27\,374.26) = 20\,530.70 \text{ J} \cdot \text{mol}^{-1}$$

The heat transfer is again calculated by the first law:

$$Q_{34} = \Delta U_{34} + W_{34} = -27\,374.26 + 20\,530.70 = -6843.56 \text{ J} \cdot \text{mol}^{-1}$$

Steps $2 \rightarrow 3$ and $4 \rightarrow 1$ remain as before, because at constant volume no work is done. The net work of the cycle is

$$W_{cycle} = W_{12} + W_{34} = -10\,797.67 + 20\,530.70 = 9733.03 \text{ J} \cdot \text{mol}^{-1}$$

and, by (2.4),

$$\eta = \frac{W_{cycle}}{Q_{23}} = \frac{9733.03}{35\,758.10} = 0.272$$

Table 2-1

	(a) Reversible; $\eta = 0.539$	(b) Irreversible; $\eta = 0.272$
Q_{12}	0 kJ $\cdot$ mol^{-1}	-2.7 kJ $\cdot$ mol^{-1}
W_{12}	-8.1	-10.8
Q_{23}	35.8	35.8
W_{23}	0	0
Q_{34}	0	-6.8
W_{34}	27.4	20.5
Q_{41}	-16.5	-16.5
W_{41}	0	0
W_{cycle}	19.3	9.7

Only the basic definition for thermal efficiency η can be used here, because the derived formulas of Example 2.8 depend upon the assumption of reversibility. It is worthy of note that use of an efficiency of 75% for each step of the cycle has reduced the thermal efficiency of the engine by a factor of about 2. Thus, for the irreversible engine, only 27.2% of the heat taken into the engine is delivered as work, and the remaining 72.8% is discarded as waste heat to the surroundings. These figures are typical of heat engines in actual operation.

Table 2-1 compares the (rounded) results of (a) and (b).

2.6   Derive an expression for the thermal efficiency of a reversible heat engine operating on the Diesel cycle with an ideal gas of constant heat capacity as the working medium.

As shown in Fig. 2-9, the Diesel cycle consists of a constant-pressure and a constant-volume step, during both of which heat is transferred, connected by reversible adiabatic (isentropic) steps.

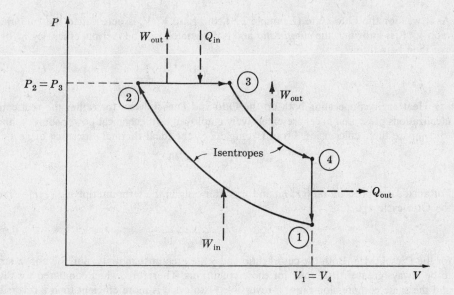

Fig. 2-9

Step 1 → 2 $Q_{12} = 0$ $W_{12} = -\Delta U_{12} = -C_V(T_2 - T_1)$

Step 2 → 3 $W_{23} = P_2(V_3 - V_2) = P_3(V_3 - V_2)$ $\Delta U_{23} = C_V(T_3 - T_2)$

$Q_{23} = W_{23} + \Delta U_{23} = P_2(V_3 - V_2) + C_V(T_3 - T_2)$

Step 3 → 4 $Q_{34} = 0$ $W_{34} = -\Delta U_{34} = -C_V(T_4 - T_3)$

Step 4 → 1 $W_{41} = 0$ $Q_{41} = \Delta U_{41} = C_V(T_1 - T_4)$

The thermal efficiency of the cycle is

$$\eta = \frac{W_{\text{cycle}}}{Q_{\text{in}}} = \frac{W_{12} + W_{23} + W_{34}}{Q_{23}} = \frac{-C_V(T_2 - T_1) + P_2(V_3 - V_2) - C_V(T_4 - T_3)}{P_2(V_3 - V_2) + C_V(T_3 - T_2)}$$

But $P_2 = P_3$, $P_2V_2 = RT_2$, and $P_3V_3 = RT_3$; whence

$$\eta = \frac{-C_V(T_2 - T_1) + R(T_3 - T_2) - C_V(T_4 - T_3)}{R(T_3 - T_2) + C_V(T_3 - T_2)} = 1 - \frac{C_V(T_4 - T_1)}{(C_V + R)(T_3 - T_2)}$$

$$= 1 - \frac{1}{\gamma}\frac{T_4 - T_1}{T_3 - T_2} = 1 - \frac{1}{\gamma}\frac{T_1}{T_2}\frac{(T_4/T_1) - 1}{(T_3/T_2) - 1}$$

For the isentropic steps, Example 2.6 gives

$$\frac{T_1}{T_2} = \left(\frac{V_2}{V_1}\right)^{\gamma-1} \qquad \frac{P_2}{P_1} = \left(\frac{V_1}{V_2}\right)^{\gamma}$$

$$\frac{P_4}{P_3} = \left(\frac{V_3}{V_4}\right)^{\gamma} = \left(\frac{V_2}{V_4}\frac{V_3}{V_2}\right)^{\gamma} = \left(\frac{V_2}{V_1}\frac{V_3}{V_2}\right)^{\gamma}$$

Thus

$$\frac{T_4}{T_1} = \frac{P_4}{P_1}\frac{V_4}{V_1} = \frac{P_4}{P_1} = \frac{P_4}{P_3}\frac{P_3}{P_2}\frac{P_2}{P_1} = \frac{P_4}{P_3}\frac{P_2}{P_1} = \left(\frac{V_2}{V_1}\frac{V_3}{V_2}\right)^{\gamma}\left(\frac{V_1}{V_2}\right)^{\gamma} = \left(\frac{V_3}{V_2}\right)^{\gamma}$$

and

$$\frac{T_3}{T_2} = \frac{P_3}{P_2}\frac{V_3}{V_2} = \frac{V_3}{V_2}$$

It follows that

$$\eta = 1 - \frac{1}{\gamma}\left(\frac{V_2}{V_1}\right)^{\gamma-1}\frac{(V_3/V_2)^{\gamma}-1}{(V_3/V_2)-1}$$

As it was for the Otto cycle (Example 2.8), the ratio V_1/V_2 is here called the *compression ratio* r; the ratio V_3/V_2 is known as the *cutoff ratio* and is designated r_c. The thermal efficiency η_D of the Diesel cycle can then be written

$$\eta_D = 1 - \left(\frac{1}{r}\right)^{\gamma-1}\frac{r_c^{\gamma}-1}{\gamma(r_c-1)}$$

Heat engines operating both on the Otto and Diesel cycles (or rather the real counterparts of the idealizations presented here) are extensively employed as mechanical power sources, and it is of interest to compare their efficiencies. From Example 2.8, the ideal thermal efficiency of an Otto cycle is

$$\eta_O = 1 - \left(\frac{1}{r}\right)^{\gamma-1}$$

For a fixed compression ratio r, η_D and η_O differ only in the term multiplying $(1/r)^{\gamma-1}$, which is unity for the Otto cycle and

$$\frac{r_c^{\gamma}-1}{\gamma(r_c-1)}$$

for the Diesel cycle. Both the cutoff ratio and γ are greater than unity, and the above term can be shown to be always greater than unity for these conditions. Therefore, when compared for the same ideal gas and the same compression ratio, a reversible Otto cycle is more efficient than a reversible Diesel cycle. However, preignition and knocking difficulties limit the compression ratios for real engines operating on the Otto cycle; so Diesel engines can be operated at greater compression ratios and resultant greater efficiencies than Otto engines.

2.7 Consider a nuclear power plant generating 750 MW; the reactor temperature is 315 °C and a river with water temperature of 20 °C is available as a heat sink. (*a*) What is the maximum possible thermal efficiency of the plant and what is the minimum rate at which heat must be discarded to the river? (*b*) If the actual thermal efficiency of the plant is 60% of the maximum, at what rate must heat be discarded to the river, and what is the temperature rise of the river if it has a flow rate of 165 m$^3 \cdot$ s^{-1}?

(*a*) The maximum thermal efficiency is given by (2.4*), which is applicable to a reversible heat engine operating between two fixed temperatures:

$$\eta^* = 1 - \frac{T_C}{T_H} = 1 - \frac{20+273.15}{315+273.15} = 0.50$$

The corresponding rate of heat transfer to the river is

$$|\dot{Q}_C| = |\dot{Q}_H| - |\dot{W}| = |\dot{W}|\left(\frac{1}{\eta^*}-1\right) = (750\text{ MW})(1.00) = 750\text{ MW}$$

(b)   For $\eta = (0.60)(0.50) = 0.30$,

$$|\dot{Q}_C| = |\dot{W}|\left(\frac{1}{\eta} - 1\right) = (750 \text{ MW})(2.33) = 1750 \text{ MW}$$

For a density of 1000 kg $\cdot$ m^{-3}, the mass flow rate of the river is

$$\dot{m} = (165)(1000) = 165\,000 \text{ kg} \cdot \text{s}^{-1}$$

Since $\dot{Q}_C = \dot{m}C_P\,\Delta T$, we have

$$\Delta T = \frac{\dot{Q}_C}{\dot{m}C_P}$$

Taking the heat capacity of water as 4.18 kJ $\cdot$ kg$^{-1} \cdot$ °C^{-1}, we get

$$\Delta T = \frac{1750 \times 10^6}{(165 \times 10^3)(4.18 \times 10^3)} = 2.53 \text{ °C}$$

This is the temperature rise (thermal pollution) of an entire river of moderate size, occasioned by a power plant of typical size.

ENTROPY CALCULATIONS   (Sections 2.3 and 2.4)

2.8   Heat is transferred directly from a heat reservoir at 280 °C to another heat reservoir at 5 °C. If the amount of heat transferred is 100 kJ, what is the total entropy change as a result of this process?

The temperatures of the two heat reservoirs are

$$T_H = 280 + 273.15 = 553.15 \text{ K} \qquad \text{and} \qquad T_C = 5 + 273.15 = 278.15 \text{ K}$$

The heat transferred is 100 kJ, and therefore

$$Q_H = -100 \text{ kJ} \qquad \text{(heat out of reservoir } H\text{)}$$
$$Q_C = +100 \text{ kJ} \qquad \text{(heat into reservoir } C\text{)}$$

Since for heat reservoirs $\Delta S = Q/T$,

$$\Delta S_H = \frac{Q_H}{T_H} = \frac{-100}{553.15} = -0.181 \text{ kJ} \cdot \text{K}^{-1} \qquad \text{and} \qquad \Delta S_C = \frac{Q_C}{T_C} = \frac{100}{278.15} = 0.360 \text{ kJ} \cdot \text{K}^{-1}$$

and $\Delta S_{\text{total}} = \Delta S_H + \Delta S_C = -0.181 + 0.360 = +0.179 \text{ kJ} \cdot \text{K}^{-1}$.

This same result is obtained from the general formula derived for this kind of process in Example 2.1:

$$\Delta S_{\text{total}} = Q_C\left(\frac{T_H - T_C}{T_H T_C}\right) = 100\left[\frac{553.15 - 278.15}{(553.15)(278.15)}\right] = 0.179 \text{ kJ} \cdot \text{K}^{-1}$$

2.9   For any closed PVT system show that (a) $dS = C_V\,dT/T$ for a process at constant volume; (b) $dS = C_P\,dT/T$ for a process at constant pressure.

(a)   At constant volume the fundamental property relation (2.6) for a closed PVT system becomes $dU = T\,dS$. However, (1.10) shows that for such a system at constant volume, $dU = C_V\,dT$. Combination of these two equations gives $dS = C_V\,dT/T$ (const. V).

(b)   For a constant-pressure process (1.7) becomes $dH = dU + P\,dV$. Substitution for dU by (2.6) gives $dH = T\,dS - P\,dV + P\,dV = T\,dS$. However, (1.11) shows that for a closed PVT system at constant pressure, $dH = C_P\,dT$. Combination of these last two equations provides $dS = C_P\,dT/T$ (const. P).

2.10   If 2 kg of liquid water at 90 °C is mixed adiabatically and at constant pressure with 3 kg of liquid water at 10 °C, what is the total entropy change resulting from this process? For simplicity, take the heat capacity of water to be constant at $C_P = 4184 \text{ J} \cdot \text{kg}^{-1} \cdot \text{K}^{-1}$.

The two masses of water in effect exchange heat under constant-pressure conditions until they reach the same final temperature T. Thus, for constant C_P,

$$Q_C = m_C C_P (T - T_C) = -Q_H = -m_H C_P (T - T_H)$$

where the subscripts C and H denote the initial conditions for the cooler and hotter masses of water. Thus

$$m_C(T - T_C) + m_H(T - T_H) = 0 \qquad \text{or} \qquad T = \frac{m_C T_C + m_H T_H}{m_C + m_H}$$

Substituting $T_C = 10 + 273.15 = 283.15$ K, $T_H = 90 + 273.15 = 363.15$ K, we find

$$T = \frac{(3)(283.15) + (2)(363.15)}{5} = 315.15 \text{ K}$$

It was shown in Problem 2.9 that for a constant-pressure process, $dS = C_P\, dT/T$. Integration with C_P constant gives

$$\Delta S = C_P \ln \frac{T_f}{T_i}$$

Application of this equation for each of the two masses of water provides

$$\Delta S_C = C_P \ln \frac{T}{T_C} = 4184 \ln \frac{315.15}{283.15} = 448.0 \text{ J} \cdot \text{kg}^{-1} \cdot \text{K}^{-1}$$

$$\Delta S_H = C_P \ln \frac{T}{T_H} = 4184 \ln \frac{315.15}{363.15} = -593.2 \text{ J} \cdot \text{kg}^{-1} \cdot \text{K}^{-1}$$

Since the process is adiabatic, there is no entropy change of the surroundings, and

$$\Delta S_{\text{total}} = m_C \, \Delta S_C + m_H \, \Delta S_H = (3)(448.0) + (2)(-593.2) = +157.6 \text{ J} \cdot \text{K}^{-1}$$

2.11 In Example 2.5 we showed that for an ideal gas with constant heat capacities the entropy can be expressed as a function of P and V by the equation

$$S = C_V \ln P + C_P \ln V + S_0''$$

(a) If the molar entropy of an ideal gas for which $C_V = (3/2)R$ and $C_P = (5.2)R$ is taken as zero at 1 bar and 0 °C, evaluate the constant S_0''. What units must be used for P and V in the resulting equation for S? (b) Calculate S for 1000 cm^3 of the ideal gas at 20 bar and 50 °C.

(a) At the conditions $P = 1$ bar and $T = 0$ °C or 273.15 K, the ideal-gas equation gives V as

$$V = \frac{RT}{P} = \frac{(83.14 \text{ cm}^3 \cdot \text{bar} \cdot \text{mol}^{-1} \cdot \text{K}^{-1})(273.15 \text{ K})}{1 \text{ bar}} = 22\,710 \text{ cm}^3 \cdot \text{mol}^{-1}$$

Substitution of values in the equation for S at the conditions for which S is zero gives

$$0 = (3/2)(8.314) \ln 1 + (5/2)(8.314) \ln 22\,710 + S_0''$$

from which

$$S_0'' = -208.48 \text{ J} \cdot \text{mol}^{-1} \cdot \text{K}^{-1} \qquad \text{and} \qquad S = C_V \ln P + C_P \ln V - 208.48$$

In this equation we have committed ourselves to units of bars for P, cm$^3 \cdot$ mol^{-1} for V and J $\cdot$ mol$^{-1} \cdot$ K^{-1} for S, C_V, and C_P.

(b) For 1000 cm^3 at 20 bar and 50 °C or 323.15 K we calculate the number of moles by the ideal-gas equation

$$n = \frac{PV'}{RT} = \frac{(20 \text{ bar})(1000 \text{ cm}^3)}{(83.14 \text{ cm}^3 \cdot \text{bar} \cdot \text{mol}^{-1} \cdot \text{K}^{-1})(323.15 \text{ K})} = 0.7444 \text{ mol}$$

The molar volume is

$$V = \frac{V'}{n} = \frac{1000}{0.7444} = 1343.36 \text{ cm}^3 \cdot \text{mol}^{-1}$$

The molar entropy as given by the equation of part (a) is

$$S = (3/2)(8.314) \ln 20 + (5/2)(8.314) \ln 1343.36 - 208.48$$
$$= -21.41 \text{ J} \cdot \text{mol}^{-1} \cdot \text{K}^{-1}$$

Thus the entropy of 1000 cm^3 or 0.7444 mol is

$$S^t = (-21.41)(0.7444) = -15.94 \text{ J} \cdot \text{K}^{-1}$$

2.12 One kilomole of an ideal gas is compressed isothermally at 400 K from 100 kPa to 1000 kPa in a piston-and-cylinder arrangement. Calculate the entropy change of the gas, the entropy change of the surroundings, and the total entropy change resulting from the process, if (a) the process is mechanically reversible and the surroundings consist of a heat reservoir at 400 K; (b) the process is mechanically reversible and the surroundings consist of a heat reservoir at 300 K; (c) the process is mechanically irreversible, requiring 20% more work than the mechanically reversible compression, and the surroundings consist of a heat reservoir at 300 K.

(a) The work of a mechanically reversible, isothermal *expansion* of an ideal gas was shown in Example 1.4 to be given by $W = nRT \ln (P_i/P_f)$. Exactly the same expression applies to a mechanically reversible, isothermal *compression* of an ideal gas. Since $T = 400$ K,

$$W = (1 \text{ kmol})(8.314 \text{ kJ} \cdot \text{kmol}^{-1} \cdot \text{K}^{-1})(400 \text{ K}) \ln (100/1000) = -7657 \text{ kJ}$$

For an ideal gas the internal energy is a function of temperature only, and since T is constant, $\Delta U = 0$. Thus the first law requires that $Q = W = -7657$ kJ. This heat leaves the system (the gas) and flows to the surroundings (the heat reservoir); hence $Q_{\text{surr}} = -Q = +7657$ kJ. The temperature of the system and of the surroundings is 400 K and is constant. Moreover, the process is reversible, and therefore

$$\Delta S = \frac{Q}{T} = \frac{-7657}{400} = -19.144 \text{ kJ} \cdot \text{K}^{-1}$$

$$\Delta S_{\text{surr}} = \frac{Q_{\text{surr}}}{T_{\text{surr}}} = \frac{7657}{400} = 19.144 \text{ kJ} \cdot \text{K}^{-1}$$

and $\Delta S_{\text{total}} = \Delta S + \Delta S_{\text{surr}} = 0$. Clearly, this result is required by the second law for a completely reversible process. Not only is the compression mechanically reversible, but also the heat transfer from system to surroundings, both at 400 K, is reversible.

(b) The only difference between this process and that of part (a) is that the heat reservoir is at 300 K rather than 400 K, the system temperature. This means that the process is no longer completely reversible, because heat transfer is now across a finite temperature difference. However, the compression is still accomplished by a mechanically reversible process, and the system changes along exactly the same path as before. Thus ΔU, W, Q, Q_{surr}, and ΔS are unchanged from the values of part (a). However, ΔS_{surr} and ΔS_{total} are different:

$$\Delta S_{\text{surr}} = \frac{Q_{\text{surr}}}{T_{\text{surr}}} = \frac{7657}{300} = 25.523 \text{ kJ} \cdot \text{K}^{-1}$$

and $\Delta S_{\text{total}} = \Delta S + \Delta S_{\text{surr}} = -19.144 + 25.523 = 6.379 \text{ kJ} \cdot \text{K}^{-1}$. Since the process is irreversible, the second law requires that $\Delta S_{\text{total}} > 0$; our result is clearly in accord with this requirement.

(c) The process described includes the same irreversibility as in part (b), but in addition the compression step is now mechanically irreversible, so that the work required is greater by 20%:

$$W = (-7657)(1.20) = -9189 \text{ kJ}$$

However, the process accomplishes exactly the same change of state in the gas as before, and as a result the property changes of the gas are the same. Thus $\Delta U = 0$, and the first law still requires that $Q = W = -9189$ kJ and $Q_{\text{surr}} = -Q = +9189$ kJ. The entropy change of the heat reservoir (surroundings) is given by

$$\Delta S_{\text{surr}} = \frac{Q_{\text{surr}}}{T_{\text{surr}}} = \frac{9189}{300} = 30.630 \text{ kJ} \cdot \text{K}^{-1}$$

The entropy change of the system is a property change like ΔU, and must have the same value as before; i.e., $\Delta S = -19.144 \text{ kJ} \cdot \text{K}^{-1}$. Thus

$$\Delta S_{\text{total}} = \Delta S + \Delta S_{\text{surr}} = -19.144 + 30.630 = 11.486 \text{ kJ} \cdot \text{K}^{-1}$$

Note that the entropy change of the system is *not* given as Q/T, because the process is mechanically irreversible, which means it is internally irreversible. The heat reservoir, however, suffers no internal irreversibilities. The irreversibility of heat transfer is in fact external to both the system and surroundings. For the system,

$$\frac{Q}{T} = \frac{-9189}{400} = -22.972 \text{ kJ} \cdot \text{K}^{-1}$$

and this *is* the entropy change in the system *caused by the heat transfer*. However, it is not the *entire* entropy change of the system, because the mechanical irreversibility within the system causes an increase in entropy that partially compensates the entropy decrease caused by heat transfer. Recognition that this is the case does not, however, provide a means for the calculation of entropy changes of systems subjected to mechanically irreversible processes. Fortunately, there is no need, because entropy is a state function, and its values are independent of the process causing a change of state.

The results of this problem are summarized in Table 2-2. The degree of irreversibility of the three processes is reflected in the values of ΔS_{total}. For the completely reversible process this is, of course, zero. For parts (*b*) and (*c*) it is positive, with the value for (*c*) being larger than for (*b*), because (*c*) includes the irreversibility of (*b*) and a further irreversibility as well.

Table 2-2

	$W/\text{kJ} = Q/\text{kJ}$	$\Delta U/\text{kJ}$	$\Delta S/\text{kJ} \cdot \text{K}^{-1}$	$\Delta S_{\text{surr}}/\text{kJ} \cdot \text{K}^{-1}$	$\Delta S_{\text{total}}/\text{kJ} \cdot \text{K}^{-1}$
(*a*)	-7657	0	-19.144	$+19.144$	0
(*b*)	-7657	0	-19.144	$+25.523$	$+6.379$
(*c*)	-9189	0	-19.144	$+30.630$	$+11.486$

2.13 One mole of an ideal gas expands isothermally from 2 bar to 1 bar in the piston-and-cylinder device shown in Fig. 2-10. The device is surrounded by the atmosphere, which exerts a constant pressure of 1 bar on the outer face of the piston. Moreover, the device is always in thermal equilibrium with the atmosphere, which constitutes a heat reservoir at 300 K. During the expansion process, a frictional force is exerted on the piston and this force varies in such a way that it always almost balances the net pressure forces on the piston. Thus the piston moves very slowly and experiences negligible acceleration. The piston and cylinder are good heat conductors. Determine the entropy change of the gas, the entropy change of the atmosphere, and the total entropy change brought about by the process.

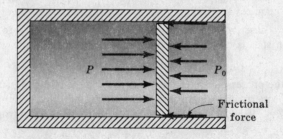

Fig. 2-10

If one takes as the system not only the gas but also the piston and cylinder, then clearly the only external force on the system is caused by P_0, the pressure of the surrounding atmosphere. The work done by the system is therefore $W = P_0 \, \Delta V$, where ΔV is given by the ideal-gas equation:

$$\Delta V = V_2 - V_1 = \frac{RT}{P_2} - \frac{RT}{P_1} = RT\left(\frac{P_1 - P_2}{P_1 P_2}\right)$$

Thus, with $P_0 = P_2$,

$$W = \frac{P_0 RT}{P_1 P_2} (P_1 - P_2) = \frac{RT}{P_1} (P_1 - P_2)$$

This work is literally work done in pushing back the atmosphere. The first-law equation is $\Delta U = Q - W$. But $\Delta U = 0$, because the process is isothermal and for an ideal gas U is a function of T only. The properties of the piston and cylinder are also presumed to be constant at constant T. Hence

$$Q = W \qquad \text{and} \qquad Q_{\text{surr}} = -Q = -W$$

Also

$$\Delta S_{\text{surr}} = \frac{Q_{\text{surr}}}{T} = \frac{-W}{T} = \frac{-R}{P_1} (P_1 - P_2) = -\frac{1}{2} R$$

The entropy change of the gas is given by (2.10), which, for constant T, is

$$\Delta S = -R \ln \frac{P_2}{P_1} = -R \ln \frac{100}{200} = 0.6932 \, R$$

The total entropy change is therefore

$$\Delta S_{\text{total}} = \Delta S + \Delta S_{\text{surr}} = 0.6932 \, R - 0.5000 \, R = 0.1932 \, R$$

The process is irreversible on account of the frictional force acting on the piston, and this is clearly reflected in the positive value obtained for ΔS_{total}, even though the frictional force was never explicitly considered in the problem solution. The basic reason that it all comes out right is that entropy is a property, and the entropy change of the gas is computed by an equation that is independent of the process.

It is instructive to rework the problem with the gas alone taken as the system. Since the process occurs very slowly, the pressure is uniform throughout the system, and is the pressure acting on the inner face of the piston. Processes such as this are sometimes said to be *quasistatic* (almost static); the notion of *internal equilibrium* is equivalent. Note that the frictional force does not act on the gas, which is the system, so that we may use the formula of Example 1.4 for the work of a reversible isothermal expansion of an ideal gas:

$$W = RT \ln \frac{P_1}{P_2} = RT \ln \frac{200}{100} = RT \ln 2$$

This is the work done by the gas on the piston, and by the first law $Q = W = RT \ln 2$. The entropy change of the gas is then

$$\frac{Q}{T} = R \ln 2 = 0.6932 \, R$$

just as before. Now, the heat which flows to the gas comes from two sources. The frictional force acting between the piston and cylinder converts part of the work done by the gas into heat which in effect flows into the gas. This heat does not come from the surrounding atmosphere and does not affect the entropy of the surroundings. Only the *additional* heat required to maintain constant temperature comes from the surroundings, and this is equal to the work done by the piston against the atmosphere, as previously calculated.

MISCELLANEOUS APPLICATIONS

2.14 Consider an ideal gas for which $C_V = (5/2)R$ and $C_P = (7/2)R$. The initial state is 1 bar and 20 °C. (*a*) 1 kmol is heated at constant volume to 80 °C. Calculate ΔU, ΔH, ΔS, Q, and W. (*b*) 1 kmol is heated at constant pressure to 80 °C. Calculate ΔU, ΔH, ΔS, Q, and W if the process is reversible. (*c*) 1 kmol is changed from its initial state to a final state of 10 bar and

80 °C by several different reversible processes. Which of the following quantities are the same for all processes: ΔU, ΔH, ΔS, Q, and W? (d) Repeat part (c) for irreversible processes.

For an ideal gas with constant heat capacities, regardless of the process,

$$\Delta U = C_V \, \Delta T \qquad\qquad \text{from } (1.13)$$
$$\Delta H = C_P \, \Delta T \qquad\qquad \text{from } (1.15)$$
$$\Delta S = C_V \ln \frac{T_2}{T_1} + R \ln \frac{V_2}{V_1} \qquad \text{from } (2.8)$$
$$= C_P \ln \frac{T_2}{T_1} - R \ln \frac{P_2}{P_1} \qquad \text{from } (2.10)$$

In the following calculations we take $R = 8.314 \text{ kJ} \cdot \text{kmol}^{-1} \cdot \text{K}^{-1}$.

(a)   Substitution into the first of the above equations gives for 1 kmol of gas:

$$\Delta U = (5/2)(8.314)(80 - 20) = 1247.1 \text{ kJ}$$
$$\Delta H = (7/2)(8.314)(80 - 20) = 1745.9 \text{ kJ}$$
$$\Delta S = \left(\frac{5}{2}\right)(8.314) \ln \frac{80 + 273.15}{20 + 273.15} = 3.87 \text{ kJ} \cdot \text{K}^{-1}$$

Since the volume is constant, $W = 0$, whence, by the first law, $Q = \Delta U = 1247.1$ kJ.

(b)   As in part (a), $\Delta U = 1247.1$ kJ and $\Delta H = 1745.9$ kJ. By the second equation for ΔS,

$$\Delta S = (7/2)(8.314) \ln 1.2047 = 5.42 \text{ kJ} \cdot \text{K}^{-1}$$

From Example 1.7, $Q = \Delta H = 1745.9$ kJ; then, by the first law,

$$W = Q - \Delta U = 1745.9 - 1247.1 = 498.8 \text{ kJ}$$

(c) ΔU, ΔH, and ΔS, which are property changes, are the same for all processes.

(d)   The answers here are identical to those of part (c). Property changes depend only on the initial and final states of the system, independent of the process, reversible or irreversible.

2.15   One mole of an ideal gas for which $C_V = 25.10 \text{ J} \cdot \text{mol}^{-1} \cdot \text{K}^{-1}$ expands adiabatically from an initial state at 340 K and 5 bar to a final state where its volume has doubled. Find the final temperature of the gas, the work done, and the entropy change of the gas, for (a) a reversible expansion and (b) a free expansion of the gas into an evacuated space (*Joule expansion*).

(a)   For a reversible, adiabatic expansion of an ideal gas with constant heat capacities, we have from Example 2.5:

$$T_2 = T_1 \left(\frac{V_1}{V_2}\right)^{\gamma - 1}$$

Here $\gamma = C_P/C_V = (25.10 + 8.314)/25.10 = 1.331$ and $V_1/V_2 = 0.5$. Thus

$$T_2 = (340)(0.5)^{0.331} = 270.3 \text{ K}$$

By the first law for an adiabatic process, and for an ideal gas with constant heat capacities,

$$W = -\Delta U = -C_V \, \Delta T = -(25.10)(270.3 - 340) = 1749.5 \text{ J}$$

By the second law, the entropy change for a reversible, adiabatic process is zero.

(b)   Expansion of the gas into an evacuated space produces no work. Since the process is also adiabatic, both Q and W are zero. Thus the first law requires that $\Delta U = 0$, which, in view of $U = U(T)$, implies that $\Delta T = 0$. Hence $T_2 = T_1$. The entropy change for 1 mol is given by (2.8) as

$$\Delta S = R \ln \frac{V_2}{V_1} = 8.314 \ln 2 = 5.763 \text{ J} \cdot \text{K}^{-1}$$

The value of ΔS for the gas is also ΔS_{total}, because $\Delta S_{\text{surr}} = 0$ for an adiabatic process. The positive value of ΔS_{total} is required by the second law for this irreversible process.

2.16 Extend (3) of Example 2.2 to a heat engine whose properties show a net change during the process.

As in Example 2.2, the entropy changes of the reservoirs are $\Delta S_H = Q_H/T_H$ and $\Delta S_C = Q_C/T_C$, where Q_H and Q_C refer to the reservoirs. There is in addition an entropy change ΔS for the engine. The total entropy change accompanying the process is then

$$\Delta S_{\text{total}} = \frac{Q_H}{T_H} + \frac{Q_C}{T_C} + \Delta S \tag{1}$$

The first law, written for the engine, $\Delta U = Q - W$, together with $Q = -Q_H - Q_C$, implies

$$W = -Q_H - Q_C - \Delta U \tag{2}$$

Combination of (1) and (2) to eliminate Q_H gives the desired expression:

$$W = -\Delta U - T_H(\Delta S_{\text{total}} - \Delta S) + Q_C\left(\frac{T_H}{T_C} - 1\right) \tag{3}$$

[For a cyclic engine, $\Delta U = \Delta S = 0$, and we retrieve (3) of Example 2.2.]

2.17 A mass of liquid water, $m = 5$ kg, initially in thermal equilibrium with the atmosphere at 20 °C, is cooled at constant pressure to 4 °C by means of heat pumps operating between the water and the atmosphere. What is the minimum work required? For water take $C_P = 4.184$ kJ $\cdot$ kg^{-1} $\cdot$ °C^{-1}.

The minimum work is required if the process is reversible, and we can imagine a series of reversible heat pumps operating so as to remove heat from the water at various temperature levels as the water cools from 20 to 4 °C, and discharging heat to the atmosphere at $T_0 = 20 + 273.15 = 293.15$ K. Each heat pump removes a differential amount of heat δQ and reduces the water temperature differentially. For a reversible heat pump, (2.5) and (2.5*) give

$$\frac{|W|}{|Q_C|} = \frac{T_H - T_C}{T_C} = \frac{T_H}{T_C} - 1$$

In the present notation T_H becomes T_0 and T_C becomes T, the water temperature; in addition, the removal of a differential amount of heat δQ with the performance of a differential amount of work δW for a given temperature T, requires that the equation be cast in differential form. Thus

$$\frac{|\delta W|}{|\delta Q|} = \frac{T_0}{T} - 1$$

The process is depicted in Fig. 2-11.

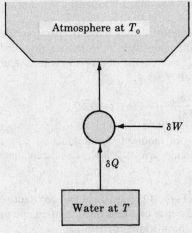

Fig. 2-11

We may remove the absolute-value signs in the above equation by noting that δW is negative (work done on the process) and that δQ is also negative *provided that δQ is taken to refer to the water*. With this understanding, $\delta W/\delta Q$ is positive, consistent with the condition $T_0/T > 1$. Thus we write

$$\frac{\delta W}{\delta Q} = \frac{T_0}{T} - 1 \quad \text{or} \quad \delta W = T_0 \frac{\delta Q}{T} - \delta Q$$

However, $\delta Q/T = dS^t$, where dS^t is the entropy change of the water (since both δQ and T refer to the water). Moreover $\delta Q = dH^t$, where dH^t is the enthalpy change of the water. This last equation follows from the result of Example 1.7, since the water is a closed PVT system cooled at constant pressure. Therefore, $\delta W = T_0\, dS^t - dH^t$; and integration over the entire process, with T_0 held constant, gives

$$W = T_0\,\Delta S^t - \Delta H^t \tag{1}$$

where ΔS^t and ΔH^t are property changes of the water. These are evaluated by $\Delta H^t = mC_P\,\Delta T$, which follows from (1.11) when C_P is constant, and

$$\Delta S^t = mC_P \ln \frac{T_2}{T_1}$$

which follows similarly from the result of Problem 2.9(b). Finally, we have by substitution into (1):

$$W = mC_P \left[T_0 \ln \frac{T_2}{T_1} - (T_2 - T_1) \right]$$

where subscripts 1 and 2 denote the initial and final temperatures. Since

$$T_1 = 20 + 273.15 = 293.15 \text{ K} \qquad T_2 = 4 + 273.15 = 277.15 \text{ K}$$

the required minimum work is

$$W = (5)(4.184)\left[\left(293.15 \ln \frac{277.15}{293.15} \right) + 16 \right] = -9.481 \text{ kJ}$$

Supplementary Problems

HEAT ENGINES AND HEAT PUMPS   (Sections 2.1 and 2.2)

2.18 A reversible engine operates on the Carnot cycle between extreme temperatures 460 °C and 25 °C. What percent of the heat taken is converted into work?   *Ans.*   59.3%

2.19 (a) Infer from (3) of Example 2.2 that the Carnot refrigerator has the greatest possible coefficient of performance. (b) At what compression ratio must a reversible Otto engine be run to obtain the same thermal efficiency as the Carnot engine of Problem 2.18? Assume that the working fluid is an ideal gas of constant heat capacity, with $\gamma = 1.4$.

Ans.   (a) $\dfrac{\omega}{\omega^*} = 1 - \dfrac{T_H\,\Delta S_{\text{total}}}{|W|}$ (b) $r = 9.5$

2.20 Calculate the minimum work required to manufacture 5 kg of ice cubes from water initially at 0 °C. Assume that the surroundings are at 298 K. The latent heat of fusion of water at 0 °C is 333.5 kJ $\cdot$ kg^{-1}.   *Ans.*   152.6 kJ

2.21 The Joule cycle, shown in Fig. 2-12, consists of two constant-pressure steps connected by two adiabatics. Determine the thermal efficiency of a reversible heat engine operating on this cycle, with an ideal gas of constant heat capacity as the working medium.
Ans. $\eta = 1 - r_P^{(1-\gamma)/\gamma}$, where $r_P = P_2/P_1 = P_3/P_4$ is the *pressure ratio*.

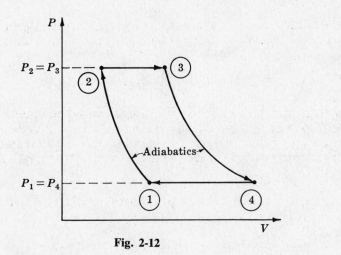

Fig. 2-12

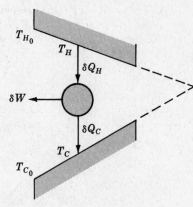

Fig. 2-13

2.22 Figure 2-13 depicts a cyclic engine exchanging heat with two constant-pressure "reservoirs" of *finite* total heat capacities C_{PH} and C_{PC}. At the beginning of the process, the temperatures of the hot and cold "reservoirs" are T_{H0} and T_{C0}, respectively, and as the process progresses T_H decreases and T_C increases. (a) Find an equation for the maximum work obtainable by decreasing the hot "reservoir" temperature from T_{H0} to any final temperature T_H. Express your results in terms of T_H, C_{PH}, C_{PC}, T_{H0}, and T_{C0}. (b) What is the lowest possible T_H consistent with the solution to part (a)?

Ans. (a) $W_{\max} = C_{PH}(T_{H0} - T_H) - C_{PC}T_{C0}[(T_{H0}/T_H)^{C_{PH}/C_{PC}} - 1]$

(b) $T_{H,\min} = (T_{C0}^{C_{PC}} T_{H0}^{C_{PH}})^{1/(C_{PC}+C_{PH})}$

ENTROPY CALCULATIONS (Sections 2.3 and 2.4)

2.23 (a) If the system in Problem 1.23 consists of 0.4 kmol of an ideal gas with constant heat capacity $C_V = (5/2)R$, calculate ΔS^t of the gas for steps 12, 23, and 31. (b) What would ΔS^t be for step 12 if this path were instead an isotherm?

Ans. (a) $\Delta S^t_{12} = +3.047 \text{ kJ} \cdot \text{K}^{-1}$ $\Delta S^t_{23} = -10.665 \text{ kJ} \cdot \text{K}^{-1}$ $\Delta S^t_{31} = +7.618 \text{ kJ} \cdot \text{K}^{-1}$

(b) the same

2.24 (a) The temperature of an ideal gas with constant heat capacity is changed from T_1 to T_2. Show that ΔS of the gas is greater if the change in state occurs at constant pressure than if at constant volume. (b) The pressure of an ideal gas is changed from P_1 to P_2 by an isothermal process and by a constant-volume process. Show that the entropy changes in the gas have opposite signs for the two processes.

2.25 Three pairs of diagrams are given in Fig. 2-14. The PV diagrams are for an ideal gas of constant heat capacity undergoing (a) a reversible Carnot cycle, (b) a reversible Otto cycle (Example 2.8), and (c) a reversible Joule cycle (Problem 2.21). Paths 12 and 34 are isentropes in all three cases. Are the TS diagrams qualitatively consistent with the corresponding PV diagrams?

Ans. (a) yes (b) no (c) yes

2.26 Two bodies, A and B, with constant total heat capacities, C_{PA} and C_{PB}, and different initial temperatures, T_{A0} and T_{B0}, exchange heat with each other under isobaric conditions. No heat is transferred to or from the surroundings, and the temperatures of both bodies are assumed uniform at all times. (a) Derive an expression for the total entropy change ΔS of the two bodies as a function of T_A, the temperature of body A. (b) Demonstrate that the process is irreversible by showing that $\Delta S > 0$. (c) What is the equilibrium temperature of the system? (*Hint:* Follow Problem 2.10.)

Ans. (a) $\Delta S = C_{PA} \ln\left(\dfrac{T_A}{T_{A0}}\right) + C_{PB} \ln\left[\dfrac{T_{B0} + (C_{PA}/C_{PB})(T_{A0} - T_A)}{T_{B0}}\right]$

(c) $T_A = T_B = (C_{PA}T_{A0} + C_{PB}T_{B0})/(C_{PA} + C_{PB})$

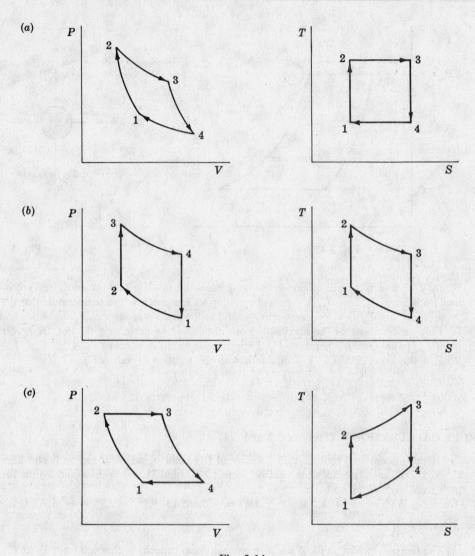

Fig. 2-14

MISCELLANEOUS APPLICATIONS

2.27   An ideal gas for which $C_V = (3/2)R$ expands adiabatically and reversibly from 300 kPa to 100 kPa in a piston-and-cylinder apparatus. If $T_1 = 315$ °C, determine T_2, ΔU, ΔH, and W.
Ans. $T_2 = 379.00$ K $\Delta U = -W = -2.61$ kJ·mol^{-1} $\Delta H = -4.35$ kJ·mol^{-1}

2.28   Obtain the overall coefficient of performance for the heat pumps of Problem 2.17.

$$Ans. \quad \omega = \frac{T_1 - T_2}{T_0 \ln (T_1/T_2) - (T_1 - T_2)}$$

2.29   Show that the same result obtains for Example 2.1 if, instead of exchanging heat directly with the cold reservoir, the hot reservoir exchanges heat with an intermediate reservoir of temperature T_M, which in turn exchanges heat with the cold reservoir.

2.30   Figure 2-15 shows the relation of temperature to entropy for a closed PVT system during a reversible process. Calculate the heat added to the system for each of the three steps 12, 23, and 31, and for the entire process 1231.   *Ans.* $Q_{12} = 1600$ J $Q_{23} = -1000$ J $Q_{31} = 0$ $Q_{\text{total}} = 600$ J

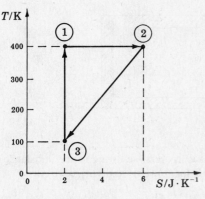

Fig. 2-15   **Fig. 2-16**

2.31   A horizontal cylinder, closed at both ends, is divided in half by a free piston. The cylinder neither absorbs nor conducts heat, and the piston is perfectly lubricated in the cylinder. The left half of the cylinder contains 1 kmol of an ideal gas at 200 kPa and 15 °C. The gas has constant heat capacities, with $C_V = (5/2)R$. The right half of the cylinder is evacuated and the piston is temporarily restrained by latches. The following operations are carried out: (1) the latches are removed, allowing the gas on the left to expand as it pushes the piston to the right-hand end of the cylinder; (2) a small rod is inserted through the right-hand end of the cylinder and exerts a force on the piston, pushing it slowly back to its initial position. (a) What is the gas temperature after process 1 once internal equilibrium is established? (b) What is the gas temperature at the end of process 2, assuming this process to be reversible and adiabatic? (c) What is the gas pressure at the end of process 2? (d) What is W for process 2? (e) What is ΔS for process 1 and for the combined processes 1 and 2?
Ans.   (a) 15 °C   (b) 107.07 °C   (c) 263.9 kPa   (d) −1914 kJ   (e) 5.763 kJ·K⁻¹

2.32   Figure 2-16 shows a closed cylinder divided into two unequal chambers A and B by a piston that is free to move, except that it is initially kept from moving by a force F applied to a piston rod that extends through the right-hand end of the cylinder. Both the cylinder wall and the piston are perfect heat insulators, so that no heat is exchanged either between the chambers or with the surroundings. The two chambers contain fixed amounts of the same ideal gas, which may be assumed to have constant heat capacities with $C_V = (5/2)R$. The piston has an area of 0.1 m² and the cylinder has a free length of 1 m. The initial location of the piston is such that chamber A represents one-third of the total volume. The initial conditions are as follows:

$$P_{A1} = 400 \text{ kPa} \qquad P_{B1} = 200 \text{ kPa}$$

$$T_{A1} = 400 \text{ K} \qquad T_{B1} = 300 \text{ K}$$

The force F is slowly reduced so as to allow the piston to move to the right until the pressures in the two chambers are equal. Assuming the process to be reversible, determine $P_{A2} = P_{B2} \equiv P_2$, the temperatures T_{A2} and T_{B2}, and the work done by the system against the external force F.
Ans. $P_2 = 262.25$ kPa $T_{A2} = 354.55$ K $T_{B2} = 324.15$ K $W = 1104$ J

2.33   A closed cylinder is divided into two unequal chambers A and B by a piston that is free to move, except that it is initially kept from moving by a stop (see Fig. 2-17). Both the cylinder wall and the piston are perfect heat insulators, so that no heat is exchanged either between the chambers or with the surroundings. The two chambers contain fixed amounts of the same ideal gas, which may be assumed to have constant heat capacities. Chamber A initially contains n_A moles at T_{A1} and P_{A1}; chamber B initially contains n_B moles at T_{B1} and P_{B1}, where $P_{A1} > P_{B1}$. The stop is removed so that the piston is free to move until the pressures in the two chambers equalize. (a) Will the change in entropy as a result of the process be less than, equal to, or greater than zero in (i) the entire system, consisting of A and B?   (ii) chamber A?   (iii) chamber B? (b) Can the final pressure $P_{A2} = P_{B2} \equiv P_2$ be determined by the methods of thermodynamics? (c) Can the final temperatures T_{A2} and T_{B2} be determined by the methods of thermodynamics? How would the problem be changed if the piston were a heat conductor?

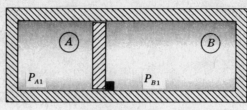

Fig. 2-17

Ans. (*a*) However one chooses the system, the process is irreversible owing to the imbalance of forces (local pressures) on the piston. Hence $\Delta S > 0$ for (i), (ii), and (iii).

(*b*)
$$P_2 = \frac{n_A T_{A1} + n_B T_{B1}}{(n_A T_{A1}/P_{A1}) + (n_B T_{B1}/P_{B1})}$$

(*c*) The final temperatures *cannot* be determined when the piston is nonconducting. When the piston is a conductor, the problem becomes equivalent to Problem 1.16.

2.34   A system consisting of gas contained in a piston-cylinder device undergoes an *irreversible* process between initial and final equilibrium states that causes its internal energy to increase by 30 kJ. During the process the system receives heat in the amount of 100 kJ from a heat reservoir at 600 K. The system is then restored to its initial state by a *reversible* process, during which the only heat transfer is between the system and the heat reservoir at 600 K. The entropy change of the heat reservoir as a result of *both* processes is $+0.026$ kJ $\cdot$ K^{-1}. Calculate (*a*) the work done by the system during the first (irreversible) process, (*b*) the heat transfer with respect to the system during the second (reversible) process, (*c*) the work done by the system during the second process.
Ans.   (*a*) 70 kJ   (*b*) -115.6 kJ   (*c*) -85.6 kJ

Chapter 3

Mathematical Formulations
of Thermodynamics

The quantitative relationships provided by the laws of thermodynamics find use in the solution of two quite different types of problems. The first is concerned with *processes*, and the equations employed deal with the relations between property changes of a system and the quantity of energy transferred between the system and its surroundings.

A second and equally important use of thermodynamics is in the elucidation of relationships among the equilibrium *properties* of a system. Derivation of these equations starts with a consideration of processes, because the laws of thermodynamics include the quantities Q and W, which are not properties but manifestations of processes. However, for reversible processes Q and W can be replaced by expressions involving properties only, and the resulting equations then become general relationships among equilibrium properties, no longer limited by the special kind of process initially chosen for the derivation. These properties are often called *state functions*. Purely mathematical considerations allow the derivation of a large number of equations interconnecting the state functions. This chapter is devoted to the development of such equations, and is a repository for numerous equations employed in later chapters.

3.1 EXACT DIFFERENTIALS AND STATE FUNCTIONS

Mathematical descriptions of the changes which occur in physical systems often lead to differential expressions of the form:

$$C_1 \, dX_1 + C_2 \, dX_2 + \cdots + C_n \, dX_n \equiv \sum C_i \, dX_i \qquad (3.1)$$

where the X_i are independent variables, and the C_i are functions of the X_i. When it is possible to set the differential expression (3.1) equal to dY, the differential of a *function* Y, where

$$Y = Y(X_1, X_2, \ldots, X_n)$$

then the differential expression (3.1) is said to be *exact*, and we can write

$$dY = C_1 \, dX_1 + \cdots + C_n \, dX_n \equiv \sum C_i \, dX_i \qquad (3.2)$$

Mathematics provides a *definition* for the differential of a multivariate function:

$$dY = \left(\frac{\partial Y}{\partial X_1}\right)_{X_j} dX_1 + \left(\frac{\partial Y}{\partial X_2}\right)_{X_j} dX_2 + \cdots + \left(\frac{\partial Y}{\partial X_n}\right)_{X_j} dX_n \equiv \sum \left(\frac{\partial Y}{\partial X_i}\right)_{X_j} dX_i$$

where the subscript X_j on the partial derivatives indicates that all X_i are held constant except the one in the derivative considered. Since the X_i are independent, this last equation and (3.2) may be equated term by term to give

$$C_1 = \left(\frac{\partial Y}{\partial X_1}\right)_{X_j}, \ldots, C_n = \left(\frac{\partial Y}{\partial X_n}\right)_{X_j} \quad \text{or} \quad C_i = \left(\frac{\partial Y}{\partial X_i}\right)_{X_j} \qquad (3.3)$$

In (3.2) each C_i and its corresponding X_i are said to be *conjugate* to each other.

If Y and its derivatives are continuous, then for any pair of independent variables X_k and X_l, we have the mathematical requirement that

$$\frac{\partial^2 Y}{\partial X_k \, \partial X_l} = \frac{\partial^2 Y}{\partial X_l \, \partial X_k}$$

or

$$\boxed{\left(\frac{\partial C_l}{\partial X_k}\right)_{X_j} = \left(\frac{\partial C_k}{\partial X_l}\right)_{X_j}}$$

(3.4)

This equation holds for any two conjugate pairs (C_l, X_l) and (C_k, X_k) in an exact differential expression. The set of $n(n-1)/2$ equations (3.4) represents a condition that is both necessary and sufficient for the exactness of (3.1).

In property relation (2.6), $dU = T\,dS - P\,dV$, U is known to be a function of S and V, so that $T\,dS - P\,dV$ must be exact. For such property relations, (3.4) is used not to test exactness but rather to provide additional thermodynamic relationships. Thus, from (2.6) we infer

$$\left(\frac{\partial T}{\partial V}\right)_S = -\left(\frac{\partial P}{\partial S}\right)_V$$

EXAMPLE 3.1   Apply (3.4) to the exact differential expression $dw = L\,dx + M\,dy - N\,dz$.
The three compatibility equations are

$$\left(\frac{\partial L}{\partial y}\right)_{x,z} = \left(\frac{\partial M}{\partial x}\right)_{y,z} \qquad \left(\frac{\partial L}{\partial z}\right)_{x,y} = -\left(\frac{\partial N}{\partial x}\right)_{z,y} \qquad \left(\frac{\partial M}{\partial z}\right)_{y,x} = -\left(\frac{\partial N}{\partial y}\right)_{z,x}$$

There are two other characteristics of exact differentials that are important in thermodynamics. If $dY = \Sigma\, C_i\, dX_i$ is an exact differential expression, then

I.   The value of the integral $\int_A^B dY \equiv \Delta Y$ is independent of the path followed from point A to point B. *Equivalently*:   The integral around any closed path $\oint dY$ is identically zero.

II.   The function Y is determined by its differential only to within an additive constant.

EXAMPLE 3.2   What applications have already been made in Chapters 1 and 2 of the characteristics of exact differentials just listed?

Axioms 1 and 3 of Section 2.1 formally identify the internal energy and the entropy as intrinsic properties of a system. Thus U and S are functions of the state variables and their differentials must be exact. Earlier, in Section 1.6, it was stated that ΔU should depend only on the end states of a system and should therefore be independent of path. In Example 2.2 use was made of the fact that ΔS_{engine} and ΔU_{engine} are zero, because the engine operates in a cycle, returning periodically to its initial state.

It was also shown that integration of (2.7) or (2.9) leads to an expression for S that contains an integration constant S_0, for which no value can be determined.

The attribute of U and S characterized in item I is shared by all proper thermodynamic functions, and the differentials of all such functions are known from experiment and experience to be exact. These functions are variously called *state functions*, *state properties*, *variables of state*, or *point functions*. Item II sets the limit to which purely mathematical considerations can aid in the deduction of values for a thermodynamic function Y defined by means of (3.2). It does not rule out the existence of absolute values for Y, but implies that such values must be otherwise obtained.

EXAMPLE 3.3   Differential expressions of the form (3.1) can be written which do *not* satisfy the exactness criterion (3.4). For example, consider $y\,dx - x\,dy$. Equation (3.4) is not satisfied because $1 \neq -1$. Hence there is *no* function of x and y whose differential is given by the original expression.

On the other hand, for the expression $y\,dx + x\,dy$, (3.4) *is* satisfied, and the expression is therefore exact. Indeed, $y\,dx + x\,dy$ is the differential of the function $z(x, y) = yx$.

EXAMPLE 3.4   Certain differentials of thermodynamics are inexact. For example, the equation

$$\delta Q_{\text{rev}} = dU + P\,dV$$

(3.5)

expresses the first law for a PVT system undergoing a reversible process. Although the variables on the right are state functions, it will be demonstrated that Q_{rev} is not.

If U is considered a function of T and V, then

$$dU = \left(\frac{\partial U}{\partial V}\right)_T dV + \left(\frac{\partial U}{\partial T}\right)_V dT \qquad (3.6)$$

Combination of (3.5) and (3.6) gives

$$\delta Q_{rev} = \left[\left(\frac{\partial U}{\partial V}\right)_T + P\right] dV + \left(\frac{\partial U}{\partial T}\right)_V dT \equiv M\, dV + N\, dT$$

Differentiation of M with respect to T and of N with respect to V results in

$$\left(\frac{\partial M}{\partial T}\right)_V = \frac{\partial^2 U}{\partial T\, \partial V} + \left(\frac{\partial P}{\partial T}\right)_V \qquad \left(\frac{\partial N}{\partial V}\right)_T = \frac{\partial^2 U}{\partial V\, \partial T}$$

Since U is a function of T and V, the second derivatives must be equal. However, $(\partial P/\partial T)_V$ is known from experiment to be nonzero in general; so (3.4) is not satisfied. Thus δQ_{rev} is not an exact differential, and Q_{rev} is not a state function. The sign δ is used with Q expressly to draw attention to this fact.

EXAMPLE 3.5 Often an inexact differential expression can be made exact by multiplying the expression by some function of the independent variables. Such a function is called an *integrating factor*; it is known that for the case of two independent variables an integrating factor always exists.

In Example 3.3 the expression $y\, dx - x\, dy$ was shown not to be exact. However, multiplication by $1/x^2$ makes it exact:

$$dw = \frac{y}{x^2}\, dx - \frac{1}{x}\, dy$$

The comparison implied by (3.4) leads to $1/x^2 = 1/x^2$, which is obviously correct. It is readily confirmed that $w = -y/x$.

It was shown in Example 3.4 that the expression $\delta Q_{rev} = dU + P\, dV$ is not exact. However, division by the absolute temperature T gives

$$dS = \frac{\delta Q_{rev}}{T} = \frac{dU}{T} + \frac{P}{T}\, dV \qquad (3.7)$$

where use has been made of (2.1). Axiom 3 of Section 2.1 affirms that S is a property, and as such it should be functionally related to U and V, provided these variables characterize the system. Thus dS should be an exact differential and (3.7) should satisfy the exactness criterion (3.4).

Elimination of dU in (3.7) by means of (3.6) gives

$$dS = \left[\frac{1}{T}\left(\frac{\partial U}{\partial V}\right)_T + \frac{P}{T}\right] dV + \frac{1}{T}\left(\frac{\partial U}{\partial T}\right)_V dT \equiv M'\, dV + N'\, dT$$

from which

$$\left(\frac{\partial M'}{\partial T}\right)_V = \frac{1}{T}\left(\frac{\partial^2 U}{\partial T\, \partial V}\right) - \frac{1}{T^2}\left(\frac{\partial U}{\partial V}\right)_T + \frac{1}{T}\left(\frac{\partial P}{\partial T}\right)_V - \frac{P}{T^2}$$

$$\left(\frac{\partial N'}{\partial V}\right)_T = \frac{1}{T}\left(\frac{\partial^2 U}{\partial V\, \partial T}\right)$$

If (3.4) is to be satisfied, then

$$-\frac{1}{T^2}\left(\frac{\partial U}{\partial V}\right)_T + \frac{1}{T}\left(\frac{\partial P}{\partial T}\right)_V - \frac{P}{T^2} = 0$$

or

$$\left(\frac{\partial U}{\partial V}\right)_T = T\left(\frac{\partial P}{\partial T}\right)_V - P \qquad (3.8)$$

Equation (3.8) is in fact a standard equation for single-phase PVT systems, and we have here derived it on the basis that the entropy S for such a system is a state function and that dS is an exact differential in the variables V and T. The validity of (3.8) is subject to experimental verification, and is known to be correct for such systems. (It is trivially true for an ideal gas.) This is simply another test of the validity of the fundamental axioms that form the basis for thermodynamics.

3.2 TRANSFORMATION RELATIONSHIPS FOR SYSTEMS WITH TWO INDEPENDENT VARIABLES

This section is devoted to the development of some useful relationships among first partial derivatives for the thermodynamically important case of a system which can be fully specified by fixing two state variables. If these variables are designated y and z, then any other state function x is related to y and z by an equation of the form

$$f(x, y, z) = 0$$

Since any pair of these variables may be selected as independent, this fundamental relationship may be expressed in three alternative forms: $x = x(y, z)$, $y = y(x, z)$, $z = z(x, y)$. Arbitrarily selecting the first two of these, we may write expressions for the total differential of dx and dy:

$$dx = \left(\frac{\partial x}{\partial y}\right)_z dy + \left(\frac{\partial x}{\partial z}\right)_y dz \tag{3.9}$$

$$dy = \left(\frac{\partial y}{\partial x}\right)_z dx + \left(\frac{\partial y}{\partial z}\right)_x dz \tag{3.10}$$

Elimination of the differential dy between (3.9) and (3.10) gives

$$\left[\left(\frac{\partial x}{\partial y}\right)_z\left(\frac{\partial y}{\partial x}\right)_z - 1\right] dx + \left[\left(\frac{\partial x}{\partial y}\right)_z\left(\frac{\partial y}{\partial z}\right)_x + \left(\frac{\partial x}{\partial z}\right)_y\right] dz = 0 \tag{3.11}$$

Since x and z are independently variable, the coefficients of dx and dz must be identically zero if (3.11) is to be generally valid. Hence

$$\boxed{\left(\frac{\partial x}{\partial y}\right)_z = \left(\frac{\partial y}{\partial x}\right)_z^{-1}} \tag{3.12}$$

and

$$\left(\frac{\partial x}{\partial z}\right)_y = -\left(\frac{\partial x}{\partial y}\right)_z\left(\frac{\partial y}{\partial z}\right)_x$$

which in view of (3.12) can be written

$$\boxed{\left(\frac{\partial x}{\partial y}\right)_z = -\left(\frac{\partial x}{\partial z}\right)_y\left(\frac{\partial z}{\partial y}\right)_x} \tag{3.13}$$

Another useful formula is the *chain rule* for partial derivatives,

$$\boxed{\left(\frac{\partial x}{\partial y}\right)_z = \left(\frac{\partial x}{\partial w}\right)_z\left(\frac{\partial w}{\partial y}\right)_z} \tag{3.14}$$

If x is now taken to be a function of y and w, then

$$dx = \left(\frac{\partial x}{\partial y}\right)_w dy + \left(\frac{\partial x}{\partial w}\right)_y dw$$

Division of this equation by dy with restriction to constant z gives

$$\boxed{\left(\frac{\partial x}{\partial y}\right)_z = \left(\frac{\partial x}{\partial y}\right)_w + \left(\frac{\partial x}{\partial w}\right)_y\left(\frac{\partial w}{\partial y}\right)_z} \tag{3.15}$$

EXAMPLE 3.6   If for a PVT system V is a function of T and P, then

$$dV = \left(\frac{\partial V}{\partial T}\right)_P dT + \left(\frac{\partial V}{\partial P}\right)_T dP \tag{3.16}$$

The partial differential coefficients in this equation are directly related to two properties commonly tabulated for pure substances:

(*a*)   The *volume expansivity* β, where

$$\boxed{\beta \equiv \frac{1}{V}\left(\frac{\partial V}{\partial T}\right)_P} \tag{3.17}$$

(*b*)   The *isothermal compressibility* κ, where

$$\boxed{\kappa \equiv -\frac{1}{V}\left(\frac{\partial V}{\partial P}\right)_T} \tag{3.18}$$

Substitution into (*3.16*) gives

$$\frac{dV}{V} = \beta \, dT - \kappa \, dP \tag{3.19}$$

Since $d \ln V$ is an exact differential, (*3.4*) must be satisfied, and therefore

$$\left(\frac{\partial \beta}{\partial P}\right)_T = -\left(\frac{\partial \kappa}{\partial T}\right)_P \tag{3.20}$$

For the special case of an ideal gas, $PV = RT$, and by differentiation it is found that

$$\beta = \frac{1}{T} \quad \text{and} \quad \kappa = \frac{1}{P} \quad \text{(ideal gas)}$$

In this case (*3.19*) becomes

$$\frac{dV}{V} = \frac{dT}{T} - \frac{dP}{P} \quad \text{(ideal gas)} \tag{3.21}$$

By (*3.12*) and (*3.13*),

$$\left(\frac{\partial P}{\partial T}\right)_V = -\frac{(\partial V/\partial T)_P}{(\partial V/\partial P)_T}$$

or, in view of (*3.17*) and (*3.18*),

$$\boxed{\left(\frac{\partial P}{\partial T}\right)_V \equiv \frac{\beta}{\kappa}} \tag{3.22}$$

The derivative $(\partial P/\partial T)_V$ is called the *thermal pressure coefficient*; it is sometimes given the symbol α. For an ideal gas, $\alpha = P/T$, as may be confirmed by direct differentiation of $PV = RT$.

EXAMPLE 3.7   If we apply (*3.15*) to a PVT system and let

$$\left(\frac{\partial x}{\partial y}\right)_z = \left(\frac{\partial T}{\partial V}\right)_S$$

and if we identify w with U, then (*3.15*) becomes

$$\left(\frac{\partial T}{\partial V}\right)_S = \left(\frac{\partial T}{\partial V}\right)_U + \left(\frac{\partial T}{\partial U}\right)_V\left(\frac{\partial U}{\partial V}\right)_S$$

According to (*1.8*), $C_V = (\partial U/\partial T)_V$ and therefore

$$\left(\frac{\partial T}{\partial U}\right)_V = \left(\frac{\partial U}{\partial T}\right)_V^{-1} = \frac{1}{C_V}$$

Furthermore, from (2.6) we have

$$\left(\frac{\partial U}{\partial V}\right)_S = -P$$

Therefore $\qquad \left(\frac{\partial T}{\partial V}\right)_S = \left(\frac{\partial T}{\partial V}\right)_U - \frac{P}{C_V} \quad$ or $\quad \left(\frac{\partial T}{\partial V}\right)_U = \left(\frac{\partial T}{\partial V}\right)_S + \frac{P}{C_V}$

This equation gives the derivative dT/dV for a process at constant internal energy. An example of such a process is the Joule expansion, in which a gas initially confined to a portion of a rigid, insulated tank is allowed to expand to fill the entire tank. In this process Q and W are zero, and therefore $\Delta U = 0$. The quantity $(\partial T/\partial V)_S$ is the temperature/volume derivative for an isentropic process.

For an ideal gas U is a function of temperature only. Therefore if U is constant, so is T, making $(\partial T/\partial V)_U = 0$ and

$$\left(\frac{\partial T}{\partial V}\right)_S = -\frac{P}{C_V}$$

which for an isentropic process may be written $dT = -(P/C_V)\,dV$. Division by T and substitution of R/V for P/T gives

$$\frac{dT}{T} = -\frac{R}{C_V}\frac{dV}{V}$$

This same equation was derived in Example 1.10 and again in Example 2.5 by entirely different methods, and leads to the relation $TV^{\gamma-1} = $ constant, for the reversible, adiabatic compression or expansion of an ideal gas with constant heat capacities.

3.3 LEGENDRE TRANSFORMATIONS

The fundamental property relation for a closed PVT system was developed in Section 2.3:

$$dU = T\,dS - P\,dV \tag{2.6}$$

This equation implies that U is always a function of the variables S and V in any closed system. However, the choice of S and V as variables is not always convenient; other pairs of variables are often advantageously employed. It is therefore useful to define new functions whose total differentials are consistent with (2.6), but for which the natural variables are pairs other than S and V. Thus in Section 1.7 it was found convenient to define the enthalpy as

$$H \equiv U + PV \tag{1.6}$$

Differentiation of (1.6) provides

$$dH = dU + P\,dV + V\,dP$$

which may be combined with (2.6) to give

$$dH = T\,dS + V\,dP$$

from which it is seen that S and P are the natural or special variables for the function H as it applies to a closed PVT system.

New thermodynamic functions cannot in general be defined by random combination of variables; there is, for example, the requirement of dimensional consistency. Fortunately, a standard mathematical method exists for the systematic definition of functions of the kind required, the *Legendre transformation*.

Recall the exact differential expression presented in Section 3.1:

$$dY = C_1\,dX_1 + C_2\,dX_2 + \cdots + C_n\,dX_n \tag{3.2}$$

Legendre transformations define Y-related functions for which the sets of variables contain one or more of the C_i in place of the conjugate X_i. For a total differential expression exhibiting n variables, there are $2^n - 1$ possible Legendre transformations, namely:

$$\mathcal{T}_1 \equiv \mathcal{T}_1(C_1, X_2, X_3, \ldots, X_n) = Y - C_1 X_1$$
$$\mathcal{T}_2 \equiv \mathcal{T}_2(X_1, C_2, X_3, \ldots, X_n) = Y - C_2 X_2$$
$$\cdots\cdots\cdots\cdots\cdots\cdots\cdots\cdots\cdots\cdots$$
$$\mathcal{T}_n \equiv \mathcal{T}_n(X_1, X_2, X_3, \ldots, C_n) = Y - C_n X_n$$
$$\mathcal{T}_{1,2} \equiv \mathcal{T}_{1,2}(C_1, C_2, X_3, \ldots, X_n) = Y - C_1 X_1 - C_2 X_2 \qquad (3.23)$$
$$\mathcal{T}_{1,3} \equiv \mathcal{T}_{1,3}(C_1, X_2, C_3, \ldots, X_n) = Y - C_1 X_1 - C_3 X_3$$
$$\cdots\cdots\cdots\cdots\cdots\cdots\cdots\cdots\cdots\cdots\cdots$$
$$\mathcal{T}_{1,\ldots,n} \equiv \mathcal{T}_{1,\ldots,n}(C_1, C_2, C_3, \ldots, C_n) = Y - \sum C_i X_i$$

Each $\mathcal{T}$ in (3.23) represents a new *function*, and in each case the independent variables, shown in parentheses, are the *canonical*[1] variables for that function. Thus (3.23) provides a recipe for the definition of a set of new functions consistent with a particular exact differential expression, and it identifies the variables which are unique to each function. These have the following special property: When a transformation function $\mathcal{T}$ is known as a function of its n canonical variables, then the remaining n variables among those appearing in the original exact differential expression (the X_i and their conjugate C_i) can be recovered by differentiation of $\mathcal{T}$. This is not true for arbitrarily chosen sets of variables.

EXAMPLE 3.8 If $Y = Y(X_1, X_2, X_3)$, then

$$dY = C_1 \, dX_1 + C_2 \, dX_2 + C_3 \, dX_3 \qquad (1)$$

From (3.23), $\mathcal{T}_1(C_1, X_2, X_3) = Y - C_1 X_1$; to simplify notation, call this function Z:

$$Z \equiv Y - C_1 X_1 \qquad (2)$$

Differentiating $Z = Z(C_1, X_2, X_3)$, we have

$$dZ = \left(\frac{\partial Z}{\partial C_1}\right)_{X_2, X_3} dC_1 + \left(\frac{\partial Z}{\partial X_2}\right)_{C_1, X_3} dX_2 + \left(\frac{\partial Z}{\partial X_3}\right)_{C_1, X_2} dX_3 \qquad (3)$$

The differential dZ can also be found from (2):

$$dZ = dY - C_1 \, dX_1 - X_1 \, dC_1$$

and substitution for dY by (1) gives

$$dZ = -X_1 \, dC_1 + C_2 \, dX_2 + C_3 \, dX_3 \qquad (4)$$

Comparison of (3) and (4) shows that

$$X_1 = -\left(\frac{\partial Z}{\partial C_1}\right)_{X_2, X_3} \qquad C_2 = \left(\frac{\partial Z}{\partial X_2}\right)_{C_1, X_3} \qquad C_3 = \left(\frac{\partial Z}{\partial X_3}\right)_{C_1, X_2}$$

Thus the variables X_1, C_2, and C_3 are given by the partial derivatives of Z with respect to its canonical variables.

EXAMPLE 3.9 Consider the fundamental property relation for a closed PVT system:

$$dU = -P \, dV + T \, dS \qquad (2.6)$$

In applying (3.23), we identify C_1 and C_2 with $-P$ and T, and X_1 and X_2 with V and S. There are $2^2 - 1 = 3$ possible U-related Legendre transforms, given by

---

[1]The adjective *canonical* means that the variables conform to a scheme that is both simple and clear.

$$H = H(P, S) \equiv U + PV$$
$$A = A(V, T) \equiv U - TS \qquad (3.24)$$
$$G = G(P, T) \equiv U + PV - TS$$

The first of these three functions is the enthalpy, already introduced. The functions A and G are the *Helmholtz* and *Gibbs energy*, respectively. (Both are sometimes called *free energies*, and in this case the modifiers *Helmholtz* and *Gibbs* are still appropriate to distinguish between them. A common European practice is to call A the *free energy* and G the *free enthalpy*.)

It is the particular form of (2.6) that makes T, S, P, and V the variables of choice in the classical thermodynamics of PVT systems, and makes U, H, A, and G the natural energy functions. One could, however, proceed differently by rearranging the fundamental property relation as

$$dS = \frac{P}{T}\, dV + \frac{1}{T}\, dU$$

This equation, used in Example 2.8, expresses the *dependent* variable S as a function of V and U. As in the case of U, there are but three possible Legendre transformations:

$$\Omega = \Omega\left(\frac{P}{T}, U\right) = S - \frac{P}{T} V$$

$$\Psi = \Psi\left(V, \frac{1}{T}\right) = S - \frac{1}{T} U$$

$$\Phi = \Phi\left(\frac{P}{T}, \frac{1}{T}\right) = S - \frac{P}{T} V - \frac{1}{T} U$$

The functions defined by these equations are called *Massieu functions*, and Φ is often called the *Planck function*. A complete network of thermodynamic equations can be developed from this formulation. However, the natural intensive variables (see Section 1.8) in this system are $1/T$ and P/T. These variables, as well as the Massieu functions, arise naturally in statistical mechanics and in irreversible thermodynamics, but they are less directly related to experience and are less useful in the description of real processes than are T and P, the intensive variables more commonly employed.

3.4 PRIMARY PROPERTY RELATIONSHIPS FOR PVT SYSTEMS OF VARIABLE COMPOSITION

As an important application of the principles of the preceding sections, we consider a homogeneous PVT system containing m chemical species present in mole numbers $n_1, n_2, \ldots, n_m$. The internal energy, entropy, and volume are extensive properties, so the total system properties are nU, nS, and nV, where U, S, and V are molar properties and n is the total number of moles of all chemical species. For the particular case of a reversible process in which *all n_i are constant* (a *constant-composition system*), we have

$$\delta Q_{\text{rev}} = T\, d(nS) \qquad \delta W_{\text{rev}} = P\, d(nV)$$

and by the first law
$$d(nU) = T\, d(nS) - P\, d(nV) \qquad (3.25)$$

Equation (3.25) is an exact differential expression and, according to (3.3),

$$\left(\frac{\partial(nU)}{\partial(nS)}\right)_{nV,n} = T \qquad (3.26)$$

$$\left(\frac{\partial(nU)}{\partial(nV)}\right)_{nS,n} = -P \qquad (3.27)$$

The additional subscript n indicates that all the n_i are held constant.

For the *general* case, however, nU must be considered a function of the n_i as well as of nS and nV. We therefore write formally for the total differential of nU

$$d(nU) = \left(\frac{\partial(nU)}{\partial(nS)}\right)_{nV,n} d(nS) + \left(\frac{\partial(nU)}{\partial(nV)}\right)_{nS,n} d(nV) + \sum_{i=1}^{m} \left(\frac{\partial(nU)}{\partial n_i}\right)_{nS,nV,n_j} dn_i \qquad (3.28)$$

where the subscript n_j means that all mole numbers except n_i are held constant. Combining (3.26), (3.27), and (3.28), we obtain

$$\boxed{d(nU) = T\,d(nS) - P\,d(nV) + \sum \mu_i\, dn_i} \qquad (3.29)$$

where μ_i is the *chemical potential*, defined by

$$\mu_i \equiv \left(\frac{\partial(nU)}{\partial n_i}\right)_{nS,nV,n_j}$$

and the plain summation symbol implies summation over all chemical species i.

Equation (3.29) is the fundamental property relation for a *homogeneous PVT system of variable composition*, and is the basis for all derived property relations for such systems. The system may be open or closed, and composition changes may result either from chemical reaction or from the transport of matter or both. Equation (3.25) is a special case of (3.29), valid for systems of fixed mole numbers. In addition, (3.25) [or (2.6)] has separate validity for *all processes connecting equilibrium states in any closed PVT system*, whether homogeneous or heterogeneous and regardless of changes in the mole numbers on account of chemical reaction.

The equations of this section and those which follow could equally well be based on unit-mass (*specific*) properties rather than on molar properties. The symbol n would then represent mass rather than number of moles.

There are $m + 2$ variables in (3.29), and therefore $2^{m+2} - 1$ possible nU-related Legendre transformations. However, only three of these find widespread use. They are the enthalpy, the Helmholtz energy, and the Gibbs energy, all of which were earlier defined in Example 3.9.

$$(nH) = (nU) + P(nV) \qquad (3.30)$$

$$(nA) = (nU) - T(nS) \qquad (3.31)$$

$$(nG) = (nU) + P(nV) - T(nS) \qquad (3.32)$$

By taking the total differentials of (3.30), (3.31), and (3.32), and employing (3.29) to eliminate $d(nU)$, one obtains $d(nH)$, $d(nA)$, and $d(nG)$ in forms which display the canonical variables for the three functions:

$$\boxed{\begin{aligned} d(nH) &= T\,d(nS) + (nV)\,dP + \sum \mu_i\, dn_i \qquad &(3.33)\\[4pt] d(nA) &= -(nS)\,dT - P\,d(nV) + \sum \mu_i\, dn_i \qquad &(3.34)\\[4pt] d(nG) &= -(nS)\,dT + (nV)\,dP + \sum \mu_i\, dn_i \qquad &(3.35) \end{aligned}}$$

A number of useful relationships follow from the fact that (3.29), (3.33), (3.34), and (3.35) are exact differential expressions. According to (3.3),

$$T = \left(\frac{\partial U}{\partial S}\right)_{V,x} = \left(\frac{\partial H}{\partial S}\right)_{P,x} \qquad (3.36)$$

$$P = -\left(\frac{\partial U}{\partial V}\right)_{S,x} = -\left(\frac{\partial A}{\partial V}\right)_{T,x} \qquad (3.37)$$

$$V = \left(\frac{\partial H}{\partial P}\right)_{S,x} = \left(\frac{\partial G}{\partial P}\right)_{T,x} \qquad (3.38)$$

$$S = -\left(\frac{\partial A}{\partial T}\right)_{V,x} = -\left(\frac{\partial G}{\partial T}\right)_{P,x} \qquad (3.39)$$

$$\mu_i = \left(\frac{\partial(nU)}{\partial n_i}\right)_{nS,nV,n_j} = \left(\frac{\partial(nH)}{\partial n_i}\right)_{nS,P,n_j} = \left(\frac{\partial(nA)}{\partial n_i}\right)_{nV,T,n_j} = \left(\frac{\partial(nG)}{\partial n_i}\right)_{T,P,n_j} \qquad (3.40)$$

Equations (3.36) through (3.39) are written in terms of molar, rather than total, properties. This is permissible because all the n_i, and hence n, are held constant in the definitions of the partial differential coefficients. Here, subscript x denotes constant composition.

We may apply the exactness criterion (3.4) to (3.29), (3.33), (3.34), and (3.35), to obtain

$$\left(\frac{\partial T}{\partial V}\right)_{S,x} = -\left(\frac{\partial P}{\partial S}\right)_{V,x} \qquad (3.41)$$

$$\left(\frac{\partial T}{\partial P}\right)_{S,x} = \left(\frac{\partial V}{\partial S}\right)_{P,x} \qquad (3.42)$$

$$\left(\frac{\partial P}{\partial T}\right)_{V,x} = \left(\frac{\partial S}{\partial V}\right)_{T,x} \qquad (3.43)$$

$$\left(\frac{\partial V}{\partial T}\right)_{P,x} = -\left(\frac{\partial S}{\partial P}\right)_{T,x} \qquad (3.44)$$

$$\left(\frac{\partial \mu_i}{\partial T}\right)_{P,x} = -\left(\frac{\partial(nS)}{\partial n_i}\right)_{T,P,n_j} \qquad (3.45)$$

$$\left(\frac{\partial \mu_i}{\partial P}\right)_{T,x} = \left(\frac{\partial(nV)}{\partial n_i}\right)_{T,P,n_j} \qquad (3.46)$$

$$\left(\frac{\partial \mu_l}{\partial n_k}\right)_{T,P,n_j} = \left(\frac{\partial \mu_k}{\partial n_l}\right)_{T,P,n_j} \qquad (3.47)$$

Equations (3.41) through (3.44) are called the *Maxwell equations*, and we have chosen to write them in terms of molar properties; as with (3.36) through (3.39), they apply to constant-composition systems. Of the twelve possible equations of the form (3.4) involving μ_i, we have written only the three given as (3.45), (3.46), and (3.47), which result from (3.35). We will see in Chapter 7 that they have special significance in the treatment of properties of solutions.

EXAMPLE 3.10   One important use of the basic property relations given by (3.29), (3.33), (3.34), and (3.35) is in the development of expressions for a large number of partial derivatives. We illustrate the general technique with (3.29) applied to a closed system of fixed composition. For such a system, (3.29) becomes

$$dU = T\,dS - P\,dV \qquad (2.6)$$

We now develop an expression for $(\partial U/\partial V)_T$. The procedure is to divide the equation through by dV and to restrict it to constant T. This gives

$$\left(\frac{\partial U}{\partial V}\right)_T = T\left(\frac{\partial S}{\partial V}\right)_T - P$$

The conversion to partial derivatives is based on the presumption that U and S are functions of T and V, and indeed this is so for single-phase PVT systems. By (3.43),

$$\left(\frac{\partial S}{\partial V}\right)_T = \left(\frac{\partial P}{\partial T}\right)_V$$

Substitution recovers (3.8),

$$\left(\frac{\partial U}{\partial V}\right)_T = T\left(\frac{\partial P}{\partial T}\right)_V - P$$

which gives the required derivative in terms of P, V, and T, all measurable quantities.

As a further illustration, we determine an expression for $(\partial U/\partial T)_V$. Division of (2.6) by dT and restriction to constant V gives immediately

$$\left(\frac{\partial U}{\partial T}\right)_V = T\left(\frac{\partial S}{\partial T}\right)_V$$

By definition (1.8), the derivative on the left is C_V, which is measurable.

3.5 PROPERTY RELATIONSHIPS FOR CONSTANT-COMPOSITION PVT SYSTEMS

The heat capacities C_V and C_P were defined in Chapter 1 by

$$C_V \equiv \left(\frac{\partial U}{\partial T}\right)_V \tag{1.8}$$

$$C_P \equiv \left(\frac{\partial H}{\partial T}\right)_P \tag{1.9}$$

For a constant-composition PVT system, (3.29) reduces to

$$dU = T\,dS - P\,dV \tag{2.6}$$

and (3.33) becomes

$$dH = T\,dS + V\,dP \tag{3.48}$$

From these, the derivatives appearing in (1.8) and (1.9) are directly obtained (see Example 3.10), and provide alternative expressions for the heat capacities:

$$C_V = T\left(\frac{\partial S}{\partial T}\right)_V \tag{3.49}$$

$$C_P = T\left(\frac{\partial S}{\partial T}\right)_P \tag{3.50}$$

Several equations will now be derived to demonstrate how heat-capacity data are used in conjunction with PVT data to evaluate changes in U, H, and S for changes in state of single-phase, constant-composition PVT systems. If U is considered a function of T and V, then

$$dU = \left(\frac{\partial U}{\partial T}\right)_V dT + \left(\frac{\partial U}{\partial V}\right)_T dV \tag{3.51}$$

Use of

$$\left(\frac{\partial U}{\partial V}\right)_T = T\left(\frac{\partial P}{\partial T}\right)_V - P \tag{3.8}$$

together with (1.8) and (3.51) gives

$$dU = C_V\,dT + \left[T\left(\frac{\partial P}{\partial T}\right)_V - P\right]dV \tag{3.52}$$

An analogous equation can be derived for dH in terms of dT and dP:

$$dH = C_P\, dT + \left[V - T\left(\frac{\partial V}{\partial T}\right)_P\right] dP \qquad (3.53)$$

Two useful (and equivalent) expressions can be derived for the total differential of the entropy, depending on whether T and V or T and P are selected as independent variables. In the former case,

$$dS = \left(\frac{\partial S}{\partial T}\right)_V dT + \left(\frac{\partial S}{\partial V}\right)_T dV$$

and substitution for the partial derivatives by (3.49) and (3.43) gives

$$dS = \frac{C_V}{T}\, dT + \left(\frac{\partial P}{\partial T}\right)_V dV \qquad (3.54)$$

With T and P as variables,

$$dS = \left(\frac{\partial S}{\partial T}\right)_P dT + \left(\frac{\partial S}{\partial P}\right)_T dP$$

and substitution by (3.50) and (3.44) yields

$$dS = \frac{C_P}{T}\, dT - \left(\frac{\partial V}{\partial T}\right)_P dP \qquad (3.55)$$

For a given change in state, (3.54) and (3.55) must give the same value for dS, and they may therefore be equated. The resulting expression, upon rearrangement, becomes

$$(C_P - C_V)\, dT = T\left(\frac{\partial V}{\partial T}\right)_P dP + T\left(\frac{\partial P}{\partial T}\right)_V dV$$

Division by dT and restriction to either constant pressure or constant volume provides an equation for the *difference* in heat capacities:

$$C_P - C_V = T\left(\frac{\partial P}{\partial T}\right)_V\left(\frac{\partial V}{\partial T}\right)_P \qquad (3.56)$$

An expression for the *ratio* of heat capacities follows directly from division of (3.50) by (3.49):

$$\frac{C_P}{C_V} = \left(\frac{\partial S}{\partial T}\right)_P\left(\frac{\partial T}{\partial S}\right)_V$$

Application of (3.13) to each of the partial derivatives gives

$$\frac{C_P}{C_V} = \left[-\left(\frac{\partial S}{\partial P}\right)_T\left(\frac{\partial P}{\partial T}\right)_S\right]\left[-\left(\frac{\partial T}{\partial V}\right)_S\left(\frac{\partial V}{\partial S}\right)_T\right] = \left[\left(\frac{\partial V}{\partial S}\right)_T\left(\frac{\partial S}{\partial P}\right)_T\right]\left[\left(\frac{\partial P}{\partial T}\right)_S\left(\frac{\partial T}{\partial V}\right)_S\right]$$

or, as a result of (3.14),

$$\frac{C_P}{C_V} = \left(\frac{\partial V}{\partial P}\right)_T\left(\frac{\partial P}{\partial V}\right)_S \qquad (3.57)$$

Equations (3.56) and (3.57) provide alternative ways of relating C_P to C_V. The partial derivatives in both equations are obtained from PVT data, with the exception of $(\partial P/\partial V)_S$, which is related to the speed of sound (see Problem 3.14).

Application of the exactness criterion (3.4) to the exact differential expressions (3.54) and (3.55) yields expressions for derivatives of the heat capacities that depend on PVT data only:

$$\left(\frac{\partial C_V}{\partial V}\right)_T = T\left(\frac{\partial^2 P}{\partial T^2}\right)_V \qquad (3.58)$$

$$\left(\frac{\partial C_P}{\partial P}\right)_T = -T\left(\frac{\partial^2 V}{\partial T^2}\right)_P \qquad (3.59)$$

It is sometimes convenient to have property relations available in forms which explicitly involve the volume expansivity β and/or the isothermal compressibility κ (see Example 3.6). Thus we list the following expressions equivalent to (3.52) through (3.59):

$$dU = C_V\, dT + \left(\frac{\beta T}{\kappa} - P\right) dV \qquad (3.60)$$

$$dH = C_P\, dT + V(1 - \beta T)\, dP \qquad (3.61)$$

$$dS = \frac{C_V}{T}\, dT + \frac{\beta}{\kappa}\, dV \qquad (3.62)$$

$$dS = \frac{C_P}{T}\, dT - \beta V\, dP \qquad (3.63)$$

$$C_P - C_V = \frac{\beta^2 V T}{\kappa} \qquad (3.64)$$

$$\frac{C_P}{C_V} = \frac{\kappa}{\kappa_S} \qquad (3.65)$$

$$\left(\frac{\partial C_V}{\partial V}\right)_T = T\left[\left(\frac{\partial(\beta/\kappa)}{\partial T}\right)_P + \frac{\beta}{\kappa}\left(\frac{\partial(\beta/\kappa)}{\partial P}\right)_T\right] \qquad (3.66)$$

$$\left(\frac{\partial C_P}{\partial P}\right)_T = -VT\left[\beta^2 + \left(\frac{\partial\beta}{\partial T}\right)_P\right] \qquad (3.67)$$

In (3.65), quantity κ_S is the *adiabatic compressibility*, defined analogously to κ as

$$\kappa_S \equiv -\frac{1}{V}\left(\frac{\partial V}{\partial P}\right)_S \qquad (3.68)$$

EXAMPLE 3.11 The usefulness of the equations of this section depends on knowledge of a PVT relation for the particular substance to which the equations are applied. Such a relation may be provided by a specific equation of state or by tables of numerical data. For gases the simplest equation of state is the ideal-gas equation, and we use it here to illustrate application of the equations just derived.

By differentiation of the ideal-gas equation, $PV = RT$, we find

$$\left(\frac{\partial P}{\partial T}\right)_V = \frac{R}{V} = \frac{P}{T} \qquad \left(\frac{\partial^2 P}{\partial T^2}\right)_V = 0$$

$$\left(\frac{\partial V}{\partial T}\right)_P = \frac{R}{P} = \frac{V}{T} \qquad \left(\frac{\partial^2 V}{\partial T^2}\right)_P = 0$$

$$\left(\frac{\partial V}{\partial P}\right)_T = \frac{-RT}{P^2} = \frac{-V}{P}$$

Substitution for these partial derivatives in (3.52) through (3.59) yields the following results, valid for ideal gases:

$$dU = C_V\, dT \qquad (3.52\ ideal) \text{ and } (1.13)$$

$$dH = C_P \, dT \qquad\qquad (3.53 \text{ ideal}) \text{ and } (1.15)$$

$$dS = C_V \frac{dT}{T} + R \frac{dV}{V} \qquad\qquad (3.54 \text{ ideal}) \text{ and } (2.7)$$

$$dS = C_P \frac{dT}{T} - R \frac{dP}{P} \qquad\qquad (3.55 \text{ ideal}) \text{ and } (2.9)$$

$$C_P - C_V = R \qquad\qquad (3.56 \text{ ideal}) \text{ and } (1.14)$$

$$\gamma \equiv \frac{C_P}{C_V} = \frac{-V}{P}\left(\frac{\partial P}{\partial V}\right)_s \qquad\qquad (3.57 \text{ ideal})$$

$$\left(\frac{\partial C_V}{\partial V}\right)_T = 0 \qquad\qquad (3.58 \text{ ideal})$$

$$\left(\frac{\partial C_P}{\partial P}\right)_T = 0 \qquad\qquad (3.59 \text{ ideal})$$

The first five of these equations were presented earlier, as indicated by the equation numbers. Equation (3.57 ideal) also reduces to a familiar equation. For an isentropic process it may be written $\gamma \, dV/V = -dP/P$, which, for constant γ, integrates to $PV^\gamma = k$, a relation derived in Example 2.5 for the reversible, adiabatic expansion of an ideal gas with constant heat capacities. Equations (3.58 ideal) and (3.59 ideal) show again that the heat capacities of an ideal gas, like the internal energy and enthalpy, are functions of temperature only. When temperature is constant, so is the heat capacity.

EXAMPLE 3.12 The "β-and-κ formulations" (3.60) through (3.67) are often applied to property calculations for condensed phases (liquids, solids). For such phases, volume may depend only weakly on temperature and pressure, allowing one to make various simplifying assumptions about the property relations. The nature of the assumptions depends on the kind of calculation and on the type and extent of available data. We illustrate some of the ideas here.

Suppose it is required to estimate property changes ΔH and ΔS of a liquid or solid for an arbitrary change in state. Choosing T and P as independent variables, we write the integrated forms of (3.61) and (3.63) as

$$\Delta H = \langle C_P \rangle (T_2 - T_1) + \langle V(1 - \beta T) \rangle (P_2 - P_1) \qquad\qquad (3.69)$$

$$\Delta S = \langle C_P \rangle \ln \frac{T_2}{T_1} - \langle \beta V \rangle (P_2 - P_1) \qquad\qquad (3.70)$$

Here we invoke the mean-value theorem of the integral calculus to introduce averages, denoted by triangular brackets $\langle \ \rangle$; thus (3.69) and (3.70) are true *by definition*. Effective use of these equations depends upon judicious estimation of the averages.

The simplest approximation, appropriate for modest changes in T and P, is to treat C_P, V, β, and κ as *constants*. In this case, (3.69) and (3.70) reduce to

$$\Delta H \approx C_P(T_2 - T_1) + V[1 - \tfrac{1}{2}\beta(T_1 + T_2)](P_2 - P_1) \qquad\qquad (1)$$

$$\Delta S \approx C_P \ln \frac{T_2}{T_1} - \beta V(P_2 - P_1) \qquad\qquad (2)$$

While the above approximation is, strictly speaking, inconsistent (V will be nonconstant, unless β and κ have the constant value zero), (1) and (2) suffice for many applications involving liquids or solids. If sufficient data are available, the "constants" C_P, V, β, and κ can be evaluated at average values of T and P. If no data at all are available for β and κ, then both are assigned the value zero; this is the *incompressible-substance* assumption, and elaborated in the next paragraph.

An analog of (3.60), with T and P as independent variables, is

$$dU = (C_P - \beta PV) \, dT + V(\kappa P - \beta T) \, dP$$

Thus, for an incompressible substance, we get $dU = C_P \, dT$, where, by (3.67), C_P is independent of P. Hence the internal energy depends only on T. However, the enthalpy is a function of both T and P, because (3.61) reduces to $dH = C_P \, dT + V \, dP$. By (3.63), $dS = (C_P/T) \, dT$, and thus the entropy is a function of T only. Finally, we see from (3.64) that $C_P = C_V$, provided that the indeterminate ratio β/κ is *finite*.

3.6 ATTAINMENT OF EQUILIBRIUM IN CLOSED, HETEROGENEOUS SYSTEMS

The preceding sections have dealt with relationships among thermodynamic properties in homogeneous systems presumed to be in equilibrium. We consider here the approach to equilibrium in heterogeneous systems which are not initially at equilibrium with respect to the distribution of the various chemical species among coexisting but separate phases. The simplest such system is one that is closed and in which the temperature and pressure are uniform (but not necessarily constant). The system is imagined to contain an arbitrary number of phases, the composition of each phase being uniform but not necessarily the same as the composition of any other phase. We further imagine that the system exchanges heat reversibly with its surroundings and that volume changes of the system occur in such a way that work exchange with the surroundings is also reversible. The process considered results from a change in the system *from* a nonequilibrium state *toward* an equilibrium state with respect to the distribution of species among the phases. As before, a total property of the system as a whole is denoted by superscript t.

For the reversible exchange of heat δQ, the entropy change of the surroundings is

$$dS_{\text{surr}} = \frac{-\delta Q}{T} \qquad (3.71)$$

where the minus sign arises because δQ is taken with reference to the system, whereas dS_{surr} refers to the surroundings. The temperature T is that of the system *and* the surroundings, since the heat transfer is assumed reversible. By the second law, $dS^t + dS_{\text{surr}} \geqq 0$, or by (3.71),

$$dS^t \geqq \frac{\delta Q}{T} \qquad (3.72)$$

The inequality sign signifies an irreversible process, here the transfer of mass between phases, since all heat and work effects have been taken to be reversible. Once the system reaches *phase equilibrium* the equality of (3.72) applies.

The first law for a closed system is given by (1.5), $dU^t = \delta Q - \delta W$; and since the work of volume change is reversible, $\delta W = P\, dV^t$, where P is the pressure of the system. Therefore,

$$dU^t = \delta Q - P\, dV^t \qquad (3.73)$$

Combination of (3.72) and (3.73) gives

$$\boxed{dU^t - T\, dS^t + P\, dV^t \leqq 0} \qquad (3.74)$$

This relation involves properties only and must be satisfied for changes of state *in any closed PVT system whatever*, without restriction to the conditions of reversibility assumed for its derivation. The equality, which reproduces (2.6), is satisfied for *any* process that leads from one state of internal equilibrium to another or to variations around an equilibrium state. The inequality must be satisfied by *any* process that starts from an initial state of uniform T and P but otherwise not a state of internal equilibrium, and it dictates the direction of change that leads toward an equilibrium state. We are considering here nonequilibrium states with respect to the distribution of chemical species among phases, but the treatment is in no way different for nonequilibrium states with respect to chemical reaction among the species or to a combination of the two.

Special Forms of the General Relation

Equation (3.74) is of such generality that its implications are difficult to visualize. It is readily reduced to simpler forms that apply to processes restricted in various ways. Thus, for processes at constant entropy and volume, (3.74) becomes

$$(dU^t)_{S^t, V^t} \leqq 0 \qquad (3.75)$$

Similarly, for processes restricted to constant U^t and V^t,

$$(dS^t)_{U^t, V^t} \geqq 0 \tag{3.76}$$

An isolated system, as discussed in Section 2.5, is a system constrained to constant internal energy and volume, and for such a system it follows immediately from the second law that (3.76) is valid.

Additional equations representing special cases of (3.74) follow from the definitions of enthalpy, the Helmholtz energy, and the Gibbs energy. Thus

$$H^t = U^t + PV^t \qquad \text{and} \qquad dH^t = dU^t + P\,dV^t + V^t\,dP$$

Combination with (3.74) gives

$$dH^t - V^t\,dP - T\,dS^t \leqq 0$$

whence

$$(dH^t)_{P, S^t} \leqq 0 \tag{3.77}$$

In an analogous fashion one finds

$$(dA^t)_{T, V^t} \leqq 0 \tag{3.78}$$

and

$$\boxed{(dG^t)_{T, P} \leqq 0} \tag{3.79}$$

Let us consider the meaning of (3.79). The inequality requires that all irreversible processes occurring at constant T and P must proceed in such a direction as to cause the total Gibbs energy of the system to decrease. Since the Gibbs energy decreases in all changes toward equilibrium in a system at constant T and P, then the equilibrium state for a given T and P must be that state for which the Gibbs energy has its minimum value with respect to all possible variations at the given T and P. At the equilibrium state the equality of (3.79) holds, and this means that differential variations may occur in the system at constant T and P without producing a change in G. Thus a criterion of equilibrium is given by the equation

$$(dG^t)_{T, P} = 0 \tag{3.80}$$

The same reasoning applies to (3.75) through (3.78); however (3.80) is the preferred relation on which to base equilibrium calculations, because T and P (intensive variables) are more conveniently treated as constant than are the other pairs of state properties.

Our discussion has shown that there are two equivalent methods which may be used to identify equilibrium states in closed systems for a given T and P. First, one can develop an expression for $(dG^t)_{T, P}$ in terms of the composition variables of the system and set it equal to zero. Alternatively, one can develop an expression for G^t as a function of the composition variables and then find the set of values for those variables that minimizes G^t at the given T and P. The former method is usually employed for phase-equilibrium calculations, and is developed below. The latter method is used for chemical-reaction-equilibrium calculations (see the last part of Section 7.10).

The application of (3.80) to phase equilibrium requires an expression for $(dG^t)_{T, P}$ that incorporates the mole numbers of the chemical species in the individual phases. We first note that the total Gibbs energy G^t is the sum of the Gibbs energies of the phases present in the system:

$$G^t = \sum_{p=\alpha}^{\pi} n^p G^p \tag{3.81}$$

where G^p is the molar Gibbs energy of the pth phase and n^p is the total number of moles in the pth phase; the summation is over all π phases. Differentiation of (3.81) provides

$$dG^t = \sum_{p=\alpha}^{\pi} d(n^p G^p) \tag{3.82}$$

Equation (3.35) applied to phase p at constant T and P becomes

$$d(n^p G^p)_{T,P} = \sum_{i=1}^{m} \mu_i^p \, dn_i^p \qquad (3.83)$$

where the summation is over all m species. Combination of (3.82) and (3.83) now gives

$$(dG^t)_{T,P} = \sum_{p=\alpha}^{\pi} \sum_{i=1}^{m} \mu_i^p \, dn_i^p \qquad (3.84)$$

In view of (3.79) we have the important result

$$\boxed{\sum_{p=\alpha}^{\pi} \sum_{i=1}^{m} \mu_i^p \, dn_i^p \leqq 0} \qquad (3.85)$$

The equality in (3.85) corresponds to (3.80), and therefore represents a criterion of equilibrium. The usefulness of this criterion in the solution of equilibrium problems is the major justification for the introduction of the chemical potential μ_i as a thermodynamic property. The equality of (3.85) has many applications, but its use for the numerical calculation of equilibrium compositions must await the development of quantitative methods for the description of thermodynamic properties for solutions.

EXAMPLE 3.13 Application of (3.85) to a system composed of two phases α and β and containing two nonreactive chemical species 1 and 2 reduces it to

$$\mu_1^\alpha \, dn_1^\alpha + \mu_2^\alpha \, dn_2^\alpha + \mu_1^\beta \, dn_1^\beta + \mu_2^\beta \, dn_2^\beta \leqq 0$$

Since the system is closed and no chemical reaction occurs, the total number of moles of each chemical species must be constant; hence

$$dn_1^\beta = -dn_1^\alpha \qquad dn_2^\beta = -dn_2^\alpha$$

Combination of these three relations gives

$$(\mu_1^\alpha - \mu_1^\beta) \, dn_1^\alpha + (\mu_2^\alpha - \mu_2^\beta) \, dn_2^\alpha \leqq 0 \qquad (3.86)$$

Note that although the system as a whole is closed, the individual phases are not. The differentials dn_1^α and $dn_2^{\alpha'}$ arise because of the transfer of matter from one phase to the other.

Consider first the case where the inequality of (3.86) holds:

$$(\mu_1^\alpha - \mu_1^\beta) \, dn_1^\alpha + (\mu_2^\alpha - \mu_2^\beta) \, dn_2^\alpha < 0 \qquad (3.86a)$$

The process represented by this equation is one during which the system changes from a nonequilibrium state toward the equilibrium state, at constant T and P. The difference $(\mu_i^\alpha - \mu_i^\beta)$ and the differential dn_i^α can both be either positive or negative. However, $(3.86a)$ will always be satisfied if

$$(\mu_i^\alpha - \mu_i^\beta) \, dn_i^\alpha < 0 \qquad (i = 1, 2)$$

and this will be true if and only if $(\mu_i^\alpha - \mu_i^\beta)$ and dn_i^α have opposite signs. It is clear upon a little thought that if the transfer of mass of species i is always in the direction of the smaller chemical potential, then $(\mu_i^\alpha - \mu_i^\beta)$ and dn_i^α *must* have opposite signs, and $(3.86a)$ *is* satisfied, as is required by the laws of thermodynamics. Thus a difference in chemical potential for a particular species represents a driving force for the transport of that species, just as temperature and pressure differences represent driving forces for heat and momentum transfer. It is a general principle that when driving forces become zero transport processes cease, and the condition characterized by the term *equilibrium* is established. In this case the equality of (3.86) holds, and

$$(\mu_1^\alpha - \mu_1^\beta) \, dn_1^\alpha + (\mu_2^\alpha - \mu_2^\beta) \, dn_2^\alpha = 0 \qquad (3.86b)$$

This equation applies to differential variations around the equilibrium state, the particular type of equilibrium considered being phase equilibrium. In $(3.86b)$ the differentials dn_1^α and dn_2^α are independent and arbitrary; thus a necessary and sufficient condition for $(3.86b)$ to be valid is

$$\mu_1^\alpha = \mu_1^\beta \qquad \text{and} \qquad \mu_2^\alpha = \mu_2^\beta$$

Although the result of Example 3.13 holds for a two-phase, two-component system, it is readily generalized so as to apply to multiphase, multicomponent systems. For each additional component, (3.86b) would include an additional term on the left, and we would conclude immediately that

$$\mu_i^\alpha = \mu_i^\beta \qquad (i = 1, 2, \ldots, m)$$

For additional phases we would consider the equilibrium requirements for the possible pairs of phases, and would conclude that for multiple phases at the same T and P the equilibrium condition can be satisfied only when the chemical potential of each species in the system is the same in all phases. Mathematically, this is expressed by

$$\boxed{\mu_i^\alpha = \mu_i^\beta = \cdots = \mu_i^\pi} \qquad (i = 1, 2, \ldots, m) \qquad (3.87)$$

Equation (3.87) is the practical or working equation that forms the basis for phase-equilibrium calculations for PVT systems of uniform temperature and pressure.

EXAMPLE 3.14 In addition to its quantitative role in the solution of phase-equilibrium problems, (3.87) provides the basis for what is known as the *phase rule*, which allows one to calculate the number of independent variables that may be arbitrarily fixed in order to establish the *intensive* state of a PVT system. Such a state is established when its temperature, pressure, and the compositions of all phases are fixed. However, for equilibrium states these variables are not all independent, and fixing a limited number of them automatically establishes the others. This number of independent variables is called the *number of degrees of freedom* of the system, and is just the difference between the number of independent intensive variables associated with the system and the number of independent equations which may be written connecting these variables.

If m represents the number of chemical species in the system, then there are $(m-1)$ independent mole fractions for each phase. (The requirement that the sum of the mole fractions equals unity makes one mole fraction dependent.) For π phases there is a total of $(m-1)(\pi)$ composition variables. In addition, the temperature and pressure, taken to be uniform throughout the system, are phase-rule variables, and this means that the total number of independent variables is $2 + (m-1)(\pi)$. Equation (3.87) shows that one may write $(\pi - 1)$ independent phase-equilibrium equations for each species, with a total of $(\pi - 1)(m)$ such equations for a nonreacting system. Since the μ_i are functions of temperature, pressure, and the phase compositions, these equations represent relations among the phase-rule variables. Subtraction of the number of independent equations from the number of independent variables gives the number of degrees of freedom F as

$$\boxed{F = 2 - \pi + m} \qquad (3.88)$$

This is the Gibbs phase rule for a nonreacting PVT system.

Solved Problems

EXACT DIFFERENTIALS AND STATE FUNCTIONS   (Section 3.1)

3.1   The fundamental property relationship for an electrochemical cell is

$$dU^t = T \, dS^t - P \, dV^t + \varepsilon \, dq$$

where U^t, S^t, and V^t are total properties of the cell; ε is the cell emf (reversible cell voltage); and q is the charge of the cell. Write the consequences of (3.3) and of the exactness criterion (3.4).

By inspection we obtain from (3.3)

$$T = \left(\frac{\partial U^t}{\partial S^t}\right)_{V^t, q} \qquad P = -\left(\frac{\partial U^t}{\partial V^t}\right)_{S^t, q} \qquad \varepsilon = \left(\frac{\partial U^t}{\partial q}\right)_{S^t, V^t}$$

and from (3.4)

$$\left(\frac{\partial T}{\partial V^t}\right)_{S^t, q} = -\left(\frac{\partial P}{\partial S^t}\right)_{V^t, q} \qquad \left(\frac{\partial T}{\partial q}\right)_{S^t, V^t} = \left(\frac{\partial \varepsilon}{\partial S^t}\right)_{V^t, q} \qquad \left(\frac{\partial P}{\partial q}\right)_{S^t, V^t} = -\left(\frac{\partial \varepsilon}{\partial V^t}\right)_{S^t, q}$$

3.2   Find an equation which must be satisfied by any integrating factor $I(x, y)$ for the inexact differential expression $\delta z = M(x, y)\, dx + N(x, y)\, dy$.

For $I\,\delta z = IM\, dx + IN\, dy$ to be exact,

$$\frac{\partial(IM)}{\partial y} = \frac{\partial(IN)}{\partial x}$$

Expansion by the product rule and rearrangement gives the desired partial differential equation for I:

$$N\frac{\partial I}{\partial x} - M\frac{\partial I}{\partial y} = \left(\frac{\partial M}{\partial y} - \frac{\partial N}{\partial x}\right) I$$

3.3   By employing the definition of H, we can write (3.5) as $\delta Q_{\text{rev}} = dH - V\, dP$. Like (3.5), this expression is inexact. Considering H a function of T and P, find an expression for $(\partial H/\partial P)_T$. Use the fact that $1/T$ is an integrating factor for δQ_{rev}.

Multiplication of the above expression by $1/T$ gives

$$dS = \frac{\delta Q_{\text{rev}}}{T} = \frac{dH}{T} - \frac{V}{T}\, dP$$

But H is a function of T and P, so that

$$dH = \left(\frac{\partial H}{\partial P}\right)_T dP + \left(\frac{\partial H}{\partial T}\right)_P dT$$

whence

$$dS = \left[\frac{1}{T}\left(\frac{\partial H}{\partial P}\right)_T - \frac{V}{T}\right] dP + \frac{1}{T}\left(\frac{\partial H}{\partial T}\right)_P dT \equiv M\, dP + N\, dT$$

Then, since

$$\left(\frac{\partial M}{\partial T}\right)_P = \frac{1}{T}\frac{\partial^2 H}{\partial T\, \partial P} - \frac{1}{T^2}\left(\frac{\partial H}{\partial P}\right)_T - \frac{1}{T}\left(\frac{\partial V}{\partial T}\right)_P + \frac{V}{T^2} \qquad \text{and} \qquad \left(\frac{\partial N}{\partial P}\right)_T = \frac{1}{T}\frac{\partial^2 H}{\partial P\, \partial T}$$

(3.4) requires that

$$-\frac{1}{T^2}\left(\frac{\partial H}{\partial P}\right)_T - \frac{1}{T}\left(\frac{\partial V}{\partial T}\right)_P + \frac{V}{T^2} = 0 \qquad \text{or} \qquad \left(\frac{\partial H}{\partial P}\right)_T = -T\left(\frac{\partial V}{\partial T}\right)_P + V$$

This equation can be obtained more directly from (3.48) by the method of Example 3.10.

TRANSFORMATION RELATIONSHIPS FOR SYSTEMS WITH TWO INDEPENDENT VARIABLES   (Section 3.2)

3.4   The differential coefficient $(\partial T/\partial P)_H$ is called the *Joule/Thomson coefficient*, and is of importance in refrigeration engineering. Show that it can be calculated from PVT and heat-capacity data by

$$\left(\frac{\partial T}{\partial P}\right)_H = -\frac{1}{C_P}\left[V - T\left(\frac{\partial V}{\partial T}\right)_P\right]$$

Consider H a function of T and P, and apply (3.13) to get

$$\left(\frac{\partial T}{\partial P}\right)_H = -\left(\frac{\partial T}{\partial H}\right)_P\left(\frac{\partial H}{\partial P}\right)_T$$

But from (1.9) and from Problem 3.3, respectively,

$$\left(\frac{\partial H}{\partial T}\right)_P = C_P \quad \text{and} \quad \left(\frac{\partial H}{\partial P}\right)_T = V - T\left(\frac{\partial V}{\partial T}\right)_P$$

Combining these three equations, we obtain the desired result.

3.5 The thermodynamic state of an axially stressed bar of constant volume can be described by the three coordinates T, σ, and ε, where σ is the stress and ε the natural strain (see Example 1.3). Any two of these coordinates may be taken as independent. The *linear expansivity* α and *Young's modulus* E of the material are defined as

$$\alpha \equiv \left(\frac{\partial \varepsilon}{\partial T}\right)_\sigma \qquad E \equiv \left(\frac{\partial \sigma}{\partial \varepsilon}\right)_T$$

Find a relationship between α and E.

Apply (3.13), letting $x = \sigma$, $y = T$, and $z = \varepsilon$:

$$\left(\frac{\partial \sigma}{\partial T}\right)_\varepsilon = -\left(\frac{\partial \sigma}{\partial \varepsilon}\right)_T\left(\frac{\partial \varepsilon}{\partial T}\right)_\sigma$$

or

$$\left(\frac{\partial \sigma}{\partial T}\right)_\varepsilon = -E\alpha$$

3.6 *Charles' law* states that, for a gas at low pressures, the volume of the gas is directly proportional to the temperature at constant pressure. *Boyle's law* asserts that, for a gas at low pressures, the pressure of the gas is inversely proportional to the volume at constant temperature. Derive the ideal-gas equation from these two observations.

By Charles' law, $V/T = C_1(P)$; by Boyle's law, $PV = C_2(T)$. Therefore

$$\frac{PV}{T} = P\,C_1(P) = \frac{C_2(T)}{T}$$

But a function of P can equal a function of T only if both functions have the same constant value. Hence $PV/T = \text{const.} \equiv R$.

LEGENDRE TRANSFORMATIONS   (Section 3.3)

3.7 Write the U'-related Legendre transformations for an electrochemical cell.

The fundamental property relationship was given in Problem 3.1:

$$dU' = T\,dS' - P\,dV' + \varepsilon\,dq$$

According to (3.23), there are $2^3 - 1 = 7$ possible U'-related Legendre transformations:

$$\begin{aligned}
\mathcal{T}_1(T, V', q) &= U' - TS' \equiv A' \\
\mathcal{T}_2(S', P, q) &= U' + PV' \equiv H' \\
\mathcal{T}_3(S', V', \varepsilon) &= U' - \varepsilon q \\
\mathcal{T}_{1,2}(T, P, q) &= U' - TS' + PV' \equiv G' \\
\mathcal{T}_{1,3}(T, V', \varepsilon) &= U' - TS' - \varepsilon q \\
\mathcal{T}_{2,3}(S', P, \varepsilon) &= U' + PV' - \varepsilon q \\
\mathcal{T}_{1,2,3}(T, P, \varepsilon) &= U' - TS' + PV' - \varepsilon q
\end{aligned}$$

The canonical variables for each transformation are displayed in parentheses. Because of their formal identity with the functions used in the analysis of PVT systems, the three transformations designated H', A', and G' are those most commonly employed in the thermodynamic analysis of electrochemical cells.

3.8 It is sometimes convenient in the thermodynamic analysis of certain types of systems to normalize extensive properties with respect to the volume, rather than the mole number or the mass, of the system. Thus one defines an *energy density* $\tilde{U}$ and an *entropy density* $\tilde{S}$ by

$$\tilde{U} \equiv U'/V' \qquad \tilde{S} \equiv S'/V'$$

where U', S', and V' are total system properties. Rewrite the fundamental property relationship for a constant-composition PVT system, (2.6), in terms of these functions and write the $\tilde{U}$-related Legendre transformations.

Equation (2.6) is $dU' = T\,dS' - P\,dV'$. Since $U' = \tilde{U}V'$ and $S' = \tilde{S}V'$,

$$dU' = \tilde{U}\,dV' + V'\,d\tilde{U} \qquad dS' = \tilde{S}\,dV' + V'\,d\tilde{S}$$

so that (2.6) becomes $\tilde{U}\,dV' + V'\,d\tilde{U} = T\tilde{S}\,dV' + TV'\,d\tilde{S} - P\,dV'$. Rearranging, we obtain

$$d\tilde{U} = T\,d\tilde{S} - \Gamma\,d\tau$$

where $\Gamma \equiv \tilde{U} - T\tilde{S} + P$ and $d\tau \equiv dV'/V'$. The function τ is called the *volume strain* and the differential $d\tau$ the *relative dilatation*. There are three Legendre transformations, given by

$$\tilde{H}(\tilde{S}, \Gamma) = \tilde{U} + \Gamma\tau \qquad \tilde{A}(T, \tau) = \tilde{U} - T\tilde{S} \qquad \tilde{G}(T, \Gamma) = \tilde{U} - T\tilde{S} + \Gamma\tau$$

where the canonical variables are displayed in parentheses. A complete network of thermodynamic equations can be developed from this formulation.

PROPERTY RELATIONSHIPS FOR PVT SYSTEMS   (Sections 3.4 and 3.5)

3.9 The enthalpy can be related to the Gibbs energy and its temperature derivatives through the *Gibbs/Helmholtz equation*. Prove that

$$H = G - T\left(\frac{\partial G}{\partial T}\right)_P = -T^2\left[\frac{\partial(G/T)}{\partial T}\right]_P$$

From (3.30) and (3.32), $H = G + TS$. But from (3.39),

$$S = -\left(\frac{\partial G}{\partial T}\right)_P$$

and the result follows.

3.10 Statistical mechanics provides a link between the thermodynamic (macroscopic) and quantum mechanical (microscopic) descriptions of a system via a *partition function* $\mathcal{Z}$. One statistical mechanical formulation for a closed PVT system gives the following expression for the Helmholtz energy:

$$A = -RT \ln \mathcal{Z}$$

where R is the universal gas constant and $\mathcal{Z}$ is a function of T and V only. Find expressions for P, S, U, H, and G in terms of T, V, and $\mathcal{Z}$.

From (3.37),

$$P = -\left(\frac{\partial A}{\partial V}\right)_T = RT\left(\frac{\partial \ln \mathcal{Z}}{\partial V}\right)_T$$

From (3.39),

$$S = -\left(\frac{\partial A}{\partial T}\right)_V = R \ln \mathcal{Z} + RT\left(\frac{\partial \ln \mathcal{Z}}{\partial T}\right)_V$$

From the definitions (3.30), (3.31), and (3.22),

$$U = A + TS = RT^2\left(\frac{\partial \ln \mathscr{Z}}{\partial T}\right)_V$$

$$H = U + PV = RT^2\left(\frac{\partial \ln \mathscr{Z}}{\partial T}\right)_V + RTV\left(\frac{\partial \ln \mathscr{Z}}{\partial V}\right)_T$$

$$G = A + PV = -RT \ln \mathscr{Z} + RTV\left(\frac{\partial \ln \mathscr{Z}}{\partial V}\right)_T$$

3.11   Show that in a PVT system, $\mu_i = G_i$ for a pure material.

From (3.40),

$$\mu_i = \left(\frac{\partial (nG)}{\partial n_i}\right)_{T,P,n_j} = n\left(\frac{\partial G}{\partial n_i}\right)_{T,P,n_j} + G\left(\frac{\partial n}{\partial n_i}\right)_{T,P,n_j}$$

But, for a pure material, $n = n_i$ and $G = G_i$, and G_i is independent of n_i. Thus

$$\left(\frac{\partial G}{\partial n_i}\right)_{T,P,n_j} = \left(\frac{\partial G_i}{\partial n_i}\right)_{T,P,n_j} = 0$$

$$\left(\frac{\partial n}{\partial n_i}\right)_{T,P,n_j} = \left(\frac{\partial n_i}{\partial n_i}\right)_{T,P,n_j} = 1$$

and therefore $\mu_i = G_i$.

We shall see in Chapter 4 that this identity is useful in studying the phase equilibrium of single-component systems; for such systems this result, when combined with (3.87), requires that the molar Gibbs energies be the same for coexisting phases at equilibrium. Thus, $G_i^\alpha = G_i^\beta$ for equilibrium between two phases α and β of pure i.

3.12   Show that, for a constant-composition mixture, $d\mu_i = -\bar{S}_i\, dT + \bar{V}_i\, dP$, where

$$\bar{S}_i \equiv \left(\frac{\partial (nS)}{\partial n_i}\right)_{T,P,n_j} \qquad \bar{V}_i \equiv \left(\frac{\partial (nV)}{\partial n_i}\right)_{T,P,n_j}$$

For a constant-composition PVT system, μ_i can be considered a function of T and P only. Thus, for a change of state,

$$d\mu_i = \left(\frac{\partial \mu_i}{\partial T}\right)_{P,x} dT + \left(\frac{\partial \mu_i}{\partial P}\right)_{T,x} dP$$

But, by (3.45) and (3.46),

$$\left(\frac{\partial \mu_i}{\partial T}\right)_{P,x} = -\left(\frac{\partial (nS)}{\partial n_i}\right)_{T,P,n_j} = -\bar{S}_i$$

$$\left(\frac{\partial \mu_i}{\partial P}\right)_{T,x} = \left(\frac{\partial (nV)}{\partial n_i}\right)_{T,P,n_j} = \bar{V}_i$$

and the result follows. The quantities $\bar{S}_i$ and $\bar{V}_i$ are called the *partial molar entropy* and the *partial molar volume* of species i in the mixture. Partial molar properties are treated in detail in Section 7.1.

3.13   Incompressibility ($\beta = \kappa = 0$; $V = $ constant) is the simplest approximation one can make to the volumetric behavior of condensed phases; see Example 3.12. A more realistic approximation is that of a substance for which β and κ are independent of P. Develop expressions for property changes of such a substance undergoing an isothermal change of state.

For a process at constant T, (3.19) gives

$$\frac{dV}{V} = -\kappa\, dP \qquad\qquad (1)$$

in which $\kappa = \kappa(T) = \text{constant}$. Integrate (1) from the initial state P_1, V_1, T to the final state P, V, T to obtain

$$V = V_1 e^{-\kappa(P-P_1)} \tag{2}$$

From (2),

$$\Delta V_T \equiv V_2 - V_1 = V_1[e^{-\kappa(P_2-P_1)} - 1] \tag{3}$$

Because the isothermal compressibility κ is inherently positive, (1) or (2) or (3) dictates a decrease in V for an isothermal increase in P.

Equations (3.61) and (3.63) give, for constant T,

$$dH_T = V(1 - \beta T) \, dP \qquad dS_T = -\beta V \, dP$$

wherein $\beta = \beta(T) = \text{constant}$. Substituting (2) for V and integrating along the isotherm from P_1, V_1 to P_2, V_2, we obtain

$$\Delta H_T = -\frac{1}{\kappa} V_1(1 - \beta T)[e^{-\kappa(P_2-P_1)} - 1] \tag{4}$$

$$\Delta S_T = -\frac{\beta}{\kappa} V_1[e^{-\kappa(P_2-P_1)} - 1] \tag{5}$$

The isothermal change in U is related to that in H by

$$\Delta U_T = \Delta H_T - \Delta(PV)_T \tag{6}$$

where, by (2),

$$\Delta(PV)_T = P_1 V_1\left[\left(\frac{P_2}{P_1}\right)e^{-\kappa(P_2-P_1)} - 1\right] \tag{7}$$

Equations (3) through (7) reduce to the appropriate incompressible-substance results when $\beta = \kappa = 0$. For sufficiently small values of the product $\kappa(P_2 - P_1)$, the exponential may be represented by the first two terms in its series expansion: $e^{-\kappa(P_2-P_1)} \approx 1 - \kappa(P_2 - P_1)$.

3.14   At low frequencies the speed of sound c in a fluid is related to the adiabatic compressibility κ_S by

$$c = \sqrt{\frac{V}{M\kappa_S}}$$

where M is the molar mass of the fluid. Show that (a) $c = \sqrt{\gamma RT/M}$ in an ideal gas; (b) $c = \infty$ in an incompressible liquid.

(a)   From (3.68) and (3.57 ideal),

$$\kappa_S = \frac{1}{P}\frac{C_V}{C_P} = \frac{1}{P\gamma}$$

Hence

$$c = \sqrt{\frac{\gamma PV}{M}} = \sqrt{\frac{\gamma RT}{M}}$$

(b)   From (3.57), $\kappa_S = \kappa/\gamma$ and therefore

$$c = \sqrt{\frac{\gamma V}{M\kappa}}$$

for *any* fluid. But for an incompressible fluid $\kappa = 0$ and therefore $c = \infty$.

These two results are idealizations, but they are in accord with the general observations that (a) the speed of sound increases with the temperature of the medium, and (b) the sound speed is substantially greater in liquids than in gases.

ATTAINMENT OF EQUILIBRIUM IN CLOSED, HETEROGENEOUS SYSTEMS   (Section 3.6)

3.15   How many degrees of freedom are there in a PVT system consisting of a vapor phase in equilibrium with a liquid phase and containing (*a*) a single component?   (*b*) two nonreacting chemical species?

In both cases, there are two equilibrium phases and therefore $\pi = 2$. The phase rule (*3.88*) then gives $F = m$.

(*a*)   For a single component, $m = 1$ and therefore $F = 1$. Thus, specification of T automatically fixes P (or vice versa), and also determines the other intensive properties of both phases.

(*b*)   Here, $m = 2$ and therefore $F = 2$. Specification of any *two* independent intensive variables, such as T and P, determines *all* intensive properties of both phases.

3.16   Show that the maximum possible number of coexisting phases at equilibrium is *three* for a single-component PVT system.

There is one component and therefore $m = 1$. By the phase rule (*3.88*),

$$\pi = 2 + m - F = 3 - F$$

The minimum possible number of degrees of freedom is zero, because it is generally impossible to solve an algebraic system for which the number of equations exceeds the number of variables. Thus the maximum possible number of phases is $\pi = 3 - 0 = 3$. This is the condition that exists at the triple point of water, at which water vapor, liquid water, and "ice I" coexist in equilibrium at 0.01 °C and 0.006 113 bar.

3.17   Solve Problem 3.16 without appealing to (*3.88*).

The condition of phase equilibrium for a π-phase, 1-component system follows from the equality of (*3.85*):

$$\sum_{p=\alpha}^{\pi} \sum_{i=1}^{1} \mu_i^p \, dn_i^p = \sum_{p=\alpha}^{\pi} \mu^p \, dn^p = 0 \tag{1}$$

where the subscript i has been dropped. Because

$$\sum_{p=\alpha}^{\pi} n^p = n = \text{constant} \tag{2}$$

we have

$$d\left(\sum_{p=\alpha}^{\pi} n^p\right) = \sum_{p=\alpha}^{\pi} dn^p = 0 \tag{3}$$

Thus, only $\pi - 1$ of the dn^p are independent; we can always suppose these to be $dn^\alpha, dn^\beta, \ldots, dn^{\pi-1}$. Then, using

$$-\sum_{p=\alpha}^{\pi-1} dn^p = dn^\pi \tag{4}$$

in (*1*), we get

$$\sum_{p=\alpha}^{\pi-1} (\mu^p - \mu^\pi) \, dn^p = 0 \tag{5}$$

Now, the dn^p are independent and arbitrary, so (*5*) can hold only if each term $(\mu^p - \mu^\pi)$ is identically zero. Thus the single equation (*5*) is equivalent to the system of $\pi - 1$ equations

$$\mu^p = \mu^\pi \qquad (\alpha \le p \le \pi - 1) \tag{6}$$

For a single-component system, composition is not a variable and the μ^p are functions of T and P only. The maximum possible number of coexisting phases obtains when the number of equations of the form (6) just equals the number of possible variables. Thus the maximum number of coexisting phases is given by $\pi - 1 = 2$, or $\pi = 3$.

3.18   For a nonreacting PVT system at equilibrium, how many variables can be fixed arbitrarily in determining completely the state of the system, if all mole numbers n_i are specified?

The state of the system is *completely* specified if we know, in addition to T, P, and the $m - 1$ independent mole fractions in each phase, the total number of moles of each phase. The total number of variables is thus

$$
\begin{array}{rl}
(m - 1)\pi & \text{independent mole fractions} \\
\pi & \text{total mole numbers for the phases} \\
2 & (T \text{ and } P) \\
\hline
m\pi + 2 & \text{independent variables}
\end{array}
$$

Phase equilibrium is defined, as before, by the $m(\pi - 1)$ equations (3.87). In addition to these equations there are m material-balance equations

$$
\sum_{p=\alpha}^{\pi} n_i^p = n_i \qquad (i = 1, 2, \dots, m)
$$

where n_i are the given component mole numbers. The total number of constraining equations is thus

$$
\begin{array}{rl}
m(\pi - 1) & \text{equations among the } \mu_i^p \\
m & \text{constraining equations on the } n_i^p \\
\hline
m\pi & \text{independent equations}
\end{array}
$$

The number of variables that can be fixed arbitrarily is the number of independent variables less the number of independent equations, or $(m\pi + 2) - (m\pi) = 2$.

Although this result is derived for phase equilibrium, it applies as well to systems which can undergo chemical reaction, and forms the basis for *Duhem's theorem*:   For any closed PVT system formed from given initial amounts of prescribed chemical species, the equilibrium state is completely determined by any two properties of the whole system, provided that these properties are independently variable at equilibrium.   •

MISCELLANEOUS APPLICATIONS

3.19   Set up the basic thermodynamic relationships for a constant-volume, axially stressed bar.

For a process in which the bar is reversibly stressed and during which it reversibly exchanges heat with its surroundings,

$$
\delta Q_{\text{rev}} = T\, dS \qquad \delta W_{\text{rev}} = -V\sigma\, d\varepsilon
$$

where S and V are the total entropy and total volume of the bar. (The equation for W_{rev} comes from Example 1.3.) Applying the first law to this process, we get

$$
dU = \delta Q_{\text{rev}} - \delta W_{\text{rev}} = T\, dS + V\sigma\, d\varepsilon
$$

But V is constant, so the above equation can be written

$$
d\tilde{U} = T\, d\tilde{S} + \sigma\, d\varepsilon \tag{1}
$$

where $\tilde{U}$ and $\tilde{S}$ are the energy density and entropy density, respectively (see Problem 3.8). Equation (1) is the fundamental property relation for our system and will form the basis for further derived relationships. There are three $\tilde{U}$-related Legendre transformations:

$$
\tilde{H}(\tilde{S}, \sigma) = \tilde{U} - \sigma\varepsilon \tag{2}
$$

$$
\tilde{A}(T, \varepsilon) = \tilde{U} - T\tilde{S} \tag{3}
$$

$$
\tilde{G}(T, \sigma) = \tilde{U} - T\tilde{S} - \sigma\varepsilon \tag{4}
$$

the total differentials of which are

$$d\tilde{H} = T\,d\tilde{S} \; - \; \varepsilon\,d\sigma \tag{5}$$

$$d\tilde{A} = -\tilde{S}\,dT \; + \; \sigma\,d\varepsilon \tag{6}$$

$$d\tilde{G} = -\tilde{S}\,dT \; - \; \varepsilon\,d\sigma \tag{7}$$

A set of equations analogous to (3.36) through (3.39) follows from application of (3.3) to (1), (5), (6), and (7):

$$T = \left(\frac{\partial \tilde{U}}{\partial \tilde{S}}\right)_{\varepsilon} = \left(\frac{\partial \tilde{H}}{\partial \tilde{S}}\right)_{\sigma} \tag{8}$$

$$\sigma = \left(\frac{\partial \tilde{U}}{\partial \varepsilon}\right)_{\tilde{S}} = \left(\frac{\partial \tilde{A}}{\partial \varepsilon}\right)_{T} \tag{9}$$

$$\varepsilon = -\left(\frac{\partial \tilde{H}}{\partial \sigma}\right)_{\tilde{S}} = -\left(\frac{\partial \tilde{G}}{\partial \sigma}\right)_{T} \tag{10}$$

$$\tilde{S} = -\left(\frac{\partial \tilde{A}}{\partial T}\right)_{\varepsilon} = -\left(\frac{\partial \tilde{G}}{\partial T}\right)_{\sigma} \tag{11}$$

Application of the exactness criterion (3.4) to (1), (5), (6), and (7) gives

$$\left(\frac{\partial T}{\partial \varepsilon}\right)_{\tilde{S}} = \left(\frac{\partial \sigma}{\partial \tilde{S}}\right)_{\varepsilon} \tag{12}$$

$$\left(\frac{\partial T}{\partial \sigma}\right)_{\tilde{S}} = -\left(\frac{\partial \varepsilon}{\partial \tilde{S}}\right)_{\sigma} \tag{13}$$

$$\left(\frac{\partial \sigma}{\partial T}\right)_{\varepsilon} = -\left(\frac{\partial \tilde{S}}{\partial \varepsilon}\right)_{T} \tag{14}$$

$$\left(\frac{\partial \varepsilon}{\partial T}\right)_{\sigma} = \left(\frac{\partial \tilde{S}}{\partial \sigma}\right)_{T} \tag{15}$$

which are analogs of the Maxwell equations (3.41) through (3.44).

The analogs of the volume expansivity and the isothermal compressibility were introduced in Problem 3.5. They are the linear expansivity α and Young's modulus E:

$$\alpha = \left(\frac{\partial \varepsilon}{\partial T}\right)_{\sigma} \tag{16}$$

$$E = \left(\frac{\partial \sigma}{\partial \varepsilon}\right)_{T} \tag{17}$$

Two other useful quantities are the *constant-strain heat capacity* and the *constant-stress heat capacity*:

$$C_{\varepsilon} = \left(\frac{\partial \tilde{U}}{\partial T}\right)_{\varepsilon} \tag{18}$$

$$C_{\sigma} = \left(\frac{\partial \tilde{H}}{\partial T}\right)_{\sigma} \tag{19}$$

Equations (1) through (19) form a basis for the thermodynamics of constant-volume stressed bars. Other formulations are possible (see Problem 3.49).

3.20 For a reversible process in a closed PVT system, show that

$$\delta W = -\sum \mu_i\,dn_i \; + \; P\,d(nV)$$

For a closed system, not in motion, the first law requires that $\delta W = \delta Q - d(nU)$. For reversible processes, the second law requires $\delta Q = T\,d(nS)$. Thus

$$\delta W = T\,d(nS) \; - \; d(nU) = -\sum \mu_i\,dn_i \; + \; P\,d(nV)$$

where the second equality follows from (3.29). From this we see that work other than expansion work can result from changes in the numbers of moles of the chemical species present. In a closed system such changes can only result from a chemical reaction. Work of this origin is most commonly manifested as electrical work, and the system in such a case is an electrochemical cell.

Supplementary Problems

EXACT DIFFERENTIALS AND STATE FUNCTIONS   (Section 3.1)

3.21   Prove that the differential expression $M\,dx + N\,dy$ is exact if M is a function of x only and N is a function of y only.

3.22   Show that $I = Ax^{B-1}y^{-B-1}$, where A and B are arbitrary constants, is an integrating factor for $y\,dx - x\,dy$. (Compare Example 3.5.)

3.23   Prove that $I = (AT + B)^{-1}$, where A and B are constants, is an integrating factor for

$$\delta Q_{\text{rev}} = dU + P\,dV$$

if and only if B is identically zero. (*Hint*:   Use the result of Problem 3.2.)

3.24   The derivative $(\partial U/\partial V)_T$ is sometimes called the *internal pressure*; and the product $T(\partial P/\partial T)_V$, the *thermal pressure*. Rationalize these identifications. How are these quantities related to P itself? What are they for an ideal gas?   for an incompressible substance?   *Ans.*   See (3.8).

TRANSFORMATION RELATIONSHIPS FOR SYSTEMS WITH TWO INDEPENDENT VARIABLES   (Section 3.2)

3.25   Show that for a constant-volume stressed bar

$$\left(\frac{\partial \varepsilon}{\partial T}\right)_{\tilde{S}} = \alpha + E^{-1}\left(\frac{\partial \sigma}{\partial T}\right)_{\tilde{S}}$$

where $\tilde{S} = S/V$. (*Hint*:   See Problem 3.5.)

3.26   For a certain gas, β and κ are given by $\beta T = \kappa P = 1 - (b/V)$, where b is a constant. Find the equation of state of the gas, if $PV/RT = 1$ at $P = 0$.   *Ans.* $P(V - b) = RT$

3.27   Show that the Joule/Thomson coefficient defined in Problem 3.4 can be written in the alternate forms

$$\left(\frac{\partial T}{\partial P}\right)_H = \frac{T^2}{C_P}\left(\frac{\partial (V/T)}{\partial T}\right)_P = \frac{V(\beta T - 1)}{C_P}$$

Prove that $(\partial T/\partial P)_H = 0$ for an ideal gas.

3.28   The *Tait equation of state* for liquids is

$$\frac{V_0 - V}{V_0} = \frac{AP}{B + P}$$

Here V_0 is the (hypothetical) liquid volume at zero pressure, and parameters A and B are positive quantities that depend only on temperature. Demonstrate that $A \leqq 1$.

3.29   Prove that

(a) $\left(\dfrac{\partial^2 y}{\partial x^2}\right)_z = -\left(\dfrac{\partial y}{\partial x}\right)_z^3\left(\dfrac{\partial^2 x}{\partial y^2}\right)_z$

(b) $\left(\dfrac{\partial^2 y}{\partial x^2}\right)_z = \left(\dfrac{\partial z}{\partial y}\right)_x^{-1}\left[\left(\dfrac{\partial}{\partial y}\dfrac{(\partial z/\partial x)_y^2}{(\partial z/\partial y)_x}\right)_x - \left(\dfrac{\partial^2 z}{\partial x^2}\right)_y\right]$

LEGENDRE TRANSFORMATIONS   (Section 3.3)

3.30   The fundamental property relation for a PVT system may be written

$$dV = \frac{T}{P}\,dS - \frac{1}{P}\,dU$$

Write the V-related Legendre transformations and show the canonical variables.

Ans. $\mathscr{T}_1\left(\dfrac{T}{P}, U\right) = V - \dfrac{T}{P}S$ $\quad \mathscr{T}_2\left(S, \dfrac{1}{P}\right) = V + \dfrac{1}{P}U$ $\quad \mathscr{T}_{1,2}\left(\dfrac{T}{P}, \dfrac{1}{P}\right) = V - \dfrac{T}{P}S + \dfrac{1}{P}U$

3.31 Show that the Massieu functions defined in Example 3.9 are related to the usual U-related transformations by

$$\Omega = \frac{U - G}{T} \qquad \Psi = -\frac{A}{T} \qquad \Phi = -\frac{G}{T}$$

Prove that $\Omega = $ constant for an ideal gas undergoing an isentropic change of state.

3.32 Prove: The Legendre transformations of a linear function are themselves linear functions.

3.33 Consider a function $y(x)$ of a single variable x, defined by the differential expression

$$dy = M\, dx \tag{1}$$

where $M \equiv y'(x)$. For this simplest of cases there is but a single Legendre transformation L; viz., $L \equiv y - Mx$. Propose a geometric interpretation for L.

Ans. With L established as a function of the canonical variable M, then

$$y = Mx + L(M) \tag{2}$$

represents a one-parameter (M) family of straight lines in the xy-plane. The envelope of this family is found by eliminating the parameter between (2), or its differential

$$dy = M\, dx + x\, dM + L'(M)\, dM \tag{3}$$

and the partial derivative of (2) with respect to M,

$$0 = x + L'(M) \tag{4}$$

Clearly the result is (1), the original curve $y = y(x)$. Thus the Legendre transformation is a *contact transformation* between a curve $y = y(x)$ and its tangent field $L = L(M)$; see Fig. 3-1.

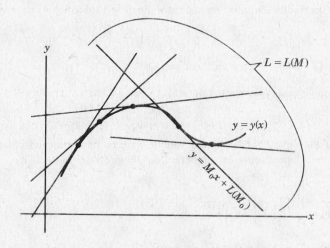

Fig. 3-1

PROPERTY RELATIONSHIPS FOR PVT SYSTEMS (Sections 3.4 and 3.5)

3.34 Prove that for a PVT system

(a)

$$U = -T^2\left(\frac{\partial (A/T)}{\partial T}\right)_V$$

(b)
$$\left(\frac{\partial U}{\partial V}\right)_T = T^2\left(\frac{\partial (P/T)}{\partial T}\right)_V$$

(c)
$$\left(\frac{\partial U}{\partial S}\right)_T = -P^2\left(\frac{\partial (T/P)}{\partial P}\right)_V$$

3.35 For a PVT system, show that if V is a function only of the ratio P/T, then U is a function only of T.

3.36 Prove that a substance for which $U = B + CV^{-R/\delta}e^{S/\delta}$ obeys the ideal-gas law, if B, C, and δ are constants.

3.37 In Problem 3.12 we defined the partial molar entropy $\bar{S}_i$ and the partial molar volume $\bar{V}_i$. Show that for a pure material $\bar{S}_i = S_i$ and $\bar{V}_i = V_i$.

3.38 Derive (3.53).

3.39 (a) Prove that
$$dS = \frac{C_V}{T}\left(\frac{\partial T}{\partial P}\right)_V dP + \frac{C_P}{T}\left(\frac{\partial T}{\partial V}\right)_P dV$$

    (b) For an ideal gas with constant heat capacities, use the above equation to derive the PV relation for an isentropic process: $PV^\gamma = $ constant.

    (c) Show that the equation of part (a) may also be written
$$T\,dS = C_V\left(\frac{\partial T}{\partial P}\right)_V\left[dP - \gamma\left(\frac{\partial P}{\partial V}\right)_T dV\right]$$

    (d) Derive (3.57) from the result of part (c).

3.40 Prove that
$$(a)\quad \left(\frac{\partial U}{\partial T}\right)_S = C_V\left(\frac{\partial \ln T}{\partial \ln P}\right)_V \qquad (b)\quad \left(\frac{\partial H}{\partial T}\right)_S = C_P\left(\frac{\partial \ln T}{\partial \ln V}\right)_P$$

3.41 Develop (3.60) through (3.67).

3.42 Prove that
$$\left(\frac{\partial U}{\partial P}\right)_T = V(\kappa P - \beta T)$$

Of the common thermodynamic properties, U is among the most insensitive to P. Why?

3.43 Equation (3.65) suggests the possibility of analogous behavior for the pairs C_P, C_V and κ, κ_S. Explore this possibility. In particular, show that, analogous to (3.64),
$$\kappa - \kappa_S = \frac{\beta^2 VT}{C_P}$$

What can be said about the *signs* of the differences $C_P - C_V$ and $\kappa - \kappa_S$? Under what conditions can we expect to have $C_P = C_V$ and $\kappa = \kappa_S$?

3.44 The enthalpy of an ideal-gas mixture depends on temperature and composition only. Hence, H^{ig} is said to be "independent of pressure," and one writes
$$\left(\frac{\partial H^{ig}}{\partial P}\right)_{T,x} = 0$$

Why is it then that
$$(a)\quad \left(\frac{\partial H^{ig}}{\partial P}\right)_{V,x} \neq 0 \qquad (b)\quad \left(\frac{\partial H^{ig}}{\partial P}\right)_{s,x} \neq 0$$

Determine expressions for these derivatives.

Ans.   (*a*) $\left(\dfrac{\partial H^{\text{ig}}}{\partial P}\right)_{V,x} = C_P^{\text{ig}} \dfrac{T}{P}$ (*b*) $\left(\dfrac{\partial H^{\text{ig}}}{\partial P}\right)_{S,x} = R \dfrac{T}{P}$

3.45   Explain the "squeaky-voice syndrome" exhibited by ocean divers breathing helium-laced gas mixtures. (*Hint*:   See Problem 3.14.)

ATTAINMENT OF EQUILIBRIUM IN CLOSED, HETEROGENEOUS SYSTEMS   (Section 3.6)

3.46   Derive (*3.78*) and (*3.79*).

3.47   Deduce (*3.85*) from (*3.74*) *without* specifically considering a process at constant T and P.

3.48   Which of the following phase-equilibrium situations are compatible with the phase rule for nonreacting systems? (*a*) Equilibrium among three allotropic forms of ice. (*b*) Equilibrium among water vapor, liquid water, and two allotropic forms of ice. (*c*) Equilibrium between two dense gas phases in a system with an arbitrary number of components. (*d*) Equilibrium among an A-rich liquid phase, a B-rich liquid phase, and a vapor phase containing A and B in an AB-binary system. (*e*) Equilibrium among $m + 3$ phases in a system containing m components.   *Ans.*   (*a*)   (*c*)   (*d*)

MISCELLANEOUS APPLICATIONS

3.49   In Problem 3.19 we outlined the derivation of the primary thermodynamic property relationships for an axially stressed bar, and in that problem the special variables were T, $\tilde{S}$, σ, and ε. In that development, however, we could equally as well have employed the reversible-work expression $\delta W_{\text{rev}} = - F\, dl$, rather than the equivalent expression $\delta W_{\text{rev}} = -\sigma V\, d\varepsilon$ (see Example 1.3), in which case the fundamental property relation would be

$$dU = T\, dS + F\, dl$$

For this formulation, the special variables are T, S, F, and l, and a network of equations analogous to those in Problem 3.19 can be written down by formally substituting S for $\tilde{S}$, F for σ, and l for ε in equations (*2*) through (*15*). In this alternate system, quantities analogous to α, E, C_ε, and C_σ are defined as follows:

$$\alpha' \equiv \frac{1}{l}\left(\frac{\partial l}{\partial T}\right)_F \qquad E' \equiv \frac{l^2}{V}\left(\frac{\partial F}{\partial l}\right)_T$$

$$C_l \equiv \left(\frac{\partial U}{\partial T}\right)_l \qquad C_F \equiv \left(\frac{\partial H}{\partial T}\right)_F$$

Recalling that $\sigma = Fl/V$ and that V is assumed constant, derive the following equations relating the expansivities, Young's moduli, and heat capacities for the two formulations:

$$\alpha' = \frac{\alpha E}{E - \sigma} \qquad E' = E - \sigma$$

$$C_l = VC_\varepsilon \qquad C_F = V\left[C_\sigma + \frac{E\alpha\sigma}{E - \sigma}(\alpha T - \varepsilon)\right]$$

3.50   For some applications involving the PVT equation of state, it is convenient to work with the *compressibility factor* Z $(\equiv PV/RT)$, the *coldness* τ $(\equiv T^{-1})$, and the *molar density* ρ $(\equiv V^{-1})$. Here, natural energies are the scaled internal energy u $(\equiv U/RT)$ and the scaled Helmholtz energy a $(\equiv A/RT)$.

(*a*)   For a constant-composition PVT system, show that

$$da = u\, \frac{d\tau}{\tau} + Z\, \frac{d\rho}{\rho} \qquad \text{(constant } x\text{)} \tag{1}$$

and thus that

$$u = \tau \left(\frac{\partial a}{\partial \tau} \right)_{\rho, x}$$

$$Z = \rho \left(\frac{\partial a}{\partial \rho} \right)_{\tau, x}$$

$$\rho \left(\frac{\partial u}{\partial \rho} \right)_{\tau, x} = \tau \left(\frac{\partial Z}{\partial \tau} \right)_{\rho, x}$$

Here, as usual, x stands for the set of all mole fractions.

(b) If we take (1) as "fundamental," then we can use it as the basis for defining three Legendre transformations related to a; viz.,

$$\mathcal{T}_1 \equiv a - u \qquad \mathcal{T}_2 \equiv a - Z \qquad \mathcal{T}_{1,2} \equiv a - u - Z$$

Verify these transformations, and show that

$$d\mathcal{T}_1 = -\tau \, d\left(\frac{u}{\tau} \right) + Z \, \frac{d\rho}{\rho} \qquad \text{(constant } x\text{)} \tag{2}$$

$$d\mathcal{T}_2 = u \, \frac{d\tau}{\tau} - \rho \, d\left(\frac{Z}{\rho} \right) \qquad \text{(constant } x\text{)}$$

$$d\mathcal{T}_{1,2} = -\tau \, d\left(\frac{u}{\tau} \right) - \rho \, d\left(\frac{Z}{\rho} \right) \qquad \text{(constant } x\text{)}$$

Demonstrate that (2) is equivalent to (2.6), the usual fundamental property relation for a constant-composition PVT system.

Chapter 4

Properties of Pure Substances

It is evident from the preceding chapter that thermodynamics provides a multitude of equations interrelating the properties of substances. The properties themselves differ from one substance to the next. Thermodynamics is in no sense a model or description of the behavior of matter; rather, it depends for its usefulness on the experimental or theoretical evaluation of a minimal number of properties (e.g., P, V, T, C_V). From these data, thermodynamics allows the development of a complete set of thermodynamic property values from which one can subsequently calculate the heat and work effects of various processes and determine equilibrium conditions in a variety of systems. This chapter therefore treats qualitatively the general behavior of substances in equilibrium states and quantitatively the methods used for correlation of experimental data and for calculation of property values.

4.1 PVT BEHAVIOR OF A PURE SUBSTANCE

The relationship of specific or molar volume to temperature and pressure for a pure substance in equilibrium states can be represented by a surface in three dimensions, as shown in Fig. 4-1. The surfaces marked S, L, and G represent respectively the solid, liquid, and gas regions of the diagram. In addition, there are three regions of coexistence of two phases in equilibrium: solid-gas (S-G), solid-liquid (S-L), and liquid-gas (L-G). Heavy lines separate the various regions and form boundaries of the surfaces representing the individual phases. The heavy line passing through points A and B marks the intersections of the two-phase regions, and is the three-phase line, along which solid, liquid, and gas phases exist in three-phase equilibrium. According to the phase rule such systems have zero degrees of freedom; they exist for a given pure substance at but one temperature and one pressure. For this reason the projection of this line on the *PT* plane (shown to the left of the main diagram of Fig. 4-1) is a point, known as the *triple point*. The phase rule also requires that systems made up of two phases in equilibrium have just one degree of freedom, and therefore the two-phase regions must project as lines on the *PT* plane, forming a PT diagram of three lines—fusion (or melting), sublimation, and vaporization—which meet at the triple point. The *PT* projection of Fig. 4-1 provides no information about the volumes of the systems represented. This is, however, given explicitly by the other projection of Fig. 4-1, where all surfaces of the three-dimensional diagram show up as areas on the *PV* plane. The *PT* projection of Fig. 4-1 is shown to a larger scale in Fig. 4-2, and the liquid and gas regions of the *PV* projection are shown in more detail by Fig. 4-3.

The solid lines of Fig. 4-2 clearly represent phase boundaries. The fusion curve (line 2-3) normally has a positive slope, but for a few substances (water is the best known) it has a negative slope. The two curves 1-2 and 2-C represent the *vapor pressures* of solid and liquid, respectively. The terminal point C is the *critical point*, which represents the highest pressure and highest temperature at which liquid and gas can coexist in equilibrium. The termination of the vapor-pressure curve at point C means that at higher temperatures and pressures no clear distinction can be drawn between what is called liquid and what is called gas. Thus there is a region extending indefinitely upward from, and indefinitely to the right of, the critical point that is called simply the *fluid region*. It is bounded by dashed lines that do not represent phase transitions, but which conform to arbitrary definitions of what is considered a liquid and what is considered a gas. The region designated as *liquid* in Fig. 4-2 lies above the vaporization curve. Thus a liquid can always be vaporized by a sufficient reduction in pressure at constant temperature. The gas region of Fig. 4-2 lies to the right of the sublimation and vaporization curves. Thus a gas can always be condensed by a sufficient reduction in temperature at constant pressure. A substance at a temperature above its critical

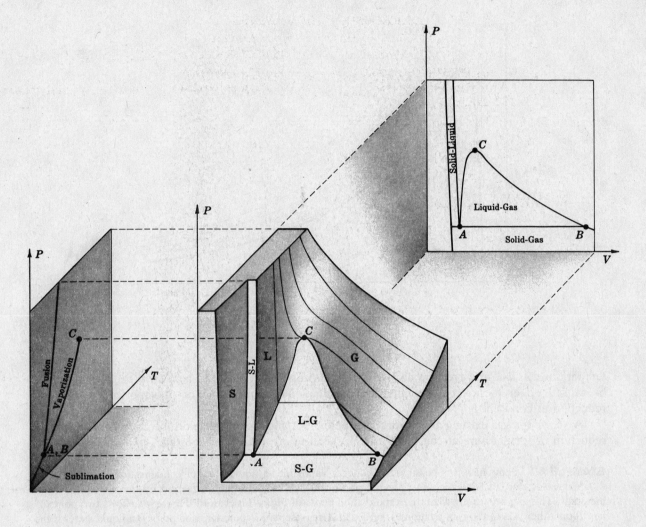

Fig. 4-1

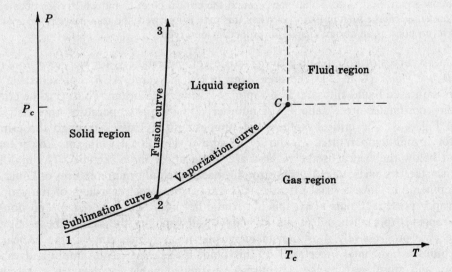

Fig. 4-2

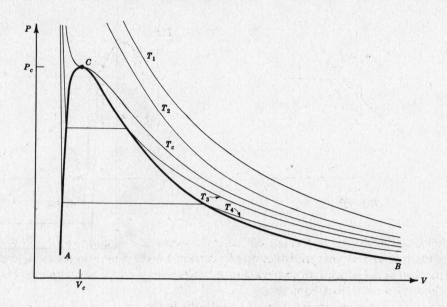

Fig. 4-3

temperature T_c *and* at a pressure above its critical pressure P_c is said to be a *fluid* because it cannot be caused either to liquefy by temperature reduction at constant P or to vaporize by pressure reduction at constant T.

A *vapor* is a gas existing at temperatures below T_c, and it may therefore be condensed either by reduction of temperature at constant P or by increase of pressure at constant T.

EXAMPLE 4.1   Show how it is possible to change a vapor into a liquid without condensation.

When a vapor condenses it undergoes an abrupt change of phase, characterized by the appearance of a meniscus. This can occur only when the vaporization curve of Fig. 4-2 is crossed. Figure 4-4 shows two points A and B on either side of the vaporization curve. Point A represents a vapor state and point B a liquid state. If the pressure on the vapor at A is raised at constant T, then the vertical path from A to B is followed, and this path crosses the vaporization curve, at which point condensation occurs at a fixed pressure. However, it is also possible to follow a path from A to B that goes around the critical point C and which then does not cross the vaporization curve, as can be seen in Fig. 4-4. When this path is followed, the transition from vapor to liquid is gradual, and at no point is an abrupt change in properties observed.

The primary curves of Fig. 4-3 give the pressure-volume relations for *saturated liquid* (A to C) and for *saturated vapor* (C to B). The area lying below the curve ACB represents the two-phase region where saturated liquid and saturated vapor coexist at equilibrium. Point C is the critical point, for which the coordinates are P_c and V_c. A number of constant-temperature lines (isotherms) are included on Fig. 4-3. The critical isotherm at temperature T_c passes through the critical point. Isotherms for higher temperatures T_1 and T_2 lie above the critical isotherm. Lower-temperature isotherms lie below the critical isotherm, and are made up of three sections. The middle section, which traverses the two-phase region, is horizontal, because equilibrium mixtures of liquid and vapor at a fixed temperature have a fixed pressure, the vapor pressure, regardless of the proportions of liquid and vapor present. Points along this horizontal line represent different proportions of liquid and vapor, ranging from all liquid at the left end to all vapor at the right. These horizontal line segments become progressively shorter with increasing temperature, until at the critical point the isotherm exhibits a horizontal inflection. As this point is approached the liquid and vapor phases become more and more nearly alike, eventually becoming unable to sustain a meniscus between them, and ultimately becoming indistinguishable.

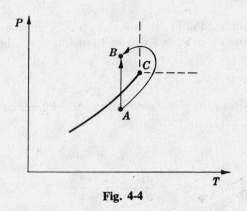

Fig. 4-4

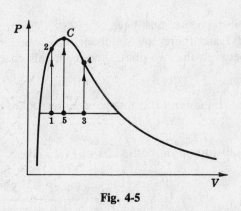

Fig. 4-5

The sections of isotherms to the left of line AC traverse the liquid region, and are very steep, because liquids are not very compressible; that is, it takes a large change in pressure to effect a small volume change. A liquid which is not saturated (not at its boiling point, along line AC) is often called *subcooled* liquid or *compressed* liquid.

The sections of isotherms to the right of line CB lie in the vapor region. Vapor in this region is called *superheated* vapor to distinguish it from saturated vapor as represented by the line CB.

EXAMPLE 4.2 Describe what would be observed if a pure substance were heated in a sealed tube as indicated by the three paths shown by Fig. 4-5.

Points 1, 3, and 5 represent three different fillings of the tube. Point 1 indicates the filling for which the most liquid is present and point 3, the least. The remainder of the tube contains vapor. (We are considering *pure* substances, and there must be no impurities such as air.)

Heating at constant volume along line 1-2 causes the meniscus to rise until at point 2 the meniscus reaches the top of the tube, and the tube is filled with liquid. Thus, starting from point 1, we see that heating causes the liquid to expand sufficiently to fill the tube, causing the vapor to condense in the process.

Heating at constant volume along line 3-4 causes the meniscus to fall until at point 4 it reaches the bottom of the tube, and the tube is filled with vapor. Thus, in this process, the liquid evaporates more rapidly than it expands, and eventually the tube contains only vapor.

Heating from the appropriate intermediate point along line 5-C produces a different result. The liquid expands and evaporates at compensating rates, and the meniscus does not move much either way. As the critical point is approached the meniscus becomes indistinct, then hazy, and finally disappears. The tube is then filled with neither liquid nor vapor, but with fluid.

4.2 PHASE CHANGES OF PURE SUBSTANCES. CLAPEYRON'S EQUATION

A phase change occurs at constant T and P whenever one of the curves of Fig. 4-2 is crossed, resulting in discrete changes in most thermodynamic properties. Thus, the molar or specific volume of a saturated liquid is very different from that of the saturated vapor at the same T and P. Discrete changes resulting from phase transition also occur in the internal energy, enthalpy, and entropy. The exception is the Gibbs energy; it does not change during melting, vaporization, sublimation, or an allotropic transformation that occurs at constant T and P.

Since a phase change represents a transition between two phases that coexist in equilibrium, (*3.80*) applies to the process of change:

$$(dG')_{T,P} = 0 \qquad\qquad\qquad (3.80)$$

For a pure substance T and P are indeed constant, and according to (*3.80*) the total Gibbs energy for the closed system in which the transition occurs must also be constant. This can be true only if the molar Gibbs energy is the same for both phases. Thus for two coexisting phases α and β of a pure substance we must have $G^\alpha = G^\beta$. (This result was obtained in Problem 3.11 by a different method.) If T and P are now simultaneously changed by dT and dP in such a way as to maintain equilibrium between the two phases, then for this change

$$dG^\alpha = dG^\beta \tag{4.1}$$

Equation (3.35), applied to one mole of a homogeneous pure material, reduces immediately to

$$dG = -S\,dT + V\,dP$$

Substitution on both sides of (4.1) gives

$$-S^\alpha\,dT + V^\alpha\,dP = -S^\beta\,dT + V^\beta\,dP \quad\text{or}\quad \frac{dP}{dT} = \frac{S^\beta - S^\alpha}{V^\beta - V^\alpha}$$

Since the pressure in this equation is always a saturation pressure on the appropriate phase boundary, we write P^{sat} in place of P to make this explicit. Furthermore, $S^\beta - S^\alpha$ and $V^\beta - V^\alpha$, representing property changes of phase transition, are most conveniently written $\Delta S^{\alpha\beta}$ and $\Delta V^{\alpha\beta}$, and our equation is expressed most simply as

$$\frac{dP^{\text{sat}}}{dT} = \frac{\Delta S^{\alpha\beta}}{\Delta V^{\alpha\beta}} \tag{4.2}$$

A phase transition occurring at constant T and P requires an exchange of heat between the substance taken as the system and its surroundings. When the transition is carried out reversibly, this heat, known as the *latent heat*, is equal to the enthalpy change; that is, $Q = \Delta H^{\alpha\beta}$. This result for a constant-pressure process was found in Example 1.7. In addition, for a reversible process at constant temperature, $\Delta S^{\alpha\beta} = Q/T$. Since the phase transition is carried out at constant T and P, we may combine these two equations to get $\Delta S^{\alpha\beta} = \Delta H^{\alpha\beta}/T$. This equation involves properties only and therefore does not depend on the assumption of reversibility made in its derivation. Combination with (4.2) gives the *Clapeyron equation*,

$$\boxed{\frac{dP^{\text{sat}}}{dT} = \frac{\Delta H^{\alpha\beta}}{T\,\Delta V^{\alpha\beta}}} \tag{4.3}$$

This applies to any phase change of a pure substance and relates the slope of the appropriate phase-boundary curve in the PT plane, at a given T and P, to the enthalpy change (latent heat) and the volume change of phase transition at the same T and P. The following superscripts designate the common phases and phase transitions:

$s \equiv$ saturated solid $\qquad sl \equiv$ fusion
$l \equiv$ saturated liquid $\qquad lv \equiv$ vaporization
$v \equiv$ saturated vapor $\qquad sv \equiv$ sublimation

The Clapeyron equation (4.3) may now be written for each of the three phase transitions:

$$\textbf{fusion}\qquad \frac{dP^{\text{sat}}}{dT} = \frac{\Delta H^{sl}}{T\,\Delta V^{sl}} \equiv \frac{H^l - H^s}{T(V^l - V^s)} \tag{4.3a}$$

$$\textbf{vaporization}\qquad \frac{dP^{\text{sat}}}{dT} = \frac{\Delta H^{lv}}{T\,\Delta V^{lv}} \equiv \frac{H^v - H^l}{T(V^v - V^l)} \tag{4.3b}$$

$$\textbf{sublimation}\qquad \frac{dP_{\text{sat}}}{dT} = \frac{\Delta H^{sv}}{T\,\Delta V^{sv}} \equiv \frac{H^v - H^s}{T(V^v - V^s)} \tag{4.3c}$$

EXAMPLE 4.3 The Riedel equation (*Chem. Ing. Tech.*, **26**: 679, 1954) is useful for estimation of latent heats of vaporization of pure substances at their normal boiling points, when experimental data are lacking:

$$\frac{\Delta H_n^{lv}/T_n}{R} = \frac{(1.092)(\ln P_c - 1.013)}{0.930 - T_{r,n}}$$

where $T_n \equiv$ normal boiling point

$\Delta H_n^{lv} \equiv$ molar latent heat of vaporization at T_n

$P_c \equiv$ critical pressure, bar

$T_{r,n} \equiv$ reduced temperature at $T_n = T_n/T_c$

Since $\Delta H_n^{lv}/T_n$ has the dimensions of the gas constant R, the units of this ratio are governed by the choice of units for R.

Applied to *n*-octane, for which $T_n = 398.83$ K, $T_c = 568.8$ K, and $P_c = 24.8$ bar, the Riedel equation gives:

$$\frac{\Delta H_n^{lv}}{T_n} = R\left[\frac{(1.092)(\ln 24.8 - 1.013)}{0.930 - 0.701}\right] = (10.49)R$$

or $\Delta H_n^{lv} = (10.49)(0.008\,314)(398.83) = 34.780$ kJ $\cdot$ mol^{-1}. The experimental value is 34.410 kJ $\cdot$ mol^{-1}.

EXAMPLE 4.4 For sublimation and vaporization processes at low pressure one may introduce reasonable approximations into the Clapeyron equation by assuming that the vapor phase is an ideal gas and that the molar volume of the condensed phase is negligible compared with the molar volume of the vapor phase. What form does the Clapeyron equation assume under these approximations?

Consider the case of vaporization. If $V^v \gg V^l$, then $V^v - V^l \approx V^v$, and by the ideal-gas equation, $V^v = RT/P^{sat}$. Then (*4.3b*) becomes

$$\frac{dP^{sat}}{dT} = \frac{\Delta H^{lv}}{RT^2/P^{sat}}$$

or
$$\Delta H^{lv} = \frac{RT^2}{P^{sat}}\frac{dP^{sat}}{dT} = -R\frac{d(\ln P^{sat})}{d(1/T)} \qquad (4.4)$$

An identical expression is obtained for ΔH^{sv}, the latent heat of sublimation. This equation is known as the *Clausius/Clapeyron equation*, and it allows an approximate value for the latent heat to be determined from data for the saturation pressure alone. No volumetric data are required.

More specifically, (*4.4*) indicates that the value of ΔH^{lv} (or of ΔH^{sv}) is given by the slope of a plot of $\ln P^{sat}$ against $1/T$. It is a fact, however, that such plots of experimental data produce lines that are nearly straight, which implies through (*4.4*) that the latent heat of vaporization ΔH^{lv} is independent of temperature. This is far from true. In fact, ΔH^{lv} decreases markedly with increasing temperature, becoming zero at the critical temperature, where the phases become identical. The difficulty is that (*4.4*) is based on assumptions which have approximate validity only at low pressures and at pressures well below the critical pressure.

4.3 VAPOR PRESSURES AND LATENT HEATS

The Clapeyron equation provides a vital connection between the properties of different phases. It is usually applied to the calculation of latent heats of vaporization and sublimation from vapor-pressure and volumetric data:

$$\Delta H^{\alpha\beta} = T \,\Delta V^{\alpha\beta}\,\frac{dP^{sat}}{dT}$$

For this, we need accurate representations of P^{sat} as a function of T. As already stated, a plot of $\ln P^{sat}$ versus $1/T$ generally yields a line that is very nearly straight, suggesting a vapor-pressure equation of the form

$$\ln P^{sat} = A - \frac{B}{T} \qquad (4.5)$$

where A and B are constants. This equation is useful for many purposes, but it does not represent data sufficiently well to provide accurate values of derivatives. The *Antoine equation* is more satisfactory, and has found wide use:

$$\ln P^{\text{sat}} = A - \frac{B}{T + C} \tag{4.6}$$

where A, B, and C are constants.

Extensive vapor-pressure data of high accuracy cannot usually be represented faithfully by any simple equation. An equation that is generally satisfactory has the form

$$\ln P^{\text{sat}} = A - \frac{B}{T + C} + DT + E \ln T \tag{4.7}$$

where A, B, C, D, and E are constants.

Equations for the representation of vapor-pressure data are empirical. (Again, thermodynamics provides no model for the behavior of substances, either in general or individually.)

4.4 PROPERTIES OF TWO-PHASE SYSTEMS

When a phase change of a pure substance is brought about at constant T and P, the molar (or specific) properties of the individual phases do not change. If we start with a given amount of a substance in phase α, for which the molar (or specific) properties are V^α, U^α, H^α, S^α, etc., and gradually carry out a phase change at constant T and P to give finally the same quantity of the substance in phase β, then it has the molar (or specific) properties V^β, U^β, H^β, S^β, etc. Intermediate states of the system are made up of the two phases α and β in varying amounts, but each phase always has the same set of molar (or specific) properties. If we want an average molar (or specific) property over the entire two-phase system, we need a weighted sum of the properties of the individual phases. If we let x be the fraction of the total number of moles (or of the total mass) of the system that exists in phase β, then the average molar (or specific) properties of the two-phase system are given by

$$V = (1 - x)V^\alpha + xV^\beta \qquad U = (1 - x)U^\alpha + xU^\beta$$
$$H = (1 - x)H^\alpha + xH^\beta \qquad S = (1 - x)S^\alpha + xS^\beta$$

These equations can be represented in general by

$$\boxed{M = (1 - x)M^\alpha + xM^\beta \qquad (0 \leqq x \leqq 1)} \tag{4.8}$$

or, equivalently,

$$M = M^\alpha + x \, \Delta M^{\alpha\beta} \qquad (0 \leqq x \leqq 1) \tag{4.9}$$

where M represents any molar (or specific) thermodynamic property. The total system property is given by $nM \equiv (n^\alpha + n^\beta)M$ [or by $mM \equiv (m^\alpha + m^\beta)M$].

EXAMPLE 4.5 Specialize (4.8) and (4.9) to the case of the volume of a two-phase system of liquid and vapor.
In this case M becomes V, and the superscripts α and β become l and v. Then (4.8) and (4.9) are written as

$$V = (1 - x)V^l + xV^v \qquad \text{and} \qquad V = V^l + x \, \Delta V^{lv}$$

The second equation has a simple physical interpretation: the mixture has the volume of the saturated liquid plus the volume gained by vaporization of the fraction of the system represented by x. For liquid/vapor systems, x is the mass fraction which is vapor; it is called the *quality*.

4.5 VOLUME EXPANSIVITY AND ISOTHERMAL COMPRESSIBILITY OF SOLIDS AND LIQUIDS

The volume expansivity β and isothermal compressibility κ were defined by (3.17) and (3.18):

$$\beta \equiv \frac{1}{V}\left(\frac{\partial V}{\partial T}\right)_P \qquad \kappa \equiv -\frac{1}{V}\left(\frac{\partial V}{\partial P}\right)_T$$

Experimental values for these quantities are tabulated in handbooks of data. For solids and liquids such tabulations take the place of an equation of state, when used in conjunction with

$$\frac{dV}{V} = \beta\, dT - \kappa\, dP \tag{3.19}$$

Both β and κ are generally positive numbers; however β can take on negative values in unusual cases, as it does for liquid water betwen 0 and 4 °C. Both β and κ are in general functions of temperature and pressure. Normally, both β and κ increase with increasing temperature, though there are exceptions. The effect of pressure is usually quite small, an increase in pressure invariably causing κ to decrease and usually, but not always, causing β also to decrease.

When both β and κ are weak functions of T and P, or where the changes in T and P are relatively small, β and κ may be treated as constants, and (3.19) may be integrated to give

$$\ln \frac{V_2}{V_1} = \beta(T_2 - T_1) - \kappa(P_2 - P_1) \tag{4.10}$$

This is a different order of approximation than the assumption of incompressibility, for which β and κ are taken to be zero, as discussed in Example 3.12.

When β and κ cannot be considered constant, (3.19) must be integrated generally:

$$\ln \frac{V_2}{V_1} = \int_{T_1}^{T_2} \beta\, dT - \int_{P_1}^{P_2} \kappa\, dP \tag{4.11}$$

Since V is a state function, the value of $\ln (V_2/V_1)$ must be independent of the path of integration for given states 1 and 2. This means that one may choose any path at all leading from state 1 to state 2, and the choice is made on the basis of convenience. Different paths give different values for the individual integrals of (4.11), but the difference between the two integrals is always the same.

The two most obvious (and most useful) paths are shown by Fig. 4-6, where each path consists of two steps, one at constant T and the other at constant P. Only one term of (4.11) applies to each step, as shown. The path indicated by dashed lines consists of a first step at the constant pressure P_1 and a second step at the constant temperature T_2. The two integrals require data for β *as a function*

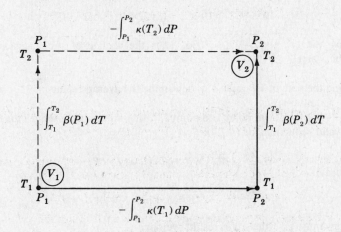

Fig. 4-6

of T at pressure P_1 and data for κ as a function of P at temperature T_2. The path described by the solid lines takes the steps in the opposite order, and requires different data for β and κ, as indicated. All other paths require data for both β and κ *as functions of both T and P*, and an exact T versus P relation would have to be mapped out. Fortunately, it is never necessary to complicate matters in this way. The choice of which rectangular path to use depends entirely upon what data are available.

EXAMPLE 4.6   The following values of β and κ for liquid water are available:

$T/°C$	P/bar	$\beta/°C^{-1}$	κ/bar^{-1}
20	1	208×10^{-6}	45.8×10^{-6}
20	1000	—	34.8×10^{-6}
30	1	304×10^{-6}	44.6×10^{-6}
30	1000	—	33.8×10^{-6}

If water undergoes a change of state from 20 °C and 1 bar to 30 °C and 1000 bar, determine the percentage change in the specific volume.

Since data are available for β just at $P_1 = 1$ bar, the only reasonable path to adopt for this calculation is the one indicated by the dashed lines of Fig. 4-6. For this path (4.11) becomes

$$\ln \frac{V_2}{V_1} = \int_{T_1}^{T_2} \beta(P_1)\, dT - \int_{P_1}^{P_2} \kappa(T_2)\, dP$$

Since no data are given for β and κ at intermediate values of T and P, we can do no better than to take β and κ out from under the integral signs as arithmetic average values:

$$\ln \frac{V_2}{V_1} = \overline{\beta(P_1)}(T_2 - T_1) - \overline{\kappa(T_2)}(P_2 - P_1)$$

where for $P_1 = 1$ bar,

$$\overline{\beta(P_1)} = \frac{208 + 304}{2} \times 10^{-6} = 256 \times 10^{-6} \ °C^{-1}$$

and for $T_2 = 30$ °C,

$$\overline{\kappa(T_2)} = \frac{44.6 + 33.8}{2} \times 10^{-6} = 39.2 \times 10^{-6} \ bar^{-1}$$

Substitution of numerical values gives

$$\ln \frac{V_2}{V_1} = (256 \times 10^{-6} \ °C^{-1})(10 \ °C) - (39.2 \times 10^{-6} \ bar^{-1})(999 \ bar) = -0.0366$$

whence $V_2/V_1 = 0.964$ and $(V_2 - V_1)/V_1 = -0.036$. Thus the change of state causes a 3.6% decrease in the specific volume of the water.

EXAMPLE 4.7   Using the data of Example 4.6, determine the average value of β for water at 1000 bar between the temperatures 20 °C and 30 °C.

Figure 4-6 shows two paths from point 1 to point 2. Using arithmetic averages for β and κ for the two paths, we get two equally valid reductions of (4.11):

$$\ln \frac{V_2}{V_1} = \overline{\beta(P_1)}(T_2 - T_1) - \overline{\kappa(T_2)}(P_2 - P_1)$$

$$\ln \frac{V_2}{V_1} = -\overline{\kappa(T_1)}(P_2 - P_1) + \overline{\beta(P_2)}(T_2 - T_1)$$

Since both expressions must give the same result, we have

$$[\overline{\beta(P_1)} - \overline{\beta(P_2)}](T_2 - T_1) = [\overline{\kappa(T_2)} - \overline{\kappa(T_1)}](P_2 - P_1)$$

where, as before, $\overline{\beta(P_1)} = 256 \times 10^{-6}$ °C^{-1} and $\overline{\kappa(T_2)} = 39.2 \times 10^{-6}$ bar^{-1}, and where, for $T_1 = 20$ °C,

$$\overline{\kappa(T_1)} = \frac{45.8 + 34.8}{2} \times 10^{-6} = 40.3 \times 10^{-6} \text{ bar}^{-1}$$

Thus

$$[256 \times 10^{-6} - \overline{\beta(P_2)}](10) = (39.2 \times 10^{-6} - 40.3 \times 10^{-6})(999) \quad \text{or} \quad \overline{\beta(P_2)} = 366 \times 10^{-6} \text{ °C}^{-1}$$

The procedure followed in this example illustrates the finite-difference counterpart of the reciprocity relation for an exact differential, given for the present case by (3.20).

4.6 HEAT CAPACITIES OF SOLIDS AND LIQUIDS

In general, heat capacities must be determined by experiment for each substance of interest. Data for solids and liquids are usually taken at atmospheric pressure and are reported as functions of temperature in either of two forms:

$$C_P = a + bT + cT^2 \tag{4.12a}$$
$$C_P = a + bT + cT^{-2} \tag{4.12b}$$

In either case the set of constants a, b, and c is peculiar to the substance. Heat capacities usually increase with increasing temperature. The effect of pressure on the heat capacities of liquids and solids is normally very small; unless the pressure is very high, it can be neglected.

The difference in heat capacities is expressed in terms of the volume expansivity and the isothermal compressibility by

$$C_P - C_V = \frac{TV\beta^2}{\kappa} \tag{3.64}$$

This difference is usually significant, except at very low temperatures.

EXAMPLE 4.8 For metallic copper at 300 K, the following values are known:

$$C_P = 24.50 \text{ J} \cdot \text{mol}^{-1} \cdot \text{K}^{-1} \qquad \kappa = 0.778 \times 10^{-6} \text{ bar}^{-1}$$
$$\beta = 50.4 \times 10^{-6} \text{ K}^{-1} \qquad V = 7.06 \times 10^{-6} \text{ m}^3 \cdot \text{mol}^{-1}$$

Determine C_V at 300 K.
    Direct substitution into (3.64), and use of 1 bar = 10^5 Pa, gives

$$C_P - C_V = \frac{(300 \text{ K})(7.06 \times 10^{-6} \text{ m}^3 \cdot \text{mol}^{-1})(50.4 \times 10^{-6} \text{ K}^{-1})^2}{0.778 \times 10^{-11} \text{ Pa}^{-1}} = 0.692 \text{ Pa} \cdot \text{m}^3 \cdot \text{mol}^{-1} \cdot \text{K}^{-1}$$

$$= 0.692 \text{ J} \cdot \text{mol}^{-1} \cdot \text{K}^{-1}$$

Thus, $C_V = C_P - 0.69 = 24.50 - 0.69 = 23.81$ J $\cdot$ mol$^{-1} \cdot$ K^{-1}.

EXAMPLE 4.9 Over the range 298–848 K, the heat capacity of quartz (SiO$_2$) at atmospheric pressure is approximated as

$$C_P = 40.50 + (44.60 \times 10^{-3})T - (8.32 \times 10^5)T^{-2}$$

where T is in kelvins and C_P is in J $\cdot$ mol$^{-1} \cdot$ K^{-1}. If 1000 kg of quartz is heated from 300 to 700 K at atmospheric pressure, how much heat is required?
    As shown in Example 1.7, $\Delta H = Q$ for a constant-pressure process. Furthermore, (1.11) gives $dH = C_P dT$ for a constant-pressure process. As a result,

$$Q = \int_{T_1}^{T_2} C_P \, dT$$

Substitution for C_P and integration provide the equation:

$$Q = 40.50(T_2 - T_1) + \frac{44.60 \times 10^{-3}}{2}(T_2^2 - T_1^2) - (8.32 \times 10^5)\left(\frac{1}{T_1} - \frac{1}{T_2}\right)$$

Substituting numerical values for the temperatures, we get $Q = 23\,535$ kJ $\cdot$ kmol^{-1}. The quantity of SiO_2 is $(1000$ kg$)/(60$ kg $\cdot$ kmol$^{-1}) = 16.67$ kmol. The total heat requirement is therefore $(16.67)(23\,535) = 392\,300$ kJ.

EXAMPLE 4.10 From the data below, determine the enthalpy change of mercury for a change of state from 1 bar and 100 °C to 1000 bar and 0 °C.

$T/°C$	P/bar	V/cm$^3 \cdot$ mol^{-1}	β/°C^{-1}	C_P/J $\cdot$ mol$^{-1} \cdot$ °C^{-1}
0	1	14.72	181×10^{-6}	28.0
0	1000	14.67	174×10^{-6}	28.0
100	1	—	—	27.5

Enthalpy is a state function and enthalpy changes are independent of path. Thus, as in Example 4.6, we may choose the most convenient calculational path, as indicated in Fig. 4-7.

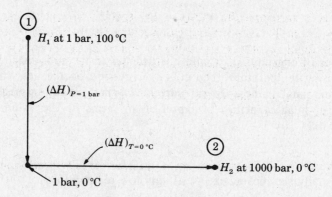

Fig. 4-7

The data show that C_P, V, and β are all very weak functions of T and P, and therefore the use of arithmetic average values should give accurate results. Thus we write (3.69) as

$$\Delta H = \underbrace{\overline{C_P}(T_2 - T_1)}_{\text{for } P = 1 \text{ bar}} + \underbrace{\bar{V}(1 - \bar{\beta}T)(P_2 - P_1)}_{\text{for } T = 0 \text{ °C}} \qquad (1)$$

where

$$\overline{C_P} = \frac{28.0 + 27.5}{2} = 27.75 \text{ J} \cdot \text{mol}^{-1} \cdot °\text{C}^{-1}$$

$$\bar{V} = \frac{14.72 + 14.67}{2} = 14.695 \text{ cm}^3 \cdot \text{mol}^{-1} = 14.695 \times 10^{-6} \text{ m}^3 \cdot \text{mol}^{-1}$$

$$\bar{\beta} = \frac{181 + 174}{2} \times 10^{-6} = 177.5 \times 10^{-6} \text{ °C}^{-1}$$

Substitution of these values into (1) gives:

$$\Delta H = (27.75 \text{ J} \cdot \text{mol}^{-1} \cdot °\text{C}^{-1})(-100 \text{ °C}) + (14.695 \times 10^{-6} \text{ m}^3 \cdot \text{mol}^{-1})$$
$$\times [1 - (177.5 \times 10^{-6} \text{ K}^{-1})(273.15 \text{ K})](999 \times 10^5 \text{ Pa})$$
$$= -1378 \text{ J} \cdot \text{mol}^{-1}$$

4.7 HEAT CAPACITIES OF GASES

The heat capacities of gases are strong functions of both temperature and pressure. However, the effect of pressure on the thermodynamic properties of gases is determined in a way that does not require knowledge of heat capacities as a function of pressure. Instead, use is made of heat capacities of gases *in the ideal-gas state*. The ideal-gas state at temperature T and pressure P for a given gas is the state that would be reached if the gas at temperature T and at a pressure approaching zero, where $PV = RT$, were to remain an ideal gas when compressed isothermally to pressure P. This state is, of course, imaginary, except at pressures approaching zero; nevertheless, it is of considerable practical use.

These ideal-gas heat capacities, designated C_P^{ig} and C_V^{ig}, are independent of pressure, and are always related by the equation

$$\frac{C_P^{ig}}{R} - \frac{C_V^{ig}}{R} = 1 \tag{1.14}$$

Just as for liquids and solids [see (4.12)], the temperature dependence of C_P^{ig} can usually be expressed by an equation of the form

$$\frac{C_P^{ig}}{R} = A + BT + CT^2 + DT^{-2} \tag{4.13}$$

where either C or D is zero and the constants are specific to a particular gas. Values for a few common gases are given in Table 4-1. Since gases at low pressures usually are approximately ideal, heat capacities for the ideal-gas state are suitable for almost all calculations for real gases at pressures up to atmospheric. Since C_P^{ig}/R is dimensionless, the units of C_P^{ig} are governed by the specification of R.

Ideal-gas heat capacities are used in the calculation of enthalpy and entropy changes. Integration of (1.15), $dH^{ig} = C_P^{ig} dT$, gives

$$\Delta H^{ig} = \int_{T_0}^{T_1} C_P^{ig} dT \tag{4.14}$$

As a matter of convenience, we introduce the *mean heat capacity with respect to T* [cf. (3.69)]:

$$\langle C_P^{ig} \rangle_T \equiv \frac{1}{T_1 - T_0} \int_{T_0}^{T_1} C_P^{ig} dT \tag{4.15}$$

Table 4-1.   Constants for Equation (4.13) for T from 298 K to T_{max}

Chemical Species		T_{max}/K	A	$B/10^{-3}K^{-1}$	$C/10^{-6}K^{-2}$	$D/10^5K^2$
Acetylene	C_2H_2	1500	6.132	1.952		−1.299
Air		2000	3.355	0.575		−0.016
Ammonia	NH_3	1800	3.578	3.020		−0.186
Benzene	C_6H_6	1500	−0.206	39.064	−13.301	
Carbon dioxide	CO_2	2000	5.457	1.045		−1.157
Carbon monoxide	CO	2500	3.376	0.557		−0.031
Chlorine	Cl_2	3000	4.442	0.089		−0.344
Ethanol	C_2H_6O	1500	3.518	20.001	−6.002	
Ethylene	C_2H_4	1500	1.424	14.394	−4.392	
Hydrogen	H_2	3000	3.249	0.422		0.083
Hydrogen chloride	HCl	2000	3.156	0.623		0.151
Hydrogen sulfide	H_2S	2300	3.931	1.490		−0.232
Methane	CH_4	1500	1.702	9.081	−2.164	
Nitrogen	N_2	2000	3.280	0.593		0.040
Oxygen	O_2	2000	3.639	0.506		−0.227
Propane	C_3H_8	1500	1.213	28.785	−8.824	
Sulfur dioxide	SO_2	2000	5.699	0.801		−1.015
Water	H_2O	2000	3.470	1.450		0.121

When (4.13) is substituted for C_P^{ig} in (4.15), integration gives:

$$\frac{1}{R} \langle C_P^{ig} \rangle_T = A + BT_{am} + \frac{C}{3}(4T_{am}^2 - T_0 T_1) + \frac{D}{T_0 T_1} \tag{4.16}$$

where $T_{am} \equiv (T_0 + T_1)/2$ is the arithmetic-mean temperature. Thus the integration required for evaluation of enthalpy changes has been accomplished, and

$$\Delta H^{ig} = \langle C_P^{ig} \rangle_T (T_1 - T_0) \tag{4.17}$$

Entropy changes for ideal gases are given by (2.10), here rewritten

$$\Delta S^{ig} = \int_{T_0}^{T_1} C_P^{ig} \frac{dT}{T} - R \ln \frac{P_1}{P_0} \tag{4.18}$$

Equation (4.13), giving the temperature dependence of the molar heat capacity C_P^{ig}, allows integration of the first term on the right. For this purpose, we define the *mean heat capacity with respect to* $\ln T$ [cf. (3.70)]:

$$\langle C_P^{ig} \rangle_{\ln T} \equiv \frac{1}{\ln(T_1/T_0)} \int_{T_0}^{T_1} C_P^{ig} \frac{dT}{T} \tag{4.19}$$

When (4.13) is substituted for C_P^{ig} in (4.19), integration gives

$$\frac{1}{R} \langle C_P^{ig} \rangle_{\ln T} = A + BT_{lm} + T_{am} T_{lm} \left[C + \frac{D}{(T_0 T_1)^2} \right] \tag{4.20}$$

where T_{am} is the arithmetic-mean temperature, and T_{lm} is the logarithmic-mean temperature, defined as

$$T_{lm} \equiv \frac{T_1 - T_0}{\ln(T_1/T_0)}$$

Eliminating the integral between (4.18) and (4.19) gives

$$\Delta S^{ig} = \langle C_P^{ig} \rangle_{\ln T} \ln \frac{T_1}{T_0} - R \ln \frac{P_1}{P_0} \tag{4.21}$$

EXAMPLE 4.11 Evaluate ΔH and ΔS for 1 mol of methane when it is heated from 25 °C to 550 °C at atmospheric pressure.

As methane is essentially ideal under the stated conditions, the superscript "ig" can be dropped. With $T_0 = 298.15$ K and $T_1 = 823.15$ K, we have

$$T_{am} = 560.65 \text{ K} \qquad T_{lm} = \frac{823.15 - 298.15}{\ln(823.15/298.15)} = 516.97 \text{ K}$$

The mean heat capacities are given by (4.16) and (4.20) as

$$\frac{1}{R} \langle C_P \rangle_T = 1.702 + (9.081 \times 10^{-3})(560.65) - \frac{2.164 \times 10^{-6}}{3} [(4)(560.65)^2 - (298.15)(823.15)] = 6.063$$

$$\frac{1}{R} \langle C_P \rangle_{\ln T} = 1.702 + (9.081 \times 10^{-3})(516.97) - (2.164 \times 10^{-6})(560.65)(516.97) = 5.769$$

With $R = 0.008\,314$ kJ $\cdot$ mol$^{-1} \cdot$ K^{-1}, (4.17) gives

$$\Delta H = (6.063)(0.008\,314)(823.15 - 298.15) = 26.464 \text{ kJ} \cdot \text{mol}^{-1}$$

and by (4.21), with $P_1 = P_0$,

$$\Delta S = (5.769)(0.008\,314) \ln \frac{823.15}{298.15} = 0.0487 \text{ kJ} \cdot \text{mol}^{-1} \cdot \text{K}^{-1}$$

4.8 RESIDUAL PROPERTIES OF PVT SYSTEMS

A general procedure for the calculation of the thermodynamic properties of gases or vapors makes use of what are called *residual properties*. These are defined by the generic equation

$$\boxed{M^R \equiv M - M^{ig}}$$
$$(4.22)$$

where the M's may represent molar values of any extensive thermodynamic property. The residual property M^R is the difference between M, representing an actual property, and M^{ig}, representing a property in the ideal-gas state (Section 4.7). It is understood that M and M^{ig} refer to the same conditions of temperature, pressure, and composition.

The fundamental property relation (3.35) may be written for the special case of 1 mol of a constant-composition gas or vapor as

$$dG = -S\,dT + V\,dP$$
$$(4.23)$$

In the differential identity

$$d\left(\frac{G}{RT}\right) \equiv \frac{1}{RT}\,dG - \frac{G}{RT^2}\,dT$$

substitute for dG by (4.23) and for G by its definition, $H - TS$; this yields

$$d\left(\frac{G}{RT}\right) = \frac{V}{RT}\,dP - \frac{H}{RT^2}\,dT$$
$$(4.24)$$

The advantage of (4.24) is that all terms are dimensionless; moreover, in contrast to (4.23), the enthalpy rather than the entropy appears on the right-hand side. Equation (4.24), written for the special case of an ideal gas, becomes

$$d\left(\frac{G^{ig}}{RT}\right) = \frac{V^{ig}}{RT}\,dP - \frac{H^{ig}}{RT^2}\,dT$$

Subtracting this equation from (4.24) gives

$$\boxed{d\left(\frac{G^R}{RT}\right) = \frac{V^R}{RT}\,dP - \frac{H^R}{RT^2}\,dT}$$
$$(4.25)$$

a *fundamental property relation* for residual properties of constant-composition fluids. From (4.25),

$$\frac{V^R}{RT} = \left[\frac{\partial(G^R/RT)}{\partial P}\right]_T$$
$$(4.26)$$

$$\frac{H^R}{RT} = -T\left[\frac{\partial(G^R/RT)}{\partial T}\right]_P$$
$$(4.27)$$

In addition, the defining equation for the Gibbs energy, $G \equiv H - TS$, written for the special case of an ideal gas, is $G^{ig} = H^{ig} - TS^{ig}$; by difference, $G^R = H^R - TS^R$, from which

$$\frac{S^R}{R} = \frac{H^R}{RT} - \frac{G^R}{RT}$$
$$(4.28)$$

According to (4.26)–(4.28), the residual Gibbs energy serves as a generating function for the other residual properties.

A direct link can be established between residual properties and the (experimentally measurable) *compressibility factor Z*, defined by $PV = ZRT$. (See Section 5.1.) First rewrite (4.26) as

$$d\left(\frac{G^R}{RT}\right) = \frac{V^R}{RT}\,dP \qquad \text{(constant } T\text{)}$$

Integration from zero pressure to arbitrary pressure P gives

$$\frac{G^R}{RT} = \int_0^P \frac{V^R}{RT}\,dP \qquad \text{(constant } T\text{)} \tag{4.29}$$

where, at the lower limit, G^R/RT has been equated to zero; justification is provided in Example 4.13. Since, by definition of Z,

$$V^R = \frac{RT}{P}\,(Z-1) \tag{4.30}$$

(4.29) becomes

$$\frac{G^R}{RT} = \int_0^P (Z-1)\,\frac{dP}{P} \qquad \text{(constant } T\text{)} \tag{4.31}$$

When (4.31) is differentiated with respect to temperature and (4.27) is applied, we get

$$\frac{H^R}{RT} = -T \int_0^P \left(\frac{\partial Z}{\partial T}\right)_P \frac{dP}{P} \qquad \text{(constant } T\text{)} \tag{4.32}$$

Combining (4.31) and (4.32) with (4.28) gives

$$\frac{S^R}{R} = -T \int_0^P \left(\frac{\partial Z}{\partial T}\right)_P \frac{dP}{P} - \int_0^P (Z-1)\,\frac{dP}{P} \qquad \text{(constant } T\text{)} \tag{4.33}$$

With $\Delta H^{ig} = H^{ig} - H_0^{ig}$, $\Delta S^{ig} = S^{ig} - S_0^{ig}$, $T_1 = T$, and $P_1 = P$, (4.17) and (4.18) yield

$$H^{ig} = H_0^{ig} + \langle C_P^{ig}\rangle_T\,(T - T_0)$$

$$S^{ig} = S_0^{ig} + \langle C_P^{ig}\rangle_{\ln T} \ln \frac{T}{T_0} - R \ln \frac{P}{P_0}$$

Combining these with (4.22) gives

$$H = H^{ig} + H^R = H_0^{ig} + \langle C_P^{ig}\rangle_T\,(T - T_0) + H^R \tag{4.34}$$

$$S = S^{ig} + S^R = S_0^{ig} + \langle C_P^{ig}\rangle_{\ln T} \ln \frac{T}{T_0} - R \ln \frac{P}{P_0} + S^R \tag{4.35}$$

Values of H^R and S^R are found from (4.32) and (4.33), where the integrals are evaluated either numerically from given PVT data or analytically when Z is expressed by an equation of state.

The reference state at T_0 and P_0 may be selected arbitrarily and values may be assigned to H_0^{ig} and S_0^{ig} arbitrarily. This is another manifestation of the characteristic of exact differentials mentioned in Section 3.1; namely, that the function (here H or S) can be defined only to within an additive constant. For the determination of numerical values of H and S the data needed are ideal-gas heat capacities as functions of T and PVT data to allow evaluation of the residual properties. Once H and S are known (along with the PVT data), then the other thermodynamic functions are calculated from (3.24): $U = H - PV$, $A = U - TS$, $G = H - TS$.

EXAMPLE 4.12 By definition,

$$V^R = V - V^{ig} = V - \frac{RT}{P}$$

Experimental data show that as $P \to 0$, the residual volume in general approaches a finite nonzero limiting value. How is this experimental evidence to be reconciled with the oft-repeated assertion that the ideal-gas equation becomes valid as $P \to 0$?

The experimental evidence rules out ordinary (Cauchy) convergence $V \to RT/P$; i.e., $V^R \to 0$. But it is consistent with *asymptotic convergence* $V \approx RT/P$; i.e., $V/(RT/P) \to 1$, or $V^R/(RT/P) \to 0$. Consider a specific case. At 283 K the limiting value of V^R for methane is -50 cm$^3 \cdot$ mol^{-1}. For this temperature and $P = 0.0001$ bar, we calculate

$$\frac{RT}{P} = \frac{(83.14 \text{ cm}^3 \cdot \text{bar} \cdot \text{mol}^{-1} \cdot \text{K}^{-1})(283 \text{ K})}{0.0001 \text{ bar}} = 235\,286\,200 \text{ cm}^3 \cdot \text{mol}^{-1}$$

Thus, while the *difference* between RT/P and V is essentially 50 cm$^3 \cdot$ mol^{-1}, the *ratio*

$$\frac{V}{RT/P} = \frac{V^{ig} + V^R}{RT/P} = 1 + \frac{V^R}{RT/P} \approx 1 - \frac{50}{235\,286\,200} \approx 1$$

We therefore ought to say: As the pressure of a real gas approaches zero, the ideal-gas equation obtains asymptotically.

EXAMPLE 4.13   As noted in Example 4.12, limiting values of V^R as $P \to 0$ are found by experiment. In this, the residual volume is unique, for the corresponding limiting values of other residual properties are not subject to experimental determination. Their behaviors as $P \to 0$ are nevertheless of interest.

The internal energy U^{ig} of an ideal gas was shown in Section 1.9 to be a function of temperature only, and therefore constant along an isotherm. For a real gas, isothermal expansion occurs with increasing internal energy U, because the molecules move apart against attractive forces. Ultimately, as $P \to 0$, expansion reduces these forces to zero, exactly as in an ideal gas; a reasonable presumption for a given temperature is that $U \to U^{ig}$, or $U^R \to 0$.

Because $H^R = U^R + PV^R$ and (Example 4.12) V^R is bounded near $P = 0$,

$$\lim_{P \to 0} H^R = \lim_{P \to 0} U^R + \lim_{P \to 0} PV^R = 0 + 0 = 0$$

With respect to the residual Gibbs energy, we assume that (4.25) is valid along the boundary isobar $P = 0$. For any isobar, $dP = 0$, and the first term on the right of (4.25) is zero. Moreover, for $P = 0$, as just shown, $H^R = 0$. Thus (4.25) becomes

$$d(G^R/RT) = 0$$

This equation shows that $G^R(P = 0)/RT$ is a constant, independent of T. Since G can be defined only to within an additive constant (see Section 3.1), we are free to set the constant value of $G^R(P = 0)/RT$ equal to zero. Equation (4.29) and subsequent equations derived from it depend on this result.

The extension of property calculations from the vapor phase to the liquid phase is made by application of the Clapeyron equation $(4.3b)$:

$$H^v - H^l \equiv \Delta H^{lv} = T \, \Delta V^{lv} \frac{dP^{\text{sat}}}{dT} \qquad (4.36)$$

In addition,

$$S^v - S^l \equiv \Delta S^{lv} = \frac{\Delta H^{lv}}{T} \qquad (4.37)$$

The additional data needed for these calculations are vapor pressures as a function of T and volumetric data for the saturated liquid and saturated vapor.

Once property values are known for the saturated liquid, the calculations may be extended into the liquid phase by application of (3.53) and (3.55), integrated at constant T from the saturation pressure to any higher pressure:

$$H - H^l = \int_{P^{\text{sat}}}^{P} \left[V - T\left(\frac{\partial V}{\partial T} \right)_P \right] dP \qquad \text{(constant } T) \qquad (4.38)$$

$$S - S^l = -\int_{P^{sat}}^{P} \left(\frac{\partial V}{\partial T}\right)_P dP \quad \text{(constant } T) \tag{4.39}$$

Liquid-phase PVT data are required for these calculations.

EXAMPLE 4.14 Experimental data for a certain substance are represented by the following equations.

Ideal-gas heat capacity:

$$C_P^{ig}/R = 12.03 + 1.2 \times 10^{-3}\, T \tag{1}$$

where T is in K.

Equation of state of vapor phase:

$$Z = 1 + \frac{P}{RT}\left(b - \frac{a}{RT}\right) \quad \text{or} \quad V = \frac{RT}{P} + b - \frac{a}{RT} \tag{2}$$

where, for P in bars, $a = 20 \times 10^6$ cm$^6 \cdot$ bar $\cdot$ mol^{-2} and $b = 140$ cm$^3 \cdot$ mol^{-1}.

Vapor pressure:

$$\ln P^{sat} = 10.0 - \frac{3550}{T} \tag{3}$$

where T is in K and P^{sat} is in bar.

Molar volume of saturated liquid:

$$V^l = 60 + 0.10\, T \tag{4}$$

where T is in K and V^l is in cm$^3 \cdot$ mol^{-1}.

Liquid expansivity and compressibility:

$$\beta = 0.0012 \text{ K}^{-1} \qquad \kappa = 0 \tag{5}$$

Determine the numerical relationships of H and S to the pressure ($1 \le P \le 10$ bar) at the fixed temperature 400 K.

For application of (4.34) and (4.35), evaluate H^R and S^R for the vapor phase by means of (4.32) and (4.33). From (2),

$$\left(\frac{\partial Z}{\partial T}\right)_P = \frac{P}{RT^2}\left(\frac{2a}{RT} - b\right)$$

Substitution for the derivative and for $Z - 1$ in (4.32) and (4.33) gives after integration

$$H^R = -\left(\frac{2a}{RT} - b\right)P \quad \text{and} \quad S^R = -\frac{aP}{RT^2}$$

For the given values of a and b, and with $T = 400$ K and $R = 83.14$ bar $\cdot$ cm$^3 \cdot$ mol$^{-1} \cdot$ K^{-1},

$$H^R = -1059P \quad \text{cm}^3 \cdot \text{bar} \cdot \text{mol}^{-1} = -105.9P \quad \text{J} \cdot \text{mol}^{-1} \quad \text{and} \quad S^R = -0.1503P \quad \text{J} \cdot \text{mol}^{-1} \cdot \text{K}^{-1}$$

With respect to (4.34) and (4.35) one must now assign values to H_0^{ig} and S_0^{ig} corresponding to some selected T_0 and P_0. Let

$$H_0^{ig} = 1000 \text{ J} \cdot \text{mol}^{-1} \qquad S_0^{ig} = 10 \text{ J} \cdot \text{mol}^{-1} \cdot \text{K}^{-1}$$

at $T_0 = 300$ K and $P_0 = 1$ bar. Applying (4.16) and (4.20) to (1), we find

$$\langle C_P^{ig} \rangle_T = 103.51 \text{ J} \cdot \text{mol}^{-1} \cdot \text{K}^{-1}$$

$$\langle C_P^{ig} \rangle_{\ln T} = 103.49 \text{ J} \cdot \text{mol}^{-1} \cdot \text{K}^{-1}$$

Substituting into (4.35) and (4.36) gives

$$H/\text{J} \cdot \text{mol}^{-1} = 1000 + 10\,350 - 105.9P = 11\,350 - 105.9P$$

$$S/\text{J} \cdot \text{mol}^{-1} \cdot \text{K}^{-1} = 10 + 29.77 - 8.314 \ln P - 0.1503P = 39.77 - 8.314 \ln P - 0.1503P$$

The initial value of P is 1 bar, and the final value is P^{sat} at 400 K, which is given by (3) as

$$\ln P^{sat} = 10.0 - \frac{3550}{400} = 1.125 \quad \text{or} \quad P^{sat} = 3.081 \text{ bar}$$

Evaluations of the vapor-phase H and S at 400 K follow.

P/bar	$H/\text{J}\cdot\text{mol}^{-1}$	$S/\text{J}\cdot\text{mol}^{-1}\cdot\text{K}^{-1}$
1	11 244	39.62
2	11 134	33.71
3	11 025	30.19
3.081 (P^{sat})	11 016 (H^v)	29.95 (S^v)

Application of the Clapeyron equation (to pass to the liquid phase) requires the derivative dP^{sat}/dT, evaluated at 400 K; from (3),

$$\frac{dP^{sat}}{dT} = \frac{3550 P^{sat}}{T^2} = \frac{(3550)(3.081)}{(400)^2} = 0.068\,36 \text{ bar}\cdot\text{K}^{-1}$$

The molar volume of saturated liquid at 400 K is, from (4),

$$V^l = 60 + (0.10)(400) = 100 \text{ cm}^3\cdot\text{mol}^{-1}$$

The molar volume of saturated vapor is computed from (2) with $T = 400$ K and $P^{sat} = 3.081$ bar:

$$V^v = \frac{(83.14)(400)}{3.081} + 140 - \frac{20\times10^6}{(8.314)(400)} = 10\,332 \text{ cm}^3\cdot\text{mol}^{-1}$$

Thus, $\Delta V^{lv} = V^v - V^l = 10\,232 \text{ cm}^3\cdot\text{mol}^{-1}$. Substitution of numerical values into (4.36) gives

$$\Delta H^{lv} = (400)(10\,232)(0.068\,36) \text{ cm}^3\cdot\text{bar}\cdot\text{mol}^{-1} \quad \text{or} \quad 27\,978 \text{ J}\cdot\text{mol}^{-1}$$

and, by (4.37),

$$\Delta S^{lv} = \frac{\Delta H^{lv}}{T} = \frac{27\,978}{400} = 69.95 \text{ J}\cdot\text{mol}^{-1}\cdot\text{K}^{-1}$$

The enthalpy and entropy of saturated liquid at 400 K and $P^{sat} = 3.081$ bar are therefore

$$H^l = H^v - \Delta H^{lv} = -16\,962 \text{ J}\cdot\text{mol}^{-1} \qquad S^l = S^v - \Delta S^{lv} = -40.00 \text{ J}\cdot\text{mol}^{-1}\cdot\text{K}^{-1}$$

Now, for the liquid phase, (3.69) and (3.70) may be written for the isothermal pressure change from P^{sat} to P as

$$H - H^l = \langle V(1-\beta T)\rangle (P - P^{sat}) \qquad S - S^l = -\langle \beta V\rangle (P - P^{sat})$$

Since V, T, and β are all constant, these become

$$H = H^l + V^l(1 - T\beta)(P - P^{sat}) \qquad S = S^l - V^l\beta(P - P^{sat})$$

and substitution of known numerical values gives

$$H = -16\,962 + \frac{(100)(1 - 400\times0.0012)(P-3.081) \quad \text{cm}^3\cdot\text{bar}\cdot\text{mol}^{-1}}{10 \text{ cm}^3\cdot\text{bar}\cdot\text{J}^{-1}} = [-16\,962 + 5.2(P-3.081)] \text{ J}\cdot\text{mol}^{-1}$$

$$S = -40.00 - \frac{(100)(0.0012)(P-3.081)}{10} = [-40.00 - 0.012(P-3.081)] \text{ J}\cdot\text{mol}^-\cdot\text{K}^{-1}$$

Values of H and S for compressed liquid are now readily calculated:

P/bar	$H/\text{J}\cdot\text{mol}^{-1}$	$S/\text{J}\cdot\text{mol}^{-1}\cdot\text{K}^{-1}$
3.081 (P^{sat})	-16 962 (H^l)	-40.00 (S^l)
4	-16 957	-40.01
6	-16 947	-40.04
8	-16 936	-40.06
10	-16 926	-40.08

Our tabulated enthalpies and entropies of course depend on the assumed zero-point values; but this is not true of the property *differences*, which alone have physical significance.

4.9 THERMODYNAMIC DIAGRAMS AND TABLES FOR PVT SYSTEMS

Calculations such as those illustrated in Example 4.14 may be carried out for a pure substance for a series of different temperatures. The results obtained for each temperature plot as an isotherm on one or another thermodynamic diagram; Example 4.14 would generate PH and PS diagrams (Fig. 4-8). Also shown is the corresponding PV diagram, which represents the data upon which the calculations are based.

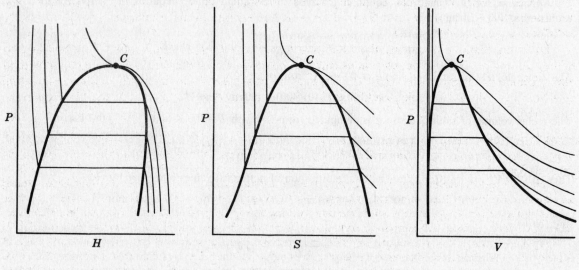

Fig. 4-8

The lighter lines in Fig. 4-8 represent isotherms at various temperatures, whereas the heavy curves drawn through the points where the isotherms abruptly change slope represent saturated liquid (to the left of C) and saturated vapor (to the right of C). Point C is in each case the *critical point*, and the isotherm with C as an inflection point is for temperature T_c. Isotherms for temperatures greater than T_c are smooth and lie above the critical isotherm.

Once diagrams such as these (or equivalent tabular data) are available, information may be taken from one graph and included on another. For example, vertical lines on the PS and PV diagrams represent lines of constant S and of constant V, and values of P and T along these lines may be used to locate lines of constant S and of constant V on the PH diagram. In this way the PH diagram may be made to include the relations among all the properties P, V, T, H, and S.

One can also use other pairs of coordinates. For example, one can take data from the PS diagram and plot a TS diagram showing lines of constant P (isobars). In addition, data from the PV and PH diagrams can be included as lines of constant V and of constant H.

The PH diagram and the TS diagram are in common use along with a third diagram, employed almost exclusively for steam and known as the *Mollier diagram*. It employs H and S as coordinates and shows lines of constant pressure. Data for volume are not usually included, but in the two-phase (liquid/vapor) region, lines of constant quality (percentage of vapor by weight) appear. In the vapor or gas region, lines of constant temperature and of constant *superheat* are shown. Superheat is a term used to designate the difference between the temperature and the saturation temperature at the same pressure. In fact, the entire vapor region exclusive of the saturation curve is often called the superheat region.

The three diagrams just discussed are exemplified in Fig. 4-9. [Note the logarithmic scale in Fig. 4-9(b).] These figures are based on data for water, and vary in detail from substance to substance. In all cases the heavy lines represent phase boundaries, and the lighter lines show the relations between coordinates for a constant value of a particular property.

The information presented by means of graphs can, of course, also be given in tables. The most widely used such tables are the *steam tables*; they represent extensive collections of data for liquid and vapor H_2O. Similar, but usually less detailed, tables for other common substances can be found in the literature.

Thermodynamic tables for water appear in Appendix D, where the equilibrium properties of saturated-liquid water and saturated-vapor steam are given for equal increments of temperature from the triple point to the critical point. Data for saturated liquid and for both saturated and superheated vapor appear in Appendix E, where the properties of steam are tabulated for equal increments of pressure at temperatures equal to and higher than the saturation value for each given pressure. The tabular spacing for V, U, H, and S (in SI units) is fine enough to make linear interpolation *usually* satisfactory (see Problem 4.18).

By international agreement, the basis for numerical values in steam tables is established by setting U^l and S^l, the specific internal energy and entropy of saturated-liquid water, equal to zero at the triple-point temperature, 0.01 °C. Since H is calculated from U by the defining equation $H = U + PV$, it is not possible in addition to assign an arbitrary value for H^l, the specific enthalpy of saturated liquid, at 0.01 °C.

EXAMPLE 4.15 Consider two tanks, each of volume 1 L. Tank A is filled with saturated-liquid water and tank B is filled with saturated vapor (steam), both at a pressure of 10 bar. If both tanks explode, which will do the greater damage?

Suppose that the tanks undergo sudden, catastrophic failure and that the contents of the tanks expand rapidly to a pressure of 1 bar in an adiabatic process. If damage is assumed to be proportional to the work done by the system, then the greatest possible damage results when the explosive expansion is also reversible; i.e., isentropic. (For purposes of comparing damages, it is permissible to contemplate a "reversible explosion.")

Figure 4-9(a) shows that the expansion of either saturated liquid or saturated vapor along a line of constant S leads to a final state in the two-phase (liquid/vapor) region, a state made up of a mixture of saturated liquid and saturated vapor. Thus the only data needed are for saturated liquid and saturated vapor at pressures of 10 bar (1000 kPa) and 1 bar (100 kPa). These data, taken from Appendix E, are as follows:

P/kPa	T/°C	V/L·kg^{-1}		U/kJ·kg^{-1}		S/kJ·kg^{-1}·K^{-1}	
		V^l	V^v	U^l	U^v	S^l	S^v
100	99.63	1.043	1693.7	417.51	2506.1	1.3027	7.3598
1000	179.88	1.127	194.29	761.48	2581.9	2.1382	6.5828

By the first law the work of an adiabatic process per unit mass is $W = -\Delta U$. The initial values of the internal energy for the two tanks are tabulated at 1000 kPa; the final values of U are determined by the condition of isentropic expansion.

Tank A. The initial entropy is given in the table as $S_1 = 2.1382$ kJ·kg^{-1}·K^{-1}. Since S is constant, (4.8) gives

$$S_1 = S_2 = (1 - x)S_2^l + xS_2^v$$

where x is the quality in the final state. Substitution of numerical values gives

$$2.1382 = (1 - x)(1.3027) + x(7.3598) \qquad \text{whence} \qquad x = 0.1379$$

We now find the final internal energy by again applying (4.8):

$$U_2 = (1 - x)U_2^l + xU_2^v = (1 - 0.1379)(417.41) + (0.1379)(2506.1) = 705.52 \text{ kJ·kg}^{-1}$$

Hence, $\Delta U = U_2 - U_1 = 705.52 - 761.48 = -55.96$ kJ·kg^{-1}. The initial specific volume of the saturated liquid is 1.127 L·kg^{-1}, whence the mass of liquid is $m = 1/1.127 = 0.8873$ kg. The total work of the explosion is therefore

$$W^t = m(-\Delta U) = (0.8873)(+55.96) = 49.65 \text{ kJ}$$

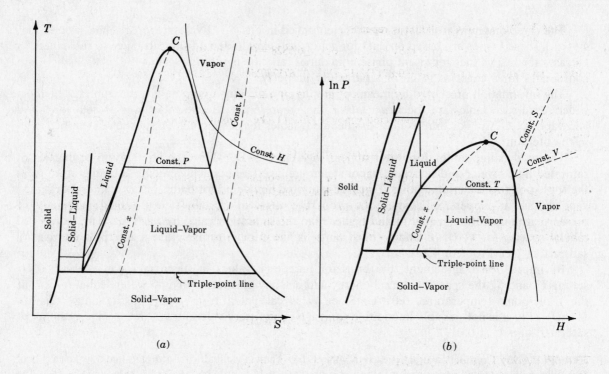

(a)

(b)

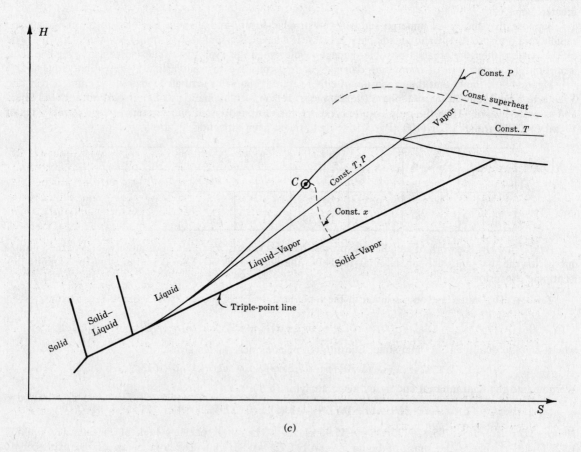

(c)

Fig. 4-9

Tank B. The above calculation is repeated.

$$S_1 = S_2 = (1-x)S_2^l + xS_2^v \quad \text{or} \quad 6.5828 = (1-x)(1.3027) + x(7.3598)$$

whence $x = 0.8717$. Then $U_2 = (1 - 0.8717)(417.41) + (0.8717)(2506.1) = 2238.2$ kJ·kg^{-1} and

$$\Delta U = U_2 - U_1 = 2238.2 - 2581.9 = -343.7 \text{ kJ·kg}^{-1}$$

The initial specific volume of the saturated vapor is 194.29 L·kg^{-1}, giving a mass $1/194.29 = 0.005\,15$ kg. The total work of the explosion is

$$W^t = m(-\Delta U) = (0.005\,15)(+343.7) = 1.77 \text{ kJ}$$

The surprising conclusion is that for a tank of a given volume the destructive potential upon explosion is nearly 30 times greater when the tank is filled with saturated-liquid water than when it is filled with saturated steam at the same pressure. The same considerations apply when one is concerned with the storage of energy in a given volume. Saturated liquid at elevated pressure is a much more effective medium than saturated vapor at the same pressure, because a far greater mass of liquid than of vapor can be put in a given container.

Solved Problems

PVT BEHAVIOR OF A PURE SUBSTANCE   (Section 4.1)

4.1   On a PV diagram show the locations of the fluid region, and the vapor region.

The regions are both indicated in Fig. 4-10. The fluid region includes all states for which the temperature *and* the pressure are above their critical values. The vapor region includes those states from which condensation can be caused *both* by compression at constant T *and* by cooling at constant P.

4.2   The PT diagram of Fig. 4-2 gives no information on volumes. For the liquid and gas regions, what would be the general appearance of constant-volume lines (*isochores* or *isometrics*) on this diagram?

An examination of Fig. 4-3 shows that vertical lines (isochores) on a PV diagram intersect the saturation line ACB over its full range. For small volumes these lines intersect the saturated-liquid curve AC, and for larger volumes, the saturated-vapor curve CB. Thus there are two constant-volume lines that intersect curve ACB at a given pressure, one for liquid and one for vapor. Of course there is an exception for the line at the critical volume, which passes through point C at the maximum of ACB. On a PT diagram the two branches of the saturation curve, AC and BC, are identical, and the coordinates (P, T) of the intersections of constant-volume lines for both liquid and vapor lie along the vaporization curve $2\text{-}C$ of Fig. 4-2. Moreover, they represent terminal points for these curves on the PT diagram. From these terminal points the constant-volume lines extend upward into the liquid region for the set of liquid isochores, and to the right into the gas region for the set of gas isochores. With respect to Fig. 4-1, the isochores in the PT-plane are the projections on that plane of the intersections of the three-dimensional surface with the planes $V = \text{const.}$

An idea of the general shape of the isochores can be obtained by consideration of two idealizations: the ideal gas and the liquid for which β and κ are constant. For the ideal gas, $PV = RT$, and

$$\left(\frac{\partial P}{\partial T}\right)_V = \frac{R}{V}$$

Thus the slope of an isochore for an ideal gas is constant and positive, and the isochores become increasingly steep as V decreases. For the liquid phase we have from (3.19) $dV/V = \beta\, dT - \kappa\, dP$, which at constant V becomes

$$\left(\frac{\partial P}{\partial T}\right)_V = \frac{\beta}{\kappa}$$

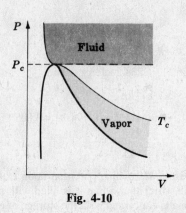

Fig. 4-10

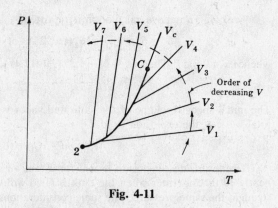

Order of decreasing V

Fig. 4-11

For constant and positive values of β and κ, the slope of an isochore is constant and positive, independent of V. These results for idealized gas and liquid phases hold qualitatively for real materials. Though isochores do in fact show curvature, it is small, and PT diagrams look essentially like Fig. 4-11.

VAPOR PRESSURES AND LATENT HEATS   (Sections 4.2 and 4.3)

4.3   If the vapor pressure P^{sat} of a certain liquid is represented by an equation of the form

$$\ln P^{sat} = A - \frac{B}{T}$$

where A and B are constants, and T is absolute temperature, show that for this material

$$\Delta S^{lv} = \Delta V^{lv}\,\frac{BP^{sat}}{T^2} \qquad\qquad (1)$$

Equation (4.2) is directly applicable:

$$\frac{dP^{sat}}{dT} = \frac{\Delta S^{lv}}{\Delta V^{lv}} \qquad\text{or}\qquad \frac{d\ln P^{sat}}{dT} = \frac{\Delta S^{lv}}{P^{sat}\,\Delta V^{lv}}$$

Differentiating the given expression for $\ln P^{sat}$, we find upon substitution:

$$\frac{B}{T^2} = \frac{\Delta S^{lv}}{P^{sat}\,\Delta V^{lv}}$$

and solution for ΔS^{lv} yields (1).

4.4   From the data that follow, estimate the latent heat of vaporization ΔH^{lv} for benzene at 340 K. At this temperature, $P^{sat} = 66.15$ kPa, $V^l = 0.0012\ \text{m}^3\cdot\text{kg}^{-1}$, and $V^v = 0.5332\ \text{m}^3\cdot\text{kg}^{-1}$. In addition, at 350 K, $P^{sat} = 91.62$ kPa.

By direct application of the Clapeyron equation (4.3b),

$$\Delta H^{lv} = T\,\Delta V^{lv}\,\frac{dP^{sat}}{dT}$$

The volume change of vaporization is $\Delta V^{lv} = 0.5332 - 0.0012 = 0.5320\ \text{m}^3\cdot\text{kg}^{-1}$. Over the small interval from 340 to 350 K, the vapor pressure should be well represented by (4.5), from which

$$\frac{dP^{sat}}{dT} = \frac{BP^{sat}}{T^2} \qquad\text{and}\qquad \Delta H^{lv} = \frac{(\Delta V^{lv})BP^{sat}}{T}$$

Application of (4.5) both at 340 K and 350 K gives

$$\ln 66.15 = A - \frac{B}{340} \qquad\qquad \ln 91.62 = A - \frac{B}{350}$$

Solving for B yields $B = 3876.1$ K, whence,

$$\Delta H^{lv} = \frac{(0.532\ \text{m}^3 \cdot \text{kg}^{-1})(3876.1\ \text{K})(66.15\ \text{kPa})}{340\ \text{K}} = 401.2\ \text{kJ} \cdot \text{kg}^{-1}$$

The reported value is $402.8\ \text{kJ} \cdot \text{kg}^{-1}$.

TWO-PHASE SYSTEMS   (Section 4.4)

4.5   Steam at a pressure of 35 bar is known to have a specific volume of $50\ \text{cm}^3 \cdot \text{g}^{-1}$. What is its specific enthalpy?

Examination of the steam tables for values of V at a pressure of 35 bar shows that the given value lies between $V^l = 1.235\ \text{cm}^3 \cdot \text{g}^{-1}$ and $V^v = 57.025\ \text{cm}^3 \cdot \text{g}^{-1}$. The steam is therefore "wet," being a mixture of saturated vapor and saturated liquid. We first determine the quality of the steam through use of (Example 4.5) $V = V^l + x\,\Delta V^{lv}$; thus,

$$x = \frac{V - V^l}{\Delta V^{lv}} = \frac{50.000 - 1.235}{57.025 - 1.235} = 0.8741$$

The specific enthalpy is given by $H = H^l + x\,\Delta H^{lv}$. From Appendix D, at 35 bar, $H^l = 1049.8\ \text{kJ} \cdot \text{kg}^{-1}$ and $\Delta H^{lv} = 1752.2\ \text{kJ} \cdot \text{kg}^{-1}$. Therefore,

$$H = 1049.8 + (0.8741)(1752.2) = 2581.3\ \text{kJ} \cdot \text{kg}^{-1}$$

4.6   A tank contains liquid water and steam in equilibrium at 700 kPa. If the liquid and the vapor each occupy half of the volume of the tank, what is the specific enthalpy of the contents of the tank?

For saturated liquid and vapor at 700 kPa, Appendix E gives

$$V^l = 1.108\ \text{cm}^3 \cdot \text{g}^{-1} \qquad V^v = 272.68\ \text{cm}^3 \cdot \text{g}^{-1}$$
$$H^l = 697.06\ \text{kJ} \cdot \text{kg}^{-1} \qquad H^v = 2762.0\ \text{kJ} \cdot \text{kg}^{-1}$$

Equating the relative volumes of vapor and liquid, $xV^v = (1-x)V^l$; thus

$$x = \frac{V^l}{V^v + V^l} = \frac{1.108}{272.68 + 1.108} = 0.004\,047$$

Then, by *(4.8)*, $H = (1-x)H^l + xH^v = (0.995\,953)(697.06) + (0.004\,047)(2762.0) = 705.42\ \text{kJ} \cdot \text{kg}^{-1}$.

4.7   A rigid tank contains $0.05\ \text{m}^3$ of saturated-liquid water and $0.95\ \text{m}^3$ of saturated-vapor water in equilibrium at 1 bar pressure. How much heat must be added to the contents so that the liquid is just vaporized?

The system is closed, and the first-law energy equation is simply $Q = \Delta U^t = m\,\Delta U$, because there is no work. Property values in the initial state are (Appendix E):

$$V_1^l = 1.043\ \text{cm}^3 \cdot \text{g}^{-1} \qquad V_1^v = 1693.7\ \text{cm}^3 \cdot \text{g}^{-1}$$
$$U_1^l = 417.41\ \text{kJ} \cdot \text{kg}^{-1} \qquad U_1^v = 2506.1\ \text{kJ} \cdot \text{kg}^{-1} \qquad \Delta U_1^{lv} = 2088.7\ \text{kJ} \cdot \text{kg}^{-1}$$

The initial masses of liquid and vapor are

$$m_1^l = \frac{V_{\text{liq},1}}{V_1^l} = \frac{0.05 \times 10^6\ \text{cm}^3}{1043\ \text{cm}^3 \cdot \text{kg}^{-1}} = 47.939\ \text{kg}$$

$$m_1^v = \frac{V_{\text{vap},1}}{V_1^v} = \frac{0.95 \times 10^6}{1693.7 \times 10^3} = 0.561\ \text{kg}$$

The final state of the system is saturated vapor occupying the entire tank:

$$m_2^v = m = 47.939 + 0.561 = 48.500 \text{ kg} \quad \text{and} \quad V_{\text{tank}} = 0.05 + 0.95 = 1.00 \text{ m}^3$$

Hence the final specific volume is

$$V_2^v = \frac{1.00}{48.500} = 20.619 \times 10^{-3} \text{ m}^3 \cdot \text{kg}^{-1} \quad \text{or} \quad 20.619 \text{ cm}^3 \cdot \text{g}^{-1}$$

Saturated vapor with this specific volume occurs at just one pressure, which may be found by interpolation in the steam tables for saturated vapor (Appendix D). Thus we find $P_2 = 89.55$ bar; at this pressure, the internal energy of saturated vapor is

$$U_2 = U_2^v = 2560.7 \text{ kJ} \cdot \text{kg}^{-1}$$

The initial internal energy is $U_1 = U_1^l + x_1 \Delta U_1^{lv}$, where x_1, the initial quality, is given by

$$x_1 = \frac{m_1^v}{m} = \frac{0.561}{48.500} = 0.01157$$

Hence, $U_1 = 417.41 + (0.011\,57)(2088.7) = 441.57 \text{ kJ} \cdot \text{kg}^{-1}$, and

$$Q = m(U_2 - U_1) = (48.5)(2560.7 - 441.6) = 1.0278 \times 10^5 \text{ kJ} \quad \text{or} \quad 102.78 \text{ MJ}$$

4.8 A rigid tank, of 0.3-m³ capacity, contains water as saturated liquid and vapor in equilibrium at 15 bar. If 99% of the mass is liquid, how much heat must be added before the tank becomes just full of liquid?

The energy equation is, as in Problem 4.7, $Q = \Delta U^t = m \Delta U$, and initial property values from the steam tables (Appendix E) are

$$V_1^l = 1.154 \text{ cm}^3 \cdot \text{g}^{-1} \qquad V_1^v = 131.7 \text{ cm}^3 \cdot \text{g}^{-1}$$
$$U_1^l = 842.9 \text{ kJ} \cdot \text{kg}^{-1} \qquad V_1^v = 2592.4 \text{ kJ} \cdot \text{kg}^{-1}$$

The initial specific volume of the contents of the tank is

$$V_1 = (1 - x_1)V_1^l + x_1 V_1^v = (0.99)(1.154) + (0.01)(131.7) = 2.46 \text{ L} \cdot \text{kg}^{-1}$$

it being given that the initial mass-fraction vapor is $x_1 = 0.01$; similarly, the initial internal energy is

$$U_1 = (0.99)(842.9) + (0.01)(2592.4) = 860.4 \text{ kJ} \cdot \text{kg}^{-1}$$

The final state as the last bubble of vapor condenses is saturated liquid, which occupies the entire volume of the tank. Since neither the volume of the tank nor the total mass of the system has changed, the final specific volume of the saturated liquid in the tank is $V_2^l = V_1 = 2.46 \text{ cm}^3 \cdot \text{g}^{-1}$. The pressure at which saturated liquid has this specific volume, found by interpolation in the table of Appendix E, is $P_2 = 217.56$ bar, and similar interpolation for its internal energy gives $U_2 = 1907.7 \text{ kJ} \cdot \text{kg}^{-1}$. Finally, the mass of the system is

$$m = \frac{V_{\text{tank}}}{V_1} = \frac{0.3 \times 10^6 \text{ cm}^3}{2.46 \times 10^3 \text{ cm}^3 \cdot \text{kg}^{-1}} = 121.95 \text{ kg}$$

Thus $Q = m(U_2 - U_1) = (121.95)(1907.7 - 860.4) = 127\,720 \text{ kJ}$ or 127.720 MJ.

4.9 The initial quality in both Problem 4.7 and Problem 4.8 is very nearly 0.01. How is it that in the one case the tank fills with vapor, while in the other it fills with liquid?

The explanation lies in the shape of the saturation curve for water. As is shown in Fig. 4-12 (not to scale), either starting point divides its isobar in the ratio 1 : 99. But this makes one isochore supercritical, and the other subcritical; consequently, the isochores terminate respectively on the vapor boundary and on the liquid boundary.

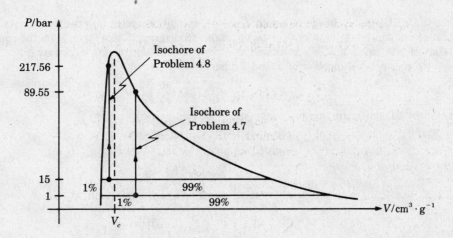

Fig. 4-12

HEAT CAPACITIES   (Sections 4.6 and 4.7)

4.10 The following data are available for liquid water at 25 °C and atmospheric pressure: $\beta = 256 \times 10^{-6} \,°C^{-1}$, $(\partial\beta/\partial T)_P = 9.6 \times 10^{-6} \,°C^{-2}$, $V = 1.003 \,cm^3 \cdot g^{-1}$. Determine the effect of pressure on C_P—i.e., calculate the value of $(\partial C_P/\partial P)_T$—for water at these conditions.

The required equation is (3.67):

$$\left(\frac{\partial C_P}{\partial P}\right)_T = -VT\left[\beta^2 + \left(\frac{\partial\beta}{\partial T}\right)_P\right]$$

Substitution of numerical values gives

$$\left(\frac{\partial C_P}{\partial P}\right)_T = -(1.003 \,cm^3 \cdot g^{-1})(298.15 \,K)[(256 \times 10^{-6} \,K^{-1})^2 + (9.6 \times 10^{-6} \,K^{-2})]$$

$$= -0.002\,89 \,cm^3 \cdot g^{-1} \cdot K^{-1} \quad \text{or} \quad -2.89 \times 10^{-4} \,kJ \cdot kg^{-1} \cdot K^{-1} \cdot bar^{-1}$$

Thus, a pressure increase of 1 bar at the stated conditions causes a very small decrease in C_P. Since the heat capacity of water at 25 °C is about $4.18 \,kJ \cdot kg^{-1} \cdot K^{-1}$, its percentage decrease for a pressure increase of 1 bar is only about 0.007%.

4.11 The general defining equation for a heat capacity is given in Section 1.9 as

$$C_X = \left(\frac{\delta Q}{dT}\right)_X$$

where X indicates a reversible and fully specified path. We have so far considered only two heat capacities, C_V and C_P, but many others could be defined. Two that are sometimes found useful are the heat capacities *of saturated liquid* and *of saturated vapor*. These represent $\delta Q/dT$ for reversible changes along the appropriate saturation curves, and we will designate them C_{sat} in general. Since the pressure, temperature, and volume all vary along a saturation curve, C_{sat} is different from C_V and C_P. Nor is it given by $(dU/dT)_{sat}$ or by $(dH/dT)_{sat}$, derivatives taken along a saturation curve. These quantities, however, are all related to one another. Derive equations that show these relationships.

Since the defining equation for C_X restricts consideration to reversible processes, we may always make the substitution $\delta Q = T\,dS$; thus

$$C_{sat} = T\left(\frac{dS}{dT}\right)_{sat}$$

where $(dS/dT)_{sat}$ is the change of S with T along a saturation curve; i.e., the reciprocal of the slope of the saturated-liquid or saturated-vapor curve on the TS diagram [Fig. 4-9(a)]. With the equation written in terms of properties only, no restriction to reversible processes is now necessary.

If we consider S a function of T and V, then

$$dS = \left(\frac{\partial S}{\partial T}\right)_V dT + \left(\frac{\partial S}{\partial V}\right)_T dV$$

Division by dT and restriction to changes along a saturation curve give

$$\left(\frac{dS}{dT}\right)_{sat} = \left(\frac{\partial S}{\partial T}\right)_V + \left(\frac{\partial S}{\partial V}\right)_T \left(\frac{dV}{dT}\right)_{sat} \tag{1}$$

By (3.49) and (3.43):

$$\left(\frac{\partial S}{\partial T}\right)_V = \frac{C_V}{T} \quad \text{and} \quad \left(\frac{\partial S}{\partial V}\right)_T = \left(\frac{\partial P}{\partial T}\right)_V$$

Hence

$$C_{sat} = T\left(\frac{dS}{dT}\right)_{sat} = T\left[\frac{C_V}{T} + \left(\frac{\partial P}{\partial T}\right)_V \left(\frac{dV}{dT}\right)_{sat}\right]$$

But, by (3.22), $(\partial P/\partial T)_V = \beta/\kappa$, and by substitution we get a final expression relating C_{sat} and C_V:

$$C_{sat} = C_V + \frac{T\beta}{\kappa}\left(\frac{dV}{dT}\right)_{sat} \tag{2}$$

An equation analogous to (1) is

$$\left(\frac{dV}{dT}\right)_{sat} = \left(\frac{\partial V}{\partial T}\right)_P + \left(\frac{\partial V}{\partial P}\right)_T \left(\frac{dP}{dT}\right)_{sat}$$

which by (3.17) and (3.18) becomes

$$\left(\frac{dV}{dT}\right)_{sat} = \beta V - \kappa V\left(\frac{dP}{dT}\right)_{sat} \tag{3}$$

Combination of (2) and (3) gives another expression relating C_{sat} and C_V:

$$C_{sat} = C_V + \frac{T\beta^2 V}{\kappa} - T\beta V\left(\frac{dP}{dT}\right)_{sat} \tag{4}$$

However, by (3.64), $T\beta^2 V/\kappa = C_P - C_V$; therefore

$$C_{sat} = C_P - T\beta V\left(\frac{dP}{dT}\right)_{sat} \tag{5}$$

and we have a relation between C_{sat} and C_P. Note that $(dP/dT)_{sat}$ is just the slope of the vaporization curve of Fig. 4-2.

Another equation analogous to (1) is

$$\left(\frac{dH}{dT}\right)_{sat} = \left(\frac{\partial H}{\partial T}\right)_P + \left(\frac{\partial H}{\partial P}\right)_T \left(\frac{dP}{dT}\right)_{sat}$$

However, $(\partial H/\partial T)_P = C_P$ and, by (3.61), $(\partial H/\partial P)_T = V(1 - \beta T)$. Therefore

$$\left(\frac{dH}{dT}\right)_{sat} = C_P - T\beta V\left(\frac{dP}{dT}\right)_{sat} + V\left(\frac{dP}{dT}\right)_{sat}$$

According to (5), the first two terms on the right equal C_{sat}; thus

$$C_{sat} = \left(\frac{dH}{dT}\right)_{sat} - V\left(\frac{dP}{dT}\right)_{sat} \tag{6}$$

which shows the relation between C_{sat} and $(dH/dT)_{sat}$.

Since $U \equiv H - PV$, we have

$$\left(\frac{dU}{dT}\right)_{sat} = \left(\frac{dH}{dT}\right)_{sat} - V\left(\frac{dP}{dT}\right)_{sat} - P\left(\frac{dV}{dT}\right)_{sat}$$

However, (6) shows the first two terms on the right to be equal to C_{sat}. Therefore

$$C_{sat} = \left(\frac{dU}{dT}\right)_{sat} + P\left(\frac{dV}{dT}\right)_{sat} \tag{7}$$

an equation which relates C_{sat} to $(dU/dT)_{sat}$.

In summary, we have from (2) and (7)

$$C_{sat} = C_V + \frac{T\beta}{\kappa}\left(\frac{dV}{dT}\right)_{sat} = \left(\frac{dU}{dT}\right)_{sat} + P\left(\frac{dV}{dT}\right)_{sat}$$

and from (5) and (6)

$$C_{sat} = C_P - T\beta V\left(\frac{dP}{dT}\right)_{sat} = \left(\frac{dH}{dT}\right)_{sat} - V\left(\frac{dP}{dT}\right)_{sat}$$

It is clear from these equations that C_{sat}, C_V, C_P, $(dU/dT)_{sat}$, and $(dH/dT)_{sat}$ are all different, although related, quantities.

4.12 Estimate C_{sat} for water at 1 atm and 100 °C from the following data:

T °C	P bar	V^l $cm^3 \cdot g^{-1}$	V^v $cm^3 \cdot g^{-1}$	H^l $kJ \cdot kg^{-1}$	H^v $kJ \cdot kg^{-1}$	S^l $kJ \cdot kg^{-1} \cdot K^{-1}$	S^v $kJ \cdot kg^{-1} \cdot K^{-1}$
99	0.9776	1.0429	1730.0	414.8	2674.4	1.2956	7.3675
100	1.0132	1.0437	1673.0	419.1	2676.0	1.3069	7.3554
101	1.0500	1.0445	1618.2	423.3	2677.6	1.3182	7.3434

The most direct calculation of C_{sat} uses the entropy data with the basic equation

$$C_{sat} = T\left(\frac{dS}{dT}\right)_{sat}$$

which we rewrite for saturated liquid and for saturated vapor as

$$C^l = T\frac{dS^l}{dT} \qquad C^v = T\frac{dS^v}{dT}$$

We assume as a reasonable approximation that $(dS/dT)_{sat}$ at 100 °C is equal to $(\Delta S/\Delta T)_{sat}$ applied to the interval between 99 and 101 °C. Thus for 100 °C,

$$\frac{dS^l}{dT} \approx \frac{1.3182 - 1.2956}{2.0} = 0.011\,30 \text{ kJ} \cdot \text{kg}^{-1} \cdot \text{K}^{-2}$$

giving

$$C^l = (373.15)(0.011\,30) = 4.217 \text{ kJ} \cdot \text{kg}^{-1} \cdot \text{K}^{-1}$$

and

$$\frac{dS^v}{dT} \approx \frac{7.3434 - 7.3675}{2.0} = -0.012\,05 \text{ kJ} \cdot \text{kg}^{-1} \cdot \text{K}^{-2}$$

giving

$$C^v = (373.15)(-0.012\,05) = -4.496 \text{ kJ} \cdot \text{kg}^{-1} \cdot \text{K}^{-1}$$

The latter result is surprising: the heat capacity of saturated-vapor water is *negative*; that is, as the temperature rises, heat must be *withdrawn*. The explanation is that both T and P change along the saturation curve, and both affect the entropy. An increase in temperature causes the entropy to increase, whereas the simultaneous pressure increase causes the entropy to decrease. When the latter effect overbalances the former, the heat capacity becomes negative.

By comparison, C_P for vapor at 100 °C and 1.0132 bar may be estimated from superheated-vapor enthalpies (Appendix E) via

$$C_P = \left(\frac{\partial H}{\partial T}\right)_P \approx \left(\frac{\Delta H}{\Delta T}\right)_P$$

The enthalpy change between the temperatures 100 °C and 125 °C at $P = 1.0132$ bar is 50.4 $kJ \cdot kg^{-1}$; thus

$$C_P \approx \frac{50.4}{25} = 2.02 \text{ kJ} \cdot \text{kg}^{-1} \cdot \text{K}^{-1}$$

The rate of enthalpy change along the saturation curve is quite different from this:

$$\frac{dH^v}{dT} \approx \frac{\Delta H^v}{\Delta T} = \frac{2677.6 - 2674.4}{2.0} = 1.60 \text{ kJ} \cdot \text{kg}^{-1} \cdot \text{K}^{-1}$$

PROPERTY VALUES FOR PVT SYSTEMS   (Section 4.8)

4.13   Property values for subcooled or compressed liquid are calculated from property values for saturated liquid at the same temperature by integration of the equations [see (3.53) and (3.55)]

$$dH = \left[V - T\left(\frac{\partial V}{\partial T}\right)_P \right] dP \quad \text{and} \quad dS = -\left(\frac{\partial V}{\partial T}\right)_P dP$$

Often data are not available for V and $(\partial V/\partial T)_P$ as functions of P in the liquid region, and the assumption is made that these values are constant at the saturation vaues. Integration then gives

$$H - H^l = \left(V^l - T\frac{dV^l}{dT} \right)(P - P^{\text{sat}}) \quad \text{(const. } T\text{)} \tag{1}$$

$$S - S^l = -\frac{dV^l}{dT}(P - P^{\text{sat}}) \quad \text{(const. } T\text{)} \tag{2}$$

(Since volumes are assumed independent of P, the total derivative is written.) Apply (1) and (2) to liquid water, using values of the saturation properties, and calculate H and S for water at $T = 100\,°C$ and (a) $P = 25$ bar, (b) $P = 200$ bar.

At 100 °C, we have from the data of Problem 4.12:

$$P^{\text{sat}} = 1.0132 \text{ bar} \qquad H^l = 419.1 \text{ kJ} \cdot \text{kg}^{-1}$$
$$S^l = 1.3069 \text{ kJ} \cdot \text{kg}^{-1} \cdot \text{K}^{-1} \qquad V^l = 1.0437 \text{ cm}^3 \cdot \text{g}^{-1}$$

$$\frac{dV^l}{dT} \approx \frac{V^l_{101} - V^l_{99}}{101 - 99} = \frac{1.0445 - 1.0429}{2} = 0.0008 \text{ cm}^3 \cdot \text{g}^{-1} \cdot \text{K}^{-1}$$

(a)   Direct substitution of values into (1) gives

$$H - H^l = \{[1.0437 - (373.15)(8 \times 10^{-4})]\text{ cm}^3 \cdot \text{g}^{-1}\}[(25 - 1.0132)\text{ bar}] = 17.9 \text{ bar} \cdot \text{cm}^3 \cdot \text{g}^{-1}$$
$$= 1.79 \text{ kJ} \cdot \text{kg}^{-1}$$

whence $H = H^l + 1.79 = 420.9 \text{ kJ} \cdot \text{kg}^{-1}$. Similarly, by (2),

$$S - S^l = (-8 \times 10^{-4} \text{ cm}^3 \cdot \text{g}^{-1} \cdot \text{K}^{-1})[(25 - 1.0132)\text{ bar}] = -0.019 \text{ bar} \cdot \text{cm}^3 \cdot \text{g}^{-1} \cdot \text{K}^{-1}$$
$$= -0.0019 \text{ kJ} \cdot \text{kg}^{-1} \cdot \text{K}^{-1}$$

whence $S = 1.3069 - 0.0019 = 1.3050 \text{ kJ} \cdot \text{kg}^{-1} \cdot \text{K}^{-1}$.

(b)   The same procedure for a pressure of 200 bar yields

$$H - H^l = 14.8 \text{ kJ} \cdot \text{kg}^{-1} \quad \text{or} \quad H = 433.9 \text{ kJ} \cdot \text{kg}^{-1}$$
$$S - S^l = -0.0159 \text{ kJ} \cdot \text{kg}^{-1} \cdot \text{K}^{-1} \quad \text{or} \quad S = 1.2910 \text{ kJ} \cdot \text{kg}^{-1} \cdot \text{K}^{-1}$$

The results of (a) and (b) are in substantial agreement with experiment. However, the procedure would be entirely inappropriate in the critical region, where liquid properties change rapidly with pressure.

4.14   Very pure liquid water can be subcooled at atmospheric pressure to temperatures well below 0 °C. Assume that a mass of water has been cooled as a liquid to −5 °C. A small crystal of ice (whose mass is negligible) is added to "seed" the subcooled liquid. If the subsequent change of state occurs adiabatically and at constant (atmospheric) pressure, (a) what fraction of the system solidifies?   (b) what is the entropy change of the system?

(a)   As data:

latent heat of fusion of water at 0 °C = 333.4 kJ · kg^{-1}

heat capacity of water between 0 and −5 °C = 4.22 kJ · kg^{-1} · °C^{-1}

The final state of the system is presumed to be the equilibrium state of a mixture of ice and water at atmospheric pressure. This state exists at 0 °C, or 273.15 K. The change of state occurs in a closed system at constant pressure, and for such a process (see Example 1.7) $Q = \Delta H^t = m \, \Delta H$. However, this process is adiabatic, and therefore $\Delta H = 0$.

Since ΔH is a property change, its value is independent of path, and therefore for purposes of calculation we may consider an arbitrary path made up of the two following steps. (1) The subcooled liquid is warmed as a liquid from -5 °C to 0 °C. (2) The heat added during step (1) is now withdrawn at 0 °C, causing part of the liquid to freeze. The amounts of heat for the two steps just compensate, so that overall $\Delta H = 0$.

For step (1), $\Delta H_1 = C_P \, \Delta T = (4.22)(5) = 21.1 \text{ kJ} \cdot \text{kg}^{-1}$; for step (2), $\Delta H_2 = -z \, \Delta H_{\text{fusion}} = -333.4z \text{ kJ} \cdot \text{kg}^{-1}$, where z is the fraction of the system that freezes. Then

$$-333.4z + 21.1 = 0 \qquad \text{or} \qquad z = 0.0633$$

Thus 6.33% of the system solidifies.

(b) The process as it occurs is obviously irreversible; thus the entropy change of the system should be positive. (Since the process is adiabatic, no entropy change occurs in the surroundings.) The entropy is a property, and its change may be calculated along the same path as the enthalpy. For step (1),

$$\Delta S_1 = C_P \ln \frac{T_2}{T_1} = 4.22 \ln \frac{273.15}{268.15} = 0.077 \, 96 \text{ kJ} \cdot \text{kg}^{-1} \cdot \text{K}^{-1}$$

For step (2), the temperature is constant, so that

$$\Delta S_2 = \frac{Q_2}{T_2} = \frac{-21.1}{273.15} = -0.077 \, 25 \text{ kJ} \cdot \text{kg}^{-1} \cdot \text{K}^{-1}$$

Hence $\Delta S_{\text{total}} = +0.000 \, 71 \text{ kJ} \cdot \text{kg}^{-1} \cdot \text{K}^{-1}$.

4.15 If 50 mol of CO_2, initially at $T_0 = 300$ °C and atmospheric pressure, is heated so as to increase its enthalpy by 1500 kJ, by how much does its entropy increase?

We first find T_1, the final temperature. Assuming that CO_2 is an ideal gas at the given conditions, we have from (4.17) that

$$\Delta H^t = n \left\langle C_P^{\text{ig}} \right\rangle_T (T_1 - T_0) \qquad \text{or} \qquad T_1 = \frac{\Delta H^t / nR}{\left\langle C_P^{\text{ig}} \right\rangle_T / R} + T_0$$

Here

$$\frac{\Delta H^t}{nR} = \frac{1500}{(50)(0.008 \, 314)} = 3608.37 \text{ K} \qquad \text{and} \qquad T_0 = 573.15 \text{ K}$$

Then

$$T_1 = \frac{3608.37}{\left\langle C_P^{\text{ig}} \right\rangle_T / R} + 573.15 \qquad \text{K} \tag{1}$$

Evaluation of $\left\langle C_P^{\text{ig}} \right\rangle_T / R$ is by (4.16), with constants for CO_2 from Table 4-1 and with an *assumed* value of T_1. Equation (1) then provides a new value of T_1, and $\left\langle C_P^{\text{ig}} \right\rangle_T / R$ is reevaluated by (4.16). This iterative process continues until successive values of T_1 differ by no more than an insignificant amount. The result here is $T_1 = 1156.45$ K, corresponding to $\left\langle C_P^{\text{ig}} \right\rangle_T / R = 6.1862$.

By (4.21), with $P_1 = P_0$,

$$\Delta S^t = n \left\langle C_P^{\text{ig}} \right\rangle_{\ln T} \ln \frac{T_1}{T_0}$$

where $\left\langle C_P^{\text{ig}} \right\rangle_{\ln T} / R$ is given by (4.20), with constants for CO_2 again from Table 4-1. For the given value of T_0 and the calculated value of T_1, the result is

$$\left\langle C_P^{\text{ig}} \right\rangle_{\ln T} / R = 6.1361 \qquad \text{or} \qquad \left\langle C_P^{\text{ig}} \right\rangle_{\ln T} = 51.0155 \text{ J} \cdot \text{mol}^{-1} \cdot \text{K}^{-1}$$

Thus

$$\Delta S^t = (50)(51.0155) \ln \frac{1156.45}{573.15} = 1790.5 \text{ J} \cdot \text{K}^{-1}$$

4.16   Sketch graphs of V, S, G, and C_P, versus T, for a pure substance along a constant-pressure path that includes a phase change from liquid to vapor.

The volume of a liquid normally increases slowly with temperature. During vaporization at constant T and P, the volume increases enormously (except in the critical region). The vapor then expands fairly rapidly with further temperature increase. Thus a plot of V versus T usually has the general characteristics of Fig. 4-13.

The change of entropy with temperature for a homogeneous system is given by (3.50) as

$$\left(\frac{\partial S}{\partial T}\right)_P = \frac{C_P}{T}$$

The heat capacity C_P is always positive, both for liquids and vapors. Therefore a plot of S versus T has a positive slope in both regions. During vaporization the entropy increases by an amount equal to the latent heat divided by the absolute temperature. Thus an S-versus-T plot looks generally like Fig. 4-14. The slope of the vapor line is indicated to be less than that of the liquid line, because the heat capacity of a vapor is generally less than that of the liquid. Since C_P normally increases as the temperature rises, C_P/T may either increase or decrease, and therefore no curvature is shown for these lines.

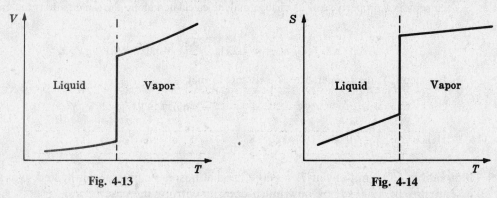

Fig. 4-13   Fig. 4-14

The Gibbs energy is related to T and P by (3.35), which for 1 mole of a pure material becomes $dG = -S\,dT + V\,dP$. At constant T and P, G must also be constant; thus we arrive again at the conclusion reached in Problem 3.11, that no change in the Gibbs energy occurs during phase change at constant T and P. For the single phases we have from the above equation

$$\left(\frac{\partial G}{\partial T}\right)_P = -S$$

From this and Fig. 4-14 we construct Fig. 4-15 for G versus T.

As already mentioned, C_P is positive and generally increases with T. Moreover, the value for saturated vapor is generally less than for saturated liquid, as shown in Fig. 4-16, which also indicates that C_P becomes infinite during the vaporization process. This follows from the defining equation

$$\left(\frac{\delta Q}{dT}\right)_P = C_P$$

and the fact that the temperature change is zero during the vaporization process.

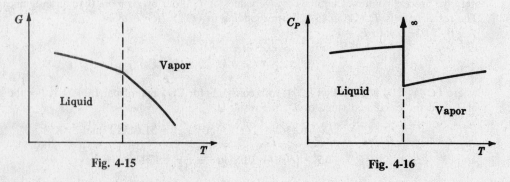

Fig. 4-15   Fig. 4-16

4.17 Develop three general expressions for the average molar properties of a pure material at its triple point (where the three phases coexist) in terms of the molar property of one phase and the property changes of fusion and vaporization.

Let z, y, and x respectively denote the molar fractions of material in the solid, liquid, and vapor phases. Then an average molar property M of the system is given by

$$M = zM^s + yM^l + xM^v \qquad (z + y + x = 1) \tag{1}$$

Substitution of $z = 1 - x - y$ into (1) gives

$$M = M^s + y(M^l - M^s) + x(M^v - M^s) = M^s + y\,\Delta M^{sl} + x\,\Delta M^{sv}$$

However, $\Delta M^{sv} \equiv \Delta M^{sl} + \Delta M^{lv}$ (an algebraic identity). Making the substitution, we have

$$M = M^s + (x + y)\,\Delta M^{sl} + x\,\Delta M^{lv} \tag{2}$$

In words, (2) says that any three-phase, pure-species system has the property of the solid phase, increased by the property change of fusion for the fraction of the system that is not solid, plus the property change of vaporization for the fraction of the system that is vapor.

Similar substitutions of $y = 1 - x - z$ and $x = 1 - z - y$ into (1) yield

$$M = M^l + x\,\Delta M^{lv} - z\,\Delta M^{sl} \tag{3}$$

$$M = M^v - (z + y)\,\Delta M^{lv} - z\,\Delta M^{sl} \tag{4}$$

4.18 Linear interpolation in tables of property values is not always accurate. The following table lists data for saturated-vapor steam at both 10 and 30 kPa:

P/kPa	$t^{\mathrm{sat}}/{}^\circ\mathrm{C}$	$V/\mathrm{cm^3 \cdot g^{-1}}$	$H/\mathrm{kJ \cdot kg^{-1}}$	$S/\mathrm{kJ \cdot kg^{-1} \cdot K^{-1}}$
10	45.83	14 670	2584.8	8.1511
30	69.12	5229.3	2625.4	7.7695

Devise interpolation methods whereby accurate values for t^{sat} and for V, H, and S for saturated vapor at 20 kPa may be obtained.

Since V, H, and S all depend strongly on temperature, first find the saturation temperature corresponding to the vapor pressure 20 kPa. An excellent interpolation formula is based on (4.5). With the two states for which property values are given labeled by subscripts 1 and 2:

$$\ln P_1 = A - \frac{B}{T_1} \tag{1}$$

$$\ln P_2 = A - \frac{B}{T_2} \tag{2}$$

and for a third (intermediate) state:

$$\ln P = A - \frac{B}{T} \tag{3}$$

Subtracting (1) from (2), and (1) from (3), and then dividing the second result by the first yields

$$\frac{\ln P - \ln P_1}{\ln P_2 - \ln P_1} = \frac{(1/T) - (1/T_1)}{(1/T_2) - (1/T_1)}$$

Solved for $1/T$, this equation becomes

$$\frac{1}{T} = \frac{1}{T_1} + \left(\frac{\ln P - \ln P_1}{\ln P_2 - \ln P_1}\right)\left(\frac{1}{T_2} - \frac{1}{T_1}\right)$$

In this form of (4.5), the three pressures may be expressed in an arbitrary unit. From the data,

$$T_1 = 45.83 + 273.15 = 318.98 \text{ K} \qquad \ln P_1 = 2.3026$$
$$T_2 = 69.12 + 273.15 = 342.27 \text{ K} \qquad \ln P_2 = 3.4012$$

and $\ln P = 2.9957$. Then

$$\frac{1}{T} = \frac{1}{318.98} + \left(\frac{2.9957 - 2.3026}{3.4012 - 2.3026}\right)\left(\frac{1}{342.27} - \frac{1}{318.98}\right) = 0.003\ 000\ 4\ \text{K}^{-1}$$

whence $T = 333.29$ K and $t = 60.14\ ^{\circ}\text{C}$.

Since the pressure is low, interpolation for V is based on the ideal-gas equation, written for each of the three states:

$$V_1 = \frac{RT_1}{P_1} \qquad V_2 = \frac{RT_2}{P_2} \qquad V = \frac{RT}{P}$$

Then

$$\frac{V - V_1}{V_2 - V_1} = \frac{(T/P) - (T_1/P_1)}{(T_2/P_2) - (T_1/P_1)}$$

which yields

$$V = V_1 + \frac{(T/P) - (T_1/P_1)}{(T_2/P_2) - (T_1/P_1)}\ (V_2 - V_1)$$

$$= 14\ 670 + \frac{(333.29/20) - (318.98/10)}{(342.27/30) - (318.98/10)}\ (5229.3 - 14\ 670) = 7650.9\ \text{cm}^3 \cdot \text{g}^{-1}$$

For an ideal gas, enthalpy is independent of P, and over a small temperature range linear interpolation with temperature provides accurate results:

$$H = H_1 + \left(\frac{t - t_1}{t_2 - t_1}\right)(H_2 - H_1)$$

$$= 2584.8 + \left(\frac{60.14 - 45.83}{69.12 - 45.83}\right)(2625.4 - 2584.8) = 2609.7\ \text{kJ} \cdot \text{kg}^{-1}$$

For an ideal gas with constant (specific) heat capacity, (2.9) integrates to give

$$S = C_P \ln T \ - \ (R/M) \ln P + I$$

where I is the constant of integration and M is molar mass. Application to this equation to states 1 and 2 gives

$$S_1 = C_P \ln T_1 \ - \ (R/M) \ln P_1 + I$$
$$S_2 = C_P \ln T_2 \ - \ (R/M) \ln P_2 + I$$

Elimination of C_P and I from these equations leads to

$$S = S_1 \ - \ (R/M) \ln \frac{P}{P_1} \ + \ \left(S_2 \ - \ S_1 + (R/M) \ln \frac{P_2}{P_1}\right) \frac{\ln T \ - \ \ln T_1}{\ln T_2 \ - \ \ln T_1}$$

With $R/M = (8.314/18.016)\ \text{kJ} \cdot \text{kg}^{-1} \cdot \text{K}^{-1}$ and known values, one obtains $S = 7.9093\ \text{kJ} \cdot \text{kg}^{-1} \cdot \text{K}^{-1}$.

A comparison of these interpolated results with values for saturated vapor at 20 kPa from Appendix E is presented in the following table. The close agreement of values is in contrast to the results obtained by simple linear interpolation on P, given in the last column of the table.

	Interpolated Values	Steam-Table Values	Arithmetic Averages
$t/^{\circ}\text{C}$	60.14	60.09	57.48
$V/\text{cm}^3 \cdot \text{g}^{-1}$	7650.9	7649.8	9949.7
$H/\text{kJ} \cdot \text{kg}^{-1}$	2609.7	2609.9	2605.1
$S/\text{kJ} \cdot \text{kg}^{-1} \cdot \text{K}^{-1}$	7.9093	7.9094	7.9603

It should be noted that direct application of the ideal-gas equation for the calculation of V yields

$$V = \frac{(R/M)T}{P} = \frac{(8314/18.016)(333.29)}{20} = 7609.3\ \text{cm}^3 \cdot \text{g}^{-1}$$

This result is significantly less accurate than the interpolated value.

4.19 Superheated steam at 32 bar and 325 °C expands to 7 bar. Determine the final state (2) of the steam if the expansion is (*a*) at constant enthalpy, (*b*) at constant entropy. Property values in the initial state (1) are, from Appendix E, $H_1 = 3052.5$ kJ·kg^{-1} and $S_1 = 6.6127$ kJ·kg^{-1}·K^{-1}.

(*a*) We find from the steam tables that superheated vapor at 7 bar has an enthalpy equal to H_1 at a temperature between 280 and 300 °C. Linear interpolation in the tables gives the value $T_2 = 296.5$ °C, at which $S_2 = 7.2867$ kJ·kg^{-1}·K^{-1}.

(*b*) We find from the steam tables that no stable vapor state at 7 bar has an entropy as low as S_1; rather, S_1 lies between the entropy of saturated liquid and saturated vapor at 7 bar. The final state must therefore be a mixture of these, or "wet" steam, of which we can determine the quality from

$$S_2 = S_1 = S_2^l + x_2 \Delta S_2^{lv} \qquad \text{or} \qquad x_2 = \frac{S_1 - S_2^l}{\Delta S_2^{lv}}$$

At 7 bar: $S_2^l = 1.9918$ and $\Delta S_2^{lv} = 4.7134$ kJ·kg^{-1}·K^{-1}; thus

$$x_2 = \frac{6.6127 - 1.9918}{4.7134} = 0.9804$$

and $H_2 = H_2^l + x_2 \Delta H_2^{lv} = 697.1 + (0.9804)(2064.9) = 2721.5$ kJ·kg^{-1}.

Note that in (*a*) the entropy increased at constant enthalpy and in (*b*) the enthalpy decreased at constant entropy. Both of these processes are easily followed on a Mollier (*H*-versus-*S*) diagram.

4.20 Five kilograms of steam, initially at pressure $P_1 = 1.5$ bar and temperature $T_1 = 150$ °C, is compressed isothermally and reversibly in a piston/cylinder device to a final pressure P_2 such that the steam is just saturated (i.e., saturated vapor). Calculate Q and W for the process.

In its initial state the steam is a superheated vapor, whereas it is a saturated vapor at the end of the process. From the appropriate steam tables we have the following values:

$$T_1 = 150 \text{ °C} \qquad\qquad T_2 = 150 \text{ °C}$$
$$P_1 = 1.5 \text{ bar} \qquad\qquad P_2 = 4.76 \text{ bar}$$
$$V_1 = 1285.2 \text{ cm}^3\cdot\text{g}^{-1} \qquad V_2 = 392.4 \text{ cm}^3\cdot\text{g}^{-1}$$
$$U_1 = 2579.7 \text{ kJ}\cdot\text{kg}^{-1} \qquad U_2 = 2558.6 \text{ kJ}\cdot\text{kg}^{-1}$$
$$S_1 = 7.4194 \text{ kJ}\cdot\text{kg}^{-1} \qquad S_2 = 6.8358 \text{ kJ}\cdot\text{kg}^{-1}$$

The value of Q is obtained from the entropy change of the system. Since the process is reversible and T is constant, we have $\Delta S^t = Q/T$, or $Q = Tm\,\Delta S$. Since $T = 150 + 273.15 = 423.15$ K,

$$Q = (423.15)(5)(6.8358 - 7.4194) = -1234.8 \text{ kJ}$$

Now the first law gives

$$W = Q - m\,\Delta U = -1234.8 - (5)(2558.6 - 2579.7) = -1129.3 \text{ kJ}$$

4.21 Select data from the steam tables to confirm that (*3.79*) holds for a phase transition from vapor to liquid.

For a finite change of state at constant T and P, (*3.79*) may be integrated to give $(\Delta G^t)_{T,P} \leqq 0$. The inequality applies to an irreversible process, whereas the equality applies to a reversible process, i.e., one where equilibrium conditions always obtain. Consider now 1 kg of water undergoing a phase change from saturated vapor to saturated liquid. At saturation the liquid and vapor phases are always at equilibrium and our equation becomes $\Delta G^{lv} = 0$.

Let us test the following data from the steam tables (Appendix D) at 2500 kPa and 223.94 °C or 497.09 K:

$$H^l = 962.0 \text{ kJ}\cdot\text{kg}^{-1} \qquad\qquad H^v = 2800.9 \text{ kJ}\cdot\text{kg}^{-1}$$
$$S^l = 2.5543 \text{ kJ}\cdot\text{kg}^{-1}\cdot\text{K}^{-1} \qquad S^v = 6.2536 \text{ kJ}\cdot\text{kg}^{-1}\cdot\text{K}^{-1}$$

Since $G = H - TS$,

$$G^l = 962.0 - (497.09)(2.5543) = -307.7 \text{ kJ} \cdot \text{kg}^{-1}$$
$$G^v = 2800.9 - (497.09)(6.2536) = -307.7 \text{ kJ} \cdot \text{kg}^{-1}$$

Thus ΔG^{lv} is zero. Now let us try another set of data, one for which G^l and G^v should not be identical. Steam-table data may be extrapolated into the subcooled-vapor region, where vapor is unstable at a temperature below its saturation temperature. For example, data are given below for subcooled vapor at 4 °C below the temperature at which vapor would normally condense. The stable state at these conditions is liquid water, data for which are also available. Thus, at 25 bar and 220 °C or 493.15 K,

$$H^{vap} = 2788.5 \text{ kJ} \cdot \text{kg}^{-1} \qquad S^{vap} = 6.2290 \text{ kJ} \cdot \text{kg}^{-1} \cdot \text{K}^{-1}$$
$$H^{liq} = 943.7 \text{ kJ} \cdot \text{kg}^{-1} \qquad S^{liq} = 2.5178 \text{ kJ} \cdot \text{kg}^{-1} \cdot \text{K}^{-1}$$

From these we calculate $G = H - TS$:

$$G^{vap} = 2788.5 - (493.15)(6.2290) = -283.3 \text{ kJ} \cdot \text{kg}^{-1}$$
$$G^{liq} = 943.7 - (493.15)(2.5178) = -298.0 \text{ kJ} \cdot \text{kg}^{-1}$$

These two states are at the same T and P, and the change that occurs from one to the other at this T and P is from the unstable vapor to the stable liquid. For this case,

$$(\Delta G')_{T,P} = G^{liq} - G^{vap} = -14.7 \text{ kJ} \cdot \text{kg}^{-1}$$

which conforms with the requirement that $(\Delta G')_{T,P} < 0$ for irreversible processes. Any change from an unstable state to a stable one is of course inherently irreversible.

4.22 Show how the saturation or vapor pressure of a spherical droplet of water at 25 °C depends on the radius of the droplet. At 25 °C the surface tension of water is $\gamma = 0.0694 \text{ N} \cdot \text{m}^{-1}$.

A free-body diagram (Fig. 4-17) showing the forces on a droplet hemisphere of radius r provides the basis for a force balance.

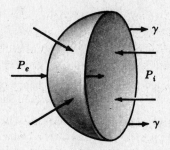

Fig. 4-17

Force acting to the left as a result of the internal pressure P_i: $P_i(\pi r^2)$

Force acting to the right as a result of the external pressure P_e: $P_e(\pi r^2)$

Force acting to the right as a result of the surface tension at the surface of the droplet: $\gamma(2\pi r)$

Thus $\qquad P_i(\pi r^2) = P_e(\pi r^2) + \gamma(2\pi r) \qquad$ or $\qquad P_i - P_e = 2\gamma/r$

From this, we calculate:

$r/\mu\text{m}$	$(P_i - P_e)/\text{kPa}$
10 000	0.014
100	1.39
1	138.8

The properties of the liquid in a droplet depend, of course, on the *internal* pressure, which we have just seen to be always higher than the external pressure. We have already developed equations which allow us to determine the influence of this pressure increase on the thermodynamic properties of the droplet. In particular, we have from (*3.35*) at constant T that

$$dG = V \, dP \qquad \text{(constant } T\text{)} \tag{1}$$

Consider a pool of liquid water with a flat surface at 25 °C in equilibrium with water vapor at its saturation pressure, $P_0^{\text{sat}} = 3.166$ kPa. If a small amount of liquid is taken from this pool and formed into a droplet of 1 μm radius at 25 °C, there would be a pressure increase $P_i - P_e = P_i - P_0^{\text{sat}}$ of 138.8 kPa, as determined above. This causes a change in the Gibbs energy, which we determine by integration of (*1*), with V taken to be constant for liquid water at 1.003 cm$^3 \cdot$ g^{-1} for $T = 25$ °C. Thus,

$$\Delta G = V(P_i - P_0^{\text{sat}}) = (1.003 \times 10^{-6})(138.8 \times 10^3) = 0.139 \text{ kJ} \cdot \text{kg}^{-1}$$

If the droplet is to remain in equilibrium with water vapor, then the Gibbs energy for the water vapor must also increase by exactly this amount, in accord with (*4.1*). Integration of (*1*) from P_0^{sat} to the new vapor pressure P^{sat}, with V for the vapor given by the ideal-gas equation, gives

$$\Delta G = \frac{R}{M} T \ln \frac{P^{\text{sat}}}{P_0^{\text{sat}}}$$

Noting that $\Delta P^{\text{sat}} \equiv P^{\text{sat}} - P_0^{\text{sat}}$ is very small, we have

$$\Delta G = \frac{R}{M} T \ln \left(1 + \frac{\Delta P^{\text{sat}}}{P_0^{\text{sat}}}\right) \approx \frac{R}{M} T \frac{\Delta P^{\text{sat}}}{P_0^{\text{sat}}}$$

Substitution of numerical values, noting that the molar mass of H_2O is 18.016 kg $\cdot$ kmol^{-1}, gives

$$0.139 = \frac{8.314}{18.016} (298.15) \frac{\Delta P^{\text{sat}}}{3.166}$$

from which $\Delta P^{\text{sat}} = 0.0032$ kPa $= 3.2$ Pa. This increase of vapor pressure with total pressure is known as the *Poynting effect*.

4.23   In accordance with the Poynting effect (Problem 4.22), the saturation pressure of water vapor in equilibrium with liquid water droplets of radius r becomes infinite as $r \to 0$. (*a*) How then do raindrops manage to form from vapor at finite pressures? (*b*) How can a fog (of higher saturation pressure) remain in equilibrium with ground water?

(*a*)   Raindrops form only on dust particles, which provide a finite radius from which the droplet can grow.

(*b*)   Vapor pressure is a strong function of temperature, and a slight reduction of temperature reduces the vapor pressure of a fog to that of the ground water. Assuming a starting temperature of 25 °C, we can calculate the required temperature decrease from (*4.4*), the Clausius/Clapeyron equation, written on a mass basis as

$$\frac{d \ln P^{\text{sat}}}{dT} = \frac{\Delta H^{lv}}{(R/M)T^2}$$

or

$$\Delta T \approx \frac{(R/M)T^2 \, \Delta P^{\text{sat}}}{(\Delta H^{lv})P_0^{\text{sat}}}$$

Here, $\Delta P^{\text{sat}} = -3.2$ Pa, the negative of the increase calculated in Problem 4.22, because we seek the temperature change that *compensates* the Poynting effect. Taking the value of ΔH^{lv} from the steam tables at 25 °C, we have

$$\Delta T = \frac{(8.314/18.016)(298.15)^2(-3.2)}{(2442.5)(3166)} = -0.017 \text{ K}$$

or a temperature decrease of 0.017 °C. We note that the vapor pressure is also lowered by impurities dissolved in the liquid droplets, and fogs stabilized in this way become *smogs*.

Supplementary Problems

VAPOR PRESSURES AND LATENT HEATS (Section 4.3)

4.24   The vapor pressure of a certain pure liquid is given by

$$\ln (P^{sat}/\text{Pa}) = 20.9042 - \frac{3456.8}{(T/\text{K} - 78.67)}$$

Under the assumptions of Example 4.4, calculate a value for ΔH^{lv} at 298 K. Are these assumptions justified in this case?
Ans. $\Delta H^{lv} = 53.030 \text{ kJ} \cdot \text{mol}^{-1}$. At 298 K, $P^{sat} = 171$ Pa, a pressure so low that the assumptions made should introduce negligible error.

TWO-PHASE SYSTEMS (Section 4.4)

4.25   A rigid tank contains 0.01 m³ of saturated-liquid water and 0.99 m³ of saturated-vapor water at 1 bar. How much heat must be added to the system so that the liquid is just vaporized?
Ans. $Q = 20.967$ MJ

4.26   One kilogram of water, 20% vapor by weight, is contained in a rigid vessel at a pressure of 2 bar. It is heated until the liquid is just vaporized. Determine the final temperature and pressure; the property changes $\Delta U'$, $\Delta H'$, and $\Delta S'$; and the heat added to the system.
Ans. $\Delta U' = 1675$ kJ $\Delta H' = 1835$ kJ $\Delta S' = 3.901 \text{ kJ} \cdot \text{K}^{-1}$ $Q = 1675$ kJ

4.27   (a) If air at 50 °C and 1 bar, flowing at the rate 20 m³·s⁻¹, is heated at constant pressure to 500 °C, what rate of heat transfer is required? (b) Heat in the amount of 1 MJ is added to 25 mol of CO_2, initially at 200 °C, in a process occurring at approximately atmospheric pressure. What is the final temperature?   *Ans.*   (a) 10.204 MW   (b) 936.31 °C

4.28   Making use of the Clapeyron equation and of (6) of Problem 4.11 as applied to both saturated liquid and to saturated vapor, show that

$$C^v - C^l = \frac{d(\Delta H^{lv})}{dT} - \frac{\Delta H^{lv}}{T}$$

PROPERTY VALUES FOR PVT SYSTEMS (Section 4.8)

4.29   A mass of saturated-liquid water at a pressure of 1 bar fills a container. The saturation temperature is 99.63 °C. Heat is added to the water until its temperature reaches 120 °C. If the volume of the container does not change, what is the final pressure? *Data:*   The average value of β between 100 and 120 °C is $80.8 \times 10^{-5} \text{ K}^{-1}$. The value of κ at 1 bar and 120 °C is $4.93 \times 10^{-5} \text{ bar}^{-1}$ and may be assumed independent of P. The specific volume of saturated-liquid water at 1 bar is $1.0434 \text{ cm}^3 \cdot \text{g}^{-1}$.
Ans.   335 bar

4.30   (a) Show that the residual properties H^R and S^R can be estimated from tabulated specific enthalpy and entropy data by the formulas

$$H^R(T, P) \approx H(T, P) - H(T, P^*)$$

$$S^R(T, P) \approx S(T, P) - S(T, P^*) + \frac{R}{M} \ln \frac{P}{P^*}$$

Here P^* is a suitably small but nonzero pressure; e.g., the lowest pressure for which properties of superheated vapor are reported at temperature T.

(b) Use steam-table data to estimate V^R, H^R, and S^R for saturated-water vapor at 200 °C.
Ans.   (b) $V^R = -13.2 \text{ cm}^3 \cdot \text{g}^{-1}$ $H^R = -89.2 \text{ kJ} \cdot \text{kg}^{-1}$ $S^R = -0.1486 \text{ kJ} \cdot \text{kg}^{-1} \cdot \text{K}^{-1}$

4.31 Data for liquid mercury at 0 °C and 100 kPa are:
$$V = 14.72 \text{ cm}^3 \cdot \text{mol}^{-1} \qquad \beta = 181 \times 10^{-6} \text{ K}^{-1} \qquad \kappa = 3.88 \times 10^{-6} \text{ bar}^{-1}$$
Taking these values to be essentially independent of P, calculate:

(a) The pressure increase above 100 kPa required to cause a 0.1% decrease in V. (For a 0.1% change, V may be considered "essentially" constant.)

(b) ΔU, ΔH, and ΔS for 1 mol of mercury for the change described in (a).

(c) Q and W, if the change is carried out reversibly and isothermally.

Ans. (a) 257 bar   (b) $\Delta U = -18.5$ J, $\Delta H = 359.5$ J, $\Delta S = -0.068\,46$ J $\cdot$ K^{-1}
(c) $Q = -18.7$ J, $W = -0.2$ J

4.32 For a single-phase PVT system consisting of one mole (or a unit mass) of a pure material, (3.48) gives $dH = T\,dS + V\,dP$. Starting with this equation, show that

(a) $(\partial H/\partial P)_S = V$

(b) $(\partial H/\partial T)_S = V(\partial P/\partial T)_S = C_P/\beta T$

(c) $(\partial H/\partial V)_S = V(\partial P/\partial V)_S = -\gamma/\kappa$

(d) $\left(\dfrac{\partial H}{\partial S}\right)_P = T$

(e) $\left(\dfrac{\partial H}{\partial S}\right)_T = T - V\left(\dfrac{\partial T}{\partial V}\right)_P = T - \dfrac{1}{\beta}$

(f) $\left(\dfrac{\partial H}{\partial S}\right)_V = T - V\left(\dfrac{\partial T}{\partial V}\right)_S = T + \dfrac{\gamma - 1}{\beta}$

(*Hint*: See Example 3.10 and Problem 4.34.)

4.33 For a PVT system consisting of one mole (or a unit mass) of a pure material, show that

(a) $\left(\dfrac{\partial^2 G}{\partial T^2}\right)_P = \dfrac{-C_P}{T}$

(b) $\left(\dfrac{\partial^2 G}{\partial P^2}\right)_T = -\kappa V$

(c) $\dfrac{\partial^2 G}{\partial P\,\partial T} = \beta V$

Show also that all three quantities become infinite during a phase transition.

4.34 For a single-phase PVT system consisting of one mole (or a unit mass) of a pure material, show that

(a) $\left(\dfrac{\partial V}{\partial T}\right)_S = -\dfrac{\kappa C_V}{\beta T}$

(b) $\left(\dfrac{\partial P}{\partial T}\right)_S = \dfrac{C_P}{V\beta T}$

(c) $\left(\dfrac{\partial V}{\partial P}\right)_S = -\dfrac{\kappa V}{\gamma}$

Show that for an ideal gas these equations lead to the results of Example 2.5.

4.35 Verify, for a single-phase PVT system consisting of one mole (or a unit mass) of a pure material, the derivatives *at constant* T, $(\partial y/\partial x)_T$, indicated in Table 4-2. In addition, show that all derivatives reduce to their proper values for an ideal gas.

Table 4-2

$\dfrac{\partial y}{\downarrow} \Big/ \partial x \rightarrow$	∂P	∂V
∂P	—	(3.18)
∂V	(3.18)	—
∂U	$-V(T\beta - P\kappa)$	$\dfrac{T\beta}{\kappa} - P$
∂H	$V(1 - \beta T)$	$\dfrac{1}{\kappa}(T\beta - 1)$
∂S	$-\beta V$	$\dfrac{\beta}{\kappa}$
∂C_V	—	$\dfrac{T}{\kappa}\left[\left(\dfrac{\partial \beta}{\partial T}\right)_V - \dfrac{\beta}{\kappa}\left(\dfrac{\partial \kappa}{\partial T}\right)_V\right]$
∂C_P	(3.67)	—

4.36   Subcooled liquid water at 6 bar and 20 °C expands to 1 atm (101.325 kPa). (a) If the expansion is isenthalpic, estimate ΔT and ΔS. Which contributes more to ΔS—the temperature change or the pressure change? (b) If the expansion is isentropic, estimate ΔT and ΔH. Which contributes more to ΔH—the temperature change or the pressure change? *Data*:   For liquid water at 20 °C and 1 atm, $V = 1.002 \text{ cm}^3 \cdot \text{g}^{-1}$, $C_P = 4.182 \text{ kJ} \cdot \text{kg}^{-1} \cdot \text{K}^{-1}$, and $\beta = 2.07 \times 10^{-4} \text{ K}^{-1}$.
   Ans.   (a) $\Delta T = +0.11 \text{ K}$, $\Delta S = 1.67 \times 10^{-3} \text{ kJ} \cdot \text{kg}^{-1} \cdot \text{K}^{-1}$,   contribution of ΔT is greater.
            (b) $\Delta T = -0.007 \text{ K}$, $\Delta H = -0.498 \text{ kJ} \cdot \text{kg}^{-1}$,   contribution of ΔP is greater.

USE OF THE STEAM TABLES

4.37   A closed rigid vessel of volume 0.5 m^3 is filled with steam at 7 bar and 325 °C. Heat is transferred from the steam until its temperature reaches 200 °C. Determine Q.   *Ans.*   −256.4 kJ

4.38   Steam at 150 °C and 1 bar is compressed isothermally in a reversible process until it reaches a final state of saturated liquid. Determine Q and W for 1 kg of steam.
   Ans. $Q = -2442.5 \text{ kJ}$ $W = -491.4 \text{ kJ}$

4.39   One kilogram of saturated-vapor steam at 200 °C undergoes a reversible, adiabatic expansion in a piston/cylinder device; the final pressure is 1 atm (101.325 kPa). Determine (a) the final state of the steam, (b) the work done.
   Ans.   (a) wet steam ($x = 0.8466$) at 1 atm and 100 °C   (b) +406.9 kJ

4.40   Water at 200 °C undergoes an isenthalpic expansion to 1.5 bar. Determine the final state, if the initial state is (a) saturated liquid, (b) steam of 50% quality, (c) saturated vapor.
   Ans.   (a) wet steam ($x = 0.1731$) at 111.37 °C and 1.5 bar   (b) wet steam ($x = 0.6085$) at 111.37 °C and 1.5 bar   (c) superheated steam at 159.1 °C and 1.5 bar.

4.41   A system consists of liquid and vapor water in mutual equilibrium. Thermodynamic analysis produces a numerical value of specific property M. Show that the value of some other specific property N is given by

$$N = N^l + \frac{\Delta N^{lv}}{\Delta M^{lv}} (M - M^l)$$

4.42   A rigid, insulated cylinder is divided in half by a rigid partition. Initially, the left half of the cylinder contains saturated-vapor steam at 200 °C and the right half is evacuated. The partition is ruptured, and the steam expands to fill the entire volume. Estimate the final condition of the steam.
   Ans.   Superheated steam at approximately 179.3 °C and 7.77 bar. (This result is based on double linear interpolation in the steam tables.)

4.43   (a) Estimate the state of minimum temperature and pressure from which an isenthalpic expansion of saturated steam can start, if the expansion path is to enter the two-phase liquid/vapor region. (b) An isenthalpic expansion of saturated-water vapor produces saturated-water vapor at 1 atm (101.325 kPa). Estimate the temperature and pressure of the saturated water vapor in its initial state.
   Ans.   (a) 236 °C, 31.2 bar   (b) 328.4 °C, 126 bar

Review Questions for Chapters 1 through 4

For each of the following statements indicate whether it is true or false.

_____1. For a closed gaseous system, the value of $\int P\,dV$ for the change of the gas from one given state to another is independent of the path so long as all processes are reversible.

_____2. All ideal gases have the same molar heat capacity at constant pressure (C_P).

_____3. The molar heat capacity at constant volume (C_V) of an ideal gas is independent of temperature.

_____4. The molar heat capacity at constant pressure (C_P) of an ideal gas is independent of pressure.

_____5. The enthalpy of an ideal gas is a function of temperature only.

_____6. The entropy of an ideal gas is a function of temperature only.

_____7. Work is *always* given by the integral $\int P\,dV$.

_____8. The first law of thermodynamics requires that the total energy of any system be conserved within the system.

_____9. For any gas at constant temperature, the product PV approaches zero as the pressure approaches zero.

_____10. The energy of an isolated system must be constant.

_____11. The entropy of an isolated system must be constant.

_____12. The equation $PV^{\gamma} = \text{const.}$ is valid for any adiabatic process in an ideal gas.

_____13. If a system undergoes a reversible adiabatic change of state, the entropy of the system does not change.

_____14. There is but a single degree of freedom at equilibrium for a three-phase PVT system made up of three nonreacting chemical species.

_____15. For wet steam, $V = V^v - x'\,\Delta V^{lv}$, where V is the specific volume of the mixture and x' is the mass fraction of liquid.

_____16. If a given amount of an ideal gas undergoes a process during which $PV^2 = k$, where k is a constant, then $T/P^{1/2} = k'$, where k' is another constant.

_____17. If a system undergoes an *irreversible* change from an initial equilibrium state i to a final equilibrium state f, the entropy change of the surroundings must be less (algebraically) than it would be if the system changed from i to f reversibly.

_____18. The heat capacity at constant volume of a single-component system consisting of liquid and vapor in equilibrium is infinite.

_____19. The heat capacity at constant pressure of a single-component system consisting of liquid and vapor in equilibrium is infinite.

_____20. At the critical point the internal energy of saturated liquid is equal to the internal energy of saturated vapor.

_____21. If a system undergoes a process during which its entropy does not change, the process is reversible and adiabatic.

_____22. Heat is *always* given by the integral $\int T \, dS$.

_____23. The equation $dH = T \, dS + V \, dP$ can be applied *only* to reversible processes.

_____24. In a phase transition, $\Delta H^{lv} = T \, \Delta S^{lv}$, where T is the absolute temperature at which ΔH^{lv} and ΔS^{lv} are evaluated.

_____25. For any process the second law of thermodynamics requires that the entropy change of the system be zero or positive.

_____26. Cyclic processes get work from heat in defiance of the qualitative expression of the second law.

_____27. If stirring work is done on an ideal gas in a closed system during a constant-volume process, then $\delta Q \neq C_V \, dT$.

_____28. The equation for a bar stressed at constant volume, $d\tilde{U} = T \, d\tilde{S} + \sigma \, d\varepsilon$, implies

$$\left(\frac{\partial \sigma}{\partial \tilde{S}}\right)_\varepsilon = \left(\frac{\partial T}{\partial \varepsilon}\right)_{\tilde{S}}$$

_____29. The change ΔG^{lv} in the Gibbs energy for vaporization at constant T and P is always positive.

_____30. When a molten salt crystallizes, the atoms arrange themselves in a highly ordered lattice structure; since increasing order is associated with decreasing entropy, we must conclude that the entropy of the universe decreases as a result of this process.

_____31. If a saturated liquid undergoes a reversible adiabatic expansion to a lower pressure, some of the liquid will vaporize.

_____32. For any single-phase PVT system at constant pressure,

$$\Delta H = \int_{T_1}^{T_2} C_P \, dT$$

_____33. The *Hilsch vortex tube* holds special interest for scientists and engineers because it operates in violation of the second law of thermodynamics.

_____34. Nuclear fission power plants present a special problem to society because they cause thermal pollution, whereas conventional steam power plants do not.

Ans.

1	2	3	4	5	6	7	8	9	10	11	12	13	14	15
F	F	F	T	T	F	F	F	F	T	F	F	T	F	T

16	17	18	19	20	21	22	23	24	25	26	27	28	29	30
T	F	F	T	T	F	F	F	T	F	F	T	T	F	F

31	32	33	34
T	T	F	F

For the following multiple-choice questions indicate your answers by the appropriate numbers: 1, 2, 3, or 4.

(a) An inventor claims to have devised an engine that produces 3 MJ of work while receiving 2.7 MJ of heat from a single heat reservoir during a complete cycle of the engine. Such an engine would violate:

    (1) the first law   (3) both the first and second laws

    (2) the second law   (4) neither the first nor the second law

(b) The inventor also claims to have constructed a device that rejects 100 kJ of heat to a single heat reservoir while absorbing 100 kJ of work during a single cycle of the device. Analyze this claim as in part (a).

(c) A system is changed from a given initial equilibrium state to a given final equilibrium state by either of two different processes, one reversible, and one irreversible. Which of the following is true, where ΔS refers to the system?

$\quad$ (1) $\Delta S_{irr} = \Delta S_{rev}$ $\qquad$ (3) $\Delta S_{irr} < \Delta S_{rev}$

$\quad$ (2) $\Delta S_{irr} > \Delta S_{rev}$ $\qquad$ (4) none of the above

(d) For any process at all, the second law requires that the entropy change of the *system* be:

$\quad$ (1) positive or zero $\qquad$ (3) negative or zero

$\quad$ (2) zero $\qquad$ (4) no requirement

(e) A hypothetical substance has the following volume expansivity and isothermal compressibility: $\beta = a/V$ and $\kappa = b/V$, where a and b are constants. The equation of state of such a substance would be:

$\quad$ (1) $V = aT + bP + \text{const.}$ $\qquad$ (3) $V = bT + aP + \text{const.}$

$\quad$ (2) $V = aT - bP + \text{const.}$ $\qquad$ (4) $V = bT - aP + \text{const.}$

(f) An exact differential expression relating thermodynamic variables is given by

$$dB = C\,dE - F\,dG + H\,dJ$$

Which of the following would *not* be a new thermodynamic function consistent with the above expression?

$\quad$ (1) $B - FG - CE$ $\qquad$ (3) $B - HJ$

$\quad$ (2) $B - CE$ $\qquad$ (4) $B - HJ + FG - CE$

(g) Given the same exact differential expression as in (f), we conclude that:

$\quad$ (1) $(\partial C/\partial G)_E = (\partial F/\partial E)_G$ $\qquad$ (3) $(\partial F/\partial G)_{E,J} = -(\partial E/\partial C)_{G,J}$

$\quad$ (2) $(\partial C/\partial J)_{E,G} = (\partial H/\partial E)_{J,G}$ $\qquad$ (4) none of the three

(h) For a PVT system, $T(\partial S/\partial T)_P - T(\partial S/\partial T)_V$ is *always* equal to

$\quad$ (1) zero $\qquad\qquad$ (3) R

$\quad$ (2) $\gamma = \dfrac{C_P}{C_V}$ $\qquad$ (4) $T\left(\dfrac{\partial P}{\partial T}\right)_V\left(\dfrac{\partial V}{\partial T}\right)_P$

(i) The product $(\partial P/\partial V)_T(\partial T/\partial P)_S(\partial S/\partial T)_P$ is equivalent to

$\quad$ (1) $(\partial S/\partial V)_T$ $\qquad$ (3) $(\partial V/\partial T)_S$

$\quad$ (2) $(\partial P/\partial T)_V$ $\qquad$ (4) $-(\partial P/\partial T)_V$

(j) A system consisting of a liquid phase and a vapor phase in equilibrium contains three chemical species: water, ethanol, and methanol. The number of degrees of freedom for the system is:

$\quad$ (1) zero $\qquad$ (3) two

$\quad$ (2) one $\qquad$ (4) three

Ans. a$\quad$ b$\quad$ c$\quad$ d$\quad$ e$\quad$ f$\quad$ g$\quad$ h$\quad$ i$\quad$ j

$\quad\quad$ (3)$\quad$ (4)$\quad$ (1)$\quad$ (4)$\quad$ (2)$\quad$ (1)$\quad$ (2)$\quad$ (4)$\quad$ (4)$\quad$ (4)

Chapter 5

Equations of State and Corresponding-States Correlations for PVT Systems

As noted in Section 4.8, the usefulness of thermodynamic property relations depends on the availability of a certain amount of data. In fact, data for a minimum number of properties must be provided as input to the network of thermodynamic equations if they are to yield any quantitative output. Such data may be stored and displayed as tables or graphs, but the most concise and generally useful kind of representation is through a mathematical *equation of state*.

For a thermodynamic system (not necessarily a PVT system) characterized by n *independent* variables, we define an equation of state (EOS) to be an algebraic expression connecting $n + 1$ state variables. Such an equation describes the inherent material behavior of a system or substance, and is an example of what in continuum mechanics is called a *constitutive equation*.

As a practical matter, the material behavior of real systems is often too complicated to be faithfully described by equations of convenient simplicity, except over limited ranges of the state variables. Thus the ideal-gas equation of state, $PV = RT$, correctly describes real-gas behavior only in the limit as $P \to 0$, although it is an exact expression for a model gas made up of molecules which have no volume and which exert no forces on one another.

Though any $n + 1$ state variables may appear in an equation of state, we restrict ourselves in this chapter to equations which apply to PVT systems and which relate the measurable variables P, V, T, and composition x. Such equations are sometimes called *thermal equations of state*.

5.1 FORMULATIONS OF THE EQUATION OF STATE

PVT equations of state are commonly classified as either *volume-explicit*,

$$V = \mathscr{V}(T, P, x) \tag{5.1}$$

or *pressure-explicit*,

$$P = \mathscr{P}(T, V, x) \tag{5.2}$$

The distinction is important in applications, because it dictates the approach used in evaluating derived properties and influences the level of complexity of numerical calculations. When they are appropriate, V-explicit equations are much preferred, because their independent variables (T, P, and x) are those quantities that are fixed in most scientific and engineering calculations. Unfortunately, nature is unkind in this respect; except for dilute gases and vapors, accurate correlation and prediction of volumetric and derived properties of fluids frequently requires the use of a P-explicit EOS.

Most equations of state contain one or more *parameters*. For pure materials, these are temperature-dependent, substance-specific quantities whose numerical values vary from one material to another. For mixtures, the parameters may depend also on composition; in fact, in most engineering equations of state, composition enters *solely* through the parameters, and is expressed through algebraic recipes called *mixing rules*. Thus an EOS parameter α has the general functional form $\alpha(T, x)$. If all the composition dependence is carried by the parameters, then the generic equations (5.1) and (5.2) may be written more precisely as

$$V = \mathscr{V}(T, P; \alpha(T, x), \beta(T, x), \ldots) \tag{5.3}$$

$$P = \mathscr{P}(T, V; \alpha(T, x), \beta(T, x), \ldots) \tag{5.4}$$

The compressibility factor Z is frequently used as the dependent variable in fluid-phase equations of state. By definition, Z is the ratio of the molar volume of a phase to that value it would have were the phase an ideal-gas mixture at the same T, P, and x: $Z \equiv V/V^{ig}$. Since $V^{ig} = RT/P$, this definition reduces to the familiar prescription

$$Z = \frac{PV}{RT}$$

(5.5)

According to Example 4.12,

$$\lim_{P \to 0} \frac{V}{RT/P} = \lim_{P \to 0} Z = 1$$

which we take to be a universal feature of real-fluid behavior. This universal limit constitutes a major motivation for the use of Z. Another advantage is that Z exhibits a much more restricted range of numerical values than does, e.g., the molar volume V. This is illustrated by Figs. 5-1 and 5-2, which depict the isothermal variation of Z with P for the fluid states of water. For the conditions shown, Z is approximately bounded by 0 and 1, whereas V for the same conditions assumes an enormous range of values.

Equations (5.3) and (5.4) have obvious analogs in which Z, rather than V or P, is the dependent variable. Thus the generic V-explicit ZPT equation of state is

$$Z = \mathscr{Z}(T, P; \alpha(T, x), \beta(T, x), \ldots)$$

(5.6)

For the P-explicit EOS with Z as dependent variable, molar *density* ρ is a more convenient independent variable than V. Thus we have, analogous to (5.4), the generic P-explicit $Z\rho T$ equation of state:

$$Z = \mathscr{Z}(T, \rho; \alpha(T, x), \beta(T, x), \ldots)$$

(5.7)

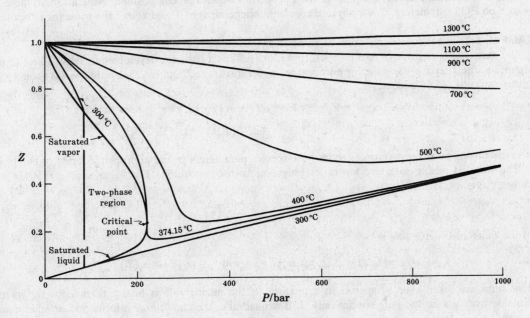

Fig. 5-1

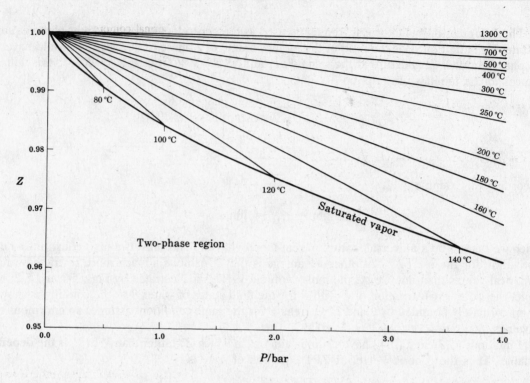

Fig. 5-2

EXAMPLE 5.1   The V-explicit and P-explicit forms of the ideal-gas EOS are

$$V = \frac{RT}{P} \quad \text{and} \quad P = \frac{RT}{V}$$

The EOS in Z, $Z = 1$, is both V-explicit and P-explicit. Note that composition is not an *equation-of-state* variable for ideal-gas mixtures, even though most of their properties depend on composition. Note also that there are in this case no EOS parameters; R is a *physical constant*, independent of T and x, and the same for all substances.

EXAMPLE 5.2   What constraints must be satisfied by *mixing rules* for EOS parameters?

We may choose any one of (5.3), (5.4), (5.6), or (5.7) as a basis for development; let us select (5.6). By assumption, function $\mathscr{Z}$ is the same for pure species as for mixtures. Then, for each pure species i composing an arbitrary mixture, (5.6) becomes

$$Z_i = \mathscr{Z}(T, P; \alpha_i(T), \beta_i(T), \ldots)$$

provided that $\qquad \lim_{x_i \to 1} \alpha(T, x) = \alpha_i(T) \qquad \lim_{x_i \to 1} \beta(T, x) = \beta_i(T) \qquad \ldots$

Thus, the mixing rules must reproduce the pure-species parameters in the appropriate limits of composition.

The simplest mixing rules are low-order polynomials in composition. The *linear mixing rule* for generic parameter α is

$$\alpha(T, x) = \sum_i x_i \alpha_{ii}(T)$$

and the *quadratic mixing rule* is

$$\alpha(T, x) = \sum_i \sum_j x_i x_j \alpha_{ij}(T) \quad \text{with} \quad \alpha_{ii}(T) \equiv \alpha_i(T)$$

where sums are taken over all species in a mixture. If the mixing rule is linear, then α for the mixture is calculable from α's for the pure species only. If it is quadratic, then additional information or approximations are required: α_{ij} for $i = j$ is a pure-species parameter, whereas for $i \neq j$ it is *not*.

EXAMPLE 5.3 Deduce expressions for volume expansivity β and isothermal compressibility κ from a ZPT equation of state.

Since by definition $PV = ZRT$, we have for a general change in state:

$$P \, dV + V \, dP = RT \, dZ + ZR \, dT$$

Dividing this by dT and restricting the result to constant P, we find

$$\left(\frac{\partial V}{\partial T}\right)_P = \frac{RT}{P}\left[\left(\frac{\partial Z}{\partial T}\right)_P + \frac{Z}{T}\right] \qquad\qquad (5.8)$$

Similar procedures yield the following equations:

$$\left(\frac{\partial V}{\partial P}\right)_T = \frac{RT}{P}\left[\left(\frac{\partial Z}{\partial P}\right)_T - \frac{Z}{P}\right] \qquad\qquad (5.9)$$

$$\left(\frac{\partial P}{\partial T}\right)_V = \frac{RT}{V}\left[\left(\frac{\partial Z}{\partial T}\right)_V + \frac{Z}{T}\right] \qquad\qquad (5.10)$$

$$\left(\frac{\partial P}{\partial V}\right)_T = \frac{RT}{V}\left[\left(\frac{\partial Z}{\partial V}\right)_T - \frac{Z}{T}\right] \qquad\qquad (5.11)$$

Equations (3.17) and (5.8) now give

$$\beta = \frac{RT}{PV}\left[\left(\frac{\partial Z}{\partial T}\right)_P + \frac{Z}{T}\right] = \frac{1}{T}\left[1 + \frac{T}{Z}\left(\frac{\partial Z}{\partial T}\right)_P\right]$$

Similarly, (3.18) and (5.9) give

$$\kappa = \frac{1}{P}\left[1 - \frac{P}{Z}\left(\frac{\partial Z}{\partial P}\right)_T\right]$$

Equations (5.10) and (5.11), not used in this exercise, find use in similar applications involving *pressure*-explicit equations of state (Problem 5.28).

A major application of PVT equations of state is to the calculation of residual properties of fluids and fluid mixtures. Residual properties were introduced in Section 4.8; we showed there how to compute them from the ZPT form of the EOS. The important results, (4.31) through (4.33), are repeated here:

$$\frac{G^R}{RT} = \int_0^P (Z-1)\,\frac{dP}{P} \qquad \text{(constant } T, x) \qquad\qquad (5.12)$$

$$\frac{H^R}{RT} = -T \int_0^P \left(\frac{\partial Z}{\partial T}\right)_{P,x} \frac{dP}{P} \qquad \text{(constant } T, x) \qquad\qquad (5.13)$$

$$\frac{S^R}{R} = -\int_0^P \left[T\left(\frac{\partial Z}{\partial T}\right)_{P,x} + Z - 1\right]\frac{dP}{P} \qquad \text{(constant } T, x) \qquad\qquad (5.14)$$

If the EOS is of $Z\rho T$ form, (5.12) through (5.14) must be modified by use of the following differential relations, valid for changes of state *at constant temperature and composition*:

$$\frac{dP}{P} = \frac{d\rho}{\rho} + \frac{dZ}{Z}$$

$$T\left(\frac{\partial Z}{\partial T}\right)_{P,x} \frac{dP}{P} = T\left(\frac{\partial Z}{\partial T}\right)_{\rho,x} \frac{d\rho}{\rho} - dZ$$

Thus we find

$$\frac{G^R}{RT} = \int_0^\rho (Z-1)\,\frac{d\rho}{\rho} + Z - 1 - \ln Z \qquad \text{(constant } T, x) \qquad\qquad (5.15)$$

$$\frac{H^R}{RT} = -T \int_0^\rho \left(\frac{\partial Z}{\partial T}\right)_{\rho,x} \frac{d\rho}{\rho} + Z - 1 \qquad \text{(constant } T, x) \qquad\qquad (5.16)$$

$$\frac{S^R}{R} = -\int_0^\rho \left[T\left(\frac{\partial Z}{\partial T}\right)_{\rho,x} + Z - 1\right]\frac{d\rho}{\rho} + \ln Z \qquad \text{(constant } T, x) \qquad\qquad (5.17)$$

EXAMPLE 5.4 The residual properties M^R represent a comparison of real-fluid and ideal-gas properties at the same temperature, *pressure*, and composition:

$$M^R \equiv M - M^{ig}(T, P, x) \tag{1}$$

One may instead make the comparison at the same temperature, *molar volume* (or molar density), and composition. In this case, we designate the residual property M', where

$$M' \equiv M - M^{ig}(T, V, x) \tag{2}$$

Determine a general relation connecting M' and M^R, and develop expressions for A'/RT, U'/RT, and S'/R appropriate for use with a $Z\rho T$ equation of state.

Subtraction of (1) from (2) yields

$$M' = M^R - [M^{ig}(T, V, x) - M^{ig}(T, P, x)] \tag{3}$$

The bracketed term involves ideal-gas properties evaluated at actual conditions; it is given by

$$M^{ig}(T, V, x) - M^{ig}(T, P, x) = \int_P^{RT/V} \left(\frac{\partial M^{ig}}{\partial P}\right)_{T,x} dP \tag{4}$$

Equations (3) and (4) produce the desired result:

$$M' = M^R - \int_P^{RT/V} \left(\frac{\partial M^{ig}}{\partial P}\right)_{T,x} dP \tag{5}$$

According to (5), the two types of residual property are identical for those M for which M^{ig} depends on temperature and composition only. Consequently,

$$U' = U^R \qquad H' = H^R \qquad C_V' = C_V^R \qquad C_P' = C_P^R \tag{6}$$

The ideal-gas entropy and quantities related to it by definition (e.g., G^{ig} and A^{ig}) depend on P, as well as on T and x. For these properties $M' \neq M^R$, and we obtain

$$S' = S^R - R \ln Z \tag{7}$$
$$G' = G^R + RT \ln Z \tag{8}$$
$$A' = A^R + RT \ln Z \tag{9}$$

(See Problem 5.32.)

To find A'/RT, we make use of previous results. Thus,

$$A = G - PV \quad \text{implies} \quad A^R = G^R - PV^R = G^R - RT(Z-1)$$

the last step following from (4.30). Now (9) gives

$$A' = G^R - RT(Z-1) + RT \ln Z$$

and combination with (5.15) yields

$$\frac{A'}{RT} = \int_0^\rho (Z-1) \frac{d\rho}{\rho} \quad \text{(constant } T, x) \tag{5.18}$$

To find U'/RT, recall that $U = H - PV$; then $U^R = H^R - PV^R = H^R - RT(Z-1)$. Hence, by (6) and (5.16),

$$\frac{U'}{RT} = -T \int_0^\rho \left(\frac{\partial Z}{\partial T}\right)_{\rho,x} \frac{d\rho}{\rho} \quad \text{(constant } T, x) \tag{5.19}$$

The expression for S'/R follows immediately from (7) and (5.17):

$$\frac{S'}{R} = -\int_0^\rho \left[T\left(\frac{\partial Z}{\partial T}\right)_{\rho,x} + Z - 1\right] \frac{d\rho}{\rho} \quad \text{(constant } T, x) \tag{5.20}$$

Comparison of (5.18) through (5.20) with (5.12) through (5.14) reveals some striking analogies between A', U', and S' (as derived from the $Z\rho T$ equation of state), and G^R, H^R, and S^R (as derived from the ZPT equation of state). The analogies are in fact expected, because T, ρ, and x are *canonical* variables for A (and hence for A'), whereas T, P, and x are canonical variables for G (and hence for G^R): see Section 3.3.

5.2 VIRIAL EQUATIONS OF STATE

At low-to-moderate pressure or density levels, the compressibility factor of a vapor or gas at constant T and x is a slowly varying, well-behaved function of P or ρ; this is illustrated for water by Fig. 5-2. Moreover, the zero-pressure (zero-density) limit of Z is known; it is always unity. These observations suggest the use of power series for representing the EOS of dilute and moderately dense gases and vapors. Two formulations are possible, depending on whether one takes pressure or molar density as the variable of expansion. Expanding Z in powers of P about $P = 0$, we write

$$Z = 1 + B'P + C'P^2 + D'P^3 + \cdots \qquad (5.21)$$

Equation (5.21) is the *virial expansion in pressure*, and parameters B', C', D', ..., are called *pressure-series virial coefficients*: B' is the *second* virial coefficient; C', the third; etc. These virial coefficients depend on temperature and (for a mixture) composition; hence (5.21) is a volume-explicit equation of state.

Alternatively to (5.21), we may expand Z in powers of ρ about $\rho = 0$, obtaining the *virial expansion in density*:

$$Z = 1 + B\rho + C\rho^2 + D\rho^3 + \cdots \qquad (5.22)$$

Quantities B, C, ..., are the second, third, ..., *density-series virial coefficients*; they too depend on temperature and composition. Series (5.22) is a pressure-explicit EOS. As it is the density-series coefficients which are normally reported and tabulated, they are normally considered "the" virial coefficients. Conversion between the two types of coefficients is through comparison of the two infinite series (5.21) and (5.22); see Problem 5.6. Results for the first three coefficients are:

$$B' = \frac{B}{RT} \qquad C' = \frac{C - B^2}{(RT)^2} \qquad D' = \frac{D + 2B^3 - 3BC}{(RT)^3} \qquad (5.23)$$

EXAMPLE 5.5 Give operational definitions of the virial coefficients.

Consider a function Y of several variables $X_1, X_2, X_3, \ldots$ The formal Taylor expansion of Y in powers of X_1 about $X_1 = 0$ is

$$Y = \sum_{n=0}^{\infty} K_n X_1^n \qquad \text{where} \qquad K_n \equiv \frac{1}{n!} \lim_{X_1 \to 0} \left(\frac{\partial^n Y}{\partial X_1^n} \right)_{X_2, X_3, \ldots}$$

Thus the pressure-series virial coefficients are defined as

$$B' \equiv \lim_{P \to 0} \left(\frac{\partial Z}{\partial P} \right)_{T,x} \qquad C' \equiv \frac{1}{2} \lim_{P \to 0} \left(\frac{\partial^2 Z}{\partial P^2} \right)_{T,x} \qquad \cdots \qquad (5.24)$$

and the density-series coefficients as

$$B \equiv \lim_{\rho \to 0} \left(\frac{\partial Z}{\partial \rho} \right)_{T,x} \qquad C \equiv \frac{1}{2} \lim_{\rho \to 0} \left(\frac{\partial^2 Z}{\partial \rho^2} \right)_{T,x} \qquad \cdots \qquad (5.25)$$

It is obvious from (5.24) and (5.25) that collection of PVT data (or $Z\rho T$ data) suitable for extraction of virial coefficients places extreme demands on the experimentalist, for the data must survive differentiations at a condition (zero pressure or density) where in fact no data can actually be taken.

EXAMPLE 5.6 With reference to Example 5.5, devise a procedure for extracting density-series virial coefficients from PVT data that formally avoids numerical differentiation of the data at zero density.

Rewrite (5.22) as $\Psi_1 \equiv (Z-1)/\rho = B + C\rho + D\rho^2 + \cdots$. Then

$$B = \lim_{\rho \to 0} \Psi_1 \qquad (5.26)$$

With B determined, the procedure may be extended, by letting $\Psi_2 \equiv (\Psi_1 - B)/\rho$. Substituting for Ψ_1 by the virial expansion and taking the zero-density limit, we obtain

$$C = \lim_{\rho \to 0} \Psi_2 \qquad (5.27)$$

Although this "Ψ-procedure" can in principle be extended indefinitely, real PVT data (however they are reduced) rarely provide meaningful values for virial coefficients higher than the third. If the data are of limited extent or of marginal precision or accuracy, even the *sign* of C can be ambiguous, particularly if the temperature level is low.

EXAMPLE 5.7   We illustrate the evaluation of virial coefficients from PVT data. The basis for this example is the set of specific volumes for water vapor at 260 °C reported in the steam tables of Keenan et al. [*Steam Tables (International Edition—Metric Units)*, Wiley, New York, 1969]. These, of course, are not actual observations; they are values generated from a comprehensive equation of state *based on* many observations. They are extremely smooth, and thus simulate a data set of very high precision.

Keenan et al. list 141 entries of specific volume V at 260 °C for pressures between 0.01 bar and the saturation pressure 46.88 bar. Values selected at reasonable intervals are converted to compressibility factors and molar densities. The corresponding values of Ψ_1 (see Example 5.6) are then computed and plotted against ρ (Fig. 5-3). On the scale of this graph, there is substantial scatter up to a ρ of about 20×10^{-5} mol $\cdot$ cm^{-3} but a much smoother trend at higher densities. The curve drawn through the points yields an intercept of -142.2 cm$^3 \cdot$ mol^{-1}, and by (5.26) this is the second virial coefficient B.

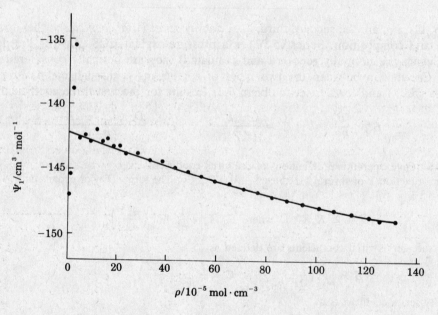

Fig. 5-3

The third virial coefficient is found from a Ψ_2-versus-ρ plot, shown in Fig. 5-4. Very significant scatter obtains up to a ρ of about 50×10^{-5} mol $\cdot$ cm^{-3}, after which a linear trend with less scatter is observed. Extrapolation of the higher-density data to $\rho = 0$ gives an intercept of -7140 cm$^6 \cdot$ mol^{-2}, which by (5.27) is C.

It should be obvious that these data cannot survive another Ψ analysis. However, we can obtain from Fig. 5-4 an *estimate* for the fourth virial coefficient D. From $\Psi_2 = C + D\rho + \cdots$,

$$\left(\frac{\partial \Psi_2}{\partial \rho} \right)_{T,x} = D + \cdots$$

Hence, from the slope of the smoothing curve on Fig. 5-4, $D \approx 1.51 \times 10^6$ cm$^9 \cdot$ mol^{-3}.

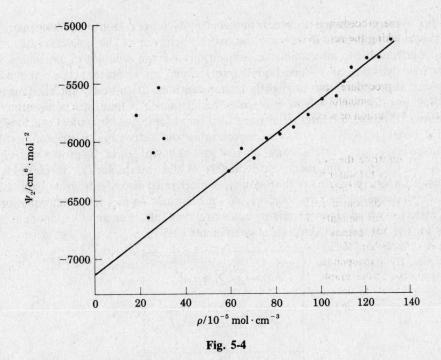

Fig. 5-4

No indication is given of the uncertainties in B, C, and D, but it is evident from Fig. 5-4 that the determination of the latter two, particularly D, is subject to rather large errors. Figure 5-4 also suggests that virial coefficients beyond D are unnecessary for water vapor at 260 °C.

It is instructive to compare the relative contributions to Z of the terms containing B, C, and D; we do this in Table 5-1. Also shown are specific volumes calculated with the derived virial coefficients and the corresponding specific volumes from the steam tables. Up to a pressure of about 5 bar, the virial equation truncated to two terms gives an essentially perfect fit of the data, and even at 25 bar the contribution of the C and D terms to Z is only about 0.4%. This is why the extrapolation to $\rho = 0$ from higher densities in Fig. 5-4 is sufficient for determination of C. The excellent agreement between calculated and steam-table volumes illustrates the appropriateness of a cubic polynomial in ρ as a curve-fitting device for this isotherm.

Table 5-1

P/bar	$B\rho$	$C\rho^2$	$D\rho^3$	Z	V_{calc}/cm$^3 \cdot$ g^{-1}	V_{table}/cm$^3 \cdot$ g^{-1}
0.01	−0.000 03	-4×10^{-10}	2×10^{-14}	0.999 97	246 046	246 046
0.1	−0.000 32	-4×10^{-8}	2×10^{-11}	0.999 68	24 598	24 598
1	−0.003 22	-4×10^{-6}	2×10^{-8}	0.996 78	2 452.6	2 453
10	−0.033 18	−0.000 39	0.000 02	0.966 45	237.80	237.8
25	−0.088 16	−0.002 74	0.000 36	0.911 93	89.511	89.51
46.88	−0.186 95	−0.012 34	0.003 43	0.804 14	42.206	42.21

Although we have justified (5.21) and (5.22) on empirical grounds, the real utility of the virial equations derives from the fact that the coefficients also have theoretical significance. If we picture a gas as comprising a large number of discrete particles (molecules), then we find that the macroscopic PVT behavior is determined on the microscopic scale by interactions among molecular force fields. [The very word *virial* derives from *vires* = strengths (of the intermolecular forces); it is due to Clausius.] At least two factors determine the effects of these molecular interactions: (1) the nature of the molecules themselves, which determines the types of forces in action, and (2) the intermolecular separations. At very large intermolecular separations (low densities), the molecules exert negligible forces on each other, and the gas approaches ideal behavior ($Z = 1$). As the density of the system is increased, the average distance between molecules decreases, and intermolecular interactions become increasingly important. If we view the overall behavior of the collection of molecules as a combination of effects due to interactions among various average *numbers* of molecules, we conclude that at very low densities only "one-body" interactions are significant (i.e., the molecules act independently of each other); that at slightly higher densities, "two-body" interactions must also be considered (i.e. interactions involving molecules taken two at a time are important); that at still higher densities, "three-body" interactions are also important; etc. Statistical mechanics interprets the successive terms 1, $B\rho$, $C\rho^2, \ldots$, as representing the successive contributions of one-body, two-body, three-body, ..., interactions to real-gas behavior; and it provides recipes for the calculation of B, C, etc., from mechanical models of the intermolecular force fields. The most important result for our purposes is that statistical mechanics also yields *exact* expressions for the composition dependence of the virial coefficients for mixtures. In fact the virial equation is the only equation of state for which *rigorous* mixing rules are available. For an *m*-component mixture, the mixing rules for the second and third virial coefficients are

$$B = \sum_{i=1}^{m} \sum_{j=1}^{m} x_i x_j B_{ij} \tag{5.28}$$

$$C = \sum_{i=1}^{m} \sum_{j=1}^{m} \sum_{k=1}^{m} x_i x_j x_k C_{ijk} \tag{5.29}$$

where the x's are mole fractions; analogous formulas hold for the higher coefficients. The coefficients B_{ij} and C_{ijk} are functions of temperature only, and have the symmetry properties

$$B_{ij} = B_{ji} \tag{5.30a}$$

$$C_{ijk} = C_{ikj} = C_{jik} = C_{jki} = C_{kij} = C_{kji} \tag{5.30b}$$

When the sums in (5.28) and (5.29) are expanded in full, two types of coefficients appear in the result: those in which the successive subscripts are identical, and those for which at least one of the subscripts differs from the others. The first kind of coefficient refers to a pure component; the second type, called a *cross-coefficient*, is a mixture property.

EXAMPLE 5.8　Write (5.28) and (5.29) in expanded form for a binary mixture.
　　Equation (5.28) becomes

$$B = \sum_{i=1}^{2} \sum_{j=1}^{2} x_i x_j B_{ij} = x_1^2 B_{11} + x_1 x_2 B_{12} + x_2 x_1 B_{21} + x_2^2 B_{22}$$

But by $(5.30a)$, $B_{12} = B_{21}$; therefore $B = x_1^2 B_{11} + 2x_1 x_2 B_{12} + x_2^2 B_{22}$.
　　Similarly, using $(5.30b)$,

$$C = \sum_{i=1}^{2} \sum_{j=1}^{2} \sum_{k=1}^{2} x_i x_j x_k C_{ijk}$$

$$= x_1^3 C_{111} + x_2 x_1^2 C_{211} + x_1^2 x_2 C_{121} + x_1 x_2^2 C_{221} + x_1^2 x_2 C_{112} + x_2^2 x_1 C_{212} + x_1 x_2^2 C_{122} + x_2^3 C_{222}$$

$$= x_1^3 C_{111} + 3x_1^2 x_2 C_{112} + 3x_1 x_2^2 C_{221} + x_2^3 C_{222}$$

EXAMPLE 5.9　The mixing rule (5.28) is quadratic in mole fraction. Rewrite this expression as the sum of two terms, one linear in mole fraction and the other quadratic, such that the linear part reduces to B_{ii} at $x_i = 1$ for all i.
　　Write (5.28) as

$$B = \sum_i x_i B_{ii} + \sum_i \sum_j x_i x_j B_{ij} - \sum_i x_i B_{ii} \tag{1}$$

The leading term on the right is linear in mole fraction and satisfies the stated boundary conditions. The identical last term must be cast into quadratic form. First note that

$$\sum_i x_i B_{ii} = \frac{1}{2} \sum_i x_i B_{ii} + \frac{1}{2} \sum_j x_j B_{jj}$$

The two terms on the right are unaltered if we multiply them by $\Sigma_j x_j$ and $\Sigma_i x_i$, respectively, because the mole fractions sum to unity. Thus

$$\frac{1}{2} \sum_i x_i B_{ii} = \frac{1}{2} \left(\sum_j x_j\right)\left(\sum_i x_i B_{ii}\right) = \frac{1}{2} \sum_i \sum_j x_i x_j B_{ii}$$

$$\frac{1}{2} \sum_j x_j B_{jj} = \frac{1}{2} \left(\sum_i x_i\right)\left(\sum_j x_j B_{jj}\right) = \frac{1}{2} \sum_i \sum_j x_i x_j B_{jj}$$

giving the *symmetric* representation

$$\sum_i x_i B_{ii} = \frac{1}{2} \sum_i \sum_j x_i x_j (B_{ii} + B_{jj}) \tag{2}$$

Combination of (1) and (2) gives the desired result:

$$B = \sum_i x_i B_{ii} + \frac{1}{2} \sum_i \sum_j x_i x_j \delta_{ij} \tag{5.31}$$

where
$$\delta_{ij} \equiv 2B_{ij} - B_{ii} - B_{jj} \tag{5.32}$$

Note that $\delta_{ij} = 0$ if $i = j$. Also, $\delta_{ji} = \delta_{ij}$ because $B_{ji} = B_{ij}$. In the special case that $\delta_{ij} = 0$ for *all* i and j, the quadratic mixing rule reduces to a linear mixing rule. We show in Problem 7.3 that linear composition dependence of B is a feature of vapor mixtures that are *ideal solutions*.

The temperature dependence of the virial coefficients must be inferred from experiment or from statistical-mechanical calculations based on models for the intermolecular potential energy. Although the quantitative dependence is different for different species, one can make qualitative generalizations for B and C that are valid for all nonreacting substances. For this purpose, the *reduced temperature* $T_r \equiv T/T_c$, where T_c is the gas/liquid critical temperature, becomes the convenient independent variable.

At low reduced temperatures, B is strongly negative. It becomes less negative with increasing T, passing through zero at $T_r \approx 2.7$. Thereafter it assumes small positive values, eventually exhibiting a very flat maximum, after which it approaches a finite positive limiting value as temperature tends to infinity. For low reduced temperatures, C is also strongly negative. It increases steeply with temperature, passes through zero at $T_r \approx 0.7$, and shows a maximum at $T_r \approx 0.9$. Thereafter it decreases monotonically with T; like B, C possesses a finite, positive, infinite-temperature limit.

Figure 5-5 shows the *reduced* second virial coefficient, $\hat{B} \equiv BP_c/RT_c$, as a function of T_r for the noble gases Ar, Kr, and Xe. This *corresponding-states* plot (Section 5.4) is based on reliable correlations of a large data base. Figure 5-6 is a more conjectural rendering of the reduced third virial coefficient, $\hat{C} \equiv CP_c^2/R^2T_c^2$, for argon. Here, the data base is meager compared with that for B; as already noted, the higher the virial coefficient, the more difficult it is to extract meaningful values for it from PVT data.

The near linearity of the isotherms at low pressures shown in Fig. 5-2 suggests that the two-term truncation of (5.21) should suffice for dilute and moderately dense gases. We therefore write $Z = 1 + B'P$, or

$$Z = 1 + \frac{BP}{RT} \qquad \text{(dilute gases)} \tag{5.33}$$

where we have eliminated B' in favor of B by (5.23). Equation (5.33) provides an excellent approximation to reality up to reduced densities $\rho_r(\equiv \rho/\rho_c)$ of $\frac{1}{4}$ to $\frac{1}{2}$, depending on the reduced temperature.

At high pressure or density levels, Z-versus-P and Z-versus-ρ isotherms show departures from linearity. Here, two-term truncations of the virial equations are inappropriate. Comparative studies

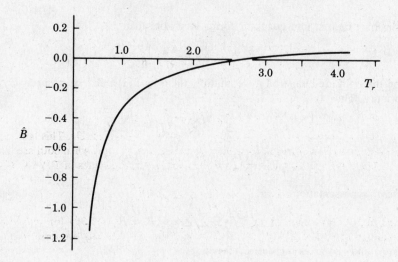

Fig. 5-5

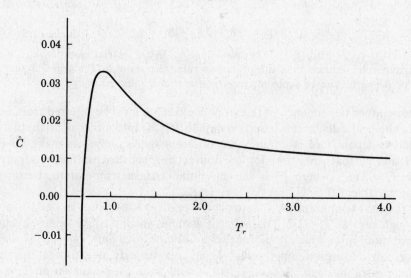

Fig. 5-6

of the three-term versions of (5.21) and (5.22) show that the latter is generally superior. Thus, for conditions where (5.33) is unsuitable, we use

$$Z = 1 + B\rho + C\rho^2 \qquad (5.34)$$

Equation (5.34) appears to be accurate up to reduced densities of about 0.75.

EXAMPLE 5.10   An alternative to (5.33) is the two-term virial equation in density:

$$Z = 1 + B\rho \qquad (5.35)$$

Compare (5.35) with (5.33).

Equation (5.33) is explicit in volume, whereas (5.35) is explicit in pressure. Comparison is facilitated if the two equations are of the same type; we choose to recast (5.35) in V-explicit form. Since $Z = P/\rho RT$, (5.35) may be written

$$\frac{P}{\rho RT} = 1 + B\rho$$

Solution for ρ gives

$$\rho = \frac{1}{2B}\left(-1 + \sqrt{1 + \frac{4BP}{RT}}\right)$$

and substitution of this expression into (5.35) yields

$$Z = \frac{1}{2} + \frac{1}{2}\sqrt{1 + \frac{4BP}{RT}} \qquad (5.36)$$

Equation (5.36), the V-explicit form of (5.35), is clearly different from (5.33). This is to be expected; a virial equation in pressure and a virial equation in density can be equivalent only when both are in their infinite-series forms. However, (5.36) and (5.33) become nearly identical for small values of BP/RT, because then

$$\sqrt{1 + \frac{4BP}{RT}} \approx 1 + \frac{2BP}{RT}$$

Comparative studies show that (5.33) is at least as accurate as (5.36) for representing PVT data at low-to-moderate pressures. Since (5.33) is the simpler, it is the preferred two-term virial equation of state.

EXAMPLE 5.11   Derive expressions for the residual Gibbs energy, residual enthalpy, and residual entropy of a gas described by the two-term virial equation (5.33). What do these results suggest about the algebraic signs of G^R, H^R, and S^R?

Equation (5.33) is explicit in volume, indicating application of (5.12) through (5.14). From (5.33),

$$Z - 1 = \frac{BP}{RT} \qquad \left(\frac{\partial Z}{\partial T}\right)_{P,x} = \frac{P}{RT}\left(\frac{dB}{dT} - \frac{B}{T}\right)$$

Substitution into (5.12) through (5.14) and integration give

$$\frac{G^R}{RT} = \frac{BP}{RT} \qquad (5.37)$$

$$\frac{H^R}{RT} = -\frac{P}{R}\left(\frac{dB}{dT} - \frac{B}{T}\right) \qquad (5.38)$$

$$\frac{S^R}{R} = -\frac{P}{R}\frac{dB}{dT} \qquad (5.39)$$

As shown by Fig. 5-5 and noted in the accompanying discussion, B is negative for temperatures below about $2.7T_c$, and dB/dT is positive up to extremely high temperatures. Thus, according to (5.37) through (5.39), the three residual properties are *negative* at normal temperatures and at pressures where (5.33) is valid. Very light gases, particularly H_2 and He, are exceptional; they have extremely low critical temperatures and hence positive second virial coefficients at normal temperatures. For these gases, G^R and H^R are positive under commonly encountered conditions.

5.3   EMPIRICAL EQUATIONS OF STATE

Although either of the virial equations (5.21) or (5.22) can in principle be used to *fit* most gas-phase isotherms to any required accuracy, *true* virial coefficients beyond the third are generally unobtainable (see Examples 5.5 through 5.7). As a result, (5.34) is for all practical purposes the best one can do with virial equations. Unfortunately, it is impossible to describe both the liquid and vapor portions of subcritical isotherms by a truncated virial equation and a single set of coefficients. Prediction and correlation of volumetric properties of fluids at high densities and in the liquid region is therefore usually accomplished with more comprehensive *empirical* equations of state. Of the scores of such equations that have been proposed, we restrict ourselves here to the simplest: those that are cubic in volume or density.

Inspection of the subcritical isotherms in Fig. 4-3 shows that an equation of state suitable for gases *and* liquids must capture at least three features of such isotherms:

(1)   "steepness," for small values of V

(2) *multivaluedness in V* at the liquid/vapor saturation pressure P^{sat}

(3) an approach to ideal-gas behavior as V approaches infinity

The simplest general class of equations of state capable of accommodating these features comprises expressions *polynomial in volume*; of these, the simplest realistic specimens are *cubic*. The general cubic EOS may be written in PVT form as

$$P = \frac{RT}{V-b} - \frac{\theta(V-\eta)}{(V-b)(V^2 + \delta V + \varepsilon)} \tag{5.40}$$

or in ZρT form as

$$Z = \frac{1}{1-b\rho} - \frac{\theta(1-\eta\rho)\rho}{RT(1-b\rho)(1+\delta\rho+\varepsilon\rho^2)} \tag{5.41}$$

Here, b, θ, δ, ε, and η are parameters which in general depend on temperature and (for a mixture) composition.

So comprehensive a cubic EOS as (5.40) or (5.41) is rarely used in practice. One deals with special cases, obtained from one or another assignment of the parameters. The simplest nontrivial assignment is $b = b(x)$, $\theta = a(x)$, $\eta = b(x)$, $\delta = \varepsilon = 0$; whence the *van der Waals equation of state*,

$$P = \frac{RT}{V-b} - \frac{a}{V^2} \tag{5.42}$$

$$Z = \frac{1}{1-b\rho} - \frac{a\rho}{RT} \tag{5.43}$$

The van der Waals EOS is the prototypical cubic EOS; deduced by van der Waals from molecular arguments, it won him the Nobel Prize in physics for 1910. Given its simplicity, it provides a remarkably faithful qualitative description of the PVT behavior of real fluids. However, it is generally inappropriate for accurate numerical work, and has been supplanted for quantitative applications by other cubic equations. An early-modern (1949) example is the *Redlich/Kwong equation*,

$$P = \frac{RT}{V-b} - \frac{a}{T^{1/2}V(V+b)} \tag{5.44}$$

$$Z = \frac{1}{1-b\rho} - \frac{a\rho}{RT^{3/2}(1+b\rho)} \tag{5.45}$$

Here, as in the van der Waals equation, a and b depend only on composition.

Although the parameters in an empirical EOS may have *interpretations* (molecular or otherwise), they have no *operational definitions*—unlike the virial coefficients. No empirical EOS is known that with a single set of parameter values can represent precisely the PVT behavior of a substance throughout the range of temperatures and pressures of practical interest. Thus, if a and b in (5.42) through (5.45) are regarded as curve-fitting coefficients, then their values depend upon the set of data used to derive them. It is possible to arrive at optimal values of a and b for a *particular* range of T and P by numerical regression techniques such as the method of least squares; one is thereby assured of optimal *local* performance. Alternatively, one can estimate a and b by subjecting the EOS to physically inspired constraints; if the constraints are chosen carefully, satisfactory *global* performance results. The most common constraints follow from the requirement that the critical isotherm have a horizontal inflection at the critical state (Section 4.1); i.e.,

$$\left(\frac{\partial P}{\partial V}\right)_{T;cr} = 0 \tag{5.46}$$

$$\left(\frac{\partial^2 P}{\partial V^2}\right)_{T;cr} = 0 \tag{5.47}$$

EXAMPLE 5.12 Find values of the van der Waals a and b which result when (5.46) and (5.47) are satisfied. From (5.42),

$$\left(\frac{\partial P}{\partial V}\right)_{T;cr} = \frac{-RT_c}{(V_c - b)^2} + \frac{2a}{V_c^3} = 0$$

$$\left(\frac{\partial^2 P}{\partial V^2}\right)_{T;cr} = \frac{2RT_c}{(V_c - b)^3} - \frac{6a}{V_c^4} = 0$$

where T_c and V_c are the critical temperature and the critical volume. Solution of these equations for a and b gives

$$a = \frac{9}{8} RT_c V_c \qquad b = \frac{1}{3} V_c \tag{5.48}$$

With a and b given by (5.48), (5.42) yields

$$P_c = \frac{3}{8} \frac{RT_c}{V_c} \tag{5.49}$$

Equation (5.49) allows us to rewrite (5.48) in terms of P_c and V_c or in terms of T_c and P_c. Thus we obtain as alternatives to (5.48)

$$a = 3P_c V_c^2 \qquad b = \frac{1}{3} V_c \tag{5.50}$$

$$a = \frac{27}{64} \frac{R^2 T_c^2}{P_c} \qquad b = \frac{1}{8} \frac{RT_c}{P_c} \tag{5.51}$$

Equation (5.49) is subject to experimental test, for it asserts that $Z_c \equiv P_c V_c / RT_c = 0.375$ for *all* fluids. But Z_c is known to vary from substance to substance, and in fact usually ranges from 0.23 to 0.29. This means that (5.48), (5.50), and (5.51), although equivalent and consistent with respect to the EOS, give different sets of values for a and b when used with measured critical coordinates. Which to use? This question arises in connection with any two-parameter EOS; practical considerations and the results of comparative numerical studies both suggest that it is best to estimate EOS parameters from T_c and P_c. Thus, for *quantitative* applications of the van der Waals EOS, use (5.51).

EXAMPLE 5.13 (a) Write the van der Waals EOS in the reduced coordinates

$$T_r \equiv \frac{T}{T_c} \qquad P_r \equiv \frac{P}{P_c} \qquad V_r \equiv \frac{V}{V_c}$$

(b) Plot in the $P_r V_r$ plane the isotherms $T_r = 0.9, 1.0, 1.1, 1.2$.

(a) When (5.42), (5.48), and (5.49) are combined, we obtain

$$P = \frac{RT}{V - \frac{V_c}{3}} - \frac{\frac{9}{8}RT_c V_c}{V^2} = \frac{RT_c}{V_c}\left[\frac{T/T_c}{(V/V_c) - \frac{1}{3}} - \frac{\frac{9}{8}}{(V/V_c)^2}\right]$$

from which

$$P_r = \frac{8T_r}{3V_r - 1} - \frac{3}{V_r^2} \tag{5.52}$$

The van der Waals equation written in the form (5.52) applies to all fluids, because both material parameters have been absorbed in the dimensionless variables.

(b) Figure 5-7 is a $P_r V_r$ diagram of (5.52) evaluated at $T_r = 0.9, 1.0, 1.1$, and 1.2. Since (5.52) is cubic in V_r, there are either one or three real roots which satisfy the equation for a given P_r at fixed T_r. If only one real root exists, it corresponds either to a gas or fluid state (e.g., to a point on the $T_r = 1.1$ or 1.2 isotherm of Fig. 5-7) or to a liquid state (e.g., to a point above B on the $T_r = 0.9$ isotherm).

When three real roots exist, there are three possible situations. If all three roots are distinct, as on $T_r = 0.9$ at P_{r2}, the smallest (at point A) is a liquid volume and the largest (at point G) is a vapor volume. The middle root for V_r (at point E) has no physical significance, as discussed later. Thus the liquid state at A and the vapor state at G exist at the same T_r (and T) and the same P_r (and P), and are therefore presumably the states of liquid and vapor which exist in vapor/liquid equilibrium at $T_r = 0.9$. However, Fig. 5-7 exhibits a *range* of values for P_r (from P_{r3} to P_{r1}) for which both liquidlike and vaporlike states are shown for the $T_r = 0.9$ isotherm. On the other hand, the phase rule applied to a system consisting of vapor and liquid in equilibrium requires that for a given T_r there be but one P_r, namely P_r^{sat}. This apparent contradiction is resolved by *Maxwell's equal-area rule* (Problem 5.15), which states that P_r^{sat} is that pressure for which the horizontal line AEG cuts off equal areas $ADEA$ and $EFGE$ below and above the isobar. The line meeting this condition on Fig. 5.7 is the one at P_{r2}. Thus points A and G correspond to saturated liquid and vapor, respectively, and exist at $P_r^{sat} = P_{r2} \approx 0.65$. The portions of the $T_r = 0.9$ isotherm above A and to the right of G represent subcooled liquid and superheated vapor, respectively.

If two of the three roots of (5.52) are equal, as at P_{r1} and P_{r3}, then a minimum or maximum in the isotherm occurs at the double root. Between these two extremes, i.e., between points D and F, $(\partial P_r/\partial V_r)_{T_r}$ is positive. However, this quantity must be *negative* (a stability requirement) for all real materials; we conclude that the portion of the isotherm between D and F has no physical significance. This is why we earlier disregarded the middle root at E. Between A and D and between F and G, $(\partial P_r/\partial V_r)_{T_r}$ is negative, as required, although the values of P_r do not equal the equilibrium pressure P_r^{sat}. These portions of the isotherm represent *metastable* liquid and vapor states, respectively, and correspond to superheated liquid and subcooled vapor, nonequilibrium physical states that can be produced and maintained in the laboratory under carefully controlled conditions.

For three equal real roots, (5.52) predicts a horizontal inflection point. This is an exceptional point and occurs only on the critical isotherm $T_r = 1$ at the critical pressure $P_r = 1$. This is the gas/liquid critical point and is designated as point C on Fig. 5-7.

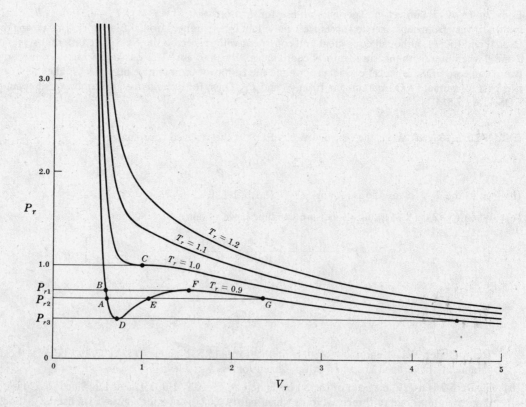

Fig. 5-7

Mixing rules for the parameters of an empirical EOS are necessarily empirical. Often they are assumed to be quadratic in mole fraction:

$$\alpha = \sum_i \sum_j x_i x_j \alpha_{ij} \qquad (5.53)$$

as in Example 5.2. As with the subscripted virial coefficients in (5.28), the α_{ij} in (5.53) are of two kinds. If $i = j$, the parameter is for a pure species; if $i \neq j$, the parameter is a mixture property, called an *interaction parameter*. Evaluation of empirical interaction parameters presents special difficulties, because the methods used to estimate pure-species parameters (see, e.g., Example 5.12) often have no convenient extension to mixtures. In any event, one would like to estimate *all* parameters from pure-species information alone. A practical approach to this problem is the use of *combining rules*.

Combining rules are empirical recipes relating an interaction parameter (say, α_{ij}) to the corresponding parameters (α_{ii} and α_{jj}) of the pure species. Frequently, a simple average is employed. Thus, the *arithmetic-mean* combining rule for α_{ij} is

$$\alpha_{ij} = \tfrac{1}{2}(\alpha_{ii} + \alpha_{jj}) \qquad (5.54)$$

and the *geometric-mean* combining rule is

$$\alpha_{ij} = (\alpha_{ii}\alpha_{jj})^{1/2} \qquad (5.55)$$

Substituting (5.54) into (5.53) and carrying out the summations, we obtain on simplification

$$\alpha = \sum_i x_i \alpha_{ii} \qquad \text{(arithmetic rule)} \qquad (5.56)$$

which is just the linear mixing rule. With α_{ij} given by (5.55),

$$\alpha = \left(\sum_i x_i \alpha_{ii}^{1/2} \right)^2 \qquad \text{(geometric rule)} \qquad (5.57a)$$

or, equivalently,

$$\alpha = \sum_i x_i \alpha_{ii} - \sum_i \sum_j x_i x_j [\tfrac{1}{2}(\alpha_{ii} + \alpha_{jj}) - (\alpha_{ii}\alpha_{jj})^{1/2}] \qquad (5.57b)$$

Because the arithmetic mean of two different numbers is larger than the geometric mean, (5.56) yields a larger estimate for the mixture α than does (5.57).

In cubic equations of state, parameters having the dimensions of molar volume are usually calculated by linear mixing rules. Thus, for parameter b in the van der Waals and Redlich/Kwong equations,

$$b = \sum_i x_i b_{ii} \qquad (5.58)$$

For parameter a in these equations, a quadratic mixing rule is used:

$$a = \sum_i \sum_j x_i x_j a_{ij} \qquad (5.59)$$

The geometric-mean combining rule is often employed for the a_{ij}:

$$a_{ij} = (a_{ii}a_{jj})^{1/2} \qquad (5.60)$$

EXAMPLE 5.14 Write the van der Waals EOS in virial form. Use the result to rationalize quadratic mixing rules for parameters a and b.

For $|b\rho| < 1$—that is, for $V > b$—

$$\frac{1}{1 - b\rho} = 1 + b\rho + b^2\rho^2 + b^3\rho^3 + \cdots$$

and hence the virial form of (5.43) is

$$Z = 1 + \left(b - \frac{a}{RT}\right)\rho + b^2\rho^2 + b^3\rho^3 + \cdots \tag{1}$$

One may interpret the coefficients of the powers of ρ in (1) as the virial coefficients *implied* by the van der Waals EOS:

$$B(\text{vdW}) = b - \frac{a}{RT} \qquad C(\text{vdW}) = b^2 \qquad D(\text{vdW}) = b^3 \qquad \cdots$$

$B(\text{vdW})$ provides a plausible qualitative description of the second virial coefficient; but $C(\text{vdW})$, $D(\text{vdW})$, . . . , are qualitatively unreasonable, being independent of T.

The *true* second virial coefficient is quadratic in mole fraction: $B = \sum_i \sum_j x_i x_j B_{ij}$. Comparison with $B(\text{vdW})$ suggests the mixing rules

$$a = \sum_i \sum_j x_i x_j a_{ij} \tag{2}$$

$$b = \sum_i \sum_j x_i x_j b_{ij} \tag{3}$$

These should be appropriate at *low* densities, where the first two terms of (1) provide a good approximation to the van der Waals EOS. For high densities, and particularly for liquid mixtures, they are at best rational first approximations which, like the van der Waals equation itself, give reasonable *qualitative* descriptions of real-fluid behavior. As was previously shown, (3) reduces to (5.58) under the combining rule (5.54).

EXAMPLE 5.15   Estimate the compressibility factor of air at 180 K and 100 bar.

For simplicity, assume air to contain 79-mole-percent nitrogen and 21-mole-percent oxygen. Use the van der Waals equation with mixing rules (5.58) and (5.59). Parameters for the pure species are determined from T_c and P_c by (5.51); the combining rule for the a_{ij} is given by (5.60). From Appendix C, with nitrogen as species 1 and oxygen as species 2,

$$T_{c1} = 126.2 \text{ K} \qquad T_{c2} = 154.5 \text{ K}$$
$$P_{c1} = 33.9 \text{ bar} \qquad P_{c2} = 50.5 \text{ bar}$$

By (5.51),

$$a_{11} = \frac{27}{64} \times \frac{(83.14)^2(126.2)^2}{33.9} = 1.370 \times 10^6 \text{ cm}^6 \cdot \text{bar} \cdot \text{mol}^{-2}$$

$$a_{22} = \frac{27}{64} \times \frac{(83.14)^2(154.5)^2}{50.5} = 1.380 \times 10^6 \text{ cm}^6 \cdot \text{bar} \cdot \text{mol}^{-2}$$

$$b_{11} = \frac{1}{8} \times \frac{(83.14)(126.2)}{33.9} = 38.69 \text{ cm}^3 \cdot \text{mol}^{-1}$$

$$b_{22} = \frac{1}{8} \times \frac{(83.14)(154.5)}{(50.5)} = 31.80 \text{ cm}^3 \cdot \text{mol}^{-1}$$

By (5.60), $a_{12} = (1.370 \times 1.380)^{1/2} \times 10^6 = 1.375 \times 10^6 \text{ cm}^6 \cdot \text{bar} \cdot \text{mol}^{-2}$. Then, by (5.59),

$$a = x_1^2 a_{11} + 2x_1 x_2 a_{12} + x_2^2 a_{22} = [(0.79)^2(1.370) + 2(0.79)(0.21)(1.375) + (0.21)^2(1.380)] \times 10^6$$
$$= 1.372 \times 10^6 \text{ cm}^6 \cdot \text{bar} \cdot \text{mol}^{-2}$$

By (5.58), $b = x_1 b_{11} + x_2 b_{22} = 37.24 \text{ cm}^3 \cdot \text{mol}^{-1}$.

Now substitute the numerical values of a and b together with $\rho = P/ZRT$ in the van der Waals equation (5.43) to find

$$Z^3 - \left(1 + \frac{bP}{RT}\right)Z^2 + \frac{aP}{R^2 T^2} Z - \left(\frac{aP}{R^2 T^2}\right)\left(\frac{bP}{RT}\right) = 0 \tag{5.61}$$

where

$$\frac{aP}{R^2 T^2} = \frac{(1.372 \times 10^6)(100)}{(83.14)^2(180)^2} = 0.6126 \qquad \frac{bP}{RT} = \frac{(37.24)(100)}{(83.14)(180)} = 0.2488$$

Thus we must solve the cubic equation $Z^3 - 1.2488 Z^2 + 0.6126 Z - 0.1524 = 0$. This equation has only one real root, $Z = 0.6762$, and this is the predicted compressibility factor for air at 180 K and 100 bar. A handbook value is $Z = 0.7084$; at these severe conditions, the van der Waals equation is in error by about 5%. As shown in Table 5-2, performance improves at lower pressures and deteriorates further as the pressure level increases. However, the qualitative features of the isotherm are reproduced.

Table 5-2

P/bar	Z for Air at 180 K	
	Handbook	van der Waals
1	0.9967	0.9964
5	0.9832	0.9816
10	0.9660	0.9629
20	0.9314	0.9243
40	0.8625	0.8432
60	0.7977	0.7611
80	0.7432	0.6965
100	0.7084	0.6762
200	0.7986	0.8907
300	1.0068	1.1758
400	1.2232	1.4609
500	1.4361	1.7418

EXAMPLE 5.16 Algebraic solution of pressure-explicit equations of state is generally inconvenient and may be impossible. (a) Outline the *fixed-point iteration method* for numerical solution of the equation $F(X) = 0$. (b) Apply this method to the van der Waals equation (5.61).

(a) The basis for solution by fixed-point iteration is a transformation of the original equation into the form $X = f(X)$, where $f(X)$ is a new function of X. A solution to the transformed equation—i.e., an argument mapped into itself by f—is called a *fixed point* of $f(X)$. Let X_0 be an initial estimate ("zeroth approximation") to a fixed point, and consider the sequence of approximations defined by

$$X_{j+1} = f(X_j) \qquad (j = 0, 1, 2, \ldots)$$

If X_0 is well chosen—a matter both of physical understanding and of mathematical experience—the sequence will converge to the desired fixed point, and iteration may be terminated when $|X_{j+1} - X_j| < \varepsilon$, where ε is a prescribed tolerance.

(b) At sufficiently low temperatures ($T < T_c$ for a pure fluid), (5.61) can have three real roots. The smallest root is for the liquid state and the largest is for the vapor state; the middle root has no physical significance (Example 5.13). We therefore seek *two* transformations $Z = f(Z)$ of (5.61), one for liquids and one for vapors. Inspiration for suitable transformations is provided by the following rearrangement of (5.61):

$$Z^2\left(Z - 1 - \frac{bP}{RT}\right) + \frac{aP}{RT^2}\left(Z - \frac{bP}{RT}\right) = 0 \tag{5.62}$$

For vapor states, $f(Z)$ should be of order unity; thus we focus on the group of terms in the first pair of parentheses in (5.62). Solution for this group and rearrangement gives

$$Z = 1 + \frac{bP}{RT} - \frac{aP}{Z^2 R^2 T^2}\left(Z - \frac{bP}{RT}\right) \tag{5.63}$$

Equation (5.63) with initial estimate $Z_0 = 1$, is a basis for finding Z for a vapor by fixed-point iteration.

For liquid states, Z is normally small, being bounded below by bP/RT (the value corresponding to $V = b$, the minimum molar volume allowed by the van der Waals equation). Thus $f(Z)$ should be dominated by bP/RT, a result obtained by solution for the terms in the second pair of parentheses in (5.62). We find, on rearrangement,

$$Z = \frac{bP}{RT} + \frac{Z^2 R^2 T^2}{aP}\left(1 + \frac{bP}{RT} - Z\right) \tag{5.64}$$

Equation (5.64), with initial estimate $Z_0 = 0$ or $Z_0 = bP/RT$, is a basis for finding Z for a liquid by fixed-point iteration.

The usefulness of (5.63) and (5.64) is not restricted to conditions for which *both* vapor and liquid states are possible. Equation (5.63) is appropriate for superheated vapors and (5.64) for subcooled liquids. Supercritical fluids are generally best treated with (5.63), although convergence may be slow if the density is high.

EXAMPLE 5.17 Derive expressions for the residual Gibbs energy, residual enthalpy, and residual entropy for a fluid described by the van der Waals EOS. How does application of these expressions differ from application of similar results (see, e.g., Example 5.11) for a *volume*-explicit EOS?

The van der Waals equation is explicit in pressure, indicating use of (5.15) through (5.17). From (5.43), we find

$$Z - 1 = \frac{b\rho}{1 - b\rho} - \frac{a\rho}{RT} \qquad \left(\frac{\partial Z}{\partial T}\right)_{\rho,x} = \frac{a\rho}{RT^2}$$

Substitution into (5.15) through (5.17) and integration yields, on rearrangement,

$$\frac{G^R}{RT} = -\frac{a\rho}{RT} + Z - 1 - \ln(1 - b\rho)Z \tag{5.65}$$

$$\frac{H^R}{RT} = -\frac{a\rho}{RT} + Z - 1 \tag{5.66}$$

$$\frac{S^R}{RT} = \ln(1 - b\rho)Z \tag{5.67}$$

Temperature and *pressure* are usually fixed in applications, whereas (5.65) through (5.67), with (5.43), give the residual properties as functions of temperature and *density*. Thus, with T and P fixed, calculation of residual properties from the van der Waals equation requires prior solution of the EOS for the appropriate Z and ρ. The fixed-point procedures just described are appropriate. This extra numerical effort is not required with a volume-explicit EOS.

5.4 CORRESPONDING-STATES CORRELATIONS

Although the PVT behavior of all fluids is qualitatively similar, gross differences may be observed when comparisons are made of the same thermodynamic property of two different fluids at, say, the same T and P. For convenience in correlation and prediction of thermodynamic properties, it is desirable to reduce available data for different substances to some common basis. The classical approach to this goal is motivated by Example 5.13, where it is shown that the van der Waals equation can be put into a form containing only pure numbers and the dimensionless variables P_r, V_r, and T_r. The resulting equation (5.52) implies that all fluids, when compared at the same T_r and P_r, have the same V_r. This is the *theorem of corresponding states*. It is found that only a few substances actually conform to this principle, but extensions of the corresponding-states concept have led to the development of correlations which unify to a reasonable degree the volumetric behavior of a large number of fluids.

The simplest of these modern correlations are based on the premise that all fluids having the same value of some dimensionless material parameter should have the same V_r when compared at the same T_r and P_r; they are sometimes called *three-parameter* corresponding-states correlations. The most obvious choice for a third parameter is Z_c (see Problem 5.16). However, determination of Z_c requires a value for V_c, a difficult quantity to measure. A parameter subject to more precise determination is *Pitzer's acentric factor* ω.

In Section 5.2 we noted that interactions among molecular force fields are manifested as PVT behavior on the macroscopic level. The simplest type of molecular force field is spherical, and we expect that the behavior of fluids consisting of molecules having fields of this type should conform to some simple scheme against which we could compare the behavior of more complex substances. Indeed it is found that these *simple fluids*, exemplified by argon, krypton, and xenon, conform very closely to the theorem of corresponding states. The basis for the definition of the acentric factor is the observation that, for Ar, Kr, and Xe,

$$P_r^{\text{sat}} \approx 0.1 \quad \text{at} \quad T_r = 0.7$$

It is also observed that $P_r^{sat} < 0.1$ at $T_r = 0.7$ for most other fluids, and that P_r^{sat} at $T_r = 0.7$ generally decreases with increasing asymmetry of the molecular force field. Accordingly, ω is defined as

$$\omega = \log_{10}(P_r^{sat})_{\substack{\text{simple fluid} \\ \text{at } T_r = 0.7}} - \log_{10}(P_r^{sat})_{T_r = 0.7}$$

Since P_r^{sat} for the simple fluids is approximately 0.1 at $T_r = 0.7$, this definition reduces to

$$\omega = -1 - \log_{10}(P_r^{sat})_{T_r = 0.7} \tag{5.68}$$

Thus ω can be determined from T_c, P_c and a single vapor-pressure measurement made at $T_r = 0.7$. A list of values for the acentric factor for some common fluids is given in Appendix C, along with data for T_c, P_c, V_c, and Z_c.

A correlation incorporating Pitzer's acentric factor and suitable for dilute gases is based on the corresponding-states form of (5.33):

$$Z = 1 + \hat{B}\frac{P_r}{T_r} \tag{5.69}$$

Here, $\hat{B}$ is the reduced second virial coefficient,

$$\hat{B} \equiv \frac{BP_c}{RT_c} \tag{5.70}$$

The second virial coefficient of a pure fluid depends on T only; hence, for a three-parameter (ω, T_r, P_r) correlation, we write $\hat{B} = \hat{B}(T_r; \omega)$. What is needed is an explicit expression for $\hat{B}(T_r; \omega)$.

Pitzer proposed the following general decomposition of Z:

$$Z = Z^0(T_r, P_r) + \omega Z^1(T_r, P_r) \tag{5.71}$$

Here Z^0 is a two-parameter corresponding-states correlation of Z for simple fluids, for which $\omega = 0$. The second term is a correction for nonsimple fluid behavior in which, by assumption, the effect of molecular asymmetry (as manifested in ω) is separable from effects of temperature and pressure (as manifested in Z^1). Consistent with (5.71) and (5.69), we therefore write

$$\hat{B} = B^0(T_r) + \omega B^1(T_r) \tag{5.72}$$

where B^0 and B^1 are dimensionless functions of T_r.

The adequacy of (5.72) as a correlating scheme is illustrated by Fig. 5-8, where there are plotted data for B^0 [$= \hat{B}$ for simple fluids ($\omega = 0$)], and also values for B^1 as deduced from data for eleven nonsimple but nonpolar, nonassociating fluids. The curves in this figure are computed from the empirical correlating equations

$$B^0 = 0.083 - \frac{0.422}{T_r^{1.6}} \qquad B^1 = 0.139 - \frac{0.172}{T_r^{4.2}} \tag{5.73}$$

which are seen to provide satisfactory representation of the data for reduced temperatures between about 0.6 and 2.0. Equation (5.69) for Z, with $\hat{B}$ defined by (5.70) and correlated by (5.72) and (5.73), is the basis for corresponding-states calculations in this Outline.

Any corresponding-states correlation for Z implies a full set of correlations for the residual properties. With respect to the present correlation, we refer to Example 5.11, where are developed the expressions for G^R, H^R, and S^R implied by the two-term virial equation (5.33). When expressed in terms of T_r, P_r, and $\hat{B}$, the results (5.37) through (5.39) become

$$\frac{G^R}{RT} = \hat{B}\frac{P_r}{T_r} \tag{5.74}$$

$$\frac{H^R}{RT} = -\left(\frac{d\hat{B}}{dT_r} - \frac{\hat{B}}{T_r}\right)P_r \tag{5.75}$$

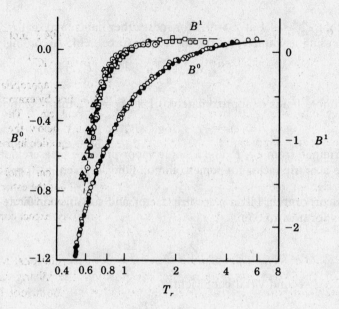

Fig. 5-8

$$\frac{S^R}{R} = -\frac{d\hat{B}}{dT_r} P_r \tag{5.76}$$

Here the derivative follows from (5.72) and (5.73):

$$\frac{d\hat{B}}{dT_r} = \frac{dB^0}{dT_r} + \omega \frac{dB^1}{dT_r} \qquad \text{with} \qquad \frac{dB^0}{dT_r} = \frac{0.675}{T_r^{2.6}} \qquad \frac{dB^1}{dT_r} = \frac{0.722}{T_r^{5.2}} \tag{5.77}$$

EXAMPLE 5.18 Estimate Z, G^R, H^R, and S^R for sulfur hexafluoride gas at 100 °C and 10 bar. For SF_6, $T_c = 318$ K, $P_c = 37.6$ bar, and $\omega = 0.286$.

For the stated conditions,

$$T_r = \frac{100 + 273.15}{318} = 1.173 \qquad P_r = \frac{10}{37.6} = 0.266$$

Equations (5.73) and (5.77) give

$$B^0 = 0.083 - \frac{0.422}{(1.173)^{1.6}} = -0.244 \qquad B^1 = 0.139 - \frac{0.172}{(1.173)^{4.2}} = 0.051$$

$$\frac{dB^0}{dT_r} = \frac{0.675}{(1.173)^{2.6}} = 0.446 \qquad \frac{dB^1}{dT_r} = \frac{0.722}{(1.173)^{5.2}} = 0.315$$

Thus, by (5.72) and (5.77),

$$\hat{B} = -0.244 + (0.286)(0.051) = -0.229$$

$$\frac{d\hat{B}}{dT_r} = 0.446 + (0.286)(0.315) = 0.536$$

and therefore, by (5.69) and (5.74) through (5.76),

$$Z = 1 - \frac{(0.229)(0.266)}{1.173} = 0.948$$

$$G^R = (8.314)(373.15) \frac{(-0.229)(0.266)}{1.173} = -161 \text{ J} \cdot \text{mol}^{-1}$$

$$H^R = -(8.314)(373.15)\left(0.536 + \frac{0.229}{1.173}\right)(0.266) = -604 \text{ J} \cdot \text{mol}^{-1}$$

$$S^R = -(8.314)(0.536)(0.266) = -1.19 \text{ J} \cdot \text{mol}^{-1} \cdot \text{K}^{-1}$$

How good are these estimates? There are two determining factors: the appropriateness of the underlying EOS (5.33) and the validity of the correlation for $\hat{B}$. We may check the first by examining the reduced density. Since $\rho = P/ZRT$, we find from the estimated Z that $\rho = 0.340 \times 10^{-3}$ mol $\cdot$ cm^{-3}. The critical density of SF$_6$ is $\rho_c = 5.05 \times 10^{-3}$ mol $\cdot$ cm^{-3}; thus $\rho_r \equiv \rho/\rho_c = 0.067$. This is substantially below the range $\rho_r = 0.25$ to 0.50 generally taken to represent the upper limit for use of the two-term virial equation in pressure. We conclude that the EOS is suitable for this application.

The value for B determined from the correlation is $\hat{B}RT_c/P_c = -161$ cm$^3 \cdot$ mol^{-1}, which is in excellent agreement with a reported value of -163 cm$^3 \cdot$ mol^{-1} for SF$_6$ at 100 °C. Thus the estimates for Z and for $G^R = RT(Z-1)$, should be quite reliable. The accuracies of H^R and S^R are more difficult to assess, because these quantities depend on the temperature *derivative* $d\hat{B}/dT_r$; generally, one must expect computed values of H^R and S^R to be less reliable than the corresponding values for Z and G^R.

The correlation just described, like all corresponding-states correlations, was developed for pure substances. Extension to mixtures is done in this case through the mixing rule for B, (5.28). We generalize the earlier definition (5.70) of a reduced second virial coefficient by writing

$$\hat{B}_{ij} \equiv \frac{B_{ij}P_{c,ij}}{RT_{c,ij}} \tag{5.78}$$

When $i = j$, the doubly subscripted corresponding-states parameters $T_{c,ij}$ and $P_{c,ij}$ are the usual pure-species properties; when $i \neq j$, they are empirical *interaction parameters*, whose values must be estimated. By (5.28) and (5.78), the mixture second virial coefficient is

$$B = \sum_i \sum_j x_i x_j \frac{\hat{B}_{ij}RT_{c,ij}}{P_{c,ij}} \tag{5.79}$$

Combination of (5.79) and (5.33) gives the equation of state for the mixture:

$$Z = 1 + \sum_i \sum_j x_i x_j \hat{B}_{ij} \frac{P_{r,ij}}{T_{r,ij}} \tag{5.80}$$

Here, $T_{r,ij} \equiv T/T_{c,ij}$ and $P_{r,ij} \equiv P/P_{c,ij}$. The corresponding expressions for residual properties are analogs of (5.74) through (5.76):

$$\frac{G^R}{RT} = \sum_i \sum_j x_i x_j \hat{B}_{ij} \frac{P_{r,ij}}{T_{r,ij}} \tag{5.81}$$

$$\frac{H^R}{RT} = -\sum_i \sum_j x_i x_j \left(\frac{d\hat{B}_{ij}}{dT_{r,ij}} - \frac{\hat{B}_{ij}}{T_{r,ij}}\right) P_{r,ij} \tag{5.82}$$

$$\frac{S^R}{R} = -\sum_i \sum_j x_i x_j \frac{d\hat{B}_{ij}}{dT_{r,ij}} P_{r,ij} \tag{5.83}$$

Equations (5.79) through (5.83) can be used with any correlation for $\hat{B}_{ij}$; one requires an explicit expression for $\hat{B}_{ij}$ as a function of $T_{r,ij}$, and also recipes for evaluating interaction parameters. Employing the Pitzer formalism, we write, analogous to (5.72),

$$\hat{B}_{ij} = B^0(T_{r,ij}) + \omega_{ij}B^1(T_{r,ij}) \tag{5.84}$$

where B^0 and B^1 are the same *functions* as in (5.73). For estimating interaction parameters, we use a procedure recommended by Prausnitz et al. (*Molecular Thermodynamics of Fluid-Phase Equilibria*, 2nd ed., Sec. 5.7, Prentice-Hall, Englewood Cliffs, N.J., 1986). The required equations are:

$$T_{c,ij} = (T_{c,ii}T_{c,jj})^{1/2} \tag{5.85}$$

$$Z_{c,ij} = \tfrac{1}{2}(Z_{c,ii} + Z_{c,jj}) \qquad (5.86)$$

$$V_{c,ij} = \tfrac{1}{8}(V_{c,ii}^{1/3} + V_{c,jj}^{1/3})^3 \qquad (5.87)$$

$$P_{c,ij} = \frac{Z_{c,ij} R T_{c,ij}}{V_{c,ij}} \qquad (5.88)$$

$$\omega_{ij} = \tfrac{1}{2}(\omega_{ii} + \omega_{jj}) \qquad (5.89)$$

When $i = j$ all of these equations reduce to the appropriate values for a pure species; furthermore, (5.84) becomes identical to (5.72). This extension of the correlation to mixtures is essentially empirical and can be expected in general to give only estimates of second virial cross-coefficients. It is most reliable for mixtures of molecules of similar size and chemical type.

Solved Problems

FORMULATIONS OF THE EOS   (Section 5.1)

5.1   Show that the heat of vaporization ΔH^{lv} of a pure fluid is related to the vapor pressure P^{sat} by

$$\Delta H^{lv} = -R \, \Delta Z^{lv} \, \frac{d \ln P^{\text{sat}}}{d(1/T)}$$

where ΔZ^{lv} is the change of the compressibility factor on vaporization.

  Start with the Clapeyron equation

$$\frac{dP^{\text{sat}}}{dT} = \frac{\Delta H^{lv}}{T \, \Delta V^{lv}} \qquad (4.3b)$$

But $V = (RT/P)Z$ implies

$$\Delta V^{lv} = \frac{RT \, \Delta Z^{lv}}{P^{\text{sat}}} \qquad (1)$$

Also, we have the mathematical identity

$$\frac{dP^{\text{sat}}}{dT} = -\frac{P^{\text{sat}}}{T^2} \frac{d \ln P^{\text{sat}}}{d(1/T)} \qquad (2)$$

Combination of (4.3b), (1), and (2) gives, after rearrangement, the desired expression. Note the similarity between this exact result and the approximate Clausius/Clapeyron equation, (4.4), which obtains under the approximation $Z^v = 1$ (vapor phase of ideal gas) and $Z^l = 0$ (liquid phase of negligible volume).

5.2   Derive expressions for the Joule/Thomson coefficient (Problem 3.4) in terms of (a) T, P, C_P, and Z and its derivatives; and (b) T, ρ, C_P, and Z and its derivatives. (c) From the results of (a) and (b), infer equations for the *inversion curve* in the TP and $T\rho$ planes.

  From Problem 3.4,

$$\left(\frac{\partial T}{\partial P}\right)_H = -\frac{1}{C_P}\left[V - T\left(\frac{\partial V}{\partial T}\right)_P\right] \qquad (1)$$

(a) By (5.8),

$$\left(\frac{\partial V}{\partial T}\right)_P = \frac{RT}{P}\left[\left(\frac{\partial Z}{\partial T}\right)_P + \frac{Z}{P}\right]$$

Also, $V = ZRT/P$. Substitution in (1) gives

$$\left(\frac{\partial T}{\partial P}\right)_H = \frac{RT^2}{C_P P}\left(\frac{\partial Z}{\partial T}\right)_P \tag{2}$$

(b) By (3.12) and (3.13),

$$\left(\frac{\partial V}{\partial T}\right)_P = -\left(\frac{\partial P}{\partial T}\right)_V \Big/ \left(\frac{\partial P}{\partial V}\right)_T$$

Expressions for $(\partial P/\partial T)_V$ and $(\partial P/\partial V)_T$ are given as (5.10) and (5.11); from them,

$$\left(\frac{\partial V}{\partial T}\right)_P = \frac{V}{T}\left[\frac{Z + T\left(\frac{\partial Z}{\partial T}\right)_V}{Z - V\left(\frac{\partial Z}{\partial V}\right)_T}\right]$$

Substitution into (1) and elimination of molar volume V in favor of molar density ρ gives finally

$$\left(\frac{\partial T}{\partial P}\right)_H = \frac{1}{C_P \rho}\left[\frac{T\left(\frac{\partial Z}{\partial T}\right)_\rho - \rho\left(\frac{\partial Z}{\partial \rho}\right)_T}{Z + \rho\left(\frac{\partial Z}{\partial \rho}\right)_T}\right] \tag{3}$$

(c) The inversion curve of a fluid is the locus of vanishing Joule/Thomson coefficient. Thus, (2) and (3) give the respective equations

$$\left(\frac{\partial Z}{\partial T}\right)_P = 0 \tag{4}$$

$$T\left(\frac{\partial Z}{\partial T}\right)_\rho - \rho\left(\frac{\partial Z}{\partial \rho}\right)_T = 0 \tag{5}$$

More often than not, the inversion curve is determined from a pressure-explicit EOS, rendering (5) the more appropriate.

5.3 Isochores of a pure fluid are observed to be approximately linear on a PT-diagram (Problem 4.2). Develop a general form of the $Z\rho T$ equation of state that is consistent with this observation.

The slope of a *linear* isochore depends only on density:

$$\left(\frac{\partial P}{\partial T}\right)_\rho = f(\rho)$$

Integration gives $P = f(\rho)T + g(\rho)$, or

$$Z = F(\rho) + \frac{G(\rho)}{RT} \tag{1}$$

where f, g, F, and G are arbitrary functions. Equation (1) is the required EOS. The simplest vapor-phase EOS producing linear isochores is the ideal-gas equation, for which $F = 1$ and $G = 0$. According to (3.22), a liquid substance of constant β and κ obeys (1) if $F(\rho) = \beta/\kappa R\rho$. The van der Waals equation of state gives linear isochores for both vapors and liquids; here $F(\rho) = (1 - b\rho)^{-1}$ and $G(\rho) = -a\rho$.

5.4 (a) Express the heat of vaporization ΔH^{lv} in terms of residual enthalpies. (b) Under what conditions is the residual enthalpy $H^{R,l}$ of a saturated liquid approximately equal to $-\Delta H^{lv}$?

(a) The definition $\Delta H^{lv} \equiv H^v(T, P^{sat}) - H^l(T, P^{sat})$ is unchanged if we subtract from H^v and H^l the ideal-gas enthalpy $H^{ig}(T, P^{sat})$. Hence,

$$\Delta H^{lv} = H^{R,v} - H^{R,l} \tag{1}$$

Analogs of (1) may be written for any property change of vaporization ΔM^{lv}.

(b) If the temperature level is sufficiently low, P^{sat} is very small and the vapor phase behaves approximately as an ideal gas; i.e., $H^{R,v} \approx 0$. Then (1) becomes

$$H^{R,l} \approx -\Delta H^{lv} \tag{2}$$

Table 5-3 gives $-\Delta H^{lv}$ and residual properties for saturated water as determined from steam tables (see Problem 4.30). For water, the approximation provided by (2) is excellent between the triple point (0.01 °C) and the normal boiling point (100 °C). As temperature and pressure increase, the absolute magnitude of $H^{R,v}$ increases, and (2) is increasingly in error. At the gas/liquid critical point (374.15 °C and 221.20 bar), the heat of vaporization is zero; here, $H^{R,l} = H^{R,v}$, and (2) becomes meaningless.

Table 5-3

$T/°C$	P^{sat}/bar	$H^{R,v}/\text{kJ} \cdot \text{mol}^{-1}$	$H^{R,l}/\text{kJ} \cdot \text{mol}^{-1}$	$-\Delta H^{lv}/\text{kJ} \cdot \text{mol}^{-1}$
0.01	0.006	−0.004	−45.083	−45.079
50	0.123	−0.043	−42.983	−42.940
100	1.013	−0.227	−40.896	−40.669
150	4.760	−0.690	−38.770	−38.080
200	15.55	−1.607	−36.541	−34.934
250	39.78	−3.195	−34.094	−30.899
300	85.93	−5.871	−31.207	−25.336
350	165.35	−10.989	−27.133	−16.144
374.15	221.20	−20.173	−20.173	0.000

5.5 The $Z\rho T$ equation of state is often modeled as

$$Z = Z_0(\rho, x) + Z_A(T, \rho, x) \tag{1}$$

Here, Z_0 is the "hard-particle," or "infinite-temperature," term; it reflects the harsh repulsive component of the intermolecular potential energy. Quantity Z_A represents the contribution to the EOS of intermolecular attractions and soft repulsions. Derive and discuss expressions for the residual properties A^r, U^r, and S^r (See Example 5.4).

From (1),

$$Z - 1 = (Z_0 - 1) - Z_A \qquad \left(\frac{\partial Z}{\partial T}\right)_{\rho,x} = \left(\frac{\partial Z_A}{\partial T}\right)_{\rho,x}$$

(The hard-particle Z, like Z itself, tends to unity as density approaches zero; hence the -1 is grouped with Z_0.) Substitution of these two expressions in (5.18) through (5.20) yields

$$\frac{A^r}{RT} = \int_0^\rho (Z_0 - 1) \frac{d\rho}{\rho} + \int_0^\rho Z_A \frac{d\rho}{\rho} \qquad \text{(constant } T, x) \tag{2}$$

$$\frac{U^r}{RT} = -T \int_0^\rho \left(\frac{\partial Z_A}{\partial T}\right)_{\rho,x} \frac{d\rho}{\rho} \qquad \text{(constant } T, x) \tag{3}$$

$$\frac{S^r}{R} = -\int_0^\rho (Z_0 - 1) \frac{d\rho}{\rho} - \int_0^\rho \left[Z_A + T\left(\frac{\partial Z_A}{\partial T}\right)_{\rho,x}\right] \frac{d\rho}{\rho} \qquad \text{(constant } T, x) \tag{4}$$

Of the three residual properties, only the internal energy is independent of the harsh repulsions. An important special case obtains when $Z_A = T^{-1}G(\rho, x)$, with G an arbitrary function of ρ and x. Then the second integral in (4) vanishes, and S^r reflects harsh repulsions only. Here one can make an intuitive contrast between repulsions and "entropic effects" on the one hand, and attractions and "energetic effects" on the other. We identify in Problem 5.18 the entropic and energetic contributions to the van der Waals equation of state.

VIRIAL EQUATIONS OF STATE   (Section 5.2)

5.6   Derive the conversions (5.23).

Solve (5.22) for P:

$$P = RT(\rho + B\rho^2 + C\rho^3 + D\rho^4 + \cdots)   \qquad (1)$$

Substitute (1) into (5.21):

$$Z = 1 + B'RT(\rho + B\rho^2 + C\rho^3 + D\rho^4 + \cdots)$$
$$+ C'(RT)^2(\rho + B\rho^2 + C\rho^3 + D\rho^4 + \cdots)^2$$
$$+ D'(RT)^3(\rho + B\rho^2 + C\rho^3 + D\rho^4 + \cdots)^3 + \cdots$$

or, collecting coefficients,

$$Z = 1 + (B'RT)\rho + [BB'RT + C'(RT)^2]\rho^2 + [CB'RT + 2BC'(RT)^2 + D'(RT)^3]\rho^3 + \cdots   \qquad (2)$$

Term-by-term comparison of (2) with (5.22) gives the desired results. In general, the kth coefficient of one of the virial expansions will depend on the first k coefficients of the other. Note that these results are strictly valid only when the two *infinite* series are compared.

5.7   Derive an expression for the residual volume V^R as a function of T, P, and the density-series virial coefficients.

Combine (4.30), (5.21), and (5.23):

$$V^R = \frac{RT}{P}(Z - 1) = \frac{RT}{P}(B'P + C'P^2 + \cdots)$$

$$= \frac{RT}{P}\left[\frac{BP}{RT} + \frac{C - B^2}{(RT)^2}P^2 + \cdots\right] = B + \frac{C - B^2}{RT}P + \cdots$$

In agreement with the discussion of Example 4.12, the last series shows that V^R remains finite and generally nonzero in the zero-pressure limit: $\lim_{P \to 0} V^R = B$. This limiting V^R clearly also depends on temperature and composition.

5.8   (a) Derive an expression for the Joule/Thomson coefficient from (5.21). (b) Compare the Joule/Thomson coefficients of an ideal and a real gas in the zero-pressure limit.

(a)   From (5.21) and (2) of Problem 5.2,

$$\left(\frac{\partial T}{\partial P}\right)_H = \frac{RT^2}{C_P P}\left(\frac{dB'}{dT}P + \frac{dC'}{dT}P^2 + \cdots\right)$$

$$= \frac{RT^2}{C_P}\left(\frac{dB'}{dT} + \frac{dC'}{dT}P + \cdots\right)   \qquad (1)$$

Note that (1) is not a true power series, for C_P is in general a function of P (and T).

(b)   The Joule/Thomson coefficient of an ideal gas is identically zero (Problem 3.27). For a real gas, as $P \to 0$, $C_P \to C_P^{ig}$, a nonzero function of T. Hence, by (1),

$$\lim_{P \to 0}\left(\frac{\partial T}{\partial P}\right)_H = \frac{RT^2}{C_P^{ig}}\frac{dB'}{dT}   \qquad (5)$$

Thus the Joule/Thomson coefficient of a real gas at zero pressure is generally nonzero, and depends on temperature and composition.

5.9 The second virial coefficient of an equimolar binary vapor mixture of methane and *n*-hexane is $-517 \text{ cm}^3 \cdot \text{mol}^{-1}$ at 50 °C. What is B at the same temperature for a mixture containing 25 mole % methane and 75 mole % *n*-hexane? At 50 °C, $B = -33 \text{ cm}^3 \cdot \text{mol}^{-1}$ for methane and $-1512 \text{ cm}^3 \cdot \text{mol}^{-1}$ for *n*-hexane?

Let methane $\equiv 1$ and *n*-hexane $\equiv 2$. B_{11} and B_{22} are given, and B_{12} must be determined from the given mixture B. For a binary mixture, by (5.28),

$$B = x_1^2 B_{11} + 2x_1 x_2 B_{12} + x_2^2 B_{22} \tag{1}$$

from which

$$B_{12} = \frac{B - x_1^2 B_{11} - x_2^2 B_{22}}{2x_1 x_2} = \frac{-517 - (0.5)^2(-33) - (0.5)^2(-1512)}{(2)(0.5)(0.5)} = -262 \text{ cm}^3 \cdot \text{mol}^{-1}$$

For the 25%/75% mixture, B can now be calculated from (1) as

$$B = (0.25)^2(-33) + (2)(0.25)(0.75)(-262) + (0.75)^2(-1512) = -951 \text{ cm}^3 \cdot \text{mol}^{-1}$$

5.10 A handbook gives $Z = 0.9764$ for carbon dioxide gas at 100 °C and 10 bar. Estimate Z for CO_2 at 100 °C and 20 bar.

This is an exercise in *extrapolation*. Only a single data point is available. We therefore require a one-parameter equation of state, and choose (5.33), written in the form

$$\frac{B}{RT} = \frac{Z - 1}{P}$$

On the isotherm $T = 100$ °C, B/RT is a constant, because for a pure fluid B depends on T only. Thus,

$$\frac{Z_1 - 1}{P_1} = \frac{Z_2 - 1}{P_2} \quad \text{or} \quad Z_2 = 1 + (Z_1 - 1)\frac{P_2}{P_1}$$

Substitution of numerical values gives

$$Z_2 = 1 + (0.9764 - 1)\left(\frac{20}{10}\right) = 0.9528$$

This is in essential agreement with the handbook value 0.9524, illustrating the suitability of (5.33) for representing the isotherm at these pressure levels. Were an extrapolation to substantially higher pressures required, the procedure would be inappropriate.

5.11 0.01 kg of steam at 260 °C expands reversibly and isothermally in a piston-and-cylinder apparatus from an initial volume of 2 L to a final volume 20 L. How much work is done by the steam?

Use (5.34) and the B and C for steam at 260 °C determined in Example 5.7. The reversible work is calculated from

$$W_{\text{rev}} = n \int_{V_1}^{V_2} P \, dV \tag{1}$$

where V is the molar volume of steam and n is the total number of moles of steam. From (5.34),

$$P = RT\left(\frac{1}{V} + \frac{B}{V^2} + \frac{C}{V^3}\right) \tag{2}$$

Substitution of (2) in (1) gives

$$W_{\text{rev}} = nRT\left[\ln\frac{V_2}{V_1} + \frac{V_2 - V_1}{V_1 V_2}\left(B + \frac{C}{2}\frac{V_1 + V_2}{V_1 V_2}\right)\right] \tag{3}$$

From the statement of the problem,

$$n = \frac{0.01}{18.02} = 0.000\,554\,9 \text{ kmol} = 0.5549 \text{ mol}$$

$$V_1 = \frac{2}{0.5549} = 3.604 \text{ L} \cdot \text{mol}^{-1} \qquad V_2 = 10V_1 = 36.04 \text{ L} \cdot \text{mol}^{-1}$$

and, from Example 5.7,

$$B = -0.1422 \text{ L} \cdot \text{mol}^{-1} \qquad C = -0.007\,14 \text{ L}^2 \cdot \text{mol}^{-2}$$

Substitution of these quantities into (3), with $R = 8.314$ J·mol^{-1}·K^{-1} and $T = 533.15$ K, yields $W_{\text{rev}} = 5575$ J.

We may check the validity of (2) for this application by calculating the initial and final pressures, with $R = 0.083\,14$ L·bar·mol^{-1}·K^{-1}:

$$P_1 = 11.80 \text{ bar} \qquad P_2 = 1.225 \text{ bar}$$

Comparison of these pressures with those in Table 5-1 shows that the three-term virial equation is entirely adequate in this range.

5.12 Derive expressions for the residual Gibbs energy, residual enthalpy, and residual entropy of a gas described by (5.34), the three-term virial equation in density.

Equation (5.34) is explicit in pressure, indicating application of (5.15) through (5.17). From (5.34), we obtain

$$Z - 1 = B\rho + C\rho^2 \qquad \left(\frac{\partial Z}{\partial T}\right)_{\rho,x} = \frac{dB}{dT}\rho + \frac{dC}{dT}\rho^2$$

Substitution in (5.15) through (5.17) and integration gives

$$\frac{G^R}{RT} = 2B\rho + \frac{3}{2}C\rho^2 - \ln Z \tag{1}$$

$$\frac{H^R}{RT} = -T\left(\frac{dB}{dT} - \frac{B}{T}\right)\rho - \frac{1}{2}T\left(\frac{dC}{dT} - 2\frac{C}{T}\right)\rho^2 \tag{2}$$

$$\frac{S^R}{R} = -T\left(\frac{dB}{dT} + \frac{B}{T}\right)\rho - \frac{1}{2}T\left(\frac{dC}{dT} + \frac{C}{T}\right)\rho^2 + \ln Z \tag{3}$$

Equations (1) through (3) may be compared with (5.37) through (5.39), their counterparts for the *two-term* virial equation in *pressure*.

5.13 One mole of methane gas is mixed with one mole of n-hexane gas under conditions of constant total volume and constant temperature. If the initial temperature and pressure of both pure gases are 75 °C and 100 kPa, what is the pressure change on mixing? For methane (1) and n-hexane (2) at 75 °C, $B_{11} = -26$ cm^3·mol^{-1}, $B_{22} = -1239$ cm^3·mol^{-1}, and $B_{12} = -180$ cm^3·mol^{-1}.

For the two pure gases we may write

$$PV_1^t = n_1 Z_1 RT \tag{1}$$

$$PV_2^t = n_2 Z_2 RT \tag{2}$$

Similarly, for the gas mixture,

$$P_f V_f^t = n Z_f RT \tag{3}$$

where the subscript f denotes the final, mixed condition and where $n = n_1 + n_2$. But by the statement of the problem, $V_f^t = V_1^t + V_2^t$, and (3) becomes

$$P_f(V_1^t + V_2^t) = n Z_f RT$$

Substitution in this last equation of V_1^t and V_2^t from (1) and (2) yields on rearrangement

$$P_f = \frac{n Z_f P}{n_1 Z_1 + n_2 Z_2} \tag{4}$$

The pressure change on mixing is $\Delta P \equiv P_f - P$, and the mole fractions of 1 and 2 in the mixture are

$x_1 = n_1/n$ and $x_2 = n_2/n$. Thus, from (4),

$$\Delta P = \left[\frac{Z_f - (x_1 Z_1 + x_2 Z_2)}{x_1 Z_1 + x_2 Z_2} \right] P \qquad (5)$$

For this problem, the gases are adequately described by the truncated virial equation (5.33); thus

$$Z_f = 1 + \frac{BP_f}{RT} = 1 + \frac{B(P + \Delta P)}{RT} \qquad Z_1 = 1 + \frac{B_{11}P}{RT} \qquad Z_2 = 1 + \frac{B_{22}P}{RT}$$

where B is the second virial coefficient of the mixture. Substitution of these expressions into (5) and solution for ΔP yields

$$\Delta P = \frac{(B - x_1 B_{11} - x_2 B_{22})P^2}{RT - (B - x_1 B_{11} - x_2 B_{22})P}$$

This equation can be abbreviated through the use of the δ_{ij}-notation of Example 5.9. Thus, by (5.31),

$$B - x_1 B_{11} - x_2 B_{22} = x_1 x_2 \delta_{12} \qquad \text{where} \qquad \delta_{12} = 2B_{12} - B_{11} - B_{22}$$

so that

$$\Delta P = \frac{x_1 x_2 \delta_{12} P^2}{RT - x_1 x_2 \delta_{12} P} \qquad (6)$$

From the statement of the problem, $\delta_{12} = 2(-180) - (-26) - (-1239) = 905 \text{ cm}^3 \cdot \text{mol}^{-1}$ and $x_1 = x_2 = 0.5$, $P = 100$ kPa, $T = 75\,°C = 348.2$ K. Therefore, with $R = 8314$ kPa $\cdot$ cm$^3 \cdot$ mol$^{-1} \cdot$ K^{-1}, we have from (6):

$$\Delta P = \frac{(0.5)(0.5)(905)(100)^2}{(8314)(348.2) - (0.5)(0.5)(905)(100)} = 0.788 \text{ kPa}$$

The small but significant pressure increase resulting from the mixing of the gases is about 1% of the initial pressure. Since all the B's, and hence δ_{12}, are zero for ideal gases, no pressure change would occur if the gases were ideal.

EMPIRICAL EQUATIONS OF STATE (Section 5.3)

5.14 Evaluate the Redlich/Kwong a and b under the critical derivative constraints (5.46) and (5.47).

We illustrate here an alternative to direct application of (5.46) and (5.47). One can show that these conditions, when applied to a cubic EOS, require that the equation have three equal roots for volume at the critical state; i.e., that the EOS have the form $(V - V_c)^3 = 0$, or

$$V^3 - 3V_c V^2 + 3V_c^2 V - V_c^3 = 0 \qquad (1)$$

At the critical state, the Redlich/Kwong equation (5.44) is, in expanded form,

$$V^3 - \frac{RT_c}{P_c} V^2 - \left(b^2 + \frac{bRT_c}{P_c} - \frac{a}{P_c T_c^{1/2}} \right) V - \frac{ab}{P_c T_c^{1/2}} = 0 \qquad (2)$$

Comparison of like coefficients in (1) and (2) yields three equations connecting the five quantities a, b, P_c, V_c, and T_c:

$$\frac{RT_c}{P_c} = 3V_c \qquad (3)$$

$$b^2 + \frac{bRT_c}{P_c} - \frac{a}{P_c T_c^{1/2}} = -3V_c^2 \qquad (4)$$

$$\frac{ab}{P_c T_c^{1/2}} = V_c^3 \qquad (5)$$

By (3),

$$V_c = \frac{1}{3} \frac{RT_c}{P_c} \qquad (6)$$

and thus, by (5),

$$a = \frac{1}{27} \frac{R^3 T_c^{3.5}}{b P_c^2} \tag{7}$$

Substitution of (6) and (7) into (4) yields, on rearrangement, the cubic equation

$$\hat{b}^3 + \hat{b}^2 + \frac{1}{3}\hat{b} - \frac{1}{27} = 0 \tag{8}$$

where $\hat{b} \equiv b P_c / R T_c$ is a reduced b. Equation (8) has one real root, given by

$$\hat{b} = \frac{2^{1/3} - 1}{3}$$

Hence,

$$b = \left(\frac{2^{1/3} - 1}{3}\right) \frac{R T_c}{P_c} \approx 0.08664 \frac{R T_c}{P_c} \tag{9}$$

and substitution of (9) in (7) gives

$$a = \left(\frac{2^{1/3} + 1}{3}\right)^3 \frac{R^2 T_c^{2.5}}{P_c} \approx 0.42748 \frac{R^2 T_c^{2.5}}{P_c} \tag{10}$$

Expressions (9) and (10) are the required results. Note that (6) implies that $Z_c = 1/3$ for all substances, a prediction that is both quantitatively and qualitatively incorrect; refer to the discussion in Example 5.12.

5.15 Derive Maxwell's *equal-area rule* for a pure-species subcritical isotherm having the general shape of a van der Waals isotherm.

The sort of isotherm to be considered is shown in Fig. 5-9. By assumption, P^{sat} is the vapor pressure implied by the equation of state. The states A and E, corresponding to saturated liquid and vapor volumes V^l and V^v, are then equilibrium states, for which (see Problem 3.11) $G^l = G^v$. But, from (3.24), $G = A + PV$. Thus

$$A^l + P^l V^l = A^v + P^v V^v$$

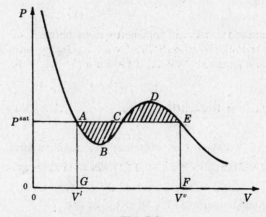

Fig. 5-9

But $P^l = P^v = P^{sat}$, and we obtain on rearrangement

$$A^l - A^v = P^{sat}(V^v - V^l) = \text{area of rectangle } AEFG \tag{1}$$

An alternative equation can be found for $A^l - A^v$ by evaluating the integral $\int dA$ along the path $EDCBA$. As this path is a segment of an isotherm, (3.34) gives $dA = -P\,dV$; thus

$$A^l - A^v = -\int_{EDCBA} P\,dV = \int_{ABCDE} P\,dV = \text{area under isotherm segment} \tag{2}$$

It follows at once from (1) and (2) that the two shaded areas in Fig. 5-9 are equal, which is Maxwell's rule.

The converse of Maxwell's rule is also true: If an isobar produces equal areas, then it corresponds to the saturation pressure for the given EOS.

5.16 The *three-parameter Clausius equation of state* is

$$P = \frac{RT}{V-b} - \frac{a}{T(V+c)^2} \tag{1}$$

Find the three-parameter corresponding-states correlation that results when (1) is required to satisfy (5.46) and (5.47), and in addition to produce the correct value of Z_c for any given fluid.

Differentiation of (1) and imposition of (5.46) and (5.47) yields:

$$\left(\frac{\partial P}{\partial V}\right)_{T;\text{cr}} = -\frac{RT_c}{(V_c-b)^2} + \frac{2a}{T_c(V_c+c)^3} = 0 \tag{2}$$

$$\left(\frac{\partial^2 P}{\partial V^2}\right)_{T;\text{cr}} = \frac{2RT_c}{(V_c-b)^3} - \frac{6a}{T_c(V_c+c)^4} = 0 \tag{3}$$

Equation (1) may be put in the form

$$Z = \frac{V}{V-b} - \frac{aV}{RT^2(V+c)^2}$$

from which

$$Z_c = \frac{V_c}{V_c-b} - \frac{aV_c}{RT_c^2(V_c+c)^2} \tag{4}$$

Combination of (2) and (3) and solution for b gives

$$b = \frac{V_c - 2c}{3} \tag{5}$$

and combination of (2) and (5) gives

$$a = \tfrac{9}{8} RT_c^2(V_c + c) \tag{6}$$

Substituting (5) and (6) in (4) and solving for c, we obtain

$$c = V_c\left(\frac{3 - 8Z_c}{8Z_c}\right) \tag{7}$$

from which

$$a = \frac{27}{64}\frac{RT_c^2 V_c}{Z_c} \quad \text{and} \quad b = V_c\left(\frac{4Z_c - 1}{4Z_c}\right) \tag{8}$$

The Clausius EOS (1) may now be written in dimensionless form. Noting that

$$P_r \equiv \frac{P}{P_c} = \frac{PV_c}{Z_c RT_c} \tag{9}$$

we obtain, after combination of (1), (7), (8), and (9),

$$P_r = \frac{4T_r}{1 + 4Z_c(V_r - 1)} - \frac{3}{T_r[1 + \tfrac{8}{3}Z_c(V_r - 1)]^2} \tag{10}$$

According to (10), all fluids having the same Z_c will have the same P_r when compared at the same V_r and T_r.

5.17 Estimate Z, G^R, H^R, and S^R for carbon dioxide gas at 100 °C and 40.53 bar (40 atm). For this purpose, use the van der Waals EOS with parameters determined from (a) T_C and V_C, (b) T_C and P_C.

Parts (*a*) and (*b*) differ only in the numerical values assigned to parameters *a* and *b*. In part (*a*), these quantities are determined from (5.48) of Example 5.12; in part (*b*), from (5.51). Critical properties for CO_2 are given in Appendix C: $T_c = 304.2$ K, $P_c = 73.8$ bar, and $V_c = 94.0$ cm$^3 \cdot$ mol^{-1}.

(*a*) By (5.48),

$$a = \tfrac{9}{8} R T_c V_c = \tfrac{9}{8}(83.14)(304.2)(94.0) = 2.675 \times 10^6 \text{ cm}^6 \cdot \text{bar} \cdot \text{mol}^{-2}$$
$$b = \tfrac{1}{3} V_c = \tfrac{1}{3}(94.0) = 31.33 \text{ cm}^3 \cdot \text{mol}^{-1}$$

The reduced temperature is $T_r = (100 + 273.15)/304.2 = 1.227$; at the given conditions, CO_2 is a supercritical fluid, and the EOS in Z has only one real root. We find Z by fixed-point iteration of (5.63). Numerical values for the dimensionless groups aP/R^2T^2 and bP/RT are

$$\frac{aP}{R^2T^2} = \frac{(2.675 \times 10^6)(40.53)}{(83.14)^2(373.15)^2} = 0.1126 \qquad \frac{bP}{RT} = \frac{(31.33)(40.53)}{(83.14)(373.15)} = 0.0409$$

Thus (5.63) becomes

$$Z = 1.0409 - 0.1126\left(\frac{Z - 0.0409}{Z^2}\right)$$

Solution by iteration gives $Z = 0.9245$, from which $\rho = P/ZRT = 1.413 \times 10^{-3}$ mol $\cdot$ cm^{-3}. With these values of Z and ρ, (5.66) and (5.67) yield

$$H^R = -612 \text{ J} \cdot \text{mol}^{-1} \qquad S^R = -1.03 \text{ J} \cdot \text{mol}^{-1} \cdot \text{K}^{-1}$$

Finally, $G^R = H^R - TS^R = -612 - (373.15)(-1.03) = -228$ J $\cdot$ mol^{-1}.

(*b*) By (5.51),

$$a = \frac{27}{64} \frac{R^2 T_c^2}{P_c} = 3.657 \times 10^6 \text{ cm}^6 \cdot \text{bar} \cdot \text{mol}^{-2} \qquad b = \frac{1}{8} \frac{RT_c}{P_c} = 42.84 \text{ cm}^3 \cdot \text{mol}^{-1}$$

Proceeding as in part (*a*), we find $Z = 0.8946$ and $\rho = 1.460 \times 10^{-3}$ mol $\cdot$ cm^{-3}; from these we can compute the residual properties.

Results of the two sets of calculations are summarized in Table 5-4, together with handbook values for comparison. Calculated compressibility factors are not too far in error, but H^R and S^R agree poorly with reported values. Part (*b*) gives the better results; this is generally to be expected.

Table 5-4

	Z	$G^R/\text{J} \cdot \text{mol}^{-1}$	$H^R/\text{J} \cdot \text{mol}^{-1}$	$S^R/\text{J} \cdot \text{mol}^{-1} \cdot \text{K}^{-1}$
(*a*)	0.9245	-228	-612	-1.03
(*b*)	0.8946	-315	-860	-1.46
Handbook	0.9012	-310	-1150	-2.25

5.18 Determine the expressions for A, S, and U implied by the van der Waals EOS.

We use the PVT form of the EOS,

$$P = \frac{RT}{V - b} - \frac{a}{V^2} \qquad (5.42)$$

along with

$$P = -\left(\frac{\partial A}{\partial V}\right)_{T,x} \qquad (3.37)$$

$$S = -\left(\frac{\partial A}{\partial T}\right)_{V,x} \qquad (3.39)$$

$$A = U - TS \qquad (3.31)$$

Integration of (3.37) with P given by (5.42) yields

$$A = -RT \ln (V - b) - \frac{a}{V} + F \qquad (1)$$

where function of integration F depends on T and x only. Differentiation of (1) according to (3.39) gives

$$S = R \ln (V - b) - \left(\frac{\partial F}{\partial T}\right)_x \qquad (2)$$

and thus we obtain by (3.31)

$$U = -\frac{a}{V} + F - T\left(\frac{\partial F}{\partial T}\right)_x \qquad (3)$$

Equations (1), (2), and (3) are the required expressions for A, S, and U. For convenience in interpretation, we examine the corresponding residual properties M^r (see Example 5.4). From (1), (2), and (3), we find that

$$A^{\text{ig}} = -RT \ln V + F \qquad S^{\text{ig}} = R \ln V - \left(\frac{\partial F}{\partial T}\right)_x \qquad U^{\text{ig}} = F - T\left(\frac{\partial F}{\partial T}\right)_x$$

so that

$$A^r = -RT \ln (1 - b\rho) - a\rho \qquad (4)$$

$$S^r = R \ln (1 - b\rho) \qquad (5)$$

$$U^r = -a\rho \qquad (6)$$

where $\rho \equiv V^{-1}$ is the molar density. Equations (5) and (6) suggest the following interpretations: The leading term in the van der Waals equation represents entropic contributions to the EOS; the second term, energetic contributions. Energy/entropy arguments are frequently used in EOS modeling; the van der Waals prescriptions for U^r and S^r are the simplest nontrivial assignments for these quantities.

CORRESPONDING-STATES CORRELATIONS   (Section 5.4)

5.19 Derive an expression for ω for a fluid obeying the vapor-pressure equation (4.5). For the same fluid, show how ω is related to the slope of the reduced-vapor-pressure curve plotted with $\log_{10} P_r$ and $1/T_r$ as coordinates.

Equation (4.5) is

$$\ln P^{\text{sat}} = A - \frac{B}{T}$$

where A and B are constants (B is not to be confused with the second virial coefficient). Since the gas/liquid critical point is the terminus of the vapor-pressure curve,

$$\ln P_c = A - \frac{B}{T_c} \qquad (1)$$

and subtraction of (1) from (4.5) gives

$$\ln P_r^{\text{sat}} = \frac{B}{T_c} \left(1 - \frac{1}{T_r}\right) \qquad (2)$$

Letting $T_r = 0.7$ and changing to common logarithms,

$$\log_{10} P_r^{\text{sat}} = -\frac{3 \log_{10} e}{7} \frac{B}{T_c} \qquad \text{at} \qquad T_r = 0.7$$

from which we obtain, according to the definition of ω, (5.68):

$$\omega = \frac{3 \log_{10} e}{7} \frac{B}{T_c} - 1 \qquad (3)$$

Equation (2) is a linear in $1/T_r$, so the slope $\mathscr{S}$ of a $(\log_{10} P_r^{\text{sat}})$-versus-$(1/T_r)$ plot is

$$\mathscr{S} = -(\log_{10} e)\, \frac{B}{T_c} \tag{4}$$

Combination of (3) and (4) yields

$$\mathscr{S} = -\frac{7}{3}\,(\omega + 1)$$

For a simple fluid ($\omega = 0$) obeying (4.5), $\mathscr{S} = -7/3$, and for nonsimple fluids $\mathscr{S}$ is increasingly negative with increasing ω. Equation (4.5) is only a first approximation to the behavior of real simple fluids, and the first of these conclusions is only in qualitative agreement with experiment; the second, however, is generally valid.

5.20 A 0.5-m^3 tank to contain propane has a bursting pressure of 30 bar. Safety considerations dictate that the tank be charged with no more propane than would exert half the bursting pressure at 400 K. With how many kilograms of propane may the tank be charged? The molar mass of propane is $0.0441\ \text{kg}\cdot\text{mol}^{-1}$.

At 400 K, propane is above its critical temperature (369.8 K), and the tank contains only gas. The allowable mass of propane is, from $PV^t = (m/M)ZRT$,

$$m = \frac{MPV^t}{ZRT} \tag{1}$$

where V^t is the volume of the tank and M is the molar mass of propane. We estimate Z from (5.33) and the generalized correlation for $\hat{B}$. From Appendix C, $T_c = 369.8$ K, $P_c = 42.5$ bar, and $\omega = 0.152$. At the stated conditions,

$$T_r = \frac{400}{369.8} = 1.082 \qquad P_r = \frac{(0.5)(30)}{42.5} = 0.353$$

From (5.73),

$$B^0 = 0.083 - \frac{0.422}{(1.082)^{1.6}} = -0.289 \qquad B^1 = 0.139 - \frac{0.172}{(1.082)^{4.2}} = -0.015$$

Thus, by (5.69) and (5.72),

$$Z = 1 + [-0.289 + (0.152)(-0.015)]\left(\frac{0.353}{1.082}\right) = 0.905$$

The mass m can now be calculated from (1):

$$m = \frac{(0.0441)(0.5 \times 30)(0.5)}{(0.905)(8.314 \times 10^{-5})(400)} = 11.0\ \text{kg}$$

5.21 Estimate the second virial coefficient of air (considered a binary mixture of nitrogen and oxygen) at 300 K.

Use (5.79), (5.84), (5.73), and the empirical combining rules (5.85) through (5.89). Take air to contain 79 mole percent nitrogen (1) and 21 mole percent oxygen (2). From Appendix C:

$T_{c,11} = 126.2$ K	$P_{c,11} = 33.9$ bar	$V_{c,11} = 89.5\ \text{cm}^3\cdot\text{mol}^{-1}$	$Z_{c,11} = 0.290$	$\omega_{11} = 0.040$
$T_{c,22} = 154.6$ K	$P_{c,22} = 50.5$ bar	$V_{c,22} = 73.4\ \text{cm}^3\cdot\text{mol}^{-1}$	$Z_{c,22} = 0.288$	$\omega_{22} = 0.021$

By (5.85): $T_{c,12} = \sqrt{(126.2)(154.6)} = 139.7$ K

By (5.86): $Z_{c,12} = \frac{1}{2}(0.290 + 0.288) = 0.289$

By (5.87): $V_{c,12} = \frac{1}{8}(89.5^{1/3} + 73.4^{1/3})^3 = 81.2\ \text{cm}^3\cdot\text{mol}^{-1}$

By (5.89): $\omega_{12} = \frac{1}{2}(0.040 + 0.021) = 0.031$

Finally, by (5.88),

$$P_{c,12} = \frac{(0.289)(83.14)(139.7)}{81.2} = 41.3\ \text{bar}$$

At 300 K, then,

$$T_{r,11} = \frac{300}{126.2} = 2.377 \qquad T_{r,22} = \frac{300}{154.6} = 1.940 \qquad T_{r,12} \equiv \frac{T}{T_{c,12}} = \frac{300}{139.7} = 2.147$$

With these values of $T_{r,ij}$, (5.73) gives

$$B^0(T_{r,11}) = -0.023 \qquad B^1(T_{r,11}) = 0.135$$
$$B^0(T_{r,22}) = -0.063 \qquad B^1(T_{r,22}) = 0.128$$
$$B^0(T_{r,12}) = -0.041 \qquad B^1(T_{r,12}) = 0.132$$

Thus, by (5.84),

$$B_{11} = \frac{\hat{B}_{11}RT_{c,11}}{P_{c,11}} = [-0.023 + (0.040)(0.135)]\frac{(83.14)(126.2)}{33.9} = -5.4 \text{ cm}^3 \cdot \text{mol}^{-1}$$

$$B_{22} = \frac{\hat{B}_{22}RT_{c,22}}{P_{c,22}} = [-0.063 + (0.021)(0.128)]\frac{(83.14)(154.6)}{50.5} = -15.4 \text{ cm}^3 \cdot \text{mol}^{-1}$$

$$B_{12} = \frac{\hat{B}_{12}RT_{c,12}}{P_{c,12}} = [-0.041 + (0.031)(0.132)]\frac{(83.14)(139.7)}{41.3} = -10.4 \text{ cm}^3 \cdot \text{mol}^{-1}$$

Finally, by (5.28):

$$B = x_1^2 B_{11} + 2x_1 x_2 B_{12} + x_2^2 B_{22}$$
$$= (0.79)^2(-5.4) + (2)(0.79)(0.21)(-10.4) + (0.21)^2(-15.4) = -7.5 \text{ cm}^3 \cdot \text{mol}^{-1}$$

The experimental value is also $-7.5 \text{ cm}^3 \cdot \text{mol}^{-1}$. One expects good agreement in this case, because N_2 and O_2 are of similar chemical type and do not differ greatly in size; hence the estimate of B_{12} should be realistic.

5.22 Propane gas undergoes a change from 105 °C and 5 bar (state 1) to 190 °C and 25 bar (state 2). Use the generalized correlation of Section 5.4 and the ideal-gas heat capacity from Table 4-1 to determine ΔH and ΔS for this change of state.

Where tabular data for thermodynamic properties are not available, one must choose a *calculational path* connecting the states of interest which allows use of available information. The property changes ΔH and ΔS being independent of path, the choice is made solely on the basis of convenience. Residual properties H^R and S^R connect the real state of a gas with the ideal-gas state at the same T and P. For the ideal-gas state, property changes are readily calculated by the equations of Section 4.7. Thus the most convenient calculational path consists of three steps:

(1) The real gas, at T_1 and P_1, is transformed into an ideal gas, at T_1 and P_1. Associated with this step are the property changes $-H_1^R$ and $-S_1^R$.

(2) The ideal gas is changed from its state at T_1 and P_1 to a state at T_2 and P_2. Associated with this step are the property changes ΔH_{12}^{ig} and ΔS_{12}^{ig}.

(3) The ideal gas, at T_2 and P_2, is transformed back into a real gas, at T_2 and P_2. Associated with this step are the property changes H_2^R and S_2^R.

The net property changes are the sums of the property changes for the three steps:

$$\Delta H = -H_1^R + \Delta H_{12}^{ig} + H_2^R \qquad (1)$$
$$\Delta S = -S_1^R + \Delta S_{12}^{ig} + S_2^R \qquad (2)$$

The required values of H^R and S^R are calculated as in Example 5.18. For propane, Appendix C gives $T_c = 369.8$ K, $P_c = 42.5$ bar, and $\omega = 0.152$. At the initial state:

$$T_{r1} = \frac{105 + 273.15}{369.8} = 1.023 \qquad P_{r1} = \frac{5}{42.5} = 0.118$$

From (5.73) and (5.77),

$$B^0 = -0.324 \qquad B^1 = -0.017$$

$$\frac{dB^0}{dT_r} = 0.636 \qquad \frac{dB^1}{dT_r} = 0.641$$

Thus, by (5.72) and (5.77),

$$\hat{B} = -0.324 + (0.152)(-0.017) = -0.327$$

$$\frac{d\hat{B}}{dT_r} = 0.636 + (0.152)(0.641) = 0.733$$

and hence, by (5.75) and (5.76),

$$\frac{H_1^R}{RT_1} = -\left(0.733 + \frac{0.327}{1.023}\right)(0.118) = -0.124 \qquad \text{or} \qquad H_1^R = -390 \text{ J} \cdot \text{mol}^{-1}$$

$$\frac{S_1^R}{R} = -(0.733)(0.118) = -0.0865 \qquad \text{or} \qquad S_1^R = -0.719 \text{ J} \cdot \text{mol}^{-1} \cdot \text{K}^{-1}$$

An identical procedure yields $H_2^R = -1292 \text{ J} \cdot \text{mol}^{-1}$ and $S_2^R = -2.004 \text{ J} \cdot \text{mol}^{-1} \cdot \text{K}^{-1}$.

It remains to determine $\Delta H_{12}^{\text{ig}}$ and $\Delta S_{12}^{\text{ig}}$. By (4.17) and (4.21),

$$\Delta H_{12}^{\text{ig}} = \langle C_P^{\text{ig}} \rangle_T (T_2 - T_1) \tag{3}$$

$$\Delta S_{12}^{\text{ig}} = \langle C_P^{\text{ig}} \rangle_{\ln T} \ln \frac{T_2}{T_1} - R \ln \frac{P_2}{P_1} \tag{4}$$

with mean heat capacities given by (4.16) and (4.20). For this problem, $T_1 = 378.15$ K, $T_2 = 463.15$ K; $T_{\text{am}} = 420.65$ K, $T_{\text{lm}} = 419.22$ K. Substitution of these temperatures and the heat-capacity parameters for propane from Table 4.1 into (4.16) and (4.20) gives

$$\frac{1}{R} \langle C_P^{\text{ig}} \rangle_T = 11.76 \qquad \frac{1}{R} \langle C_P^{\text{ig}} \rangle_{\ln T} = 11.72$$

Then, by (3) and (4),

$$\Delta H_{12}^{\text{ig}} = (8.314)(11.76)(463.15 - 378.15) = 8311 \text{ J} \cdot \text{mol}^{-1}$$

$$\Delta S_{12}^{\text{ig}} = (8.314)\left[(11.72)\ln\frac{463.15}{378.15} - \ln\frac{25}{5}\right] = 6.376 \text{ J} \cdot \text{mol}^{-1} \cdot \text{K}^{-1}$$

Finally, substituting numbers into (1) and (2), we obtain

$$\Delta H = 390 + 8311 - 1292 = 7409 \text{ J} \cdot \text{mol}^{-1}$$

$$\Delta S = 0.719 + 6.376 - 2.004 = 5.091 \text{ J} \cdot \text{mol}^{-1} \cdot \text{K}^{-1}$$

In this case, the residual-property contributions account for about 12% of the estimated ΔH and about 25% of the estimated ΔS. Notice, however, that $-M_1^R$ and M_2^R tend to cancel each other, and that the ideal-gas contributions therefore tend to dominate the estimates for ΔH and ΔS. This is a common result when initial and final temperatures and pressures are not vastly different.

5.23 Show how residual properties are determined from a comprehensive corresponding-states correlation of the form

$$Z = Z^0(T_r, P_r) + \omega Z^1(T_r, P_r) \tag{5.71}$$

Since this is a volume-explicit formulation, we determine residual properties from (5.12) through (5.14), as expressed in reduced coordinates:

$$\frac{G^R}{RT} = \int_0^{P_r} (Z-1) \frac{dP_r}{P_r} \qquad (\text{constant } T_r, x)$$

$$\frac{H^R}{RT} = -T_r \int_0^{P_r} \left(\frac{\partial Z}{\partial T_r}\right)_{P_r,x} \frac{dP_r}{P_r} \qquad \text{(constant } T_r, x)$$

$$\frac{S^R}{R} = -\int_0^{P_r} \left[T_r\left(\frac{\partial Z}{\partial T_r}\right)_{P_r,x} + Z - 1\right] \frac{dP_r}{P_r} \qquad \text{(constant } T_r, x)$$

From (5.71), we have

$$Z - 1 = (Z^0 - 1) + \omega Z^1$$

$$\left(\frac{\partial Z}{\partial T_r}\right)_{P_r,x} = \left(\frac{\partial Z^0}{\partial T_r}\right)_{P_r,x} + \omega\left(\frac{\partial Z^1}{\partial T_r}\right)_{P_r,x}$$

Notice that the -1 is associated with Z^0. The reason is that Z^0 is the compressibility factor for the *simple fluids* Ar, Kr, and Xe; hence $Z - 1$ must approach $Z^0 - 1$ as ω approaches zero. Combination of the last five equations gives the required expressions, each of which has the form of (5.71):

$$\frac{G^R}{RT} = \int_0^{P_r} (Z^0 - 1)\frac{dP_r}{P_r} + \omega \int_0^{P_r} Z^1 \frac{dP_r}{P_r} \qquad \text{(constant } T_r, x) \tag{1}$$

$$\frac{H^R}{RT} = -\int_0^{P_r} T_r\left(\frac{\partial Z^0}{\partial T_r}\right)_{P_r,x} \frac{dP_r}{P_r} - \omega \int_0^{P_r} T_r\left(\frac{\partial Z^1}{\partial T_r}\right)_{P_r,x} \frac{dP_r}{P_r} \qquad \text{(constant } T_r, x) \tag{2}$$

$$\frac{S^R}{R} = -\int_0^{P_r} \left[T_r\left(\frac{\partial Z^0}{\partial T_r}\right)_{P_r,x} + Z^0 - 1\right]\frac{dP_r}{P_r} - \omega \int_0^{P_r} \left[T_r\left(\frac{\partial Z^1}{\partial T_r}\right)_{P_r,x} + Z^1\right]\frac{dP_r}{P_r} \qquad \text{(constant } T_r, x) \tag{3}$$

Correlations for Z^0 and Z^1 in (5.71) are developed by analysis of PVT data for *pure* fluids; neither in (5.71) nor in (1) through (3) does composition appear explicitly. Empirical extension to mixtures is usually done through the *pseudoparameter* concept, according to which the corresponding-states parameters T_c, P_c, and ω are defined for mixtures as composition-weighted combinations of the pure-species corresponding-states parameters. The simplest such prescriptions are linear in mole fraction:

$$T_c = \sum_i x_i T_{c,i} \qquad P_c = \sum_i x_i P_{c,i} \qquad \omega = \sum_i x_i \omega_i$$

The first two of these are called *Kay's rules*.

MISCELLANEOUS APPLICATIONS

5.24 The equation of state of an associating gas (e.g., acetic acid vapor) is conventionally modeled by *chemical theory*, a scheme which incorporates concepts of chemical-reaction equilibrium (Sections 7.9 and 7.10). The simplest version of chemical theory produces the following EOS for a strongly dimerizing gas:

$$Z = \frac{1}{2} + \frac{1}{2}\left(1 + 4K\frac{P}{P^0}\right)^{-1/2} \tag{1}$$

Here, K is a dimensionless, positive-definite parameter (the "equilibrium constant"), a function of T alone. Quantity P^0 is a reference pressure whose value is fixed, usually at 1 bar or 1 atm. Infer from (1) expressions for (a) the residual properties G^R, H^R, and S^R; (b) the density-series virial coefficients B, C, and D.

(a) The EOS is explicit in volume. From (1),

$$Z - 1 = \frac{1}{2}\left(1 + 4K\frac{P}{P^0}\right)^{-1/2} - \frac{1}{2} \tag{2}$$

$$\left(\frac{\partial Z}{\partial T}\right)_P = -\left(1 + 4K\frac{P}{P^0}\right)^{-3/2} \frac{dK}{dT}\frac{P}{P^0} \tag{3}$$

Substitution of (2) in (5.12) and integration gives

$$\frac{G^R}{RT} = -\ln\left[\frac{1}{2} + \frac{1}{2}\left(1 + 4K\frac{P}{P^0}\right)^{1/2}\right] \tag{4}$$

Since K is positive, the group of terms in square brackets is greater than unity, and therefore G^R is negative. The residual enthalpy is found from (3) and (5.13); integration gives

$$\frac{H^R}{RT} = T \frac{d \ln K}{dT} \left[\frac{1}{2} - \frac{1}{2} \left(1 + 4K \frac{P}{P^0} \right)^{-1/2} \right] \tag{5}$$

The group of terms inside the square brackets is always positive, and $d \ln K/dT$ for a dimerization reaction is negative; thus H^R, like G^R, is negative. Since

$$\frac{S^R}{R} = \frac{H^R}{RT} - \frac{G^R}{RT}$$

the sign of S^R is undetermined. In the usual case, however, H^R is sufficiently negative to make S^R negative also.

(b)   The infinite-series representation of (1) about $P = 0$ is

$$Z = 1 - \frac{K}{P^0} P + 3 \left(\frac{K}{P^0} \right)^2 P^2 - 10 \left(\frac{K}{P^0} \right)^3 P^3 + \cdots \tag{6}$$

Hence the first three pressure-series virial coefficients are

$$B' = -\frac{K}{P^0} \qquad C' = 3 \left(\frac{K}{P^0} \right)^2 \qquad D' = -10 \left(\frac{K}{P^0} \right)^3$$

The required density-series coefficients follow by application of (5.23):

$$B = -\frac{KRT}{P^0} \qquad C = 4 \left(\frac{KRT}{P^0} \right)^2 \qquad D = -20 \left(\frac{KRT}{P^0} \right)^3$$

Series (6) converges for $P < P^0/4K$. For strongly dimerizing gases (K large), this restricts the effective use of virial expansions to very low pressure levels.

5.25   To study liquid-phase behavior with accuracy and precision, one must generally use special PVT equations of state, e.g., the Tait equation (Problem 3.28), here rewritten as

$$V = V_0 \left(1 - \frac{AP}{B + P} \right) \tag{1}$$

By means of a computerized, least-squares fitting of (1) to a set of specific volumes for liquid water at 60 °C [Keenan et al., *Steam Tables* (*International Edition—Metric Units*), Wiley, New York, 1969], one obtains the following parameter values (at 60 °C):

$$V_0 = 1.0171 \text{ cm}^3 \cdot \text{g}^{-1} \qquad A = 0.3645 \qquad B = 8393 \text{ bar} \tag{2}$$

Using (1) and (2), estimate the isothermal compressibility of liquid water at 60 °C for $P = 0.1992$ (saturation), 250, 500, 1000 bar.

By (3.18),

$$\kappa = -\left(\frac{\partial \ln V}{\partial P} \right)_T = -\frac{\partial}{\partial P} \ln \left(1 - \frac{AP}{B + P} \right) = \frac{AB}{(B + P)[B + (1 - A)P]} \tag{3}$$

Substitution of values in (3) generates Table 5-5, which includes steam-table values for comparison.

Table 5-5

P/bar	$\kappa/10^{-5}$ bar^{-1}	
	Equation (3)	Steam Tables
0.1992	4.34	4.45
250	4.14	4.15
500	3.95	3.93
1000	3.61	3.63

Supplementary Problems

FORMULATIONS OF THE EOS   (Section 5.1)

5.26 Prove:

(a)
$$C_P - C_V = \frac{\dfrac{RT^2}{Z}\left(\dfrac{\partial Z}{\partial T}\right)_V^2 + 2RT\left(\dfrac{\partial Z}{\partial T}\right)_V + RZ}{1 - \dfrac{V}{Z}\left(\dfrac{\partial Z}{\partial V}\right)_T}$$

(b)
$$\left(\frac{\partial C_V}{\partial V}\right)_T = \frac{RT^2}{V}\left(\frac{\partial^2 Z}{\partial T^2}\right)_V + \frac{2RT}{V}\left(\frac{\partial Z}{\partial T}\right)_V$$

(c)
$$\left(\frac{\partial C_P}{\partial P}\right)_T = -\frac{RT^2}{P}\left(\frac{\partial^2 Z}{\partial T^2}\right)_P - \frac{2RT}{P}\left(\frac{\partial Z}{\partial T}\right)_P$$

5.27 Derive (5.9), (5.10), and (5.11).

5.28 Use (5.10) and (5.11) to find general expressions for β and κ implied by a $Z\rho T$ equation of state.

Ans. $\beta = \dfrac{1}{T}\left[1 + \dfrac{T}{Z}\left(\dfrac{\partial Z}{\partial T}\right)_\rho\right]\left[1 + \dfrac{\rho}{Z}\left(\dfrac{\partial Z}{\partial \rho}\right)_T\right]^{-1}$ $\kappa = \dfrac{1}{Z\rho RT}\left[1 + \dfrac{\rho}{Z}\left(\dfrac{\partial Z}{\partial \rho}\right)_T\right]^{-1}$

5.29 (a) Show that in the liquid/vapor two-phase region, $Z = Z^l + x\,\Delta Z^{lv}$, where the notation is that of Section 4.4. (b) Show that at the gas/liquid critical point

$$\left(\frac{\partial Z}{\partial \rho}\right)_{T;cr} = -\frac{Z_c}{\rho_c} \qquad \left(\frac{\partial^2 Z}{\partial \rho^2}\right)_{T;cr} = 2\frac{Z_c}{\rho_c^2}$$

5.30 Derive the two equations that precede (5.15).

5.31 Show that (5) of Example 5.4 can be replaced by

$$M^r = M^R - \int_{RT/P}^{V}\left(\frac{\partial M^{ig}}{\partial V}\right)_{T,x} dV$$

5.32 Verify (7), (8), and (9) of Example 5.4.

5.33 The van der Waals hard-sphere EOS has the V-explicit form

$$Z = 1 + \frac{bP}{RT} \tag{1}$$

and the P-explicit form

$$Z = \frac{1}{1 - b\rho} \tag{2}$$

Demonstrate that residual properties computed from (1) via (5.12) through (5.14) are the same as those determined from (2) via (5.15) through (5.17). Recall that b is a function of composition only.

VIRIAL EQUATIONS OF STATE   (Section 5.2)

5.34 Prove: $B' = \lim\limits_{P\to 0}\left(\dfrac{Z-1}{P}\right)_{T,x}$ and $C' = \lim\limits_{P\to 0}\left(\dfrac{\partial\left(\dfrac{Z-1}{P}\right)}{\partial P}\right)_{T,x} = \lim\limits_{P\to 0}\left(\dfrac{\dfrac{Z-1}{P} - B'}{P}\right)_{T,x}$

5.35 Calculate B for an equimolar ternary mixture of methane (1), propane (2), and n-pentane (3) at 100 °C. At 100 °C, in units of $cm^3 \cdot mol^{-1}$,

$$B_{11} = -20 \qquad B_{22} = -241 \qquad B_{33} = -621$$
$$B_{12} = -75 \qquad B_{13} = -122 \qquad B_{23} = -399$$

Ans. $B = -230\,cm^3 \cdot mol^{-1}$

5.36   Prove: The Joule/Thomson coefficient for a gas described by (5.33) is

$$\left(\frac{\partial T}{\partial P}\right)_H = \frac{T}{C_P}\left(\frac{dB}{dT} - \frac{B}{T}\right)$$

What can be said about the sign of the coefficient? (*Hint*: See Example 5.11.)

5.37   Show that the three-term virial equation in density, (5.34), when subjected to the critical derivative conditions (5.46) and (5.47), yields the following estimates for the reduced second and third virial coefficients at the critical temperature:

$$\hat{B}_c = -\tfrac{1}{3} \qquad \hat{C}_c = \tfrac{1}{27}$$

Are these estimates reasonable?

5.38   Write the virial equation in density (5.22) as

$$Z = 1 + \sum_{n=2}^{\infty} B_n \rho^{n-1}$$

Show that this equation yields linear isochores (Problem 5.3) if, for all n,

$$B_n = a_n - \frac{b_n}{T}$$

where a_n and b_n are constants.

5.39   For nitrogen gas at 160 K:

P/bar	Z	ρ/mol·L^{-1}
40	0.8031	3.744
80	0.6304	9.540

Derive a *good* estimate for Z at 160 K and 60 bar.
Ans.   Linear interpolation on P (the least defensible approach) gives $Z = 0.7168$. Interpolation based on (5.34) gives $Z = 0.7038$. A handbook value is $Z = 0.7017$.

5.40   Show that the expressions for residual properties developed in Problem 5.12 reduce to those of Example 5.11 for sufficiently low densities. (*Hint*: $\ln Z \approx Z - 1$, for small $|Z - 1|$.)

5.41   Estimate Z, H^R, and S^R for water vapor at 300 °C and 5 bar from the following values of the second virial coefficient for H_2O:

t/°C	B/cm^3·mol^{-1}
290	−125
300	−119
310	−113

Compare your estimates with values determined from the steam tables.
Ans. $Z = 0.988$, $H^R = -231$ J·mol^{-1}, $S^R = -0.300$ J·mol^{-1}·K^{-1}. From the steam tables: $Z = 0.988$, $H^R = -216$ J·mol^{-1}, $S^R = -0.302$ J·mol^{-1}·K^{-1}.

5.42   Show that, analogous to (5.31) and (5.32),

$$C = \sum_i x_i C_{iii} + \frac{1}{3}\sum_i \sum_j \sum_k x_i x_j x_k \delta_{ijk}$$

where $\delta_{ijk} \equiv 3C_{ijk} - C_{iii} - C_{jjj} - C_{kkk}$. Demonstrate that the subscripts on δ *permute*; i.e.,

$$\delta_{ijk} = \delta_{ikj} = \delta_{jik} = \delta_{jki} = \delta_{kij} = \delta_{kji}$$

Show that for a binary mixture,

$$C = x_1 C_{111} + x_2 C_{222} + x_1 x_2 (x_1 \delta_{112} + x_2 \delta_{122})$$

EMPIRICAL EQUATIONS OF STATE   (Section 5.3)

5.43   Find G^R/RT, H^R/RT, and S^R/R, for the Redlich/Kwong equation of state.

Ans. $\dfrac{G^R}{RT} = -\dfrac{a}{2bRT^{1.5}} \ln(1 + b\rho) + (Z - 1) - \ln(1 - b\rho)Z$

$\dfrac{H^R}{RT} = -\dfrac{3a}{2bRT^{1.5}} \ln(1 + b\rho) + (Z - 1)$

$\dfrac{S^R}{R} = \ln(1 - b\rho) - \dfrac{a}{2bRT^{1.5}} \ln(1 + b\rho)$

5.44   Repeat Problem 5.17 for the Redlich/Kwong EOS. Estimate parameters from T_c and P_c, using the results of Problem 5.14.
Ans. $Z = 0.8944$ $G^R = -321 \text{ J} \cdot \text{mol}^{-1}$ $H^R = -1045 \text{ J} \cdot \text{mol}^{-1}$ $S^R = -1.94 \text{ J} \cdot \text{mol}^{-1} \cdot \text{K}^{-1}$

5.45   (*a*)   Show that the inversion curve [see Problem 5.2(*c*)] of a van der Waals fluid has the equation

$$2ab^2\rho^2 - 4ab\rho + (2a - bRT) = 0$$

(*b*)   The *Boyle curve* of a fluid is the locus of points for which $(\partial Z/\partial P)_T = 0$. Show that the Boyle curve of a van der Waals fluid has the equation

$$ab^2\rho^2 - 2ab\rho + (a - bRT) = 0$$

(*c*)   The *spinodal curve* of a fluid is the locus of points for which $(\partial P/\partial V)_T = 0$. Show that the spinodal curve of a van der Waals fluid has the equation

$$2a\rho(1 - b\rho)^2 - RT = 0$$

5.46   Determine the expressions for the second, third, and fourth virial coefficients implied by the general cubic EOS (*5.41*).
Ans. $B = b - \dfrac{\theta}{RT}$ $C = b^2 - (b - \eta - \delta)\dfrac{\theta}{RT}$ $D = b^3 - [b^2 + (\delta + \eta)(\delta - b) - \varepsilon]\dfrac{\theta}{RT}$

5.47   Justify the procedure used in Problem 5.14 to find the Redlich/Kwong a and b; i.e., prove that a cubic EOS in V has three equal roots at the critical state if and only if the critical isotherm has a horizontal inflection at the critical state. What is the generalization of this result to an EOS which is an arbitrary polynomial—or an arbitrary analytic function (see Problem 5.49)—in V? [*Hint*: If $\mathscr{F}(P, V, T) = 0$, then, by repeated implicit differentiation,

$$\left(\frac{\partial \mathscr{F}}{\partial V}\right)_{T,P} + \left(\frac{\partial \mathscr{F}}{\partial P}\right)_{T,V}\left(\frac{\partial P}{\partial V}\right)_T = 0$$

$$\left(\frac{\partial^2 \mathscr{F}}{\partial V^2}\right)_{T,P} + \left[\text{terms in }\left(\frac{\partial P}{\partial V}\right)_T \text{ or }\left(\frac{\partial^2 P}{\partial V^2}\right)_T\right] = 0$$

. . . .

Now consider the partial Taylor expansion of $\mathscr{F}$ as a function of V, about the point $V = V_c$.]

5.48   Generalize the results of Example 5.16(*b*) by developing two transformations of (*5.41*) suitable for fixed-point iteration.

Ans. $Z \text{ (vap)} = 1 + \dfrac{bP}{RT} - \dfrac{\theta P}{R^2 T^2} \dfrac{Z - \dfrac{\eta P}{RT}}{Z^2 + \dfrac{\delta P}{RT} Z + \dfrac{\varepsilon P^2}{R^2 T^2}}$

$$Z \text{ (liq)} = \dfrac{bP}{RT} + \dfrac{R^2 T^2}{\theta P} \dfrac{Z - \dfrac{bP}{RT}}{Z - \dfrac{\eta P}{RT}} \left(Z^2 + \dfrac{\delta P}{RT} Z + \dfrac{\varepsilon P^2}{R^2 T^2} \right) \left(1 + \dfrac{bP}{RT} - Z \right)$$

5.49   The *Dieterici EOS* is

$$Z = \dfrac{1}{1 - b\rho} \exp\left(-\dfrac{a\rho}{RT} \right)$$

where a and b are functions of composition only. Verify that this equation, when constrained by (5.46) and (5.47), yields

$$Z_c = \dfrac{2}{e^2} \approx 0.270\,67 \qquad \dfrac{bP_c}{RT_c} = \dfrac{1}{e^2} \approx 0.135\,34 \qquad \dfrac{aP_c}{R^2 T_c^2} \approx 0.541\,34$$

5.50   The molar density of a van der Waals liquid at zero pressure is

$$\rho = \dfrac{1}{2b} \left(1 + \sqrt{1 - \dfrac{4bRT}{a}} \right)$$

(*a*) Derive this expression from (5.43). (*b*) Sketch a subcritical isotherm on a PV diagram to illustrate the significance of the zero-pressure liquid state. (*c*) Show that for a van der Waals fluid the zero-pressure liquid exists only for $T_r \lesssim 27/32$.

CORRESPONDING-STATES CORRELATIONS   (Section 5.4)

5.51   What is the acentric factor of a fluid which obeys the reduced van der Waals EOS (5.52)? [*Hint:* Use the definition (5.68) and the results of Problem 5.15.]   *Ans.* $\omega = -0.303$

5.52   The *Boyle temperature* of a real gas is the temperature for which $(\partial Z/\partial P)_T = 0$ at $P = 0$, i.e., the temperature at which the Boyle curve [Problem 5.45(*b*)] intersects the Z axis on a ZP diagram. Show that $B = 0$ at the Boyle temperature. Estimate the Boyle temperature for oxygen from (5.72) and (5.73), the corresponding-states correlation for B.   *Ans.*   145 °C (an experimental value is 150 °C)

5.53   Develop the following corresponding-states version of the van der Waals EOS:

$$Z^3 - \left(1 + \dfrac{1}{8} \dfrac{P_r}{T_r} \right) Z^2 + \dfrac{27}{64} \dfrac{P_r}{T_r^2} Z - \dfrac{27}{512} \dfrac{P_r^2}{T_r^3} = 0$$

5.54   Estimate Z, H^R, and S^R for hydrogen sulfide gas at 10 bar and 200 °C. Use the generalized correlation for B.
Ans. $Z = 0.982$ $H^R = -238 \text{ J} \cdot \text{mol}^{-1}$ $S^R = 0.359 \text{ J} \cdot \text{mol}^{-1} \cdot \text{K}^{-1}$

5.55   Estimate B for an equimolar ternary mixture of methane, propane, and *n*-pentane at 100 °C. Compare your result with that of Problem 5.35.   *Ans.* $B = -246 \text{ cm}^3 \cdot \text{mol}^{-1}$

5.56   Show that the reduced second virial coefficient implied by the Redlich/Kwong EOS is

$$\dfrac{BP_c}{RT_c} = 0.086\,64 - \dfrac{0.427\,48}{T_r^{1.5}} \tag{1}$$

Compare (1) *numerically* with the empirical correlation for B^0 given by (5.73). What do you conclude? [*Hint:* For the origin of the numerical coefficients in (1), see Problem 5.14.]

MISCELLANEOUS APPLICATIONS

5.57 Calculate Z and $V/cm^3 \cdot mol^{-1}$ for isopropanol vapor at 200 °C and 10 bar, using (a) (5.34), with the experimentally determined virial coefficients $B = -388 \ cm^3 \cdot mol^{-1}$ and $C = -26\,000 \ cm^6 \cdot mol^{-2}$; (b) (5.33) and the generalized correlations for B; (c) the Redlich/Kwong EOS, with a and b estimated from T_c and P_c. For isopropanol, $T_c = 508.2$ K, $P_c = 50.7$ bar, and $\omega = 0.700$.

Ans.

	Z	$V/cm^3 \cdot mol^{-1}$
(a)	0.8866	3488
(b)	0.9037	3555
(c)	0.9123	3589

5.58 Steam undergoes a change of state from 475 °C and 3500 kPa to 150 °C and 275 kPa. Determine ΔH and ΔS by use of (a) the steam tables; (b) the generalized correlation for B and the method illustrated in Problem 5.22. The ideal-gas heat capacity for H_2O is given in Table 4-1.
Ans. (a) $\Delta H = -11\,385 \ J \cdot mol^{-1}$ $\Delta S = 0.6685 \ J \cdot mol^{-1} \cdot K^{-1}$
          (b) $\Delta H = -11\,258 \ J \cdot mol^{-1}$ $\Delta S = 0.9182 \ J \cdot mol^{-1} \cdot K^{-1}$

5.59 Repeat Problem 5.22, using the Redlich/Kwong EOS to estimate residual properties. Determine a and b by the procedure of Problem 5.14.
Ans. $H_1^R = -354 \ J \cdot mol^{-1}$, $S_1^R = -0.619 \ J \cdot mol^{-1} \cdot K^{-1}$; $H_2^R = -1312 \ J \cdot mol^{-1}$,
          $S_2^R = -1.857 \ J \cdot mol^{-1} \cdot K^{-1}$; $\Delta H = 7353 \ J \cdot mol^{-1}$, $\Delta S = 5.138 \ J \cdot mol^{-1} \cdot K^{-1}$

Chapter 6

Thermodynamics of Flow Processes

The first law of thermodynamics, (1.3), is rewritten here as $\Delta E^t = Q - W$. This equation is the mathematical formulation of the principle of energy conservation as applied to a process occurring in a closed system (a system of constant mass). The total energy of the system, E^t, includes energy in both its internal and external forms, and thus the total energy change ΔE^t is the sum of several terms:

$$\Delta E^t = \Delta U^t + \Delta E_K^t + \Delta E_P^t$$

Here the superscript t signifies that the various energy terms refer to the entire system.

Most of our applications of the first law have been to systems at rest, and we have therefore been able to set $\Delta E^t = \Delta U^t$, giving $\Delta U^t = Q - W$ or, in differential form,

$$dU^t = \delta Q - \delta W \qquad (6.1)$$

For a homogeneous fluid system (6.1) may also be written

$$m\, dU = \delta Q - \delta W \qquad (6.2)$$

where U is the specific internal energy of the fluid and m is the (constant) mass of the system. If instead U is the molar internal energy, then m is replaced by n, the (constant) number of moles. Inherent in the formulation of (6.2) is the assumption that the system has uniform properties throughout, and hence that each unit mass or mole of fluid experiences the same change in internal energy.

Our initial purpose in this chapter is to develop from (6.1) an equation more general than (6.2); specifically, an equation applicable to closed systems of several parts, each having its own set of uniform properties, but between which fluid may flow. Subsequently we shall develop even more general energy equations, applicable to open systems and to flow processes of all kinds. In addition we shall consider flow processes from the standpoint of the second law.

6.1 ENERGY EQUATIONS FOR CLOSED SYSTEMS

A closed system is necessarily one of constant mass. But such a system may consist of several interconnected parts, none of which is itself constrained to constant mass; for example, the system of Fig. 6-1. Shown are two cylinders, A and B, separated by a partially opened valve through which gas is transferred, from A to B, by appropriate motion of the pistons. The two identifiable parts or regions of the system are the two cylinders. We assume that the properties of the gas in each cylinder are uniform throughout (but different in the two cylinders). The total internal energy of the system at

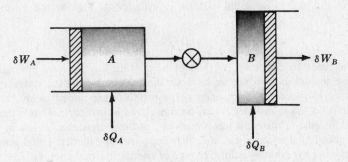

Fig. 6-1

any instant is then given by

$$U^t = m_A U_A + m_B U_B$$

and for a differential change in the system

$$dU^t = d(m_A U_A + m_B U_B)$$

The total heat and work quantities are clearly

$$\delta Q = \delta Q_A + \delta Q_B \quad \text{and} \quad \delta W = \delta W_A + \delta W_B$$

Substitution of these last three expressions in (6.1) gives

$$d(m_A U_A + m_B U_B) = \delta Q_A + \delta Q_B - (\delta W_A + \delta W_B)$$

More generally, for closed systems having an arbitrary number of interconnected regions,

$$d \sum_R (mU) = \sum \delta Q - \sum \delta W \qquad (6.3)$$

All three summations in (6.3) are over the regions of the system. However, the index R is omitted from the right-hand summations to emphasize that (6.3) remains valid when the total heat and the total work are arbitrarily partitioned. Since the differential of a sum is equal to the sum of differentials, (6.3) may equally well be written

$$\sum_R d(mU) = \sum \delta Q - \sum \delta W \qquad (6.3a)$$

EXAMPLE 6.1 With respect to the process illustrated in Fig. 6-1, expand completely the differential

$$dU^t = d(m_A U_A + m_B U_B)$$

and identify each of the resulting terms with a particular mass of gas in the system.

Expansion of the differential gives

$$dU^t = m_A \, dU_A + U_A \, dm_A + m_B \, dU_B + U_B \, dm_B \qquad (1)$$

In this expression dU_A and dU_B represent the changes in the specific internal energies of the gases in cylinders A and B, respectively; similarly dm_A and dm_B represent the changes in mass of the gas contained in the two cylinders. Equation (1) illustrates the fact that the total internal energy of a region may change for two reasons: a change in the amount of material in the region and a change in the properties of the material in the region.

Since the total mass of the system is constant, $dm_A + dm_B = 0$, and so

$$dm_B = -dm_A \equiv dm \qquad (2)$$

where dm represents the mass of gas that flows from cylinder A to cylinder B during the differential process. This mass is shown in Fig. 6.2 at its initial location in cylinder A. Making the substitutions in (1) that are indicated by (2), we get

$$dU^t = m_A \, dU_A + (U_B - U_A) \, dm + m_B \, dU_B \qquad (3)$$

Equation (3) illustrates an additional method of accounting for the total internal energy change of the system. It indicates the several masses that we have selected as comprising the system; after each such mass is multiplied by its change in specific internal energy, the products are added to give the total internal-energy change. Thus the quantity of gas in cylinder A may be considered in two parts. First is that part which remains in cylinder A after the differential process; it has a mass m_A and experiences a change in specific internal energy dU_A. Second is that differential mass of gas, dm, that flows out of cylinder A and into cylinder B during the process and which therefore experiences a finite change of *specific* internal energy, $U_B - U_A$. Finally we identify the mass of gas m_B which is initially in cylinder B, and which remains there and experiences a specific-internal-

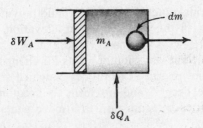

Fig. 6-2

energy change dU_B. It is always possible to determine the total internal-energy change of a closed system by such an accounting scheme.

Integration of (6.3) for a finite process gives immediately

$$\Delta \sum_R (mU) = \sum Q - \sum W \tag{6.4}$$

However, such direct integration is not always advantageous, because it may yield an equation in several unknowns. On the other hand, the original differential equation may be reducible to a simple form, which when integrated can be solved for a single unknown.

An alternative energy equation that is sometimes easier to apply results from elimination of U from (6.3) in favor of the enthalpy H. Since by definition, $H = U + PV$,

$$d \sum_R (mH) = d \sum_R (mU) + d \sum_R (mPV)$$

Combination with (6.3) gives

$$d \sum_R (mH) = \sum \delta Q - \sum \delta W + d \sum_R (mPV) \tag{6.5}$$

This equation proves to be particularly useful for application to processes where the pressure is constant in each region of the system (but not necessarily the same in the different regions), because in this case the last two terms can be made to cancel. For this to be true the process must be mechanically reversible (Section 1.5) within each region of the system. We must *assume* this is so if in fact we are to evaluate the work terms. With this assumption, each work term comes from (1.2), rewritten for constant pressure as $\delta W = d(mPV)$; (6.5) becomes

$$d \sum_R (mH) = \sum \delta Q \tag{6.6}$$

Equation (6.6) is a generalization of the result obtained in Example 1.7 for a constant-pressure, mechanically reversible process where the pressure is uniform throughout the system. The equation shows that the change in total enthalpy for a closed system is equal to the heat transfer, provided the pressure is constant in each region of the system, even if different in different regions, and provided the process is mechanically reversible within each region. Integration of (6.6) gives

$$\Delta \sum_R (mH) = \sum Q \tag{6.7}$$

or alternatively

$$\sum_R \Delta(mH) = \sum Q \tag{6.7a}$$

The equations so far developed contain no terms to account for changes in potential or kinetic energy of the fluid that flows from one region of the system to another. The absence of terms to account for changes in gravitational potential energy of fluid moving from one region of the system to another can be justified only if the differences in elevation among the various regions are inconsequential. This is almost always the case. Where it is not, the best course to follow is to employ the more general (6.10). The omission of terms for changes in kinetic energy is based on the presumption that applications are restricted to processes which start and end with the fluid at rest. Thus no kinetic-energy change occurs, and no term to account for it is necessary. Where this presumption is not fulfilled, one must again turn to a more general energy equation.

EXAMPLE 6.2 A crude method for determining the enthalpy of steam in a steam line is to bleed steam from the line through a hose into a barrel of water. The mass of the barrel with its contents and the temperature of the water are recorded at the beginning and end of the process. The condensation of steam by the water raises the temperature of the water.

Our energy equations are written for a system of constant mass. We therefore choose the system so that it includes not only the water originally in the barrel, but also the steam in the line that is condensed by the water during the process. We imagine a piston in the steam line that separates the steam in the line into two parts: that which enters the barrel and that which remains in the line. This view is represented schematically in Fig. 6-3.

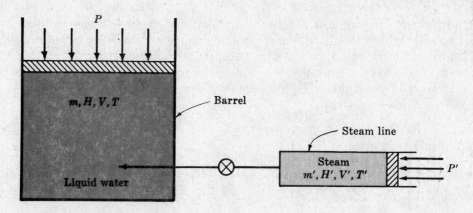

Fig. 6-3

In addition to the piston in the steam line, another piston is imagined to separate the water in the barrel from the atmosphere. Both the atmosphere and the steam in the line exert forces (shown as pressures, i.e., forces distributed over an area) which act on the system. These forces move during the process and therefore do work.

The pressure P exerted by the atmosphere on the water in the barrel is taken as constant, but the properties of the water change during the process. Thus the mass changes from m_1 to m_2; the specific enthalpy, from H_1 to H_2; the specific volume, from V_1 to V_2; and the temperature, from T_1 to T_2. In the steam line, however, the intensive properties T', P', H', and V' remain constant. The mass of steam in the line goes from an initial value m' to zero. From a mass balance,

$$m_2 - m_1 = m'$$

Since the pressure P in the barrel and the pressure P' in the steam line are both constant, (6.7) is applicable, provided we assume the process to be mechanically reversible within the barrel and the steam line. We really have no choice, because this is the only assumption that allows us to solve the problem. Furthermore, in the absence of any information about heat transfer, we assume the barrel and the steam line to be well insulated, and set $\Sigma\, Q = 0$. As a result, (6.7) becomes

$$\Delta \sum_R (mH) = 0$$

or

$$\underbrace{m_2 H_2}_{\substack{\sum_R \text{final}}} - \underbrace{(m_1 H_1 + m'H')}_{\substack{\sum_R \text{initial}}} = 0$$

whence
$$H' = \frac{m_2 H_2 - m_1 H_1}{m'} \qquad (1)$$

The result (1) is crude because it rests on crude assumptions. In particular, it is quite impossible to prevent heat transfer, and even worse, there is no way to determine a value for it.

EXAMPLE 6.3   A tank of capacity $1.5\,\mathrm{m}^3$ contains 500 kg of liquid water in equilibrium with pure water vapor, which fills the rest of the tank, at a temperature of 100 °C and 1.013 bar pressure. From a water line at slightly above this pressure 750 kg of water at 70 °C is bled into the tank. How much heat must be transferred to the contents of the tank during this process if the temperature and pressure in the tank are not to change?

The system, which must be chosen to include the initial contents of the tank and the water to be added during the process, is shown in Fig. 6-4. Since the tank at all times contains liquid and vapor in equilibrium at 100 °C and 1.013 bar, the properties H^l and V^l for the liquid and H^v and V^v for the vapor in the tank are the same at the end as at the start of the process. The properties H' and V' of the liquid water in the line are also constant. As liquid is added to the tank, it displaces some of the vapor that is present initially, and this vapor must condense. Thus the tank at the end of the process contains as liquid all the 500 kg of liquid initially present in the tank. This 500 kg of liquid is present as saturated liquid at 100 °C both at the beginning and end of the process, and hence does not change in properties. In addition, the 750 kg of liquid added to the tank remains liquid, but it changes from liquid at 70 °C to saturated liquid at 100 °C. Also, the vapor which condenses adds to the liquid in the tank at the end of the process. Let the amount of vapor which condenses be y kg. This mass of material changes from saturated vapor at 100 °C to saturated liquid at 100 °C during the process. The part of the vapor which does not condense remains as saturated vapor at 100 °C and therefore does not change in properties.

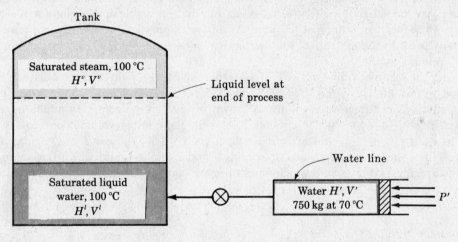

Fig. 6-4

Since the pressure is constant in both the tank and the water line, (6.7) applies, as does $(6.7a)$. Both show that $\Sigma\,Q$ is equal to the total enthalpy change of the system. However, in the above description of the process we have identified four masses which constitute the system and have described their changes during the process. This suggests that it may be advantageous to determine the total enthalpy change by summation over the changes which occur in these masses, in the manner illustrated in Example 6.1.

First note that two of the masses—the initial 500 kg of liquid in the tank and the mass of vapor which does not condense—undergo no change in properties, and may be omitted from consideration. The other two masses are the 750 kg of water added to the tank from the line and the y kg of vapor which condenses. The heat transfer is therefore equal to the enthalpy changes of these two masses:

$$\Sigma\,Q = 750(H^l - H') + y(H^l - H^v)$$

From the steam tables,

$$H^l = 419.1\ \mathrm{kJ \cdot kg^{-1}} \qquad H' = 293.0\ \mathrm{kJ \cdot kg^{-1}} \qquad H^v = 2676.0\ \mathrm{kJ \cdot kg^{-1}}$$

Thus $\Sigma\,Q = 750(419.1 - 293.0) + y(419.1 - 2676.0) = 94\,575 - 2256.9y$.

Now determine y. This is most easily done by noting that the sum of the volume changes of the masses of the system that have already been identified must be equal to the total volume change of the system. The total volume change of the system is just the volume of the 750 kg of water at 70 °C that is added to the tank. Since the total volume change is negative,

$$\sum (m\,\Delta V) = -750V'$$

The same masses that change in enthalpy also change in volume. Thus

$$750(V^l - V') + y(V^l - V^v) = -750V' \qquad \text{or} \qquad 750V^l + y(V^l - V^v) = 0$$

From the steam tables, $V^l = 1.044\ \text{cm}^3 \cdot \text{g}^{-1}$, $V^v = 1672\ \text{cm}^3 \cdot \text{g}^{-1}$; therefore

$$(750)(1.044) + y(1.04 - 1672) = 0 \qquad \text{or} \qquad y = 0.4686\ \text{kg}$$

and $\sum Q = 94\,575 - (2256.9)(0.4686) = 93\,517\ \text{kJ}$.

Analysis of this problem through identification of masses has led to a very simple solution. If we had summed over regions, we should eventually have reached the same result, but the calculations would have been much more tedious.

6.2 ENERGY EQUATION FOR STEADY-STATE FLOW

We consider now the kind of process that is referred to as "steady-state flow." Such processes are of primary importance in engineering because large-scale production of materials and energy demands continuous processing. The adjective *steady-state* implies that the inflow of mass is at all times exactly matched by the outflow of mass, so that there is no accumulation of material within the apparatus. Moreover, conditions at all points within the apparatus are steady or constant, in time. Thus, at any point in the apparatus, the thermodynamic properties are constant, although they may vary from point to point.

Consider the schematic diagram of Fig. 6-5, which shows a region of space bounded by the heavy curves and by the line segments 1-1, 2-2, and 3-3. This region is called the *control volume*. The material contained in the control volume is shown by the light shading. In addition, we recognize three other regions of space which communicate with the control volume: one to the left of 1-1, another to the right of 2-2, and a third to the right of 3-3. These we call *regions 1, 2,* and *3,* respectively. For the process considered, regions 1 and 2 contain fluid which flows into the control volume, and region 3 collects fluid which flows from the control volume. There is no limitation on the

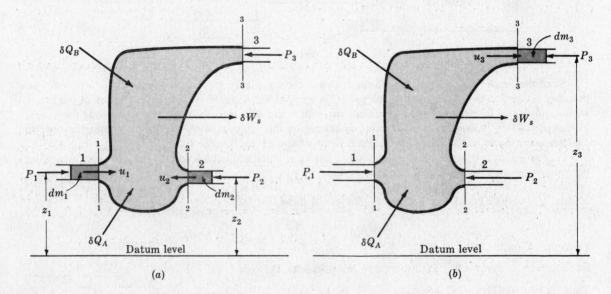

Fig. 6-5

number of inlets to and outlets from the control volume. We consider here a total of three by way of example.

Our *system* is chosen to include not only the fluid contained initially in the control volume, but in addition all fluid which enters the control volume during some arbitrarily selected time interval. Thus the initial configuration of the system as shown in Fig. 6-5(a) includes the masses dm_1 and dm_2 which flow into the control volume during a differential time interval. The final configuration of the system after time $d\tau$ is shown in Fig. 6-5(b). Here regions 1 and 2 no longer contain the masses dm_1 and dm_2; however, the mass dm_3 has collected in region 3. Since the process considered is one of steady-state flow, the mass contained in the control volume is constant, and $dm_3 = dm_1 + dm_2$.

The masses in regions 1, 2, and 3 are considered to have the properties of the fluid as measured at 1-1, 2-2, and 3-3, respectively. These properties include a velocity u and an elevation z above a datum plane, as well as the thermodynamic properties. Thus these masses have kinetic and gravitational potential energy as well as internal energy, and an energy equation for steady-state flow must contain terms for all these forms of energy. From Section 1.6 the kinetic and gravitational potential energies of a mass m are $E_K = \frac{1}{2}mu^2$ and $E_P = mgz$.

We are now able to write an energy equation for steady-state flow. Note that we have chosen our system to be one of constant mass, i.e., the mass contained in the control volume plus the mass which enters during the time interval considered. Thus we may apply the first law in a form already developed, provided that we add terms to account for changes in kinetic and potential energy. Equation (6.4) indicates a summation over regions of the system, and the major region is the control volume. However, the mass of fluid in the control volume and the properties of the fluid in the control volume do not change with time. Therefore terms to account for the energy of this region cancel in any energy equation, and may be omitted from the start. The regions that must be considered are those labeled 1, 2, and 3 in Fig. 6-5, and only differential masses leave or enter these regions during the time interval $d\tau$ required for the process that changes the system from the state shown in Fig. 6-5(a) to that of Fig. 6-5(b). Applied to this process, (6.4), with the addition of kinetic- and potential-energy terms, gives *for the system*:

$$\Delta \sum_R (U\,dm) \;+\; \Delta \sum_R (\tfrac{1}{2}u^2\,dm) \;+\; \Delta \sum_R (gz\,dm) = \sum \delta Q \;-\; \sum \delta W$$

Written out, the internal-energy term becomes

$$\Delta \sum_R (U\,dm) = U_3\,dm_3 \;-\; U_1\,dm_1 \;-\; U_2\,dm_2$$

This result can be expressed more simply by writing

$$\Delta \sum_R (U\,dm) = \sum (U\,dm)_{\text{out}} \;-\; \sum (U\,dm)_{\text{in}} \equiv \Delta(U\,dm)_{\substack{\text{flowing}\\\text{streams}}}$$

The kinetic- and potential-energy terms may be expressed similarly. Thus the energy equation may be written

$$\Delta(U\,dm)_{\substack{\text{flowing}\\\text{streams}}} \;+\; \Delta(\tfrac{1}{2}u^2\,dm)_{\substack{\text{flowing}\\\text{streams}}} \;+\; \Delta(gz\,dm)_{\substack{\text{flowing}\\\text{streams}}} = \sum \delta Q \;-\; \sum \delta W$$

The term $\sum \delta Q$ includes all heat transferred to the system. For the example of Fig. 6-5 it equals $\delta Q_A + \delta Q_B$. The term $\sum \delta W$ includes work quantities of two types. The one shown in Fig. 6-5 and designated δW_s represents *shaft work*, i.e., work transmitted across the boundaries of the system by a rotating or reciprocating shaft. The other work quantities are of the kind already considered, which result from the action of external pressures in regions 1, 2, and 3 on the moving boundaries of the system. Thus, for the process of Fig. 6-5, if we assume mechanical reversibility in regions 1, 2, and 3,

$$\sum \delta W = P_3 V_3\,dm_3 \;-\; P_1 V_1\,dm_1 \;-\; P_2 V_2\,dm_2 \;+\; \delta W_s \equiv \Delta(PV\,dm)_{\substack{\text{flowing}\\\text{streams}}} \;+\; \delta W_s$$

Substitution in the energy equation and rearrangement gives

$$\Delta(U\,dm)_{\substack{\text{flowing}\\\text{streams}}} \;+\; \Delta(PV\,dm)_{\substack{\text{flowing}\\\text{streams}}} \;+\; \Delta(\tfrac{1}{2}u^2\,dm)_{\substack{\text{flowing}\\\text{streams}}} \;+\; \Delta(gz\,dm)_{\substack{\text{flowing}\\\text{streams}}} = \sum \delta Q \;-\; \delta W_s$$

Factoring dm allows this equation to be written more simply as

$$\Delta[(U + PV + \tfrac{1}{2}u^2 + gz)\, dm]_{\substack{\text{flowing}\\\text{streams}}} = \sum \delta Q - \delta W_s$$

By definition, $H = U + PV$, and we have, finally,

$$\Delta[(H + \tfrac{1}{2}u^2 + gz)\, dm]_{\substack{\text{flowing}\\\text{streams}}} = \sum \delta Q - \delta W_s \tag{6.8}$$

This energy equation, applicable to steady-flow processes, connects the properties of the streams flowing into and out of the control volume with the heat and work quantities crossing the boundary of the control volume.

6.3 GENERAL EQUATIONS OF BALANCE

The concept of a control volume can also be applied in the description of unsteady-flow processes. It is still a bounded region of space; however, the boundaries are now considered flexible, allowing for expansion and contraction of the control volume. Furthermore, the mass contained in the control volume need no longer be constant; we deal now with transient conditions, where rates and properties vary with time. During an infinitesimal time interval $d\tau$, the mass entering the control volume may be different from the mass leaving the control volume. The difference must be accounted for by the accumulation or depletion of mass within the control volume. The general equation of mass balance is therefore

$$dm_{\substack{\text{control}\\\text{volume}}} + \Delta(dm)_{\substack{\text{flowing}\\\text{streams}}} = 0 \tag{6.9}$$

Similarly, the difference between the transport of energy out of the control volume and into it by flowing streams need no longer be accounted for solely by the heat and work terms. Energy may be accumulated or depleted within the control volume. To state the situation precisely, the energy crossing the boundaries of the control volume as heat and work, $\Sigma\, \delta Q - \Sigma\, \delta W$, must equal the change in energy of the material contained within the control volume itself, plus the net energy transport of the flowing streams. The energy change of the material in the control volume is $d(mU)_{\text{control volume}}$, presuming no gross motion of the control volume. The net energy transport of the flowing streams is shown by (6.8) to be

$$\Delta[(H + \tfrac{1}{2}u^2 + gz)\, dm]_{\substack{\text{flowing}\\\text{streams}}}$$

The energy equation therefore becomes

$$d(mU)_{\substack{\text{control}\\\text{volume}}} + \Delta[(H + \tfrac{1}{2}u^2 + gz)\, dm]_{\substack{\text{flowing}\\\text{streams}}} = \sum \delta Q - \sum \delta W \tag{6.10}$$

The work term $\Sigma\, \delta W$ may include shaft work, as well as work resulting from the expansion or contraction of the control volume itself. Height z is measured from a datum level through the (stationary) center of mass of the control volume.

In situations beyond the scope of (6.10) (e.g., an accelerating control volume) one must derive the energy equation from the conservation principle by scrupulous "bookkeeping."

In the foregoing development we have implicitly assumed that energy is conserved in all processes, whether flow or nonflow, regardless of whether the system is open or closed. In other words, we have extended the validity of the first law of thermodynamics to cover all cases. With respect to the second law, we make exactly the same extension, taking it to be a universally valid law of nature, applicable alike to flow and to nonflow processes. Thus we equate the total entropy change resulting from an unsteady-flow process to the entropy change of the material contained within the control volume, plus the net entropy transport across the boundary of the control volume by the flowing streams, plus the entropy changes of any heat reservoirs in the surroundings that serve as sources or sinks for heat transfer to or from the control volume:

$$dS_{\text{total}} = d(mS)_{\substack{\text{control} \\ \text{volume}}} + \Delta(S\,dm)_{\substack{\text{flowing} \\ \text{streams}}} - \sum \frac{\delta Q}{T_{\text{res}}}$$

The minus sign preceding the last term is required because $-\delta Q = \delta Q_{\text{res}}$. According to the second law $dS_{\text{total}} \geq 0$; therefore, our general entropy relation may be stated as an inequality between the entropy change of the system and that of the surroundings:

$$\boxed{d(mS)_{\substack{\text{control} \\ \text{volume}}} + \Delta(S\,dm)_{\substack{\text{flowing} \\ \text{streams}}} \geq \sum \frac{\delta Q}{T_{\text{res}}}} \qquad (6.11)$$

The equals sign holds in (6.11) only in the case of a *reversible* process.

Expressions (6.9) through (6.11) are written for an infinitesimal time interval $d\tau$. If we divide each through by $d\tau$ and denote the various rates by

$$\dot{m} \equiv \frac{dm}{d\tau} \qquad \dot{mU} \equiv \frac{d(mU)}{d\tau} \qquad \dot{Q} \equiv \frac{\delta Q}{d\tau} \qquad \dot{W} \equiv \frac{\delta W}{d\tau} \qquad \dot{mS} = \frac{d(mS)}{d\tau}$$

we get rate relations:

$$\dot{m}_{\substack{\text{control} \\ \text{volume}}} + \Delta \dot{m}_{\substack{\text{flowing} \\ \text{streams}}} = 0 \qquad (6.12)$$

$$(\dot{mU})_{\substack{\text{control} \\ \text{volume}}} + \Delta[\dot{m}(H + \tfrac{1}{2}u^2 + gz)]_{\substack{\text{flowing} \\ \text{streams}}} = \sum \dot{Q} - \sum \dot{W} \qquad (6.13)$$

$$(\dot{mS})_{\substack{\text{control} \\ \text{volume}}} + \Delta(\dot{m}S)_{\substack{\text{flowing} \\ \text{streams}}} \geq \sum \frac{\dot{Q}}{T_{\text{res}}} \qquad (6.14)$$

where Δ denote the difference between streams flowing out and streams flowing in. These relations apply at any instant during processes where conditions and flow rates change continuously. They may of course be integrated over a finite time interval, but this requires detailed information describing the process as a function of time.

Both the equations for nonflow and for steady-flow processes are obtained as special cases of (6.9) through (6.14). For a nonflow process, where the control volume contains the entire system, $dm = 0$, and (6.10) reduces to

$$d(mU) = \sum \delta Q - \sum \delta W$$

If the system contains several regions, we need to sum the internal-energy term over the regions:

$$\sum_R d(mU) = \sum \delta Q - \sum \delta W$$

which is (6.3a). Similarly, (6.11) becomes

$$\sum_R d(mS) \geq \sum \frac{\delta Q}{T_{\text{res}}}$$

This last relation has itself an important specialization. When the nonflow process is *periodic* (as in a cyclical heat engine), then integration over a period gives the *Clausius inequality*

$$\sum \frac{Q}{T_{\text{res}}} \leq 0$$

For a steady-flow process, (6.10) reduces to (6.8), because the first term of (6.10) is zero and $\Sigma\, \delta W$ becomes δW_s. Similarly, (6.12) through (6.14) become

$$\Delta(\dot{m})_{\substack{\text{flowing} \\ \text{streams}}} = 0 \tag{6.15}$$

$$\Delta[\dot{m}(H + \tfrac{1}{2}u^2 + gz)]_{\substack{\text{flowing} \\ \text{streams}}} = \sum \dot{Q} - \dot{W}_s \tag{6.16}$$

$$\Delta(\dot{m}S)_{\substack{\text{flowing} \\ \text{streams}}} \geqq \sum \frac{\dot{Q}}{T_{\text{res}}} \tag{6.17}$$

A special case of the steady-flow relations is often encountered. If only a single stream enters and a single stream leaves the control volume, then (6.15) requires $\dot{m}_{\text{out}} = \dot{m}_{\text{in}} \equiv \dot{m}$; (6.16) becomes

$$\dot{m}\, \Delta(H + \tfrac{1}{2}u^2 + gz) = \sum \dot{Q} - \dot{W}_s \tag{6.18}$$

Division by $\dot{m}$ gives

$$\Delta(H + \tfrac{1}{2}u^2 + gz) = \frac{\sum \dot{Q}}{\dot{m}} - \frac{\dot{W}_s}{\dot{m}} = \sum Q - W_s$$

or

$$\boxed{\Delta H + \tfrac{1}{2}\Delta u^2 + g\,\Delta z = \sum Q - W_s} \tag{6.19}$$

Each term of (6.19) refers to a *unit mass of fluid* passing through the control volume. All terms in this equation, as in all the energy equations of this chapter, must be expressed in the same energy units. Similarly, (6.17) is transformed to

$$\boxed{\Delta S \geqq \sum \frac{Q}{T_{\text{res}}}} \tag{6.20}$$

EXAMPLE 6.4 Rework Example 6.3 by application of (6.10).

We take the tank as our control volume. As there is no shaft work and no expansion work, $\Sigma\,\delta W = 0$. There is but one flowing stream, and it flows *into* the control volume. Therefore, the term in (6.10) accounting for the energy of the flowing streams reduces to

$$\Delta[(H + \tfrac{1}{2}u^2 + gz)\,dm]_{\substack{\text{flowing} \\ \text{streams}}} = -(H' + \tfrac{1}{2}u'^2 + gz')\,dm'$$

However, z' can be taken equal to zero, because the water line and the tank are at essentially the same level. We have no information on u', and can only presume it is small enough so that any contribution from a kinetic-energy term is negligible. This assumption is equivalent to the tacit assumption made in Example 6.3 that the process within the water line is mechanically reversible. This assumption is implicit in the use of (6.7) or any equivalent equation. Equation (6.10) therefore becomes

$$\sum \delta Q = d(mU)_{\text{tank}} - H'\, dm'$$

We must now integrate this equation over the entire process. Since H' is constant, the result is

$$\sum Q = \Delta(mU)_{\text{tank}} - m'H'$$

By the definition of enthalpy, $\Delta(mU)_{\text{tank}} = \Delta(mH)_{\text{tank}} - \Delta(PmV)_{\text{tank}}$. However, both the volume of the tank $(mV)_{\text{tank}}$ and the pressure in the tank are constant. Therefore $\Delta(PmV)_{\text{tank}} = 0$, and

$$\sum Q = \Delta(mH)_{\text{tank}} - m'H'$$

This equation expresses the fact that for this process the heat transferred is equal to the total enthalpy change caused by the process. The solution of Example 6.3 was based on this very idea.

6.4 APPLICATIONS TO STEADY-FLOW PROCESSES

Flow processes inevitably result from pressure gradients within the fluid. Moreover, temperature gradients, velocity gradients, and even concentration gradients may exist within the flowing fluid. Thus, in contrast to a closed system at uniform conditions throughout, we find in a system where flow occurs a distribution of conditions throughout the system or control volume, making it necessary to attribute properties to point masses of fluid. We assume that the intensive properties (density, specific enthalpy, specific entropy, etc.) at a point in a fluid are determined solely by the temperature, pressure, and composition at the point, uninfluenced by the existence of gradients in these conditions at the point. Moreover, we assume that the fluid exhibits a set of intensive properties at the point exactly the same as though the entire system were uniform, and existed at equilibrium at the same temperature, pressure, and composition. The implication is that an equation of state applies locally and instantaneously at any point in a fluid system, and that one may employ a concept of *local state*. This concept is independent of the concepts of equilibrium and reversibility. Experience shows that use of the concept of local state leads for all practical purposes to results in accord with observation. It is therefore universally accepted and employed. At the very worst it represents an acceptable approximation.

EXAMPLE 6.5 Saturated steam at 350 kPa is mixed continuously with a stream of water at 15 °C to produce hot water at 80 °C at the rate of 4 kg·s^{-1}. The inlet and outlet lines to the mixing device all have an internal diameter of 50 mm. At what rate must steam be supplied?

Equation (6.16) applies, with $\dot{W}_s = 0$. We shall assume that the apparatus is insulated so that, to a good approximation, $\Sigma \dot{Q} = 0$. Presumably, also, the device is small enough so that the inlet and outlet elevations are almost the same and the potential-energy terms may be neglected. The velocity of the outlet hot-water stream is given by $u = \dot{m}V/A$, where $\dot{m} = 4$ kg·s^{-1}, V is the specific volume of water at 80 °C, and $A = (\pi/4)D^2$. From the steam tables, $V = 1.029$ cm^3·g^{-1}; thus

$$u = \frac{(4)(1.029 \times 10^{-3})}{(\pi/4)(50 \times 10^{-3})^2} = 2.096 \text{ m·s}^{-1}$$

From this we calculate the kinetic energy of the water stream:

$$\tfrac{1}{2}\dot{m}u^2 = (\tfrac{1}{2})(4)(2.096)^2 = 8.8 \text{ J·s}^{-1} = 8.8 \text{ W}$$

which is indeed negligible compared with the other energy changes considered. (The process raises the temperature of nearly 4 kg·s^{-1} of water from 15 to 80 °C; this requires about 1 MW.) We therefore neglect the kinetic-energy terms for both water streams. Since we do not yet know the flow rate of the steam, we initially also neglect the kinetic energy of this stream. Equation (6.16) then reduces to

$$H_3\dot{m}_3 - H_2\dot{m}_2 - H_1\dot{m}_1 = 0$$

where the subscript 3 refers to the outlet stream; 2, to the inlet water; and 1, to the inlet steam.

From the steam tables: $V_1 = 524.0$ cm^3·g^{-1}, $H_1 = 2731.6$ kJ·kg^{-1}, $H_2 = 62.9$ kJ·kg^{-1}, $H_3 = 334.9$ kJ·kg^{-1}. Also, $\dot{m}_3 = 4$ kg·s^{-1} and $\dot{m}_2 = \dot{m}_3 - \dot{m}_1 = 4 - \dot{m}_1$. Substitution in the energy equation gives

$$(334.9)(4) - (62.9)(4 - \dot{m}_1) - 2731.6\dot{m}_1 = 0 \qquad \text{or} \qquad \dot{m}_1 = 0.4077 \text{ kg·s}^{-1}$$

The velocity of the steam is

$$u_1 = \frac{\dot{m}_1 V_1}{(\pi/4)D^2} = \frac{(0.4077)(524.0 \times 10^{-3})}{(\pi/4)(50 \times 10^{-3})^2} = 108.8 \text{ m·s}^{-1}$$

and its kinetic energy is

$$\tfrac{1}{2}\dot{m}_1 u_1^2 = \tfrac{1}{2}(0.4077)(108.8)^2 = 2413 \text{ J·s}^{-1} \qquad \text{or} \qquad 2.413 \text{ kW}$$

We now rewrite the energy balance to include the kinetic-energy term for the steam:

$$(334.9)(4) - (62.9)(4 - \dot{m}_1) - 2731.6\dot{m}_1 - 2.413 = 0 \qquad \text{or} \qquad \dot{m}_1 = 0.407 \text{ kg·s}^{-1}$$

The inclusion of the kinetic-energy term results in less than a 0.25% change in the answer.

EXAMPLE 6.6 A complicated process has been devised to make heat continuously available at a temperature level of 260 °C. The only source of energy is steam, saturated at 17.5 bar. Cooling water is available in large supply at 20 °C. How much heat can be transferred from the process to a heat reservoir at 260 °C per kilogram of steam condensed in the process?

We do not need to know any details about how the process is accomplished, but we do need to make a couple of basic assumptions. Therefore we assume that steam flows continuously and that it is condensed and subcooled to the cooling-water temperature by the time it emerges. The properties of the inlet steam and saturated-liquid condensate are given by the steam tables as:

$$T_1 = 205.7 \text{ °C} \qquad\qquad T_2 = 20 \text{ °C}$$
$$H_1 = 2794.1 \text{ kJ} \cdot \text{kg}^{-1} \qquad H_2 = 83.9 \text{ kJ} \cdot \text{kg}^{-1}$$
$$S_1 = 6.3853 \text{ kJ} \cdot \text{kg}^{-1} \cdot \text{K}^{-1} \qquad S_2 = 0.2963 \text{ kJ} \cdot \text{kg}^{-1} \cdot \text{K}^{-1}$$

The second law denies the possibility that the *only* effect of this process could be the transfer of heat from the steam at temperatures between 20 and 205.7 °C to a heat reservoir at 260 °C. However, in this process there is cooling water available at 20 °C. Thus we may consider that the *two* results of the process are transfer of heat from the steam to a heat reservoir at 260 °C *and* the transfer of heat to another heat reservoir at 20 °C (Fig. 6-6). Finally, in order to allow a *numerical* application of the second law, we must assume that the process is reversible. This is tantamount to assuming that it effects a *maximum* transfer of heat to the 260-°C reservoir.

With potential- and kinetic-energy changes neglected, (6.19) gives for the control volume: $\Delta H = Q_H + Q_C$, or

$$Q_H + Q_C = 83.9 - 2794.1 = -2710.2 \qquad\qquad (1)$$

Application of (6.20), with equality, gives

$$\Delta S = \frac{Q_H}{T_H} + \frac{Q_C}{T_C}$$

or, upon substitution of entropy-data and Kelvin temperatures,

$$\frac{Q_H}{533.15} + \frac{Q_C}{293.15} = -6.0890 \qquad\qquad (2)$$

Simultaneous solution of (1) and (2) yields $Q_H = -2055.3 \text{ kJ} \cdot \text{kg}^{-1}$, the minus sign merely indicating transfer out of the control volume.

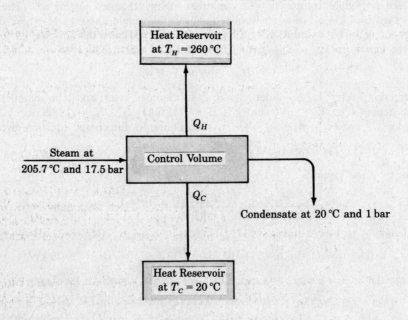

Fig. 6-6

Reversible Adiabatic Flows

The flow of real fluids is an inherently irreversible process because of the inevitable viscous dissipation. Nevertheless, one can imagine the *limiting* process of reversible flow, during which the total entropy is constant.

EXAMPLE 6.7 Consider the steady flow of fluid through a nozzle as illustrated in Fig. 6-7. We wish to consider the change of state which occurs in the fluid as it flows under the influence of the pressure drop $P_1 - P_2$ from station 1 near the nozzle inlet to station 2 at the nozzle exit. The energy equation applicable to this steady-flow process is (6.19):

$$\Delta H + \tfrac{1}{2} \Delta u^2 + g \Delta z = \sum Q - W_s$$

In the present application the potential-energy term may be dropped, because the change in elevation Δz from 1 to 2 is zero. Since no shaft work is done by the process, W_s is also zero. We must make some assumption with respect to the heat transfer $\sum Q$ between the control volume (the section of the nozzle between 1 and 2) and the surroundings. Certainly the object of the process is not heat transfer, and therefore steps are usually taken to insulate the nozzle. Moreover, the amount of heat that could be transferred to each unit mass of fluid flowing through the control volume would normally be quite small, because nozzles accommodate a high rate of flow and the area available for heat transfer is small. We therefore take the process to be adiabatic, and this allows us to write $\Delta H + \tfrac{1}{2} \Delta u^2 = 0$, or

$$\tfrac{1}{2}(u_2^2 - u_1^2) = -(H_2 - H_1) \tag{6.21}$$

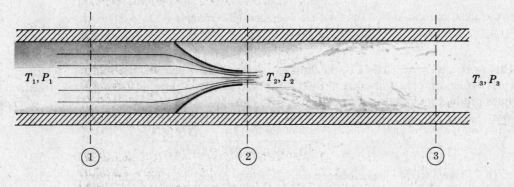

Fig. 6-7

Presuming we know the upstream conditions P_1, T_1, and u_1, as well as the exit pressure P_2, we would like to determine the conditions T_2 and u_2. In general, however, we need T_2 to find H_2, and H_2 to calculate u_2. Thus (6.21) alone does not suffice for the calculation. However, if we consider the limiting case of *reversible* adiabatic flow, then by (6.20), $\Delta S = 0$, and $S_2 = S_1$. With a path (the isentrope) specified for the process, tabular data may be used to determine $(H_2 - H_1)_S$ and T_2, and then

$$\tfrac{1}{2}(u_2^2 - u_1^2) = -(H_2 - H_1)_S \tag{6.22}$$

may be solved for u_2. This theoretical exit velocity can very nearly be attained in a properly designed nozzle. However, the design of the nozzle depends on other considerations (fluid mechanics) quite outside the scope of thermodynamics, which provides merely the limit of what can be attained, without prescribing the means of doing it.

EXAMPLE 6.8 The isentropic expansion of a fluid through a nozzle, as treated in Example 6.7, produces a fluid stream of increased kinetic energy. This stream can be made to impinge on a turbine blade so as to provide a force to move the blade. The stream thus does work on the turbine blade at the expense of its kinetic energy. This is the principle on which the operation of a turbine depends. A series of nozzles and blades is arranged to

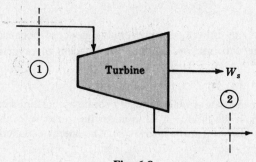

Fig. 6-8

expand the fluid in stages and to convert kinetic energy into shaft work. Thus, the overall result of the process is the expansion of a fluid from a high pressure to a low pressure with the production of work (Fig. 6-8). The appropriate energy equation is again (6.19), which applied between stations 1 and 2 reduces to

$$H_2 - H_1 = \sum Q - W_s$$

where the potential-energy term is omitted because there is negligible change in elevation, and the kinetic-energy term is omitted because the turbine is normally designed so that both the inlet and exit velocities are relatively low. There is, of course, the possibility of heat exchange between the turbine and the surroundings. However, the object of the process is not heat transfer but the production of work. Therefore, in any properly designed turbine, $\sum Q$ is entirely negligible, and we are left with

$$W_s = -(H_2 - H_1) \tag{6.23}$$

Under the additional assumption of reversible operation (cf. Example 6.7), (6.23) becomes

$$W_s = -(H_2 - H_1)_s \tag{6.24}$$

The work given by (6.24) is the limiting or *maximum* shaft work that can be produced by adiabatic expansion of a unit mass of fluid from a given initial state to a given final pressure. Actual machines produce an amount of work equal to 75% or 80% of this. Thus we can define an *expansion efficiency*

$$\eta_e \equiv \frac{W_s \text{ (actual)}}{W_s \text{ (isentropic)}} = \frac{\Delta H}{(\Delta H)_s}$$

where (6.23) and (6.24) have been used. The two processes, the actual and the reversible, are represented on a Mollier diagram in Fig. 6-9. The reversible process follows a vertical line of

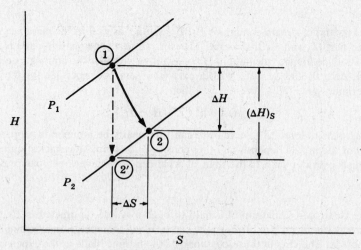

Fig. 6-9. Expansion

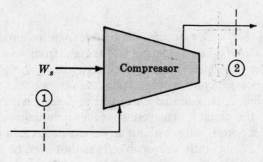

Fig. 6-10

constant entropy from point 1 at the higher pressure P_1 to point $2'$ at the discharge pressure P_2. The line representing the actual irreversible process starts again from point 1, but proceeds downward and to the right—a direction of increasing entropy. The process terminates at point 2 on the isobar for P_2. The greater the irreversibility, the further this point lies to the right of point $2'$, giving lower and lower values for the efficiency η_e of the process.

An adiabatic compression process, represented schematically in Fig. 6-10, is the opposite of an adiabatic expansion process, because work is done on a fluid so as to raise its pressure. In fact, a reversible, adiabatic expansion process can be made to retrace its path in reverse, becoming a reversible, adiabatic compression process. The energy equation is the same for both processes, because the same assumptions of negligible kinetic-energy and potential-energy changes are made. Thus (6.23) and (6.24) are applicable in either case. There is, however, a difference. In the case of adiabatic compression the reversible work is the *minimum* shaft work required for compression of a fluid from its initial state to a given final pressure. An actual irreversible process requires greater work expenditure. Thus the *compression efficiency* is defined by

$$\eta_c \equiv \frac{W_s \text{ (isentropic)}}{W_s \text{ (actual)}} = \frac{(\Delta H)_S}{\Delta H}$$

The adiabatic compression processes, actual and reversible, are represented on a Mollier diagram in Fig. 6-11. Here again the actual path 1-2 runs in a direction of increasing entropy. The figure clearly shows that in this case $(\Delta H)_S$ is the minimum possible enthalpy increase.

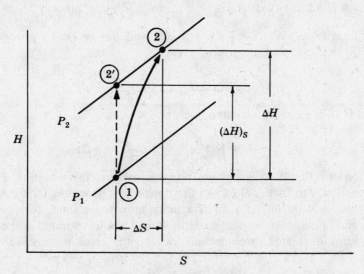

Fig. 6-11. Compression

Throttling Processes

With reference to Fig. 6-9, which depicts adiabatic expansion processes, we see that a possible path for expansion is a horizontal line extending to the right from point 1. Such a line represents an adiabatic expansion process for which there is no change in enthalpy. For this case (6.23) becomes $W_s = -\Delta H = 0$. Such a process therefore occurs adiabatically, without the production of shaft work, and with no change in kinetic or potential energy. The sole result of this kind of process is the reduction of pressure, and this result is achieved whenever a fluid flows through a restriction, such as a partially closed valve or a porous plug, without any appreciable gain in kinetic energy. It is known as a *throttling process*, and is inherently irreversible. It cannot even be imagined to be reversible; the entropy of the fluid necessarily increases along any horizontal line in Fig. 6-9.

Refer again to Fig. 6-7. The discharge stream from the nozzle, with its high kinetic energy, can be used for the production of work; however, if no work-producing device is present and if the jet from the nozzle merely discharges into the downstream pipe, then it generates turbulence downstream from the nozzle. In this case the flowing fluid swirls about, expanding to fill the pipe, and dissipates its kinetic energy as a result of the damping effects of viscosity. Kinetic energy is thereby reconverted into internal energy, and the fluid passes station 3 with a low velocity; the overall process from 1 to 3 is for all practical purposes a throttling process. Although the change from 1 to 2 can be made nearly reversible, the ensuing process from 2 to 3 cannot, and the overall process must result in an entropy increase.

EXAMPLE 6.9   What are the temperature and the entropy changes that result when an ideal gas undergoes a throttling process from 200 kPa to 100 kPa?

For a throttling process $\Delta H = 0$, and for an ideal gas H is solely a function of T (see Example 1.9). Hence, $\Delta H = 0$ implies $\Delta T = 0$. Now, the entropy change of an ideal gas is given by (2.10), which, for $T_2 = T_1$, becomes

$$\Delta S = -R \ln \frac{P_2}{P_1}$$

Thus, for $P_1 = 2P_2$, $\Delta S = R \ln 2$.

6.5   THE MECHANICAL-ENERGY BALANCE

Consider the steady flow of fluid through a control volume to which there is but one entrance and one exit. The energy equation which applies is (6.19),

$$\Delta H + \tfrac{1}{2} \Delta u^2 + g \Delta z = \sum Q - W_s$$

The property relation (3.48) is $dH = T\, dS + V\, dP$; and, *for reversible processes,* $T\, dS = \delta(\sum Q)$. Thus, $dH = \delta(\sum Q) + V\, dP$, and integration gives

$$H_2 - H_1 \equiv \Delta H = \sum Q + \int_1^2 V\, dP$$

Substituting for ΔH in the energy equation, we get

$$-W_s = \int_1^2 V\, dP + \tfrac{1}{2} \Delta u^2 + g \Delta z$$

The assumption of reversibility was made in order to derive this equation. However, the viscous nature of real fluids leads to frictional effects that make flow processes inherently irreversible. As it stands, this equation is valid only for an imaginary nonviscous fluid; for real fluids it is at best approximate. An additional, positive-definite term Ψ may be incorporated in the equation to account for mechanical energy dissipated through fluid friction; the resulting *mechanical-energy balance* is

$$-W_s = \int_1^2 V\, dP + \tfrac{1}{2} \Delta u^2 + g \Delta z + \Psi \qquad (6.25)$$

The determination of numerical values for Ψ is a problem in fluid mechanics, not thermodynamics, and will not be considered here.

The famous *Bernoulli equation* is obtained from (6.25) for a nonviscous, incompressible fluid that does not exchange shaft work with the surroundings. With $\Psi = 0$, and with $W_s = 0$, and with

$$\int_1^2 V \, dP = V \, \Delta P = \frac{\Delta P}{\rho}$$

(6.25) becomes

$$\frac{\Delta P}{\rho} + \tfrac{1}{2} \Delta u^2 + g \, \Delta z = 0 \qquad (6.26)$$

Equivalently,

$$\frac{P}{\rho} + \tfrac{1}{2} u^2 + gz = \text{const.}$$

The severe limitations on Bernoulli's equation should be carefully noted.

Solved Problems

ENERGY EQUATIONS   (Sections 6.1 through 6.3)

6.1   A 3-m³ rigid tank initially contains a mixture of saturated-vapor steam and saturated-liquid water at 3500 kPa. Of the total mass, 10% is vapor. Saturated liquid only is bled slowly from the tank until the total mass in the tank drops to half of the initial total mass. During this process the temperature of the contents of the tank is kept constant by the transfer of heat. How much heat is transferred?

We can calculate immediately the initial mass in the tank. The total volume of the tank is the sum of the liquid volume and the vapor volume:

$$V_{\text{tank}} = m_1^l V^l + m_1^v V^v = 3 \text{ m}^3$$

Since 10% of the mass is vapor and 90% is liquid, then $m_1^l/m_1^v = 9$, and

$$V_{\text{tank}} = 9m_1^v V^l + m_1^v V^v = m_1^v (9V^l + V^v) = 3$$

The steam tables provide the values $V^l = 1.235 \times 10^{-3} \text{ m}^3 \cdot \text{kg}^{-1}$, $V^v = 57.025 \times 10^{-3} \text{ m}^3 \cdot \text{kg}^{-1}$; thus,

$$m_1^v[(9)(1.235) + 57.025] \times 10^{-3} = 3 \quad \text{or} \quad m_1^v = 44.03 \text{ kg}$$

Since this represents 10% of the total mass, we have $m^t = 440.3$ kg and the mass removed from the tank is $m' = 220.15$ kg.

Since the tank is kept at constant temperature, and at all times contains saturated liquid and vapor in equilibrium, the pressure also remains constant, at 3500 kPa, and the specific properties of the phases do not change. Only the amounts of the two phases change.

If we take the tank as the control volume, and apply (6.10), we get

$$d(mU)_{\text{tank}} + H^l \, dm' = \sum \delta Q$$

The kinetic- and potential-energy terms have been assumed negligible, and since no work is involved, δW is set equal to zero. In addition, there is but a single flowing stream, which flows *out* of the control volume, and it has the specific enthalpy of the saturated liquid. Since this enthalpy remains constant during the process, integration of the energy equation gives

$$\Delta(mU)_{\text{tank}} + m'H^l = Q$$

For convenience, we eliminate U in favor of H. Since $U = H - PV$, and since P and V_{tank} are fixed,

$$\Delta(mU)_{\text{tank}} = \Delta(mH)_{\text{tank}} - \Delta(mPV)_{\text{tank}} = \Delta(mH)_{\text{tank}}$$

and the energy equation can be written

$$\Delta(mH)_{\text{tank}} + m'H^l = Q$$

Since the tank contains both saturated liquid and saturated vapor at the start and end of the process,

$$\Delta(mH)_{\text{tank}} = (m_2^l H^l - m_1^l H^l) + (m_2^v H^v - m_1^v H^v)$$

The energy equation now becomes

$$H^v(m_2^v - m_1^v) - H^l(m_1^l - m_2^l - m') = Q$$

By a mass balance, $m_2^v + m_2^l + m' = m_1^v + m_1^l$, or

$$m_2^v - m_1^v = m_1^l - m_2^l - m' \equiv y$$

The energy equation now simplifies to

$$Q = y(H^v - H^l)$$

We have merely to determine y, the mass of liquid vaporized in the tank. This is done by noting that the volume of the tank is constant:

$$m_1^l V^l + m_1^v V^v = m_2^l V^l + m_2^v V^v (= V_{\text{tank}})$$

whence $$V^v(m_1^v - m_2^v) = V^l(m_2^l - m_1^l) \qquad \text{or} \qquad V^v y = V^l(y + m')$$

Solution for y gives

$$y = \frac{m'V^l}{V^v - V^l} = \frac{(220.15)(1.235 \times 10^{-3})}{(57.025 - 1.235) \times 10^{-3}} = 4.873 \text{ kg}$$

From the steam tables, $H^l = 1049.8 \text{ kJ} \cdot \text{kg}^{-1}$, $H^v = 2802.0 \text{ kJ} \cdot \text{kg}^{-1}$; and so

$$Q = (4.873)(2802.0 - 1049.8) = 8538 \text{ kJ}$$

The above solution is rather labored, because a very general energy equation is applied to a very simple process. It is possible to reach the same result more readily if we take a different point of view at the start. We may imagine the saturated liquid that flows from the tank to be collected in a piston/cylinder device (Fig. 6-12). Since what happens to the liquid taken from the tank is not specified and does not affect the remaining contents of the tank, we are free to assume that it is collected reversibly in the cylinder as saturated liquid at the T and P of the tank. Then the process is one of constant pressure in a closed system, for which the heat transfer is equal to the total enthalpy change: $Q = \Delta H'$. The only enthalpy change is that which accompanies the change of y kg of liquid into vapor; thus

$$Q = y \, \Delta H^{lv}$$

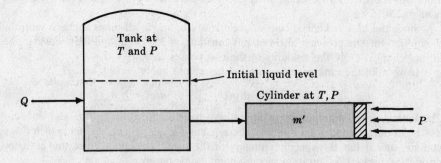

Fig. 6-12

Similarly, the volume change of the system is given by $\Delta V^t = y\,\Delta V^{lv}$. However, the total volume change is clearly also given by $\Delta V^t = m'V^l$. Thus

$$y = \frac{m'V^l}{\Delta V^{lv}}$$

These equations are the same ones derived earlier.

6.2 A standard problem solved by the methods of this chapter involves the flow of a gas or vapor from a source at constant temperature and pressure into a closed tank of known volume which is initially at a lower pressure than the source. For example, a tank containing air at pressure P_1 and temperature T_1 could be connected to a compressed-air line which supplies air at the steady conditions P' and T'. The tank is initially isolated from the line by a closed valve, and the problem is to determine the amount of gas that flows into the tank when the valve is opened long enough for a finite change to occur. Develop the general energy equation which applies to this process.

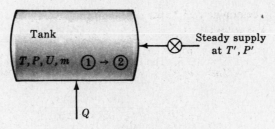

Fig. 6-13

The process is shown in Fig. 6-13. Take the tank as control volume and apply (6.10). Since there is but one flowing stream, and since it *enters* the tank, (6.10) becomes

$$d(mU)_{\text{tank}} - (H' + \tfrac{1}{2}u'^2 + gz')\,dm' = \sum \delta Q - \sum \delta W$$

However, there is no work involved in the process; there is but one heat term; and, in the absence of detailed information, the kinetic- and potential-energy terms must be taken as negligible. The energy equation therefore becomes

$$d(mU)_{\text{tank}} - H'\,dm' = \delta Q$$

A mass balance requires that $dm' = dm$; therefore,

$$d(mU)_{\text{tank}} - H'\,dm = \delta Q$$

which may be integrated immediately, because H' is constant, to give

$$m_2U_2 - m_1U_1 - H'(m_2 - m_1) = Q$$

or
$$m_2(U_2 - H') - m_1(U_1 - H') = Q \tag{1}$$

It is sometimes advantageous to express (1) in terms of molar, rather than specific, properties; thus

$$n_2(U_2 - H') - n_1(U_1 - H') = Q \tag{2}$$

For the case at hand, let us take the gas to be ideal and to have constant heat capacities. Then

$$H' = U' + P'V' = U' + RT'$$

and (2) becomes

$$n_2(U_2 - U' - RT') - n_1(U_1 - U' - RT') = Q$$

or, since $\Delta U = C_V\,\Delta T$ and $R = C_P - C_V$,

$$n_2(C_VT_2 - C_PT') - n_1(C_VT_1 - C_PT') = Q \tag{3}$$

For an ideal gas bled into an *evacuated* tank *adiabatically*, n_1 and Q are zero, and (3) gives $C_V T_2 = C_P T'$, or

$$T_2 = \gamma T' \qquad (4)$$

Thus, in the very special case considered, the temperature in the tank depends only on the supply temperature (and not on the amount of gas introduced).

6.3 A well-insulated steel tank having a volume of 13 m^3 contains air at a pressure of 100 kPa and a temperature of 295 K. This tank is attached to a compressed-air line, which supplies air at the steady conditions 500 kPa and 303 K. Initially the tank is shut off from the air line by a valve. The valve is opened and stays open until the pressure in the tank reaches 300 kPa. Calculate the temperature of the air in the tank and the amount of air the tank contains (*a*) if the process is adiabatic (no heat exchange between tank wall and air); (*b*) if the tank wall exchanges heat with the air so rapidly that the wall and the air are always at the same temperature. *Data*: Take air to be an ideal gas for which $C_P = (7/2)R$ and $C_V = (5/2)R$. The tank is of mass of 1200 kg, and steel has specific heat 448 J·kg^{-1}·K^{-1}.

The energy equation that applies is (3) of Problem 6.2,

$$n_2(C_V T_2 - C_P T') - n_1(C_V T_1 - C_P T') = Q \qquad (1)$$

in which $T_1 = 295$ K and $T' = 303$ K. The initial number of moles in the tank is given by the ideal-gas equation:

$$n_1 = \frac{P_1 V_{\text{tank}}}{RT_1} = \frac{(100 \times 10^3)(13)}{(8.314)(295)} = 530.0 \text{ mol}$$

(*a*) For $Q = 0$, the energy equation becomes

$$n_2 = n_1 \frac{C_V T_1 - C_P T'}{C_V T_2 - C_P T'} = \frac{-1.712 \times 10^5}{(5/2)T_2 - 1060.5} \qquad (2)$$

An additional relation between n_2 and T_2 is provided by the ideal-gas equation:

$$n_2 = \frac{P_2 V_{\text{tank}}}{RT_2} = \frac{4.691 \times 10^5}{T_2} \qquad (3)$$

Simultaneous solution of (2) and (3) yields $T_2 = 370.16$ K, $n_2 = 1267$ mol.

(*b*) If the tank wall exchanges heat with the air so that it is always at the air temperature, then

$$Q = -Q_{\text{wall}} = -m_{\text{wall}} C_{\text{steel}}(T_2 - T_1)$$

Combine this expression with (1), substitute the numerical data, and solve simultaneously with (3), to obtain $T_2 = 299.88$ K, $n_2 = 1564$ mol.

The great difference between the results of (*a*) and of (*b*) illustrates the importance of the assumptions made in the initial formulation of the problem. We have worked here two limiting cases. In practice, (*b*) would be much more nearly right.

6.4 A donkey engine draws its steam from a large insulated tank containing mostly high-pressure, saturated-liquid water. The water is in equilibrium with saturated-vapor steam, which occupies a relatively small space above the saturated liquid. As the engine operates, it draws off saturated vapor, causing liquid to evaporate, and this in turn lowers the temperature and pressure in the tank. Eventually, the engine must visit a power station to recharge its tank. A particular engine carries a tank having a volume of 17 m^3. For efficient operation it requires steam at pressures ranging from 10 bar to 7 bar. If the engine visits the power station when the tank pressure has fallen to 7 bar and the saturated liquid occupies 90% of the tank volume, with how much saturated steam at 11 bar from the power station must the tank be charged to raise its pressure to 10 bar? Neglect all heat transfer.

The energy equation applicable here is (1) of Problem 6.2, with Q set equal to zero:

$$\frac{m_2}{m_1} = \frac{U_1 - H'}{U_2 - H'} = \frac{H' - U_1}{H' - U_2}$$

Here m_1 and m_2 represent the total mass in the tank at the start and end of the charging process, U_1 and U_2 represent the specific internal energy of the tank contents at start and end of the process, and H' is the specific enthalpy of the steady supply of saturated steam from the power station. To find m_1, we note that $0.9V_{\text{tank}} = m_1^l V_1^l$ and $0.1V_{\text{tank}} = m_1^v V_1^v$, whence

$$m_1^l = \frac{(0.9)(17)}{1.108 \times 10^{-3}} = 13\,809 \text{ kg} \qquad m_1^v = \frac{(0.1)(17)}{272.7 \times 10^{-3}} = 6.23 \text{ kg}$$

Here values of V_1^l and V_1^v are from the steam tables at $P = 7$ bar. Therefore, $m_1 = m_1^l + m_1^v = 13\,815$ kg, and the quality of the initial two-phase mixture is

$$x_1 = \frac{m_1^v}{m_1} = \frac{6.23}{13\,815} = 0.000\,451$$

The initial internal energy U_1 is therefore given by

$$U_1 = U_1^l + x_1\,\Delta U_1^{lv} = 696.3 + (0.000\,451)(1874.8) = 697.1 \text{ kJ}\cdot\text{kg}^{-1}$$

From the steam tables, saturated vapor at 11 bar has $H' = 2779.7 \text{ kJ}\cdot\text{kg}^{-1}$. The energy equation now becomes

$$\frac{m_2}{m_1} = \frac{2779.7 - 697.1}{2779.7 - U_2} = \frac{2082.6}{2779.7 - U_2} \tag{1}$$

As both m_2 and U_2 are unknown, we require another relationship connecting the initial and final states of the system. It is provided by $V_{\text{tank}} = \text{const.}$, or

$$\frac{m_2}{m_1} = \frac{V_1}{V_2} \tag{2}$$

in which

$$V_1 = V_1^l + x_1\,\Delta V_1^{lv} = 1.108 + (0.000\,451)(271.59) = 1.230 \text{ cm}^3\cdot\text{g}^{-1}$$

Combination of (1) and (2) gives

$$\frac{1.230}{V_2} = \frac{2082.6}{2779.7 - U_2} \tag{3}$$

Now, in terms of the final quality,

$$V_2 = V_2^l + x_2\,\Delta V_2^{lv} = 1.127 + 193.16x_2$$
$$U_2 = U_2^l + x_2\,\Delta U_2^{lv} = 761.5 + 1820.4x_2$$

where data have been taken from the tables for saturated steam at $P = 10$ bar. Substituting these two expressions in (3) and solving, we obtain $x_2 = 0.000\,334\,5$; then

$$V_2 = 1.127 + (193.16)(0.000\,334\,5) = 1.1916 \text{ cm}^3\cdot\text{g}^{-1}$$

and finally, from (2),

$$m_2 = \frac{1.230}{1.1916}\,(13\,815) = 14\,260 \text{ kg}$$

We conclude that a mass $m_2 - m_1 = 445$ kg of saturated steam at 11 bar must be charged to the tank by the power station.

6.5   Another standard problem in unsteady flow involves the flow of gas or vapor *out of* a tank. One usually wishes to determine the conditions in the gas remaining in a tank after sufficient gas has been bled out to cause a finite change in state. Develop the energy equation which applies.

A differential process of the kind under consideration is shown in Fig. 6-14. Application of (6.10)

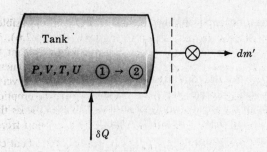

Fig. 6-14

with the tank as control volume gives $d(mU)_{tank} + H' \, dm' = \delta Q$. With $H' = H$ and $dm' = -dm$, this equation becomes

$$d(mU)_{tank} - H \, dm = \delta Q \tag{1}$$

Integration of the equation over a finite process requires a relation between the enthalpy H of the gas in the tank and mass m of the gas in the tank.

An alternative form of (1) can be developed by expansion of the first term:

$$m \, dU + U \, dm - H \, dm = \delta Q \qquad \text{or} \qquad m \, dU - (H - U) \, dm = \delta Q$$

However, $H - U = PV$, so that

$$m \, dU - PV \, dm = \delta Q \tag{2}$$

where U and V are specific properties. Alternatively,

$$n \, dU - PV \, dn = \delta Q \tag{3}$$

where now U and V are molar properties.

6.6   Suppose the gas of Problem 6.5 to be ideal, with constant heat capacities. Integrate the energy equation (3), given that the heat transfer between gas and tank wall (a) is zero; (b) is so rapid that gas and tank wall are always at the same temperature.

With $PV = RT$ and $dU = C_V \, dT$, the energy equation takes the form

$$nC_V \, dT - RT \, dn = \delta Q \tag{1}$$

(a)   For $\delta Q = 0$, (1) gives

$$\frac{dT}{T} = \frac{R}{C_V} \frac{dn}{n}$$

This equation can be integrated directly; however, it is convenient to change variables through use of the relationship $V_{tank} = nV$. Since V_{tank} is constant, differentiation gives

$$\frac{dn}{n} = -\frac{dV}{V}$$

Therefore we can write

$$\frac{dT}{T} = -\frac{R}{C_V} \frac{dV}{V}$$

Integration for constant C_V gives

$$\ln \frac{T_2}{T_1} = \frac{R}{C_V} \ln \frac{V_1}{V_2}$$

But $R/C_V = \gamma - 1$, and therefore

$$\frac{T_2}{T_1} = \left(\frac{V_1}{V_2} \right)^{\gamma - 1} \tag{2}$$

Now (2) is exactly the expression one gets for the reversible, adiabatic expansion of an ideal gas with constant heat capacities (see Examples 1.10 and 2.6). We have, of course, assumed the process to be adiabatic and the gas to be ideal with constant heat capacities, but we have not explicitly assumed reversibility. However, in the development of the energy equation, our assumptions that the stream leaving the tank has negligible velocity and has properties identical with those of the gas in the tank are equivalent to the assumption of reversibility. Thus we could have started with the assumption that for an adiabatic process the gas that remains in the tank at any instant has undergone a reversible, adiabatic expansion from its initial state.

(b) If [cf. Problem 6.3(b)] $\delta Q = -K \, dT$, where K is the total heat capacity of the tank wall, then (1) becomes

$$\frac{dT}{T} = \frac{R \, dn}{nC_V + K} = \frac{R}{C_V} \frac{d(nC_V)}{nC_V + K} = (\gamma - 1) \frac{d(nC_V + K)}{nC_V + K}$$

Integration gives

$$\frac{T_2}{T_1} = \left(\frac{n_2 C_V + K}{n_1 C_V + K} \right)^{\gamma - 1} \tag{3}$$

In addition, we have

$$P_2 V_{\text{tank}} = n_2 R T_2 \tag{4}$$

Since $n_1 = P_1 V_{\text{tank}}/RT_1$ is known, (3) and (4) constitute two equations in the two unknowns T_2 and n_2. Solution must be by trial or by some iterative procedure (e.g., Newton's method). Clearly, if $K \gg nC_V$, then $T_2 \approx T_1$.

6.7 Consider a well-insulated tank of 3-m³ capacity containing superheated steam at 200 °C and 350 kPa. A valve is opened, and steam is bled out until the pressure in the tank is reduced to 101.33 kPa. Calculate the final temperature of the steam in the tank and the amount of steam bled out, if (a) the process occurs rapidly enough so that no heat is transferred between the tank wall and the steam in the tank; (b) the process occurs slowly enough so that heat transfer between the tank wall and the steam keeps them always at the same temperature. The mass of the tank is 600 kg and its specific heat capacity is 0.448 kJ·kg⁻¹·K⁻¹.

(a) If there is no heat transfer to the steam in the tank then, as discussed in Problem 6.6(a), we can assume that the steam left in the tank at any instant has undergone a reversible, adiabatic (isentropic) expansion from its initial conditions, $T_1 = 200$ °C and $P_1 = 350$ kPa. From the tables for superheated steam,

$$S_1 = 7.2366 \text{ kJ} \cdot \text{kg}^{-1} \cdot \text{K}^{-1} \qquad V_1 = 612.31 \text{ cm}^3 \cdot \text{g}^{-1}$$

Since $S_2 = S_1 = 7.2366$ kJ·kg⁻¹·K⁻¹, we must find from the steam tables the conditions at which steam has this entropy at 101.33 kPa (1.0133 bar). The steam is found to be wet, and therefore to be at its saturation temperature, 100 °C; hence

$$S_2^l = 1.3069 \text{ kJ} \cdot \text{kg}^{-1} \cdot \text{K}^{-1} \qquad \Delta S_2^{lv} = 6.0485 \text{ kJ} \cdot \text{kg}^{-1} \cdot \text{K}^{-1}$$

The quality x_2 is

$$x_2 = \frac{S_2 - S_2^l}{\Delta S_2^{lv}} = \frac{7.2366 - 1.3069}{6.0485} = 0.9804$$

so that $V_2 = V_2^l + x_2 \Delta V_2^{lv} = 1.044 + (0.9804)(1672.0) = 1640.2 \text{ cm}^3 \cdot \text{g}^{-1}$. The initial and final masses of steam in the tank are

$$m_1 = \frac{V_{\text{tank}}}{V_1} = \frac{3}{612.31 \times 10^{-3}} = 4.899 \text{ kg} \qquad m_2 = \frac{V_{\text{tank}}}{V_2} = \frac{3}{1640.2 \times 10^{-3}} = 1.829 \text{ kg}$$

The amount of steam bled from the tank is therefore $4.899 - 1.829 = 3.07$ kg.

(b) The appropriate energy equation is (2) of Problem 6.5,

$$m \, dU \, - \, PV \, dm = \delta Q$$

The following development of this equation is intended to put it into a form most convenient for integration. Since $V_{tank} = mV = \text{constant}$, differentiation gives $-V\,dm = m\,dV$, and the energy equation becomes

$$m(dU + P\,dV) = \delta Q$$

However, by (2.6), $dU + P\,dV = T\,dS$. Also, $m = V_{tank}/V$ and $\delta Q = -\delta Q_{tank} = -(mC)_{tank}\,dT$. Combining these with the energy equation gives

$$\frac{dS}{dT} = -\left(\frac{mC}{V}\right)_{tank}\frac{V}{T} = -89.6\frac{V}{T} \qquad (1)$$

for V in $m^3 \cdot kg^{-1}$, T in K, and S in $kJ \cdot kg^{-1} \cdot K^{-1}$. It is this equation that must be integrated. Since S, T, and V are related for any given state of superheated steam by numerical data in the steam tables, we are able to carry out a numerical integration as follows.

The initial conditions T_1 and V_1 for the steam in the tank are known. From these (1) allows calculation of an initial value for dS/dT. Using this and an incremental temperature change ΔT, we can calculate the incremental change in S:

$$\Delta S = \left(\frac{dS}{dT}\right)\Delta T$$

The new values $S = S_1 + \Delta S$ and $T = T_1 + \Delta T$ identify a new state of superheated steam in the steam tables, for which there is a new value of V (and also P). The new values T and V yield a new value for dS/dT by (1), and the process may now be repeated for another increment ΔT. Thus we calculate successive values of S and V (and the corresponding P) for successive increments ΔT. The process continues until the given final pressure $P_2 = 101.33$ kPa is reached. The calculations are made increasingly accurate by decreasing the increment ΔT. A reasonable set of final values is

$$T_2 = 470.05 \text{ K} \qquad S_2 = 7.8151 \text{ kJ} \cdot kg^{-1} \cdot K^{-1} \qquad V_2 = 2.129 \text{ } m^3 \cdot kg^{-1}$$

Thus, $m_2 V_{tank}/V_2 = 1.409$ kg, and the amount of steam bled from the tank is $4.899 - 1.409 = 3.490$ kg.

Comparison of the results of (a) and (b) reveals the considerable effect of the initial assumptions made in setting up the solution.

	(a)	(b)
Final state of the steam	wet	superheated
Final temperature of the steam	100 °C	196.9 °C
Mass of steam bled from the tank	3.060 kg	3.490 kg

6.8   A steam boiler of capacity $30 \text{ } m^3$ contains saturated-liquid water and saturated steam in equilibrium at 7 bar. Initially the liquid and vapor occupy equal volumes. During a certain time interval saturated vapor is withdrawn from the boiler, while simultaneously 15 000 kg of liquid water at 40 °C is added to the boiler. During the process the boiler pressure is held at 7 bar by the addition of heat. At the end of the process the saturated liquid remaining in the boiler occupies one-quarter of the boiler volume, and saturated vapor fills the remainder. How much heat is added to the boiler during the process?

First we determine the masses of material contained in the boiler and flowing from the boiler. Within the boiler the properties of saturated liquid and saturated vapor remain constant throughout the process. From the tables for saturated steam, we have at 7 bar

$$V^l = 1.108 \text{ } cm^3 \cdot g^{-1} \qquad V^v = 272.7 \text{ } cm^3 \cdot g^{-1}$$

Thus the initial masses of liquid and vapor in the boiler are given by

$$m_1^l = \frac{(30)(0.5)}{1.108 \times 10^{-3}} = 13\,538 \text{ kg} \qquad m_1^v = \frac{(30)(0.5)}{272.7 \times 10^{-3}} = 55 \text{ kg}$$

Therefore, $m_1 = m_1^l + m_1^v = 13\,593$ kg. Similarly, the final masses are

$$m_2^l = \frac{(30)(0.25)}{1.108 \times 10^{-3}} = 6769 \text{ kg} \qquad m_1^v = \frac{(30)(0.75)}{272.7 \times 10^{-3}} = 82.5 \text{ kg}$$

and $m_2 = 6851.5$ kg. Thus the depletion of mass in the boiler amounts to $m_1 - m_2 = 6741.5$ kg. The mass of vapor withdrawn from the boiler is the sum of this depleted mass and the mass of liquid added:

$$m' = 6741.5 + 15\,000 = 21\,741.5 \text{ kg}$$

We now apply (6.10) to the boiler, taken as the control volume. There is no work associated with the process, and we assume the kinetic- and potential-energy terms negligible. A saturated-vapor stream flows out, with specific enthalpy H', and a liquid stream flows in, with specific enthalpy H''. Equation (6.10) is therefore written

$$d(mU)_{\text{boiler}} + H'\,dm' - H''\,dm'' = \delta Q$$

Since H' and H'' are constant, integration gives

$$(m_2 U_2 - m_1 U_1) + (H'm' - H''m'') = Q \tag{1}$$

The first term in (1) represents the change in total internal energy of the contents of the tank; the second term represents the net enthalpy-transport of the flowing streams. From the tables for saturated steam at 7 bar:

$$m_2 U_2 = m_2^l U^l + m_2^v U^v = (6769)(696.3) + (82.5)(2571.1) = 4.925 \times 10^6 \text{ kJ}$$
$$m_1 U_1 = m_1^l U^l + m_2^v U^v = (13\,538)(696.3) + (55)(2571.1) = 9.568 \times 10^6 \text{ kJ}$$
$$H' = 2762.0 \text{ kJ} \cdot \text{kg}^{-1}$$

and the table for saturated liquid at 40 °C gives $H'' = 167.5$ kJ $\cdot$ kg^{-1}. Substitution in (1) then yields $Q = 52.89 \times 10^6$ kJ.

STEADY-FLOW PROCESSES   (Section 6.4)

6.9   Wet steam at 15 bar is throttled adiabatically in a steady-flow process to 2 bar. The resulting stream has a temperature of 130 °C. What are the temperature and quality of the wet steam? Calculate ΔS of the steam as a result of the process.

For a throttling process, $\Delta H = 0$, or $H_1 = H_2$. At the final conditions 130 °C and 2 bar, we find by interpolation in the tables for superheated steam: $H_2 = 2726.8$ kJ $\cdot$ kg^{-1}. Since the steam in its initial condition is wet, it is at its saturation temperature, which is found from the tables for saturated steam to be $T_1 = 198.3$ °C. In addition, $H_1^l = 844.7$ kJ $\cdot$ kg^{-1} and $\Delta H_1^{lv} = 1945.2$ kJ $\cdot$ kg^{-1}. Therefore,

$$H_1 = H_1^l + x_1 \Delta H_1^{lv} = 844.7 + 1945.2 x_1 = H_2 = 2726.8 \text{ kJ} \cdot \text{kg}^{-1}$$

Solution gives $x_1 = 0.9676$; the wet steam is 96.76% vapor on a mass basis.

The initial entropy of the steam is determined as

$$S_1 = S_1^l + x_1 \Delta S_1^{lv} = 2.3145 + (0.9676)(4.1261) = 6.3069 \text{ kJ} \cdot \text{kg}^{-1} \cdot \text{K}^{-1}$$

For the final state, $S_2 = 7.1777$ kJ $\cdot$ kg^{-1} $\cdot$ K^{-1}, and

$$\Delta S = S_2 - S_1 = 7.1777 - 6.3069 = 0.8708 \text{ kJ} \cdot \text{kg}^{-1} \cdot \text{K}^{-1}$$

6.10   A thermometer is inserted in a pipe through which is flowing a steady stream of gas at temperature T_1 and velocity u_1. Will the indicated temperature be higher than, lower than, or the same as T_1?

The experimental arrangement is shown in Fig. 6-15. We choose as a control volume an imaginary "tube" (shown by the dashed lines) whose cross-section is the projected area of the thermometer bulb. The bounding planes 2-2 and 1-1 are drawn, respectively, through the centerline of the thermometer, and at a position far enough upstream so that conditions are unaffected by the presence of the

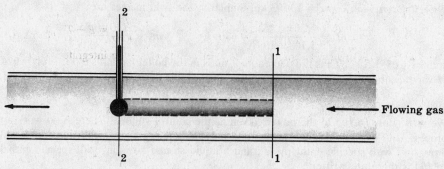

Fig. 6-15

thermometer. It is assumed that gas reaching 2-2 in the control volume is decelerated to a low velocity and that it enters the control volume at 1-1. The flow is steady, and we consider only the single stream leaving the control volume at 2-2 and entering at 1-1. Clearly, $W_s = \Delta z = 0$, and we also assume that no heat is transferred to or from the gas. Thus (6.19) reduces to (6.21), which was previously derived for flow through an adiabatic nozzle:

$$H_2 - H_1 = \tfrac{1}{2}(u_1^2 - u_2^2) \tag{6.21}$$

For an ideal gas with constant heat capacity, $H_2 - H_1 = C_P(T_2 - T_1)$. Substitution of this expression and $u_2 = 0$ into (6.21) gives $C_P(T_2 - T_1) = \tfrac{1}{2}u_1^2$, whence

$$T_2 = T_1 + \frac{u_1^2}{2C_P} \tag{1}$$

The measured temperature is therefore *higher than* the gas temperature T_1; the temperature T_2 defined by (1) is sometimes called the *stagnation temperature*. Many of the fluid-mechanical assumptions implicit in the solution of this problem are of limited validity, however; actual thermometer readings for such arrangements are only approximately equal to the stagnation temperature.

6.11 Propane at 10 bar and 320 K expands in a steady-flow process through an orifice to a pressure of 1 bar. Heat is added so as to maintain the downstream temperature also at 320 K. Upstream from the orifice the propane flows with a velocity of $2\ \mathrm{m \cdot s^{-1}}$ and downstream from the orifice its velocity is approximately $20\ \mathrm{m \cdot s^{-1}}$. How much heat must be added per kilogram of propane flowing through the orifice? What percentage error would be introduced if the kinetic-energy change of the propane were neglected? The volumetric properties of propane at these conditions are well represented by (5.33), the simplest form of the virial equation, and the second virial coefficient may be determined from the generalized correlation of Section 5.4.

The appropriate energy equation is (6.19), which reduces to

$$\Delta H + \tfrac{1}{2}\Delta u^2 = Q \tag{1}$$

because there is no work, and the change in potential energy of the propane is evidently negligible. To determine ΔH, we make use of the equation

$$\left(\frac{\partial H}{\partial P}\right)_T = -T\left(\frac{\partial V}{\partial T}\right)_P + V \tag{2}$$

which was derived in Problem 3.3. Because H and V in (1) and (2) are specific properties, we write the gas constant for propane as

$$R^* \equiv \frac{R}{M} = \frac{8.314\ \mathrm{kJ \cdot kmol^{-1} \cdot K^{-1}}}{44\ \mathrm{kg \cdot kmol^{-1}}} = 0.189\ \mathrm{kJ \cdot kg^{-1} \cdot K^{-1}}$$

Then, substitution of $Z = PV/R^*T$ in (5.33) and solution for V gives

$$V = \frac{R^*T}{P} + B \qquad [B = B(T)]$$

CHAP. 6]   THERMODYNAMICS OF FLOW PROCESSES   201

from which

$$\left(\frac{\partial V}{\partial T}\right)_P = \frac{R^*}{P} + \frac{dB}{dT} \quad \text{and} \quad \left(\frac{\partial H}{\partial P}\right)_T = B - T\frac{dB}{dT}$$

For a process occurring at constant T, this last equation may be integrated to give

$$\Delta H = \left(B - T\frac{dB}{dT}\right)\Delta P \qquad (3)$$

The generalized correlation of Section 5.4 provides the equations

$$\frac{BP_c}{R^*T_c} = B^0 + \omega B^1 \qquad (4)$$

$$\frac{P_c}{R^*}\frac{dB}{dT} = \frac{dB^0}{dT_r} + \omega\frac{dB^1}{dT_r} \qquad (5)$$

and Appendix C provides the values

$$T_c = 369.8 \text{ K} \qquad P_c = 42.5 \text{ bar} \qquad \omega = 0.152$$

Thus $T_r = 320/369.8 = 0.865$, and, from (5.73) and (5.77),

$$B^0 = -0.449 \qquad B^1 = -0.177$$
$$dB^0/dT_r = 0.983 \qquad dB^1/dT_r = 1.532$$

Therefore, from (4) and (5),

$$B = -0.476\frac{R^*T_c}{P_c} \quad \text{and} \quad \frac{dB}{dT} = 1.216\frac{R^*}{P_c}$$

Substitution in (3) gives

$$\Delta H = \left(-0.476\frac{R^*T_c}{P_c} - 1.216\frac{R^*T}{P_c}\right)\Delta P = -(0.476T_c + 1.216T)\left(\frac{\Delta P}{P_c}\right)R^*$$

$$= -[(0.476)(369.8) + (1.216)(320)]\left(\frac{-9}{42.5}\right)(0.189) = 22.61 \text{ kJ}\cdot\text{kg}^{-1}$$

We now evaluate the kinetic-energy term:

$$\tfrac{1}{2}\Delta u^2 = \tfrac{1}{2}(u_2^2 - u_1^2) = \tfrac{1}{2}(20^2 - 2^2) = 198 \text{ J}\cdot\text{kg}^{-1} \quad \text{or} \quad 0.198 \text{ kJ}\cdot\text{kg}^{-1}$$

Equation (1) now gives $Q = 22.61 + 0.198 = 22.808 \text{ kJ}\cdot\text{kg}^{-1}$. The error caused by neglect of the kinetic-energy term is

$$\frac{0.198}{22.808}(100) = 0.87\%$$

6.12 Methane gas is throttled in a steady-flow process from the upstream conditions 40 °C and 20 bar to the downstream pressure 5 bar. What is the gas temperature on the downstream side of the throttling device?

For the conditions of this problem, the volumetric properties of methane are adequately described by the truncated virial equation (5.33), used in conjunction with the generalized correlation of Section 5.4. An expression for the molar heat capacity of methane in the ideal-gas state comes from the data of Table 4-1:

$$\frac{C_P^{ig}}{R} = 1.702 + (9.081 \times 10^{-3})T - (2.164 \times 10^{-6})T^2 \qquad (1)$$

As shown in Section 6.4, (6.19) applied to a throttling process gives $\Delta H = 0$. Thus the downstream gas temperature takes on a value such that there is no enthalpy change across the throttling device.

In order to find the final temperature, we set up a calculational path similar to that of Problem 5.22. Since $\Delta H = 0$, we see from Fig. 6-16 that an equivalent requirement is $-H_1^R + \Delta H_{12}^{ig} + H_2^R = 0$, or

$$-\frac{H_1^R}{R} + \frac{\langle C_P^{ig}\rangle_T}{R}(T_2 - T_1) + \frac{H_2^R}{R} = 0 \qquad (2)$$

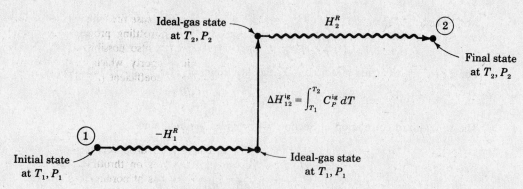

Fig. 6-16

Since both $\langle C_P^{ig} \rangle_T$ and H_2^R depend on T_2, an iterative solution for T_2 is indicated.

For methane, from Appendix C,

$$T_c = 190.6 \text{ K} \qquad P_c = 46.0 \text{ bar} \qquad \omega = 0.008$$

For the initial state, $T_{r1} = 313.15/190.6 = 1.643$ and $P_{r1} = 20.0/46.0 = 0.435$; and, from (5.73) and (5.77) for $T_{r1} = 1.643$,

$$B^0 = -0.108 \qquad B^1 = 0.118$$

$$\frac{dB^0}{dT_r} = 0.186 \qquad \frac{dB^1}{dT_r} = 0.055$$

By (5.72), (5.75), and (5.77), with T replaced by $T_c T_r$, we get

$$\frac{H^R}{RT_c} = P_r \left[B^0 - T_r \frac{dB^0}{dT_r} + \omega \left(B^1 - T_r \frac{dB^1}{dT_r} \right) \right] \tag{3}$$

Therefore,

$$\frac{H_1^R}{RT_c} = (0.435)\{-0.108 - (1.643)(0.186) + 0.008[0.118 - (1.643)(0.055)]\} = -0.180$$

whence $H_1^R/R = (0.435)(-0.4134)(190.6) = -34.13$ K.

Anticipating that the temperature change caused by throttling is small, we initially take $\langle C_P^{ig} \rangle_T$ to be equal to C_P^{ig} at the initial temperature, 313.15 K. Also, we initially evaluate H_2^R at this temperature. Thus by (1)

$$\langle C_P^{ig} \rangle_T /R = 1.702 + (9.081 \times 10^{-3})(313.15) - (2.164 \times 10^{-6})(313.15)^2 = 4.333$$

and with $P_{r2} = 5/46.0 = 0.109$,

$$\frac{H_2^R}{RT_c} = (0.109)(-0.4134) = -0.0451 \qquad \text{and} \qquad \frac{H_2^R}{R} = (-0.0451)(190.6) = -8.6 \text{ K}$$

Solving (2) for $T_2 - T_1$, we get

$$T_2 - T_1 = \frac{(H_1^R/R) - (H_2^R/R)}{\langle C_P^{ig} \rangle_T /R} = \frac{-34.1 + 8.6}{4.333} = -5.9 \text{ K} \qquad \text{or} \qquad -5.9 \text{ °C}$$

The temperature drop is therefore about 6 °C, making T_2 about 34 °C.

We may refine this result by reevaluating $\langle C_P^{ig} \rangle_T$, taking it equal to C_P^{ig} at the arithmetic mean temperature, 310.15 K; this gives $\langle C_P^{ig} \rangle_T /R = 4.310$. Further, we recalculate H_2^R at 307.15 K, finding $H_2^R/R = -8.9$ K. Then,

$$T_2 - T_1 = \frac{-34.1 + 8.9}{4.310} = -5.85 \text{ °C}$$

Clearly, further refinement of this result is pointless; thus, $T_2 = 34.2$ °C.

The effect of the throttling process of this problem is to *decrease* the temperature of the flowing gas stream by about 6 °C, and in fact a usual purpose of a throttling process (e.g., in refrigeration engineering) is to achieve a temperature drop. However it is also possible to obtain a temperature *increase*, under certain conditions. The thermodynamic property which determines the sign and magnitude of the temperature change is the Joule/Thomson coefficient $(\partial T/\partial P)_H$ [see Problems 3.4, 5.2, 5.8, and 5.36]. Thus for a change of state at constant H,

$$\Delta T = \int_{P_1}^{P_2} \left(\frac{\partial T}{\partial P}\right)_H dP$$

Clearly, if $P_2 < P_1$, as in a throttling process, a temperature drop occurs if $(\partial T/\partial P)_H$ is positive. Conversely, if $(\partial T/\partial P)_H$ is negative, the temperature increases on throttling. Certain gases, notably hydrogen and helium, have negative Joule/Thomson coefficients at normal conditions.

6.13 A test made on a stand-by turbine power unit produced the following results. With steam supplied to the turbine at 1350 kPa and 375 °C the exhaust from the turbine at 10 kPa was saturated vapor only. What is the efficiency of the turbine?

Assuming the turbine to be properly designed, we can neglect heat transfer and kinetic- and potential-energy terms. Under these circumstances (see the discussion following Example 6.8), $\eta_e = \Delta H/(\Delta H)_S$. Thus we need to determine from the steam tables the actual enthalpy change of the steam and the enthalpy change that would occur were expansion isentropic to the final pressure of 10 kPa.

From the table for superheated vapor we find, at $P_1 = 1350$ kPa and $T_1 = 375$ °C, that $H_1 = 3205.4$ kJ · kg^{-1} and $S_1 = 7.2410$ kJ · kg^{-1} · K^{-1}. For saturated steam at 10 kPa, $H_2 = 2584.8$ kJ · kg^{-1}; thus,

$$\Delta H = H_2 - H_1 = 2584.8 - 3205.4 = -620.6 \text{ kJ} \cdot \text{kg}^{-1}$$

We must now find the specific enthalpy of steam at 10 kPa for which the specific entropy is $S_2' = S_1 = 7.2410$ kJ · kg^{-1} · K^{-1} (see Fig. 6-9). We note that the entropy of saturated vapor at 10 kPa is 8.1511 kJ · kg^{-1} · K^{-1}, which is higher than the required S_2'. Thus the required state is wet steam, and we must determine its quality. At 10 kPa,

$$S_2' = 7.2410 \text{ kJ} \cdot \text{kg}^{-1} \cdot \text{K}^{-1} \qquad S^l = 0.6493 \text{ kJ} \cdot \text{kg}^{-1} \cdot \text{K}^{-1} \qquad \Delta S^{lv} = 7.5018 \text{ kJ} \cdot \text{kg}^{-1} \cdot \text{K}^{-1}$$

Since $S_2' = S^l + x\,\Delta S^{lv}$, $7.2410 = 0.6493 + 7.5018x$, or $x = 0.8787$. The required enthalpy H_2' is now given by

$$H_2' = H^l + x\,\Delta H^{lv} = 191.8 + (0.8787)(2393.0) = 2294.5 \text{ kJ} \cdot \text{kg}^{-1}$$

and $(\Delta H)_S = H_2' - H_1 = 2294.5 - 3205.4 = -910.9$ kJ · kg^{-1}. Finally,

$$\eta_e = \frac{\Delta H}{(\Delta H)_S} = \frac{-620.6}{-910.9} = 0.681$$

6.14 (*a*) For the steady-flow, adiabatic compression or expansion of an ideal gas from initial conditions T_1 and P_1 to final pressure P_2, develop a procedure for calculating the work W_s and the final temperature T_2. The efficiency η_c or η_e of the process is given. (*b*) Specialize to the case of constant heat capacities.

(*a*) For an ideal gas, (*4.21*) may be written as

$$\Delta S = \langle C_P \rangle_{\ln T} \ln \frac{T_2}{T_1} - R \ln \frac{P_2}{P_1} \qquad (1)$$

If compression or expansion is *isentropic*, then $\Delta S = 0$, and (*1*) becomes

$$\langle C_P' \rangle_{\ln T} \ln \frac{T_2'}{T_1} - R \ln \frac{P_2}{P_1} = 0 \qquad (2)$$

where T_2' is the final temperature for the *isentropic* process and where $\langle C_P' \rangle_{\ln T}$ is the mean heat capacity for the temperature range from T_1 to T_2'. Given, for example, (*4.13*) for the temperature dependence of C_P, we find $\langle C_P' \rangle_{\ln T}$ as a function of T_2' and solve (2) numerically for T_2' by the method of Example 5.16(a). Thus for isentropic compression or expansion,

$$(\Delta H)_S = \langle C_P' \rangle_T (T_2' - T_1) \tag{3}$$

The work is then given by

$$W_s(\text{comp}) = -\eta_c^{-1}(\Delta H)_S \quad \text{and} \quad W_s(\text{exp}) = -\eta_e(\Delta H)_S$$

Combining these with (6.23) and (4.17), we have,

Compression: $\qquad\qquad \langle C_P \rangle_T (T_2 - T_1) - \eta_c^{-1}(\Delta H)_S = 0 \tag{4}$

Expansion: $\qquad\qquad\quad \langle C_P \rangle_T (T_2 - T_1) - \eta_e(\Delta H)_S = 0 \tag{5}$

Solution of (4) or (5) for T_2, the actual final temperature, is again done numerically, exactly as (2) was solved for T_2'.

(b) With $\langle C_P' \rangle_{\ln T} = \langle C_P' \rangle_T = \langle C_P \rangle_T = C_P = \text{const.}$, (2) may be solved explicitly:

$$T_2' = T_1 \left(\frac{P_2}{P_1}\right)^{R/C_P} = T_1 \left(\frac{P_2}{P_1}\right)^{(\gamma-1)/\gamma}$$

(cf. Example 2.6). Then, from (3), (4), and (5),

$$W_s(\text{comp}) = -\eta_c^{-1} C_P T_1 \left[\left(\frac{P_2}{P_1}\right)^{(\gamma-1)/\gamma} - 1 \right]$$

$$\tag{6}$$

$$W_s(\text{exp}) = -\eta_e C_P T_1 \left[\left(\frac{P_2}{P_1}\right)^{(\gamma-1)/\gamma} - 1 \right]$$

Since $W_s = -\Delta H = C_P(T_2 - T_1)$, we have by (6) that

$$T_2(\text{comp}) = T_1 \left\{ 1 + \eta_c^{-1} \left[\left(\frac{P_2}{P_1}\right)^{(\gamma-1)/\gamma} - 1 \right] \right\}$$

$$\tag{7}$$

$$T_2(\text{exp}) = T_1 \left\{ 1 + \eta_e \left[\left(\frac{P_2}{P_1}\right)^{(\gamma-1)/\gamma} - 1 \right] \right\}$$

6.15 Methane gas is compressed at the rate of 3×10^4 m$^3 \cdot$ h^{-1} (as measured at 0.1013 MPa and 15 °C) from 0.69 MPa and 26.7 °C to 3.45 MPa. The compressor operates adiabatically and has an efficiency of 80% compared with isentropic compression. After compression, the methane is cooled at a constant pressure of 3.45 MPa to a temperature of 38 °C. What is the power requirement of the compressor and what is the rate of heat removal in the cooler? *Data for methane*:

At 0.69 MPa and 26.7 °C: $H = 946.7$ kJ $\cdot$ kg^{-1}, $S = 6.071$ kJ $\cdot$ kg$^{-1} \cdot$ K^{-1}

At 3.45 MPa:

T/°C	H/kJ $\cdot$ kg^{-1}	S/kJ $\cdot$ kg$^{-1} \cdot$ K^{-1}
38	949.6	5.264
150	1236.7	6.041
155	1250.3	6.075
160	1263.9	6.108
165	1277.5	6.139
170	1291.1	6.169
175	1304.8	6.199
180	1318.4	6.227
185	1332.0	6.253

At 0.1013 MPa and 15 °C methane is essentially an ideal gas, and the molar rate of compression is calculated as

$$\dot{n} = \frac{P\dot{V}^t}{RT} = \frac{(0.1013 \times 10^6)(3 \times 10^4)}{(8.314)(288)(3600)} = 352.6 \text{ mol} \cdot \text{s}^{-1}$$

Since the molar mass of methane is 16, the mass flow rate is

$$\dot{m} = (0.3526 \text{ kmol} \cdot \text{s}^{-1})(16 \text{ kg} \cdot \text{kmol}^{-1}) = 5.642 \text{ kg} \cdot \text{s}^{-1}$$

For reversible, adiabatic compression the process would occur at constant entropy. Therefore (see Fig. 6-11), $S_2' = S_1 = 6.071 \text{ kJ} \cdot \text{kg}^{-1} \cdot \text{K}^{-1}$. At 3.45 MPa the state having this entropy occurs at temperature 154.4 °C and enthalpy $H_2 = 1248.7 \text{ kJ} \cdot \text{kg}^{-1}$; therefore

$$(\Delta H)_S = 1248.7 - 946.7 = 302.0 \text{ kJ} \cdot \text{kg}^{-1}$$

The efficiency of a compression process is given by $\eta_c = (\Delta H)_S / \Delta H$, whence the actual (adiabatic) enthalpy change is

$$\Delta H = \frac{(\Delta H)_S}{\eta_c} = \frac{302}{0.8} = 377.5 \text{ kJ} \cdot \text{kg}^{-1}$$

and since $W_s = -\Delta H$, the work input to the compressor is 377.5 kJ · kg^{-1}. The power requirement is calculated as

$$\text{Power} = \dot{m}W_s = (5.642)(377.5) = 2130 \text{ kW}$$

The actual enthalpy of the compressed methane is

$$H_2 = H_1 + \Delta H = 946.7 + 377.5 = 1324.2 \text{ kJ} \cdot \text{kg}^{-1}$$

and this value is obtained at a temperature of 182 °C (interpolate in the data).

In the cooler no work is done, and the kinetic- and potential-energy terms should be negligible. Therefore (6.18) becomes

$$\dot{Q} = \dot{m} \, \Delta H = (5.642)(949.6 - 1324.2) = -2113 \text{ kW}$$

6.16 Ethylene gas undergoes continuous adiabatic compression from the initial state 1 bar and 294 K to the final pressure 18 bar. If compression is 75% efficient compared with an isentropic process, find (a) the work requirement and (b) the final temperature of the ethylene. Assume that the virial equation of state (5.33) adequately describes the volumetric behavior of ethylene, and that the second virial coefficient is given by the generalized correlation of Section 5.4. The molar heat capacity of ethylene in the ideal-gas state is found from the data of Table 4-1 as

$$\frac{C_P^{\text{ig}}}{R} = 1.424 + (14.394 \times 10^{-3})T - (4.392 \times 10^{-6})T^2 \tag{1}$$

This problem is similar to Problem 6.12 in that the property changes between initial and final states are determined for a particular calculational path. For the data available, the most convenient path is that shown in Fig. 6-17.

(a) For the *isentropic* compression of ethylene, $\Delta S = -S_1^R + \Delta S_{12}^{\text{ig}} + S_2^R = 0$, or

$$\frac{S_2^R}{R} + \frac{1}{R} \langle C_P' \rangle_{\ln T} \ln \frac{T_2'}{T_1} - \ln \frac{P_2}{P_1} - \frac{S_1^R}{R} = 0 \tag{2}$$

where T_2' is the final temperature for isentropic compression and where the mean heat capacity (notated without the superscript "ig") is for the temperature range T_1 to T_2'.

Equations (5.76) and (5.77) give

$$\frac{S^R}{R} = -P_r \left(\frac{dB^0}{dT_r} + \omega \frac{dB^1}{dT_r} \right) \tag{3}$$

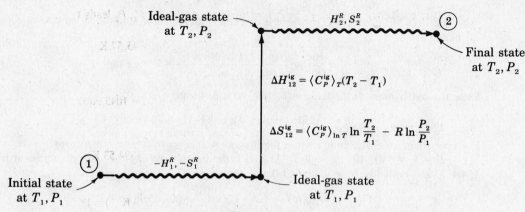

Fig. 6-17

For ethylene, from Appendix C, $T_c = 282.4$ K, $P_c = 50.4$ bar, $\omega = 0.085$; thus, for the fixed initial state,

$$T_{r1} = \frac{294}{282.4} = 1.041 \qquad P_{r1} = \frac{1}{50.4} = 0.020$$

From (5.77), for $T_{r1} = 1.041$, $dB^0/dT_r = 0.608$ and $dB^1/dT_r = 0.586$. Substitution in (3) now yields

$$\frac{S_1^R}{R} = (-0.020)[0.608 + (0.085)(0.586)] = -0.0132$$

To determine T_2' by fixed-point iteration (Example 5.16), rewrite (2) as

$$T_2' = T_1 \exp\left(\frac{F}{\langle C_P' \rangle_{\ln T}/R}\right) \tag{4}$$

where

$$F \equiv \ln\frac{P_2}{P_1} - \frac{S_2^R}{R} + \frac{S_1^R}{R}$$

Choosing an initial value of T_2', evaluate $\langle C_P' \rangle_{\ln T}/R$ by (4.20) (adjust the notation from T_0, T_1 to T_1, T_2); dB^0/dT_r and dB^1/dT_r by (5.77); and S_2^R/R by (3). Equation (4) then yields a new value of T_2', for which we reevaluate $\langle C_P' \rangle_{\ln T}/R$ and S_2^R/R for resubstitution into (4). This iterative procedure is readily programmed for execution by calculator or computer. It continues until there is no significant change from one iteration to the next.

The initial value of T_2' should be somewhat above T_1; say, $T_2' = 400$ K. The final temperature for isentropic compression, as found by iteration, is $T_2' = 442.22$ K.

The enthalpy change for isentropic compression is $(\Delta H)_S = -H_1^R + \Delta H_{12}^{ig} + H_2^R$, or

$$\frac{(\Delta H)_S}{R} = \frac{-H_1^R}{R} + \frac{1}{R}\langle C_P' \rangle_T (T_2' - T_1) + \frac{H_2^R}{R} \tag{5}$$

Evaluating the right side of (5), we have from (4.16) (involving T_1 and T_2'):

$$\frac{1}{R}\langle C_P' \rangle_T (T_2' - T_1) = (7.326)(442.22 - 294) = 1085.83 \text{ K}$$

By (3) of Problem 6.12,

$$\frac{H^R}{RT_c} = P_r\left[B^0 - T_r\frac{dB^0}{dT_r} + \omega\left(B^1 - T_r\frac{dB^1}{dT_r}\right)\right] \tag{6}$$

For the initial conditions, all quantities on the right of (6), save B^0 and B^1, have already been computed; by (5.73),

$$B^0 = -0.313 \qquad B^1 = -0.006$$

Hence

$$\frac{H_1^R}{R} = \left(\frac{H_1^R}{RT_c}\right)T_c = (-0.0200)(282.4) = -5.64 \text{ K}$$

In like manner, evaluation of (6) at the final conditions T'_2, P_2 leads to

$$\frac{H_2^R}{R} = (-0.1614)(282.4) = -45.57 \text{ K}$$

Thus, by (5),

$$\frac{(\Delta H)_S}{R} = 5.64 + 1085.83 - 45.57 = 1045.90 \text{ K}$$

and, for a compression efficiency of 75%,

$$\frac{\Delta H}{R} = \frac{(\Delta H)_S/R}{0.75} = \frac{1045.9}{0.75} = 1394.53 \text{ K}$$

The work requirement is then

$$-W_s = \Delta H = (1394.53 \text{ K})(0.008\,314 \text{ kJ} \cdot \text{mol}^{-1} \cdot \text{K}^{-1}) = 11.594 \text{ kJ} \cdot \text{mol}^{-1}$$

(b) With H_1^R/R and $\Delta H/R$ known from (a), the basic equation

$$\frac{\Delta H}{R} = \frac{-H_1^R}{R} + \frac{1}{R} \langle C_P^{\text{ig}} \rangle_T (T_2 - T_1) + \frac{H_2^R}{R} \tag{7}$$

can be solved for T_2 by iteration. Transform (7) into

$$T_2 = T_1 + \frac{J}{\langle C_P^{\text{ig}} \rangle_T/R} \tag{8}$$

where

$$J \equiv \frac{\Delta H}{R} + \frac{H_1^R}{R} - \frac{H_2^R}{R} = 1394.53 + 5.64 - \frac{H_2^R}{R} = 1400.17 - \frac{H_2^R}{R} \quad \text{(in K)}$$

With the starting value $T_2 = T'_2$, evaluate $\langle C_P^{\text{ig}} \rangle_T/R$ by (4.16); B^0, B^1, dB^0/dT_r, and dB^1/dT_r by (5.73) and (5.77); and H_2^R/R by (6). Equation (8) then yields a new value of T_2, for the next iteration. The process converges to the solution $T_2 = 481.37$ K.

6.17 (a) For steady flow in an adiabatic nozzle of local cross-sectional area A, show that

$$u^2 = -V \frac{dH}{dV} \left[\frac{d(\ln A)}{d(\ln u)} + 1 \right]$$

(b) If in addition the flow is reversible, show that

$$u^2 = -V^2 \left(\frac{\partial P}{\partial V} \right)_S \left[\frac{d(\ln A)}{d(\ln u)} + 1 \right]$$

$$\mathbf{M}^2 - 1 = \frac{d(\ln A)}{d(\ln u)}$$

where $\mathbf{M}$ is the local *Mach number*, defined as the ratio of the local velocity u to the local sound speed c; i.e., $\mathbf{M} = u/c$.

(a) The energy equation for an adiabatic nozzle is given by (6.21), which in differential form becomes

$$\frac{dH}{du} = -u$$

from which we see that dH/du *is always negative*. We may also write

$$dH = -u \, du = -u^2 \, d(\ln u) = -u^2 \frac{d(\ln u)}{d(\ln V)} \frac{dV}{V}$$

Solving for u^2, we have

$$u^2 = -V \frac{dH}{dV} \frac{d(\ln V)}{d(\ln u)} \tag{1}$$

For steady flow, the conservation of mass is expressed by

$$\frac{uA}{V} = \dot{m} = \text{constant}$$

By logarithmic differentiation, $d(\ln u) + d(\ln A) - d(\ln V) = 0$, or

$$\frac{d(\ln V)}{d(\ln u)} = \frac{d(\ln A)}{d(\ln u)} + 1$$

Substitution of this relation in (1) gives the required result:

$$u^2 = -V\frac{dH}{dV}\left[\frac{d(\ln A)}{d(\ln u)} + 1\right] \tag{2}$$

(b) If the flow is isentropic, then (2) may be written

$$u^2 = -V\left(\frac{\partial H}{\partial V}\right)_S\left[\frac{d(\ln A)}{d(\ln u)} + 1\right] \tag{3}$$

Division of the general property relation $dH = T\,dS + V\,dP$ by dV and restriction to constant S provides

$$\left(\frac{\partial H}{\partial V}\right)_S = V\left(\frac{\partial P}{\partial V}\right)_S$$

Substitution in (3) yields the required result:

$$u^2 = -V^2\left(\frac{\partial P}{\partial V}\right)_S\left[\frac{d(\ln A)}{d(\ln u)} + 1\right] \tag{4}$$

We also note that the property relation gives $(\partial H/\partial P)_S = V$. Thus, for isentropic flow, dH/dP is *always positive*.

From physics we have the following equation for the sound speed c:

$$c^2 = -V^2\left(\frac{\partial P}{\partial V}\right)_S$$

Combining this equation with (4) we get

$$\frac{u^2}{c^2} = \frac{d(\ln A)}{d(\ln u)} + 1$$

or, equivalently,

$$\mathbf{M}^2 - 1 = \frac{d(\ln A)}{d(\ln u)} \tag{5}$$

6.18 Discuss the qualitative consequences of (5) of Problem 6.17.

There are three cases to consider: $\mathbf{M} < 1$ (subsonic flow), $\mathbf{M} = 1$ (sonic flow), $\mathbf{M} > 1$ (supersonic flow). We bear in mind that the logarithmic function is strictly increasing, and that, as found in Problem 6.17,

$$\frac{dH}{du} = \ominus \quad \text{and} \quad \frac{dH}{dP} = \oplus$$

When $\mathbf{M} < 1$, (5) shows that $d(\ln A)/d(\ln u)$ is negative. Therefore, either (1) A decreases in the direction of flow, while u increases; or (2) A increases in the direction of flow, while u decreases. If $\mathbf{M} = 1$, $d(\ln A)/d(\ln u) = 0$, which implies that (3) A is stationary (as at a throat of the nozzle). When $\mathbf{M} > 1$, we have that $d(\ln A)/d(\ln u)$ is positive: either (4) A increases as u increases; or (5) A decreases as u decreases. We summarize in Table 6-1.

Case (1) is the familiar *converging* nozzle, through which a fluid flows with steadily increasing velocity as the pressure drops [Fig. 6-18(a)]. The limiting velocity is the sound speed, and for isentropic flow this can be attained only at the exit of the nozzle, where area no longer changes with length.

Case (2) applies to a device that receives a high-velocity (but subsonic) stream and decreases its velocity, converting kinetic energy into internal energy, thus causing an increase in enthalpy and

Table 6-1

(1)	(2)	(3)	(4)	(5)
$\mathbf{M} < 1$ $\dfrac{d(\ln A)}{d(\ln u)} = \ominus$ Flow is subsonic		$\mathbf{M} = 1$ $\dfrac{d(\ln A)}{d(\ln u)} = 0$ Flow is sonic	$\mathbf{M} > 1$ $\dfrac{d(\ln A)}{d(\ln u)} = \oplus$ Flow is supersonic	
A decreases u increases H decreases P decreases *Converging Nozzle*	A increases u decreases H increases P increases *Diverging Diffuser*	A is stationary u may increase or decrease H and P may decrease or increase *End Condition or Transition between Subsonic and Supersonic*	A increases u increases H decreases P decreases *Diverging Nozzle*	A decreases u decreases H increases P increases *Converging Diffuser*

pressure [Fig. 6-18(b)]. Such a device, called a *diffuser*, finds use in the compression of the intake air for a jet airplane engine.

Case (4) is again a nozzle, but a *diverging* nozzle. Once flow has reached the sound speed, further increases in velocity can be attained only if the nozzle area increases. Thus the diverging nozzle is almost always found as part of a *converging-diverging* nozzle [Fig. 6-18(c)]. At the throat, where the converging and diverging sections join, the flow is sonic and we have a transition from subsonic to supersonic flow, represented by case (3).

Case (5) is again a diffuser, one that *converges* and operates at supersonic velocities. Just as a diverging cross section is required to increase the velocity of a supersonic flow, so a converging cross section is required to decrease the velocity of a supersonic flow in an isentropic process. Such a diffuser is used on a supersonic jet aircraft for the compression of the intake air. To bring the air to subsonic velocities (relative to the aircraft) case (5) is combined with case (2) to make a converging-diverging diffuser. The throat of such a device is again represented by case (3), where the transition is from supersonic to subsonic flow.

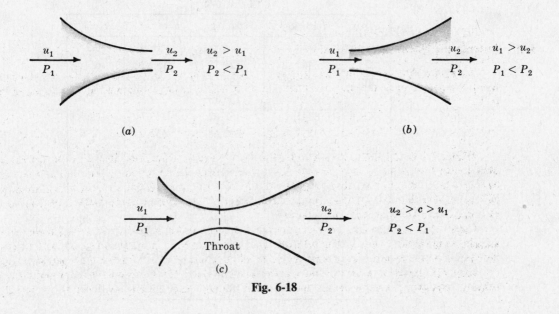

Fig. 6-18

6.19 Steam flows through an isentropic nozzle at the rate of $2 \text{ kg} \cdot \text{s}^{-1}$. Inlet conditions are 500 °C and 28 bar, and the entrance velocity is $10 \text{ m} \cdot \text{s}^{-1}$. The discharge pressure is 1.5 bar. Plot V, u, and A against P for the full pressure range of the nozzle.

The appropriate expression of the first law is (6.22):

$$\frac{u^2 - u_1^2}{2} = -(H - H_1)_S$$

where the unsubscripted symbols u and H apply to any point downstream from the nozzle inlet. Solution for u gives

$$u = \sqrt{u_1^2 + 2(H_1 - H)_S} \qquad (1)$$

In addition we have the continuity equation, $\dot{m} = uA/V$, or

$$A = \frac{\dot{m}V}{u} \qquad (2)$$

For the initial state we find the following properties from the tables for superheated steam:

$$V_1 = 124.58 \text{ cm}^3 \cdot \text{g}^{-1} \qquad H_1 = 3458.4 \text{ kJ} \cdot \text{kg}^{-1} \qquad S_1 = 7.2685 \text{ kJ} \cdot \text{kg}^{-1} \cdot \text{K}^{-1}$$

Application of (2) at the nozzle entrance gives

$$A_1 = \frac{\dot{m}V_1}{u_1} = \frac{(2)(124.6 \times 10^{-3})}{1000} = 249.2 \text{ cm}^2$$

Consider now a point in the nozzle where $P = 27$ bar. Since S is constant throughout the nozzle, we locate the state in the tables for superheated steam for which $P = 27$ bar and $S = 7.2685 \text{ kJ} \cdot \text{kg}^{-1} \cdot \text{K}^{-1}$. This state occurs at a temperature just below 494 °C, and by interpolation we find

$$H = 3445.8 \text{ kJ} \cdot \text{kg}^{-1} \qquad V = 128.2 \text{ cm}^3 \cdot \text{g}^{-1}$$

The velocity at this point is given by (1) as

$$u = \sqrt{10^2 + (2)(3458.4 - 3445.8) \times 10^3} = 159.1 \text{ m} \cdot \text{s}^{-1}$$

From (2) the corresponding value of the area is

$$A = \frac{(2)(128.2 \times 10^{-3})}{15\,910} = 16.1 \text{ cm}^2$$

We continue in like fashion to successively lower pressures. The results are given in Table 6-2 and are displayed in Fig. 6-19.

Table 6-2

P/bar	H/kJ·kg^{-1}	V/cm^3·g^{-1}	u/m·s^{-1}	A/cm^2
28	3458.4	124.58	10.0	249.2
27	3445.8	128.18	159.1	16.1
26	3432.8	132.01	226.5	11.6
25	3419.4	136.13	279.5	9.74
20	3345.3	162.02	475.7	6.81
15	3255.0	202.68	637.9	6.36
12.5	3200.6	233.49	718.1	6.50
10	3137.2	277.52	801.6	6.92
8	3076.7	329.76	873.8	7.54
6	3003.4	411.63	954.0	8.63
5	2959.2	473.56	999.3	9.48
4	2907.7	562.03	1049.5	10.71
3	2845.2	700.38	1107.5	12.65
2	2764.0	954.12	1178.5	16.19
1.5	2710.9	1187.1	1222.7	19.42

Shaping this isentropic nozzle is a problem in fluid dynamics, not thermodynamics.

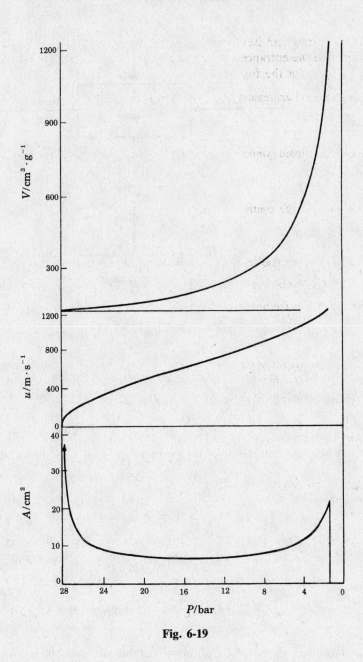

Fig. 6-19

6.20   A stream of air at atmospheric pressure is cooled continuously from 38 °C to 15 °C for purposes of air-conditioning a building. The volumetric flow is $0.5 \text{ m}^3 \cdot \text{s}^{-1}$ (as measured at 101.33 kPa and 25 °C). The temperature of the ambient air to which heat is discarded is 38 °C. What is the minimum power requirement of a mechanical refrigeration system designed to accomplish the necessary cooling?

The process is diagramed in Fig. 6-20. The molar flow rate of air is

$$\dot{n} = \frac{P \dot{V}^t}{RT} = \frac{(101.33 \times 10^3)(0.5)}{(8.314)(298.15)} = 20.44 \text{ mol} \cdot \text{s}^{-1}$$

The energy equation (6.18) can be written for this process as

$$\dot{n}(\Delta H) = \dot{Q} - \dot{W}_s \tag{1}$$

where the potential- and kinetic-energy terms are omitted as negligible. For minimum work, the process

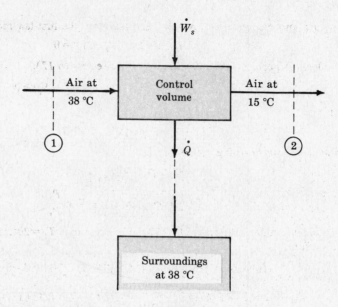

Fig. 6-20

must be reversible, and (6.20) gives

$$\Delta S = \frac{Q}{T'} \quad \text{or} \quad \dot{Q} = T' \dot{n} \, \Delta S$$

where T', the absolute temperature of the surroundings is 311.15 K. Substitution in (1) yields

$$\dot{W}_s = \dot{n}(T' \, \Delta S \, - \, \Delta H) \tag{2}$$

For air as an ideal gas with $C_P = (7/2)R$,

$$\Delta H = C_P(T_2 - T_1) = (7/2)(8.314)(288.15 - 311.15) = -669.3 \text{ J} \cdot \text{mol}^{-1}$$

$$\Delta S = C_P \ln \frac{T_2}{T_1} = (7/2)(8.314) \ln \frac{288.15}{311.15} = -2.2346 \text{ J} \cdot \text{mol}^{-1} \cdot \text{K}^{-1}$$

Therefore

$$\dot{W}_s = (20.44)[(311.15)(-2.2346) + 669.3] = -531.4 \text{ W}$$

6.21 Determine whether or not the following continuous process is possible: Nitrogen gas at 600 kPa and 21 °C enters a device which has no moving parts and which is thoroughly insulated from its surroundings. Half of the nitrogen flows out of the device at 82 °C, and the other half at −40 °C, both halves at 100 kPa. Assume nitrogen at these conditions behaves as an ideal gas for which $C_P = (7/2)R$.

If neither the first nor the second law of thermodynamics is violated by the process, then it is possible. The first law is, for this steady-flow process, given by (6.16). Since the device is well insulated, we may suppose the heat transfer to be zero. Since the device has no moving parts, the shaft work is zero. If, in addition, we neglect the kinetic- and potential-energy terms, (6.16) on a molar basis becomes

$$\Delta(\dot{n}H)_{\substack{\text{flowing} \\ \text{streams}}} = 0 \tag{1}$$

If we let $\dot{n}'$ represent the molar flow rate of the colder exit stream and $\dot{n}''$ the flow rate of the warmer exit stream, then we may write (1) as $\dot{n}'H_2' + \dot{n}''H_2'' - (\dot{n}' + \dot{n}'')H_1 = 0$, or

$$\dot{n}'(H_2' - H_1) + \dot{n}''(H_2'' - H_1) = 0$$

For the particular process described, $\dot{n}' = \dot{n}''$, and therefore the first-law requirement is simply

$$(H_2' - H_1) + (H_2'' - H_1) = 0 \tag{2}$$

For an adiabatic process, the second law is expressed by (6.17):

$$\Delta(\dot{n}S)_{\substack{\text{flowing} \\ \text{streams}}} = \dot{n}'(S_2' - S_1) + \dot{n}''(S_2'' - S_1) \geqq 0$$

Since $\dot{n}' = \dot{n}''$,

$$(S_2' - S_1) + (S_2'' - S_1) \geqq 0 \tag{3}$$

For an ideal gas we have the equations

$$H_2 - H_1 = C_P(T_2 - T_1)$$

$$S_2 - S_1 = C_P \ln \frac{T_2}{T_1} - R \ln \frac{P_2}{P_1}$$

From the problem statement: $P_1 = 600$ kPa, $P_2 = 100$ kPa; $T_1 = 294.15$ K, $T_2' = 233.15$ K, $T_2'' = 355.15$ K. Therefore,

$$H_2' - H_1 = \tfrac{7}{2}(8.314)(233.15 - 294.15) = -1775 \text{ J} \cdot \text{mol}^{-1}$$

$$S_2' - S_1 = \frac{7}{2}(8.314) \ln \frac{233.15}{294.15} - 8.314 \ln \frac{100}{600} = 8.134 \text{ J} \cdot \text{mol}^{-1} \cdot \text{K}^{-1}$$

Similarly,

$$H_2'' - H_1 = \frac{7}{2}(8.314)(355.15 - 294.15) = 1775 \text{ J} \cdot \text{mol}^{-1}$$

$$S_2'' - S_1 = \frac{7}{2}(8.314) \ln \frac{355.15}{294.15} - 8.314 \ln \frac{100}{600} = 20.380 \text{ J} \cdot \text{mol}^{-1} \cdot \text{K}^{-1}$$

Checking (2): $(H_2' - H_1) + (H_2'' - H_1) = -1775 + 1775 = 0$.

Checking (3): $(S_2' - S_1) + (S_2'' - S_1) = 8.134 + 20.380 = 28.514 > 0$.

The process is possible.

6.22 Water at 40 °C and 10 kPa is pumped adiabatically from the condenser of a power plant to the boiler pressure of 7200 kPa. If the pump is 80% efficient compared with isentropic operation, find (a) the work required and (b) the temperature rise of the water.

(a) This is a steady-flow process to which (6.19) is applicable, reducing in this case to $W_s = -\Delta H$, because there is no heat transfer and the kinetic- and potential-energy terms are negligible for any properly designed pump. Equation (3.48), $dH = T\,dS + V\,dP$, implies

$$W_s \text{ (isentropic)} = -(\Delta H)_S = -\int_1^2 V\,dP$$

As water at 40 °C is only slightly compressible, we evaluate the integral by taking V constant at its initial, steam-table value:

$$W_s \text{ (isentropic)} \approx -V_1(P_2 - P_1) = -(1.008 \times 10^{-3} \text{ m}^3 \cdot \text{kg}^{-1})(7190 \text{ kPa}) = -7.248 \text{ kJ} \cdot \text{kg}^{-1}$$

Hence, $W_s \text{ (actual)} = W_s \text{ (isentropic)}/\eta_c = -7.248/0.8 = -9.060 \text{ kJ} \cdot \text{kg}^{-1}$.

(b) The temperature rise of the water is found from (3.69), which implies

$$-W_s \text{ (actual)} = \Delta H = C_P\,\Delta T + V_1(1 - \beta T_1)\,\Delta P$$

Substituting $\beta = 385 \times 10^{-6} \text{ K}^{-1}$, $C_P = 4.178 \text{ kJ} \cdot \text{kg}^{-1} \cdot \text{K}^{-1}$, and values from (a), we obtain $\Delta T = 0.64$ K.

MECHANICAL-ENERGY BALANCE   (Section 6.5)

6.23 Water is drained from a 3-m-diameter tank through a 50-mm hole in the side located 250 mm above the bottom of the tank (see Fig. 6-21). The top of the tank is open to the atmosphere. What is the velocity of the exiting stream when the water level in the tank is 2.5 m?

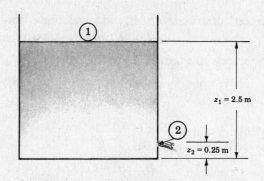

Fig. 6-21

Although the liquid level in the tank slowly recedes, the ratio of the tank to hole diameters is so large that the velocity of the water surface in the tank is essentially zero. We can therefore treat the drainage process as a quasisteady flow for which the control volume, taken as the total volume of water in the tank, remains virtually constant. There is no shaft work done on or by the fluid; so, assuming the water to be nonviscous and incompressible, we apply the Bernoulli equation (6.26) between locations 1 (the liquid surface in the tank) and 2 (the exiting jet of liquid). Thus we write

$$\frac{P_2 - P_1}{\rho} + \tfrac{1}{2}(u_2^2 - u_1^2) + g(z_2 - z_1) = 0$$

Since $u_1 \approx 0$ by assumption, and $P_1 = P_2 = $ atmospheric pressure, solution of the above equation for u_2 gives

$$u_2 = \sqrt{2g(z_1 - z_2)}$$

By the statement of the problem, $z_2 = 0.25$ m and $z_1 = 2.5$ m; therefore,

$$u_2 = \sqrt{(2)(9.8066)(2.25)} = 6.64 \text{ m} \cdot \text{s}^{-1}$$

The actual surface velocity u_1 can be estimated from the continuity equation applied between the drainage hole and a plane parallel to and just below the liquid surface:

$$\frac{u_1 A_1}{V_1} = \frac{u_2 A_2}{V_2}$$

But $V_1 = V_2$, by the incompressibility assumption; whence

$$u_1 = u_2\left(\frac{A_2}{A_1}\right) = (6.64)\left(\frac{50 \times 10^{-3}}{3}\right)^2 = 1.84 \times 10^{-3} \text{ m} \cdot \text{s}^{-1}$$

and the surface velocity is indeed negligible compared to the exit velocity.

6.24 Air at a pressure of 690 kPa and a temperature of 20 °C enters a horizontal steam-jacketed pipe with a velocity of 30 m · s^{-1}. Measurements at the exit end of the pipe show that the air is heated to 200 °C and that the pressure is 669 kPa. What percentage of the pressure drop in the pipe can be attributed to fluid friction? Assume that the air behaves as an ideal gas.

The mechanical-energy balance (6.25) may be written in differential form as

$$-\delta W_s = V\, dP + u\, du + g\, dz + \delta\Psi$$

Since there is no shaft work and no change in elevation, this reduces to

$$dP = \frac{-u}{V}\, du - \frac{\delta\Psi}{V}$$

By the continuity equation,

$$\frac{u_1 A_1}{V_1} = \frac{u_2 A_2}{V_2}$$

or, since $A_1 = A_2$,

$$\frac{u_1}{V_1} = \frac{u_2}{V_2} = \frac{u}{V} = \text{constant}$$

The expression for dP may therefore be written

$$dP = -\left(\frac{u_1}{V_1}\right) du - \frac{\delta \Psi}{V}$$

and integration gives

$$\Delta P = -\left(\frac{u_1}{V_1}\right) \Delta u - \int_1^2 \frac{\delta \Psi}{V}$$

The total pressure drop is given as the sum of two terms, and we may associate the first of these terms with the pressure drop that results in a velocity increase and the second term with the pressure drop attributed to fluid friction. The desired fraction is therefore equal to

$$1 + \left(\frac{u_1}{V_1}\right) \frac{\Delta u}{\Delta P} \tag{1}$$

Now, the initial molar volume of the air is given by the ideal-gas equation as

$$\frac{RT_1}{P_1} = \frac{(8.314)(293.15)}{690 \times 10^3} = 3.532 \times 10^{-3} \text{ m}^3 \cdot \text{mol}^{-1}$$

The molar mass of air is approximately $29 \times 10^{-3} \text{ kg} \cdot \text{mol}^{-1}$; so $V_1 = 0.1218 \text{ m}^3 \cdot \text{kg}^{-1}$. Also,

$$u_2 = u_1 \frac{V_2}{V_1} = u_1 \left(\frac{T_2}{T_1}\right)\left(\frac{P_1}{P_2}\right) = (30)\left(\frac{473.15}{293.15}\right)\left(\frac{690}{669}\right) = 49.94 \text{ m} \cdot \text{s}^{-1}$$

so that $\Delta u = u_2 - u_1 = 49.94 - 30 = 19.94 \text{ m} \cdot \text{s}^{-1}$. Finally, $\Delta P = -21 \times 10^3$ Pa. Substitution of these values in (1) gives 0.766, or 76.6%.

Supplementary Problems

MISCELLANEOUS APPLICATIONS   (Sections 6.1–6.5)

6.25   A steam turbine operates adiabatically and produces 4000 kW. Steam is fed to the turbine at 21 bar and 475 °C. The exhaust from the turbine is saturated steam at 0.1 bar and it enters a condenser where it is condensed and cooled to 30 °C. What is the steam rate (mass flow) for the turbine, and at what rate must cooling water be supplied to the condenser, if the water enters at 15 °C and is heated to 25 °C?   *Ans.* $\dot{m}_{\text{steam}} = 4.840 \text{ kg} \cdot \text{s}^{-1}$ $\dot{m}_{\text{water}} = 284 \text{ kg} \cdot \text{s}^{-1}$

6.26   Saturated steam at 1 bar is taken continuously into a compressor at low velocity and compressed to 3 bar. It then enters a nozzle where it expands back to a pressure of 1 bar and its initial saturated condition; however, it now has a velocity of $600 \text{ m} \cdot \text{s}^{-1}$. The steam rate is $2.5 \text{ kg} \cdot \text{s}^{-1}$. It is found necessary to cool the compressor at the rate of 150 kW. What is the power requirement of the compressor?   *Ans.*   600 kW

6.27   A well-insulated closed tank is of volume 70 m³. Initially, it contains 25 000 kg of water distributed between liquid and vapor phases at 30 °C. Saturated steam at 11 bar is admitted to the tank until the pressure reaches 7 bar. What mass of steam is added?   *Ans.*   6960 kg

6.28 A well-insulated tank of 3-m^3 capacity contains 1400 kg of liquid water in equilibrium with its vapor, which fills the remainder of the tank. The initial temperature is 280 °C. Liquid water at 70 °C in the amount of 900 kg flows into the tank, and nothing is removed. How much heat must be added during the process if the temperature in the tank is to be unchanged?   *Ans.*   784.54 MJ

6.29 A tank of volume 1 m^3 is initially evacuated. Atmospheric air leaks into the tank through a weld imperfection. The process is slow, and heat transfer with the surroundings keeps the tank and its contents at the ambient temperature, 27 °C. Calculate the amount of heat exchanged with the surroundings during the time it takes for the pressure in the tank to reach atmospheric pressure (101.3 kPa).   *Ans.* $Q = -101.3$ kJ

6.30 A continuous process for liquefying nitrogen includes a heat-exchanger system and throttle valve, with no moving parts. The nitrogen enters the system at 101.3 bar (100 atm) and 300 K. Two streams leave the system: first, unliquefied nitrogen gas at 1.013 bar and 294.5 K; and second, saturated liquid nitrogen at 1.013 bar. The apparatus is well insulated, and it is estimated that only 60 J of heat leaks into the system for each mole of entering nitrogen. What fraction of the entering nitrogen is liquefied under these conditions?

Available data: When nitrogen at 101.3 bar and 300 K expands adiabatically through a throttle to 1.013 bar, the temperature drops to 280.6 K. The latent heat of vaporization of nitrogen at its normal boiling point, 77.35 K, is 5590 J·mol^{-1}. The heat capacity of nitrogen gas may be taken as constant at 28.1 J·mol^{-1}·K^{-1}.

Ans.   2.83%

6.31 A frictionless piston, having a cross-sectional area of 750 cm^2 slides in a cylinder, as shown in Fig. 6-22. The piston works to compress a spring, which exerts a force on the piston $F = 440x$, where x is in meters and F is in kilonewtons. Steam from a line at 42 bar and 425 °C is admitted slowly to the cylinder through the valve, and is allowed to flow until the pressure in the cylinder is 14 bar. Assuming that no heat is exchanged with the steam and that $x_1 = 0$, determine: (*a*) $H_2 - H_1$ as a function of P_2 and V_2, where subscript 2 denotes steam in the cylinder, and subscript 1 refers to steam in the line; (*b*) the specific enthalpy and total mass of the steam in the cylinder at the end of the process.

Ans.   (*a*) $H_2 - H_1 = P_2 V_2/2$. Note that this equation must be solved in conjunction with the steam tables, which provide a second relation connecting P_2, V_2, and H_2. (*b*) $H_2 = 3443.9$ kJ·kg^{-1}, $m_2 = 0.0724$ kg

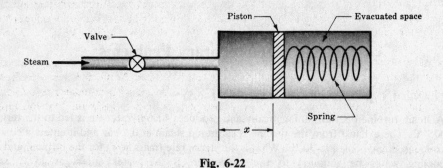

Fig. 6-22

6.32 A portable power-supply system consists of a 30-L bottle of compressed helium, charged to 140 bar at 20 °C, connected to a small turbine. The helium drives the turbine continuously until the pressure in the bottle drops to 7 bar. For a particular use at high altitude, the turbine exhausts at 0.3 bar. Neglecting all heat transfer with the gas, calculate the maximum possible shaft work obtainable from the device. Assume helium to be an ideal gas with $C_V = (3/2)R$.   *Ans.* $W_s = 523.5$ kJ

6.33 Two perfectly insulated tanks A and B of equal volume contain equal quantities of the same ideal gas at the same pressure and temperature. Tank A is connected to the inlet of a small reversible adiabatic turbine that drives an electric generator, and tank B discharges through an insulated valve. Both the turbine and the valve discharge gas to the atmosphere. Both devices are allowed to operate until gas discharge ceases. Regarding these operations, there are three alternative choices for completion of each of the following statements, namely: *equal to*, *greater than*, or *less than*. Indicate in each case which is correct. Assume no heat transfer to or from the gas.

(a) When discharge ceases in each system, the temperature in tank A is _____ the temperature in B.

(b) When the pressures in both tanks have fallen to one-half the initial pressure, the temperature of the gas discharged from the *turbine* will be _____ the temperature of the gas issuing from the *valve*.

(c) During the discharge process, the temperature of the gas leaving the *turbine* will be _____ the temperature of the gas leaving *tank A* at the same instant.

(d) During the discharge process, the temperature of the gas leaving the *valve* will be _____ the temperature of the gas leaving *tank B* at the same instant.

(e) When the tank pressures have reached atmospheric, the quantity of gas in *tank A* is _____ the quantity in *tank B*.

Ans. (a) equal to   (b) less than   (c) less than   (d) equal to   (e) equal to

6.34 Determine the efficiency (with respect to isentropic operation) of the turbine of Problem 6.25.   *Ans.* 76.1%

6.35 For an ideal gas with constant heat capacities that flows reversibly and adiabatically through a nozzle from an initial pressure P_1 and a negligible initial velocity u_1 to a final pressure P_2, show that

$$u_2 = \sqrt{2 C_P T_1 \left[1 - \left(\frac{P_2}{P_1} \right)^{(\gamma-1)/\gamma} \right]}$$

where C_P is the *specific* heat capacity of the fluid.

6.36 Air expands through a nozzle from a negligible initial velocity to the final velocity $340 \text{ m} \cdot \text{s}^{-1}$. Calculate the temperature drop of the air, assuming it an ideal gas for which $C_P = (7/2)R$. The molar mass of air is $0.029 \text{ kg} \cdot \text{mol}^{-1}$.   *Ans.* 57.6 K

6.37 (a) Saturated steam at 7 bar expands reversibly and adiabatically through a turbine to pressure 0.34 bar. What is the work of the turbine? (b) If, for the same initial conditions and the same final pressure as in (a), the steam expands adiabatically but irreversibly through a turbine and produces 80% of the work of (a), what is the final state of the steam?   *Ans.* (a) $W_s = 484.6 \text{ kJ} \cdot \text{kg}^{-1}$ (b) wet: $x = 0.8908$

6.38 A steady-flow expander (or gas turbine) is powered by a stream of hot compressed gases. It is required that the gases discharge from the turbine at 101.3 kPa and 20 °C. It is also required that the turbine produce 930 kW for a gas flow of $380 \text{ mol} \cdot \text{s}^{-1}$. Estimate the initial temperature and pressure required for the gas stream. Assume that the turbine operates isentropically and that the gases are ideal, with $C_P = 44 \text{ J} \cdot \text{mol}^{-1} \cdot \text{K}^{-1}$.   *Ans.* $T_1 = 75.6 \text{ °C}$ $P_1 = 254 \text{ kPa}$

6.39 Saturated steam is compressed continuously from 125 kPa to 575 kPa in a centrifugal compressor which operates adiabatically. For an efficiency of 75% compared with isentropic compression, determine the required work and the final state of the steam.
Ans. $W_s = -422.1 \text{ kJ} \cdot \text{kg}^{-1}$; superheated steam at 575 kPa and 321.4 °C

6.40 A power plant employs two adiabatic steam turbines in series. Steam enters the first turbine at 48 bar and 600 °C, and discharges from the second turbine at 10 kPa. The system is designed so that equal work is done by the two turbines, and the design is based on an efficiency of 80% compared with isentropic operation *in each turbine separately*. If the turbines perform according to these design conditions, what are the temperature and pressure of the steam between the turbines? What is the overall efficiency of the two turbines considered together, compared with isentropic expansion from the initial state to the final pressure?
Ans. $T = 316.0 \text{ °C}$; $P = 4.83 \text{ bar}$; $\eta = 83.6\%$. Note that solution for T and P is by trial.

6.41 A tank contains 0.5 kg of steam at a pressure of 20 bar and a temperature of 375 °C. It is connected through a valve to a vertical cylinder which contains a piston, as shown in Fig. 6-23. The piston is of such mass that a pressure of 7 bar is required to support it. Initially, the piston rests on the bottom of the cylinder. The valve is cracked so as to allow steam to flow slowly into the cylinder until the pressure is uniform throughout the system. Assuming that no heat is transferred from the steam to the surroundings and that no heat is exchanged between the two parts of the system, determine the final temperatures in the tank and in the cylinder. *Ans.* $T_{tank} = 235.7$ °C $\quad T_{cyl} = 305.5$ °C

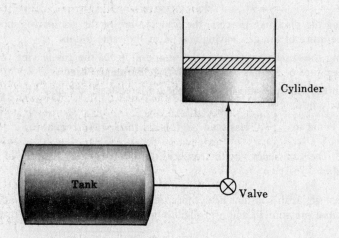

Fig. 6-23

6.42 Flake ice at 0 °C is to be produced continuously from liquid water at 20 °C, at the rate of 500 kg·h^{-1}. The process is to employ exhaust steam, available saturated at 175 kPa, as the sole source of energy. Heat rejection is to be to the atmosphere. On a day when the ambient temperature is 25 °C, what is the minimum steam rate required? The latent heat of solidification of ice at 0 °C is 333.5 kJ·kg^{-1}. *Ans.* 30.8 kg·h^{-1}

6.43 A gas turbine operates between 7.5 bar and 1.2 bar. If the exhaust temperature is $T_2 = 750$ K, what is T_1, the temperature of the inlet stream? Treat the gas as ideal, with C_P constant at 33.25 J·mol^{-1}·K^{-1}. The turbine efficiency with respect to isentropic operation is 80%. *Ans.* 1062.4 K

6.44 A steam turbine operates between $P_1 = 7000$ kPa and $P_2 = 10$ kPa. If the exhaust steam is to be no less than 95% vapor (by mass), what is the minimum degrees of superheat required in the inlet stream? The turbine efficiency with respect to isentropic operation is 80%. *Ans.* $\Delta T = 262.9$ K

Chapter 7

Chemical Thermodynamics

Preceding chapters dealt primarily with systems containing a single chemical species. However, many systems of practical interest contain two or more distinct chemical species. In addition, a system may be made up of several phases, and the distribution of species among the phases is rarely uniform. Moreover, some of the species of a system may participate in chemical reactions, which alter the composition of the system. In this chapter we treat such systems with respect to phase and chemical equilibria.

Any quantitative treatment of these topics requires methods for the description of the thermodynamic properties of mixtures. Thus Sections 7.1 through 7.6 constitute an introduction to the thermodynamics of solutions. The remaining sections deal with phase equilibrium (7.7 and 7.8) and chemical equilibrium (7.9 and 7.10). Because of the introductory nature of this outline, the examples and problems of these last four sections are limited to the simpler applications of theory; however, sufficient material is included to serve as a basis for solving more complex problems.

7.1 PARTIAL PROPERTIES

It was shown in Section 3.6 that the chemical potential μ_i plays a central role in the description of phase equilibrium. According to (3.40), μ_i can be defined as a mole-number derivative of the Gibbs energy:

$$\mu_i \equiv \left(\frac{\partial (nG)}{\partial n_i} \right)_{T,P,n_j} \tag{3.40}$$

For a constant-composition mixture, the temperature and pressure derivatives of μ_i are given by

$$\left(\frac{\partial \mu_i}{\partial T} \right)_{P,x} = - \left(\frac{\partial (nS)}{\partial n_i} \right)_{T,P,n_j} \tag{3.45}$$

$$\left(\frac{\partial \mu_i}{\partial P} \right)_{T,x} = \left(\frac{\partial (nV)}{\partial n_i} \right)_{T,P,n_j} \tag{3.46}$$

Equations (3.40), (3.45), and (3.46) suggest that derivatives of the form

$$\boxed{ \overline{M}_i \equiv \left(\frac{\partial (nM)}{\partial n_i} \right)_{T,P,n_j} } \tag{7.1}$$

will be of importance in the treatment of thermodynamic properties of solutions, particularly as regards phase equilibrium. The functions $\overline{M}_i$, called *partial molar properties*, in fact have a general significance to solution thermodynamics, whether or not one is concerned with phase-equilibrium calculations. We show in the next few paragraphs how partial molar properties arise naturally in the description of systems of variable composition.

It is a matter of experience that the thermodynamic properties of a homogeneous phase depend on temperature, pressure, and composition. Thus the total property nM of a phase may be considered a function of T, P, and the mole numbers $n_1, n_2, \ldots,$ of the individual species:

$$nM = \mathcal{M}(T, P, n_1, n_2, \ldots)$$

For an arbitrary change of state undergone by the phase, the total differential of nM is then

$$d(nM) = \left(\frac{\partial (nM)}{\partial T} \right)_{P,n} dT + \left(\frac{\partial (nM)}{\partial P} \right)_{T,n} dP + \sum_i \left(\frac{\partial (nM)}{\partial n_i} \right)_{T,P,n_j} dn_i$$

219

When all n_i are constant, so are n and all x_i. Hence

$$d(nM) = n\left(\frac{\partial M}{\partial T}\right)_{P,x} dT + n\left(\frac{\partial M}{\partial P}\right)_{T,x} dP + \sum \overline{M}_i \, dn_i \qquad (7.2)$$

where the partial molar properties $\overline{M}_i$ enter via the definition (7.1).

Equation (7.2) can be further developed. We have

$$dn_i = d(nx_i) = n \, dx_i + x_i \, dn \qquad \text{and} \qquad d(nM) = n \, dM + M \, dn$$

which combined with (7.2) yield

$$\left[dM - \left(\frac{\partial M}{\partial T}\right)_{P,x} dT - \left(\frac{\partial M}{\partial P}\right)_{T,x} dP - \sum \overline{M}_i \, dx_i\right]n + \left[M - \sum x_i \overline{M}_i\right] dn = 0$$

We are considering here an arbitrary change of state undergone by an open phase of arbitrary nominal extent. Hence quantities dn and n are arbitrary and independent of each other, and the two collections of terms enclosed by brackets must separately be zero. We therefore conclude that

$$dM = \left(\frac{\partial M}{\partial T}\right)_{P,x} dT + \left(\frac{\partial M}{\partial P}\right)_{T,x} dP + \sum \overline{M}_i \, dx_i \qquad (7.3)$$

and

$$\boxed{M = \sum x_i \overline{M}_i} \qquad (7.4)$$

Equation (7.3) is merely a special case of (7.2), corresponding to $n = 1$ (and thus to $n_i = x_i$). Equation (7.4) is new, however. Whereas the defining equation (7.1) provides a prescription for finding the "parts" ($\overline{M}_i$) from the "whole" (M), (7.4) shows how to compute the "whole" from the "parts". The *inverse* nature of (7.1) and the "summability" relation (7.4) becomes more obvious when (7.4) is written in the equivalent form

$$nM = \sum n_i \overline{M}_i$$

Equations (7.1) and (7.4) also suggest alternative interpretations of what a partial molar property actually "is." According to (7.1), $\overline{M}_i$ can be viewed as a *response function*, a numerical measure of the change in total mixture property nM resulting from addition of a minute quantity of species i to a phase. On the other hand, (7.4) suggests that $\overline{M}_i$ represents the *contribution* of species i to molar mixture property M. Depending on the application, both interpretations are useful; the second is not unique, however, for one can rationally define many quantities that satisfy summability relations like (7.4).

If one chooses to deal with the masses (rather than with the mole numbers) of the species in a system, then equations analogous to (7.1) through (7.4) follow with n_i replaced by m_i, and n by m. In this case x_i is the mass ("weight") fraction of species i, and the $\overline{M}_i$ are called partial *specific* properties. All equations of this section are developed on a molar basis, but conversion to a specific basis is simply a matter of making the above formal substitutions.

EXAMPLE 7.1 Laboratory alcohol contains 96 mass-percent ethanol and 4 mass-percent water. As an experiment a student decides to convert 2000 cm^3 of this material into vodka, having a composition of 56 mass-percent ethanol and 44 mass-percent water. Wishing to perform the experiment carefully, she searches the literature and finds the following partial-specific-volume data for ethanol/water mixtures at 25° C and 1 atm:

	In 96% ethanol	In vodka
$\overline{V}_{H_2O}/cm^3 \cdot g^{-1}$	0.816	0.953
$\overline{V}_{EtOH}/cm^3 \cdot g^{-1}$	1.273	1.243

The specific volume of water at 25 °C is 1.003 cm$^3 \cdot$ g^{-1}. How much water should be added to the 2000 cm^3 of laboratory alcohol, and how much vodka results?

Let m_w be the mass of water added to mass m_a of laboratory alcohol to produce mass m_v of vodka. The overall material balance is $m_a + m_w = m_v$, and a material balance on the water gives

$$0.04m_a + m_w = 0.44m_v$$

whence $m_w = 0.7143m_a$ and $m_v = 1.7143m_a$.

The *total* volumes V_w^t, V_a^t, and V_v^t are related to the corresponding masses through the specific volumes:

$$V_w^t = m_w V_w \qquad V_a^t = m_a V_a \qquad V_v^t = m_v V_v$$

and thus the equations for calculation of V_w^t and V_v^t are

$$V_w^t = 0.7143 \frac{V_w V_a^t}{V_a} \qquad V_v^t = 1.7143 \frac{V_v V_a^t}{V_a} \qquad (1)$$

According to (7.4), the specific volume of a binary solution is given in terms of the partial specific volumes of the components as $V = x_1 \overline{V}_1 + x_2 \overline{V}_2$. For the laboratory alcohol and the vodka, then,

$$V_a = (0.04)(0.816) + (0.96)(1.273) = 1.255 \text{ cm}^3 \cdot \text{g}^{-1}$$
$$V_v = (0.44)(0.953) + (0.56)(1.243) = 1.115 \text{ cm}^3 \cdot \text{g}^{-1}$$

and, by the statement of the problem, $V_w = 1.003$ cm$^3 \cdot$ g^{-1} and $V_a^t = 2000$ cm^3. Substitution of these values into (1) gives

$$V_w^t = (0.7143) \frac{(1.003)(2000)}{(1.255)} = 1142 \text{ cm}^3$$

$$V_v^t = (1.7143) \frac{(1.115)(2000)}{(1.255)} = 3046 \text{ cm}^3$$

Observe that $V_a^t + V_w^t \neq V_v^t$ (liquid volumes don't add).

It is convenient to rewrite the definition (7.1) in terms of the intensive variables M and x_i. First, expand the derivative:

$$\left(\frac{\partial (nM)}{\partial n_i} \right)_{T,P,n_j} = M \left(\frac{\partial n}{\partial n_i} \right)_{T,P,n_j} + n \left(\frac{\partial M}{\partial n_i} \right)_{T,P,n_j}$$

But $(\partial n/\partial n_i)_{T,P,n_j} = 1$, and (7.1) becomes

$$\overline{M}_i = M + n \left(\frac{\partial M}{\partial n_i} \right)_{T,P,n_j} \qquad (7.5)$$

Now, the intensive property M of an m-component mixture is a function of T, P, and $m-1$ independent mole fractions. In what follows we choose these mole fractions to be $x_1, x_2, \ldots, x_{i-1}, x_{i+1}, \ldots, x_m$; that is, we eliminate the mole fraction x_i of the component of interest. We can then write, at constant T and P,

$$dM = \sum_{k \neq i} \left(\frac{\partial M}{\partial x_k} \right)_{T,P,x_l} dx_k \qquad \text{(constant } T, P\text{)}$$

where the subscript x_l indicates that all mole fractions in the basic set other than x_k are held constant. Division of this equation by dn_i and restriction to constant n_j gives

$$\left(\frac{\partial M}{\partial n_i} \right)_{T,P,n_j} = \sum_{k \neq i} \left(\frac{\partial M}{\partial x_k} \right)_{T,P,x_l} \left(\frac{\partial x_k}{\partial n_i} \right)_{n_j} \qquad (7.6)$$

But, by definition, $x_k = n_k/n$, from which

$$\left(\frac{\partial x_k}{\partial n_i} \right)_{n_j} = -\frac{n_k}{n^2} = -\frac{x_k}{n} \qquad (k \neq i) \qquad (7.7)$$

Combination of (7.5), (7.6), and (7.7) gives

$$\overline{M}_i = M - \sum_{k \neq i} x_k \left(\frac{\partial M}{\partial x_k}\right)_{T,P,x_l} \qquad (l \neq k, i) \tag{7.8}$$

Equation (7.8) is, of course, merely an alternate form of (7.1), from which it is derived. However, it finds greater use in the treatment of experimental data, because compositions are generally given in terms of the x_i, rather than the n_i, and because the molar properties M, rather than M^t, are usually of interest. In addition, (7.8) allows a third interpretation of a partial molar property: $\overline{M}_i$ is the Legendre transformation (Section 3.3) of the molar property M with respect to the independent composition-variables $x_1, x_2, \ldots, x_{i-1}, x_{i+1}, \ldots, x_m$.

EXAMPLE 7.2   Write (7.8) for the components of a binary mixture and show how the $\overline{M}_i$ for a binary system can be found from a plot of M versus x_1 at constant T and P.

From (7.8),
$$\overline{M}_1 = M - x_2 \frac{dM}{dx_2}$$

or
$$\overline{M}_1 = M + (1 - x_1) \frac{dM}{dx_1} \tag{7.9}$$

where we have used the fact that $x_1 + x_2 = 1$, and that $dx_2 = -dx_1$. Since conditions of constant T and P are understood, there is only one independent variable, which we have chosen as x_1; the derivative dM/dx_1 is therefore written as a total derivative. Similarly,

$$\overline{M}_2 = M - x_1 \frac{dM}{dx_1} \tag{7.10}$$

A representative plot of M against x_1 is shown in Fig. 7-1. At any composition x_1 the derivative dM/dx_1 can be found by constructing a tangent to the curve. Call the intersections of this tangent with the M-axes I_2 (at $x_1 = 0$) and I_1 (at $x_1 = 1$). From the geometry of the figure, we can write two expressions for dM/dx_1:

$$\frac{dM}{dx_1} = \frac{M - I_2}{x_1} \qquad \text{and} \qquad \frac{dM}{dx_1} = I_1 - I_2$$

Solving for I_2 and I_1, we obtain

$$I_1 = M + (1 - x_1) \frac{dM}{dx_1} \qquad I_2 = M - x_1 \frac{dM}{dx_1}$$

Comparison of these two equations with (7.9) and (7.10) gives

$$I_1 = \overline{M}_1 \qquad I_2 = \overline{M}_2$$

Thus the $\overline{M}_i$ values for the two components of a binary solution are equal to the M-intercepts of the tangent drawn to the M versus x_1 curve at the composition of interest, as shown by Fig. 7-1. It is clear from this construction that a tangent drawn at $x_1 = 1$ gives $\overline{M}_1 = M_1$ and one drawn at $x_1 = 0$ ($x_2 = 1$) gives $\overline{M}_2 = M_2$. This is seen from Fig. 7-2, and is in accord with the requirement that for a pure material $\overline{M}_i = M_i$. The other ends of the tangents shown in Fig. 7-2 intercept the opposite axes, and give the partial molar property of the component present at *infinite dilution*, designated $\overline{M}_i^{\infty}$. Thus, for $x_1 = 1$ and $x_2 = 0$, $\overline{M}_2 = \overline{M}_2^{\infty}$, and when $x_1 = 0$ and $x_2 = 1$, $\overline{M}_1 = \overline{M}_1^{\infty}$.

EXAMPLE 7.3   Suppose for a given T and P that the composition dependence of the liquid molar volume, in $\text{cm}^3 \cdot \text{mol}^{-1}$, for a particular binary system is given by

$$V = 100x_1 + 80x_2 + 2.5x_1x_2 \tag{1}$$

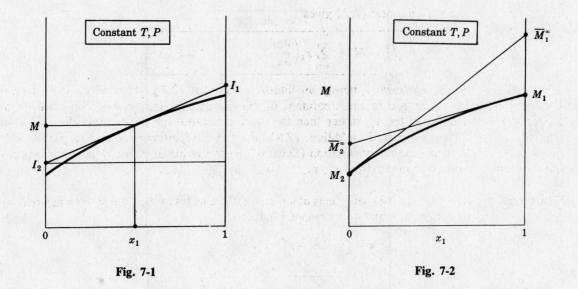

Fig. 7-1   **Fig. 7-2**

Determine the corresponding expressions for $\overline{V}_1$ and $\overline{V}_2$ from (a) (7.1), and (b) (7.9) and (7.10).

(a) Equation (7.1) here becomes

$$\overline{V}_1 = \left(\frac{\partial(nV)}{\partial n_1}\right)_{T,P,n_2} \qquad \overline{V}_2 = \left(\frac{\partial(nV)}{\partial n_2}\right)_{T,P,n_1} \tag{2}$$

Multiply (1) by n, and clear mole fractions in favor of mole numbers:

$$nV = 100n_1 + 80n_2 + 2.5\,\frac{n_1 n_2}{n} \tag{3}$$

Apply (2) to (3), noting that $n = n_1 + n_2$, and recast the results in terms of x_1 and x_2:

$$\overline{V}_1 = 100 + 2.5x_2^2 \qquad \overline{V}_2 = 80 + 2.5x_1^2 \tag{4}$$

(b) With $M = V$, (7.9) and (7.10) become

$$\overline{V}_1 = V + (1 - x_1)\frac{dV}{dx_1} \qquad \overline{V}_2 = V - x_1\frac{dV}{dx_1} \tag{5}$$

By (1), $V = 100x_1 + 80(1 - x_1) + 2.5x_1(1 - x_1) = 80 + 22.5x_1 - 2.5x_1^2$. Thus

$$\frac{dV}{dx_1} = 22.5 - 5.0x_1 \tag{6}$$

Combine (5) with (6) and the expression for $V(x_1)$:

$$\overline{V}_1 = 100 + (2.5)(1 - 2x_1 + x_1^2) = 100 + 2.5x_2^2 \qquad \overline{V}_2 = 80 + 2.5x_1^2 \tag{7}$$

Figure 7-3 shows $\overline{V}_1$, $\overline{V}_2$, and V plotted against x_1 (the solid lines), together with the *approximations* $\overline{V}_1 = V_1$, $\overline{V}_2 = V_2$, and $V = x_1V_1 + x_2V_2$ (the dashed lines). These approximations are the simplest that can be made for the composition dependence of mixture volumes; they are features of *ideal-solution* behavior (Section 7.4).

A final important partial-property equation follows from (7.3) and (7.4). By (7.4),

$$dM = \sum x_i\,d\overline{M}_i + \sum \overline{M}_i\,dx_i$$

This equation, like (7.3), is valid for an arbitrary change of state. Combining the two gives, on rearrangement,

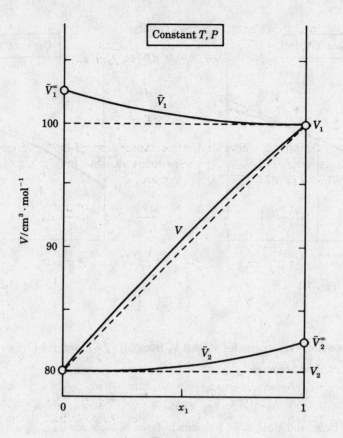

Fig. 7-3

$$\sum x_i\, d\overline{M}_i = \left(\frac{\partial M}{\partial T}\right)_{P,x} dT + \left(\frac{\partial M}{\partial P}\right)_{T,x} dP \qquad (7.11)$$

Equation (7.11) is the most general form of the *Gibbs/Duhem equation*; it implies that variations in the partial properties of the species composing a phase *are not independent*; whatever their cause (most commonly, variations in composition), the variations must satisfy (7.11). In other words, one cannot arbitrarily specify expressions for the composition dependence of the $\overline{M}_i$ for *all* the species in a phase.

Because the molar Gibbs energy G and its relatives play a central role in classical solution thermodynamics, it is useful to specialize the equations just developed to the case $M = G$. From Section 3.4,

$$\mu_i = \left(\frac{\partial(nG)}{\partial n_i}\right)_{T,P,n_j} \qquad (3.40)$$

Comparison with (7.1) shows that

$$\mu_i = \overline{G}_i \qquad (7.12)$$

That is, the chemical potential μ_i is just the partial molar Gibbs energy. By (3.38), (3.39), (7.12), and (7.2), with $M = G$, we retrieve the fundamental property relation

$$d(nG) = -(nS)\, dT + (nV)\, dP + \sum \mu_i\, dn_i \qquad (3.35)$$

Equations (7.3), (7.4), and (7.11) become respectively,

$$dG = -S\, dT + V\, dP + \sum \mu_i\, dx_i \qquad (7.13)$$

$$G = \sum x_i \mu_i \qquad (7.14)$$

$$\sum x_i\, d\mu_i = -S\, dT + V\, dP \qquad (7.15)$$

The procedure of Section 4.8 yields alternative expressions of the last four equations, in which the dimensionless Gibbs energy G/RT is the dependent variable for the mixture, corresponding to the dimensionless chemical potentials μ_i/RT. They are:

$$d\left(\frac{nG}{RT}\right) = -\frac{nH}{RT^2}\, dT + \frac{nV}{RT}\, dP + \sum \frac{\mu_i}{RT}\, dn_i \qquad (7.16)$$

$$d\left(\frac{G}{RT}\right) = -\frac{H}{RT^2}\, dT + \frac{V}{RT}\, dP + \sum \frac{\mu_i}{RT}\, dx_i \qquad (7.17)$$

$$\frac{G}{RT} = \sum x_i \frac{\mu_i}{RT} \qquad (7.18)$$

$$\sum x_i\, d\left(\frac{\mu_i}{RT}\right) = -\frac{H}{RT^2}\, dT + \frac{V}{RT}\, dP \qquad (7.19)$$

The usefulness of these equations results in part from the fact that the enthalpy, rather than the entropy, appears in the coefficient of dT. Moreover, as we shall see, the scaled properties G/RT and μ_i/RT (and properties closely related to them) have special significance in our subject.

EXAMPLE 7.4 The importance of G derives from the fact that its *canonical variables* (Section 3.3) are temperature, pressure, and composition. Classical solution thermodynamics is an experimental science, and T, P, and x (the set of mole fractions x_i) are the variables most susceptible to measurement and control in the laboratory. Moreover, these are the independent variables favored by engineers for use in the design of chemical processes. We illustrate here the derivation of thermodynamic properties from a canonical expression of the form

$$G = G(T, P, x) \qquad (1)$$

From (7.13), we obtain by inspection the molar mixture entropy and molar mixture volume:

$$S = -\left(\frac{\partial G}{\partial T}\right)_{P,x} \qquad (2)$$

$$V = \left(\frac{\partial G}{\partial P}\right)_{T,x} \qquad (3)$$

Equations (2) and (3) were developed earlier as (3.39) and (3.38). Other mixture properties are found from definitions. For example, since $H = G + TS$,

$$H = G - T\left(\frac{\partial G}{\partial T}\right)_{P,x} \qquad (4)$$

Also, since $C_P = T(\partial S/\partial T)_{P,x}$, we find from (2) that

$$C_P = -T\left(\frac{\partial^2 G}{\partial T^2}\right)_{P,x} \qquad (5)$$

Equation (1) also implies a full set of expressions for partial molar properties. With the chemical potential μ_i given by (3.40), we find partial entropy and volume from (3.45) and (3.46), written here as

$$\overline{S}_i = -\left(\frac{\partial \mu_i}{\partial T}\right)_{P,x} \tag{6}$$

$$\overline{V}_i = \left(\frac{\partial \mu_i}{\partial P}\right)_{T,x} \tag{7}$$

Notice that (6) and (7) are analogs of (2) and (3). By similar analogy with (4) and (5), we write

$$\overline{H}_i = \overline{G}_i - T\left(\frac{\partial \mu_i}{\partial T}\right)_{P,x} \tag{8}$$

$$(\overline{C}_P)_i = -T\left(\frac{\partial^2 \mu_i}{\partial T^2}\right)_{P,x} \tag{9}$$

Finally, noting that a mixture property becomes a pure-species property in the limit as $x_i \to 1$, we have from (1) through (5) that

$$G_i = \lim_{x_i \to 1} G \tag{10}$$

$$S_i = \lim_{x_i \to 1} S = -\left(\frac{\partial G_i}{\partial T}\right)_P \tag{11}$$

$$V_i = \lim_{x_i \to 1} V = \left(\frac{\partial G_i}{\partial P}\right)_T \tag{12}$$

$$H_i = \lim_{x_i \to 1} H = G_i - T\left(\frac{\partial G_i}{\partial T}\right)_P \tag{13}$$

$$(C_P)_i = \lim_{x_i \to 1} C_P = -T\left(\frac{\partial^2 G_i}{\partial T^2}\right)_P \tag{14}$$

7.2 THE IDEAL-GAS MIXTURE

In earlier chapters we considered the ideal-gas model and its application to the estimation of thermodynamic properties and property changes of pure substances. There, the model served several purposes: as an approximation to reality (for real gases at low pressures and densities); as a basis for process calculations in schemes incorporating the thermodynamic residual properties; and as a limiting law to be incorporated into formulations of the PVT equation of state. In this chapter we shall use the ideal-gas model to set the stage for the concepts of *fugacity* and *ideal solution*. For this purpose, the ideal-gas equation of state is by itself insufficient; what we need is a comprehensive statement of ideal-gas-*mixture* behavior that incorporates all known features of classical ideal gases. The methods of molecular physics and statistical mechanics provide such a statement, phrased in classical language by the canonical formulation

$$G^{ig} = \sum x_i \Gamma_i(T) + RT \sum x_i \ln x_i P \tag{7.20}$$

Here, quantities Γ_i are properties of the substances composing the mixture. Numerical values of the Γ_i (which do not concern us here) reflect the structural and energetic characteristics of individual molecular species; as signaled by the notation, the Γ_i are functions of temperatures only.

Because it is of canonical form, (7.20) yields a *complete* classical description of ideal gases and their mixtures. Properties are found by derivation, by procedures laid out in Example 7.4. Important for our purposes is the expression for the chemical potential (partial molar Gibbs energy), which is found by application of (3.40) to (7.20). The result is

$$\mu_i^{ig} = \Gamma_i(T) + RT \ln x_i P \tag{7.21}$$

Equation (7.21) is noteworthy for its simplicity: effects of temperature, pressure, and composition are clearly isolated. Note also that (7.14) is satisfied. The molar Gibbs energy of pure i is found from

(7.21) as the limit of μ_i^{ig} as $x_i \to 1$:

$$G_i^{ig} = \Gamma_i(T) + RT \ln P$$

Elimination of Γ_i between the last two equations produces an alternative expression for the chemical potential,

$$\mu_i^{ig} = G_i^{ig} + RT \ln x_i \tag{7.22}$$

whence, by (7.14),

$$G^{ig} = \sum x_i G_i^{ig} + RT \sum x_i \ln x_i \tag{7.23}$$

Let us summarize. Of the four equations (7.20) through (7.23), any *one* completely describes ideal-gas-mixture behavior. Equations (7.20) and (7.21), which show explicitly how G and μ_i depend on pressure, are those needed for applications relating to PVT equations of state; e.g., the development of fugacity concepts (Section 7.3). On the other hand, (7.22) and (7.23), which connect the mixture properties μ_i and G with the pure-component properties G_i, serve as a basis for the definition of the ideal solution (Section 7.4).

EXAMPLE 7.5 Develop expressions for ideal-gas properties implied by (7.20).

Use expressions developed in Example 7.4, and consider only the properties treated there. Beginning with the pure ideal gas i, for which

$$G_i^{ig} = \Gamma_i(T) + RT \ln P$$

apply (11)–(14) of Example 7.4 to obtain

$$S_i^{ig} = -\frac{d\Gamma_i}{dT} - R \ln P \tag{1}$$

$$V_i^{ig} = \frac{RT}{P} \tag{2}$$

$$H_i^{ig} = \Gamma_i - T\frac{d\Gamma_i}{dT} \tag{3}$$

$$(C_P^{ig})_i = -T\frac{d^2\Gamma_i}{dT^2} \tag{4}$$

Equation (2) is the familiar ideal-gas equation of state. Equations (3) and (4) assert that the enthalpy and constant-pressure heat capacity of a pure ideal gas depend on T only; according to (1), the ideal-gas entropy depends on P as well as on T. All the other well-known features of pure ideal gases are easily found from these equations.

Next consider ideal-gas mixture properties. Application of (2)–(5) of Example 7.4 to (7.23) yields

$$S^{ig} = \sum x_i S_i^{ig} - R \sum x_i \ln x_i \tag{5}$$

$$V^{ig} = \sum x_i V_i^{ig} \tag{6}$$

$$H^{ig} = \sum x_i H_i^{ig} \tag{7}$$

$$C_P^{ig} = \sum x_i (C_P^{ig})_i \tag{8}$$

In (5)–(8), the appropriate pure-species properties M_i^{ig} are given by (1)–(4). Substitution of (2) in (6) yields the remarkable result

$$V^{ig} = \frac{RT}{P} = V_i^{ig}$$

that is to say, molar volumes of ideal-gas mixtures are independent of composition. Equations (7) and (8) illustrate another important feature of such mixtures: certain mixture properties are simple linear combinations of the corresponding pure-species properties.

Finally, (6)–(9) of Example 7.4 operate on (7.22) to produce

$$\bar{S}_i^{ig} = S_i^{ig} - R \ln x_i \tag{9}$$

$$\bar{V}_i^{ig} = V_i^{ig} \tag{10}$$

$$\overline{H}_i^{\mathrm{ig}} = H_i^{\mathrm{ig}} \qquad\qquad\qquad (11)$$

$$(\overline{C}_P^{\mathrm{ig}})_i = (C_P^{\mathrm{ig}})_i \qquad\qquad\qquad (12)$$

Thus, certain of the ideal-gas-mixture partial properties (those not related to the entropy) are identical to the corresponding property of the pure ideal gas. This is the simplest approximation one can make to a partial property: $\overline{M}_i = M_i$. Even for ideal-gas mixtures, it does not apply for G, S, and A; for *real* mixtures of *real* fluids, it rarely applies for the other properties.

7.3 FUGACITY AND FUGACITY COEFFICIENT

The chemical potentials μ_i play a natural role in the formulation of criteria for phase and chemical-reaction equilibria. For use in solving practical problems, however, they suffer from several deficiencies:

(1) They are related, as derivatives of G, to the primitive quantities U and S. Since thermodynamics only provides prescriptions for finding *changes* in U and S, unequivocal absolute values for the μ_i are unknown.

(2) As the pressure of a phase approaches zero, the chemical potentials of all species in the phase approach $-\infty$. This may be seen (for ideal gases) from (7.21).

(3) As the concentration of a particular species, i, in a phase approaches zero, μ_i approaches $-\infty$. This behavior too is suggested by (7.21).

Although none of these features actually precludes use of μ_i, everyday work is made easier through use of a *surrogate* chemical potential, a quantity which retains the important attributes of μ_i but which (by contrivance) exhibits none of the deficiencies noted above. Such a quantity is the *fugacity*.

The origins of the fugacity concept reside in (7.20) and (7.21), expressions for G and μ_i for ideal gases. Consider first (7.21), which implies

$$d\mu_i^{\mathrm{ig}} = RT\, d(\ln x_i P) \qquad \text{(constant } T) \qquad\qquad (7.24)$$

Equation (7.24) holds only for a species in an ideal-gas mixture; we extend it to real mixtures by artful definition. Thus we write, analogous to (7.24),

$$\boxed{d\mu_i \equiv RT\, d(\ln \hat{f}_i) \qquad \text{(constant } T)} \qquad\qquad (7.25)$$

and call the defined quantity $\hat{f}_i$ the *fugacity of species i in solution*. Since (7.25) is a differential equation, it provides only a partial definition for $\hat{f}_i$. The definition is completed by imposing the boundary condition

$$\boxed{\lim_{P \to 0} \left(\frac{\hat{f}_i}{x_i P} \right) \equiv 1} \qquad\qquad (7.26)$$

Condition (7.26) forces equivalence of (7.25) and (7.24) in the zero-pressure limit, where real-gas behavior is described by the ideal-gas model.

It is now possible to evaluate $\hat{f}_i$ by integration of (7.25). By (3.46),

$$d\mu_i = \overline{V}_i\, dP \qquad \text{(constant } T, x) \qquad\qquad (7.27)$$

Equating the right sides of (7.25) and (7.27), we get

$$RT\, d(\ln \hat{f}_i) = \overline{V}_i\, dP \qquad \text{(constant } T, x)$$

At constant T and x, we also have the identity

$$RT\, d(\ln x_i P) = RT\, \frac{dP}{P} \qquad \text{(constant } T, x\text{)}$$

Subtraction of the last equation from the one preceding it gives, on rearrangement,

$$d\left[\ln\left(\frac{\hat{f}_i}{x_i P}\right)\right] = (\overline{Z}_i - 1)\, \frac{dP}{P} \qquad \text{(constant } T, x\text{)}$$

where $\overline{Z}_i$ is the partial molar compressibility factor: $\overline{Z}_i \equiv P\overline{V}_i/RT$. Integrating from $P = 0$ to arbitrary P we obtain, using (7.26),

$$\boxed{\ln\left(\frac{\hat{f}_i}{x_i P}\right) = \int_0^P (\overline{Z}_i - 1)\, \frac{dP}{P} \qquad \text{(constant } T, x\text{)}} \tag{7.28}$$

Equation (7.28) can be used to calculate the fugacity of a species in any phase. One requires a volume-explicit equation of state for mixtures, valid from zero pressure to the pressure (and *for the phase*) of interest. Thus, (7.28) is most frequently applied to the evaluation of fugacities of species in gas mixtures at low-to-moderate pressures.

Sometimes one merely needs $\hat{f}_i$ for a species in a particular phase relative to $\hat{f}_i$ in the same phase at a different pressure. Here again one requires volumetric data, but only for the phase of interest. Applying (7.28) separately to pressures P_2 and P_1, subtracting, and simplifying, we find on rearrangement that

$$\boxed{\frac{\hat{f}_i(P_2)}{\hat{f}_i(P_1)} = \exp\left(\int_{P_1}^{P_2} \frac{\overline{V}_i}{RT}\, dP\right) \qquad \text{(constant } T, x\text{)}} \tag{7.29}$$

where again we have used the relationship $\overline{Z}_i = P\overline{V}_i/RT$. The pressure effect on a fugacity (or related quantity) is called the *Poynting effect*, and the expression on the right side of (7.29) is an example of a *Poynting factor*. Poynting factors are most commonly applied to condensed phases (liquids and solids), where molar volumes and partial molar volumes are small; for such applications, Poynting factors are of order unity.

We introduce here a final cosmetic simplification through definition of a dimensionless function called the *fugacity coefficient*:

$$\boxed{\hat{\phi}_i \equiv \frac{\hat{f}_i}{x_i P}} \tag{7.30}$$

With this definition, (7.26) and (7.28) become

$$\lim_{P \to 0} \hat{\phi}_i = 1 \tag{7.31}$$

$$\ln \hat{\phi}_i = \int_0^P (\overline{Z}_i - 1)\, \frac{dP}{P} \qquad \text{(constant } T, x\text{)} \tag{7.32}$$

Clearly, $\hat{\phi}_i$ is of order unity for species i in a gas mixture at low P. For a species in an *ideal*-gas mixture $\hat{\phi}_i^{ig} \equiv 1$ at *all* P.

Equations (7.25), (7.26), (7.28), (7.29), and (7.30) contain all essential information relating to $\hat{f}_i$ and $\hat{\phi}_i$, the fugacity and fugacity coefficient of a *species in solution*. Analogous sets of expressions are easily developed for f and ϕ (the fugacity and fugacity coefficient of a *mixture*) and for f_i and ϕ_i (the fugacity and fugacity coefficient of *pure species i*). Derivation of these expressions is left as an exercise (Problem 7.31); results are summarized in Table 7-1.

Table 7-1 Summary of Fugacity Formulas

For Species i in Solution		For a Mixture		For Pure Species i	
$d\mu_i \equiv RT\,d(\ln \hat{f}_i)$ (constant T)	(7.25)	$dG \equiv RT\,d(\ln f)$ (constant T, x)	(7.33)	$dG_i \equiv RT\,d(\ln f_i)$ (constant T)	(7.38)
$\displaystyle\lim_{P\to 0}\left(\frac{\hat{f}_i}{x_i P}\right) \equiv 1$	(7.26)	$\displaystyle\lim_{P\to 0}\left(\frac{f}{P}\right) \equiv 1$	(7.34)	$\displaystyle\lim_{P\to 0}\left(\frac{f_i}{P}\right) \equiv 1$	(7.39)
$\hat{\phi}_i \equiv \dfrac{\hat{f}_i}{x_i P}$	(7.30)	$\phi \equiv \dfrac{f}{P}$	(7.35)	$\phi_i \equiv \dfrac{f_i}{P}$	(7.40)
$\ln \hat{\phi}_i = \displaystyle\int_0^P (\overline{Z}_i - 1)\frac{dP}{P}$ (constant T, x)	(7.32)	$\ln \phi = \displaystyle\int_0^P (Z - 1)\frac{dP}{P}$ (constant T, x)	(7.36)	$\ln \phi_i = \displaystyle\int_0^P (Z_i - 1)\frac{dP}{P}$ (constant T)	(7.41)
$\dfrac{\hat{f}_i(P_2)}{\hat{f}_i(P_1)} = \exp\left(\displaystyle\int_{P_1}^{P_2} \frac{\overline{V}_i}{RT}dP\right)$ (constant T, x)	(7.29)	$\dfrac{f(P_2)}{f(P_1)} = \exp\left(\displaystyle\int_{P_1}^{P_2} \frac{V}{RT}dP\right)$ (constant T, x)	(7.37)	$\dfrac{f_i(P_2)}{f_i(P_1)} = \exp\left(\displaystyle\int_{P_1}^{P_2} \frac{V_i}{RT}dP\right)$ (constant T)	(7.42)

The fugacity coefficient (and thus the fugacity) is closely related to the residual Gibbs energy. Consider first ϕ for a mixture. By (7.36),

$$\ln \phi = \int_0^P (Z-1)\frac{dP}{P} \qquad \text{(constant } T, x)$$

But, by (5.12),

$$\frac{G^R}{RT} = \int_0^P (Z-1)\frac{dP}{P} \qquad \text{(constant } T, x)$$

so that

$$\ln \phi = \frac{G^R}{RT} \qquad (7.43)$$

Similarly, for pure species i,

$$\ln \phi_i = \frac{G_i^R}{RT} \qquad (7.44)$$

From (5.12), the dimensionless partial molar residual Gibbs energy is

$$\frac{\overline{G}_i^R}{RT} = \int_0^P (\overline{Z}_i - 1)\frac{dP}{P} \qquad \text{(constant } T, x)$$

Comparing this equation with (7.32), we conclude that

$$\ln \hat{\phi}_i = \frac{\overline{G}_i^R}{RT} \qquad (7.45)$$

Equations (7.43) and (7.45) together imply an important pair of partial-molar-property relations. Since $\overline{G}_i^R$ is the partial molar property with respect to G^R, we can make the immediate identification

$$\ln \hat{\phi}_i = \overline{(\ln \phi)}_i \qquad (7.46)$$

That is, the *logarithm* of the fugacity coefficient of a species in solution is the partial molar property with respect to the *logarithm* of the mixture fugacity coefficient. From (7.46) and the definitions (7.30) and (7.35) of $\hat{\phi}_i$ and ϕ, it follows that

$$\ln (\hat{f}_i/x_i) = \overline{(\ln f)}_i \qquad (7.47)$$

Both (7.46) and (7.47) are nonintuitive partial-molar-property relations. Note that $\hat{\phi}_i$ is not a partial molar property with respect to ϕ, nor is $\hat{f}_i$ a partial molar property with respect to f.

Equations (7.32), (7.36), and (7.41) are appropriate for calculation of fugacity coefficients from a volume-explicit equation of state. With a *pressure*-explicit equation of state, for which temperature, density, and composition are the independent variables, one must employ alternative expressions. For ϕ and ϕ_i, we use the connections with G^R and G_i^R provided by (7.43) and (7.44), together with the expression (5.15) for the residual Gibbs energy. Thus we have

$$\ln \phi = \int_0^\rho (Z-1) \frac{d\rho}{\rho} + Z - 1 - \ln Z \qquad \text{(constant } T, x) \qquad (7.48)$$

$$\ln \phi_i = \int_0^\rho (Z_i-1) \frac{d\rho}{\rho} + Z_i - 1 - \ln Z_i \qquad \text{(constant } T) \qquad (7.49)$$

The fugacity coefficients are obtained in Problem 7.4 as

$$\ln \hat{\phi}_i = \int_0^\rho (\tilde{Z}_i-1) \frac{d\rho}{\rho} + Z - 1 - \ln Z \qquad \text{(constant } T, x) \qquad (7.50)$$

where $\tilde{Z}_i$ is the *constant-*(T, ρ) partial molar compressibility factor (see Problem 7.2):

$$\tilde{Z}_i \equiv \left(\frac{\partial (nZ)}{\partial n_i} \right)_{T,\rho,n_j} \qquad (7.51)$$

Unsubscripted compressibility factors in (7.50) are *mixture* properties.

EXAMPLE 7.6 The criterion for phase equilibrium in PVT systems of uniform T and P is concisely stated in terms of the chemical potential:

$$\mu_i^\alpha = \mu_i^\beta = \cdots = \mu_i^\pi \qquad (i = 1, 2, \ldots, m) \qquad (3.87)$$

Show that uniformity of chemical potentials implies uniformity of the fugacities of the species in solution.

We start with

$$d\mu_i \equiv RT\, d(\ln \hat{f}_i) \qquad \text{(constant } T) \qquad (7.25)$$

This expression may be integrated between any two equilibrium states, provided only that the temperature is constant. We choose for the present application a hypothetical change of state in which each species i is initially in the πth phase at equilibrium conditions. Integrating between this state and the equilibrium state in each of the remaining $\pi - 1$ phases, we obtain

$$\mu_i^\alpha = \mu_i^\pi + RT \ln (\hat{f}_i^\alpha/\hat{f}_i^\pi)$$
$$\mu_i^\beta = \mu_i^\pi + RT \ln (\hat{f}_i^\beta/\hat{f}_i^\pi)$$
$$\cdots\cdots\cdots\cdots\cdots\cdots\cdots\cdots$$
$$\mu_i^{\pi-1} = \mu_i^\pi + RT \ln (\hat{f}_i^{\pi-1}/\hat{f}_i^\pi)$$

It then follows from (3.87) that

$$\boxed{\hat{f}_i^\alpha = \hat{f}_i^\beta = \cdots = \hat{f}_i^\pi \qquad (i = 1, 2, \ldots, m)} \qquad (7.52)$$

Equation (7.52) is a practical basis for formulating solutions to practical problems in phase equilibrium. Inspection of Table 7-1 reveals that

$$\lim_{x_i \to 1} \hat{f}_i = f_i$$

That is, the fugacity of a species in solution approaches the pure-species fugacity as the species approaches purity. Hence the analog of (7.52) for *pure-species* phase equilibrium is

$$f_i^\alpha = f_i^\beta = \cdots = f_i^\pi \qquad (i = 1, 2, \ldots, m) \tag{7.53}$$

Equation (7.53) is equivalent to the equilibrium criterion (see Problem 3.11)

$$G_i^\alpha = G_i^\beta = \cdots = G_i^\pi$$

from which it can be derived by a procedure similar to that used in developing (7.52) from (3.87).

EXAMPLE 7.7 Determine expressions for the fugacity coefficients of a gas described by the two-term virial equation in pressure, (5.33). Calculate and plot as functions of composition $\hat{\phi}_i$ and $\hat{f}_i$ for methane ($i = 1$) and n-hexane ($i = 2$) in a binary mixture at 75 °C and 1 atm (1.01325 bar).

We find ϕ and the ϕ_i from (7.43) and (7.44), employing the expression for G^R/RT given as (5.37); thus,

$$\ln \phi = \frac{BP}{RT} \tag{7.54}$$

$$\ln \phi_i = \frac{B_{ii}P}{RT} \tag{7.55}$$

where B is the mixture second virial coefficient and B_{ii} is the second virial coefficient of pure species i. The expression for $\hat{\phi}_i$ can be found from (7.32) or, more directly, from (7.54) and (7.46); we choose the latter route. By the definition (7.1) of a partial molar property, (7.46) implies that

$$\ln \hat{\phi}_i = \left(\frac{\partial (n \ln \phi)}{\partial n_i} \right)_{T,P,n_j} = \frac{P}{RT} \left(\frac{\partial (nB)}{\partial n_i} \right)_{T,P,n_j} \equiv \frac{P}{RT} \overline{B}_i \tag{7.56}$$

By definition, $\overline{B}_i$ is the *partial molar second virial coefficient*; note that restriction of the derivative to constant P is, strictly, superfluous, because B depends only on temperature and composition. Evaluation of $\overline{B}_i$ requires the *mixing rule* for B, expressed by (5.31) and (5.32). Multiply (5.31) by $n = n^2/n$, change the dummy indices, and carry out the indicated differentiation, to obtain:

$$\overline{B}_i = B_{ii} + \tfrac{1}{2} \sum_k \sum_l (2\delta_{ki} - \delta_{kl}) x_k x_l \tag{7.57}$$

Combination of (7.56) and (7.57) gives finally the required expression for $\hat{\phi}_i$:

$$\ln \hat{\phi}_i = \left[B_{ii} + \tfrac{1}{2} \sum_k \sum_l y_k y_l (2\delta_{ki} - \delta_{kl}) \right] \frac{P}{RT} \tag{7.58}$$

Although (7.58) appears complex, it is the simplest realistic expression available for the composition dependence of $\hat{\phi}_i$ in a real-gas mixture. Because $\delta_{kk} = 0$ and $\delta_{kl} = \delta_{lk}$ (see Example 5.9), the double sum reduces to relatively compact expressions even for mixtures containing large numbers of species. Thus for species 1 and 2 in a binary mixture (7.58) yields

$$\ln \hat{\phi}_1 = (B_{11} + x_2^2 \delta_{12}) \frac{P}{RT} \qquad \ln \hat{\phi}_2 = (B_{22} + x_1^2 \delta_{12}) \frac{P}{RT} \tag{7.59}$$

or, equivalently, since $\hat{f}_i = x_i \hat{\phi}_i P$,

$$\hat{f}_1 = x_1 P \exp \left[(B_{11} + x_2^2 \delta_{12}) \frac{P}{RT} \right] \qquad \hat{f}_2 = x_2 P \exp \left[(B_{22} + x_1^2 \delta_{12}) \frac{P}{RT} \right] \tag{7.60}$$

For the methane (1)/n-hexane (2) system at 75 °C,

$$B_{11} = -26 \text{ cm}^3 \cdot \text{mol}^{-1} \qquad B_{22} = -1239 \text{ cm}^3 \cdot \text{mol}^{-1} \qquad \delta_{12} = 905 \text{ cm}^3 \cdot \text{mol}^{-1}$$

The $\hat{\phi}_i$ and $\hat{f}_i$ curves at 1 atm calculated from (7.59) and (7.60) with these values are shown on Fig. 7-4. On the scale of the figure, the $\hat{f}_i$ curves appear to be straight lines; actually, they deviate slightly from linearity because of the δ_{12} terms in (7.60). When $\delta_{12} = 0$, then (7.60) reduces to the linear equations

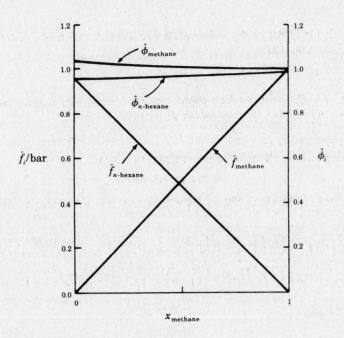

Fig. 7-4

$$\hat{f}_i = x_i P \exp\left(\frac{B_{ii}P}{RT}\right) = x_i f_i \qquad (i = 1, 2)$$

where the last equality follows from (7.55) and (7.40). When all virial coefficients are zero, then (7.60) reduces to

$$\hat{f}_i = x_i P \qquad (i = 1, 2)$$

For this very special case, the phase is an ideal-gas mixture, and the fugacities of the species in solution are equal to their *partial pressures*, $x_i P$. The departure of the $\hat{\phi}_i$ curves from straight, horizontal lines in Fig. 7-4 similarly results from the δ_{12} terms in (7.59). When $\delta_{12} = 0$, then (7.59) gives *composition-independent* values for the $\hat{\phi}_i$:

$$\hat{\phi}_i = \exp\left(\frac{B_{ii}P}{RT}\right) = \phi_i \qquad (i = 1, 2)$$

In general, mixtures of real gases for which all $\delta_{ij} = 0$ exhibit particularly simple $\hat{\phi}_i$ and $\hat{f}_i$ behavior at low pressures; we show in Problem 7.3 that such behavior is a characteristic of *ideal solutions*.

EXAMPLE 7.8 Calculate and plot ϕ and f versus P ($0 \le P \le 100$ bar) for water at 260 °C. For the vapor phase, use virial coefficients determined in Example 5.7. For the liquid phase, use the Tait equation (see Problem 5.25), with the following values for the Tait parameters:

$$V_0 = 1.2842 \text{ cm}^3 \cdot \text{g}^{-1} \qquad A = 0.2706 \qquad B = 1652 \text{ bar}$$

At 260 °C, the vapor/liquid saturation pressure of water is 46.9 bar. Therefore the required calculations are for both the vapor phase (from 0 to 46.9 bar) and the liquid phase (from 46.9 to 100 bar). In Example 5.7, we found that volumetric data for water vapor at 260 °C are fitted very precisely by

$$Z = 1 + B\rho + C\rho^2 + D\rho^3 \qquad (1)$$

Values of the virial coefficients are presented there. Since this EOS is explicit in pressure, ϕ is evaluated by (7.49), from which we find, suppressing subscript i and replacing ρ by P/ZRT,

$$\ln \phi^v = 2B\left(\frac{P}{ZRT}\right) + \frac{3}{2} C\left(\frac{P}{ZRT}\right)^2 + \frac{4}{3} D\left(\frac{P}{ZRT}\right)^3 - \ln Z \qquad (2)$$

With $\ln \phi^v$ known from (2), the relation $f^v = \phi^v P$, or

$$\ln f^v = \ln \phi^v + \ln P \tag{3}$$

gives the fugacity. All that remains is to determine Z for each P, by means of a fixed-point iteration (Example 5.16). For this purpose we write (1) as

$$Z = 1 + B\left(\frac{P}{ZRT}\right) + C\left(\frac{P}{ZRT}\right)^2 + D\left(\frac{P}{ZRT}\right)^3 \tag{1}$$

It is found that the initial estimate $Z_0 = 1$ leads to converged six-figure solutions for Z in 10 iterations or less.

The above procedure takes us to $P^{\text{sat}} = 46.9$ bar, above which pressure water is a subcooled liquid at 260 °C. According to (7.53), at the saturation conditions 260 °C and 46.9 bar,

$$\phi^l(\text{sat}) = \phi^v(\text{sat}) \equiv \phi^{\text{sat}}$$

Thus the vapor and liquid branches of the ϕ-versus-P and f-versus-P curves are continuous at the saturation pressure P^{sat}; the values of f^{sat} and ϕ^{sat} computed for the vapor phase can therefore be employed as reference values from which the liquid-phase quantities can be determined. Here we use the Poynting factor given by (7.42), which becomes for the present application (again suppress subscript i)

$$f^l = f^{\text{sat}} \exp\left(\int_{P^{\text{sat}}}^{P} \frac{V}{RT}\, dP\right) \qquad (\text{constant } T) \tag{7.61}$$

Molar volume V is given by (1) of Problem 5.25, wherein the parameters have the values given above. Carrying out the integration, we find

$$f^l = f^{\text{sat}} \exp\left\{\frac{V_0}{RT}\left[(1 - A)(P - P^{\text{sat}}) + AB \ln\left(1 + \frac{P - P^{\text{sat}}}{B + P^{\text{sat}}}\right)\right]\right\} \tag{4}$$

from which

$$\phi^l = \frac{f^l}{P} \tag{5}$$

Values calculated for f and ϕ from (2), (3), (4), and (5) are plotted against P as solid lines in Fig. 7-5. For comparison, three common approximations are shown by dashed lines. The first, labeled $\phi^v = 1$ and $f^v = P$, represents the assumption of an ideal gas for the vapor phase. For water at 260 °C, this approximation leads to increasing error with increasing pressure up to the saturation pressure, where the error is about 18%.

The second approximation, which is a natural extension of the first to the liquid state, is based on the assumption that $f^l = P^{\text{sat}} = \text{const.}$, for $P \geqq P^{\text{sat}}$. In the present case, this leads to a nearly uniform error of about 18% in both f^l and ϕ^l.

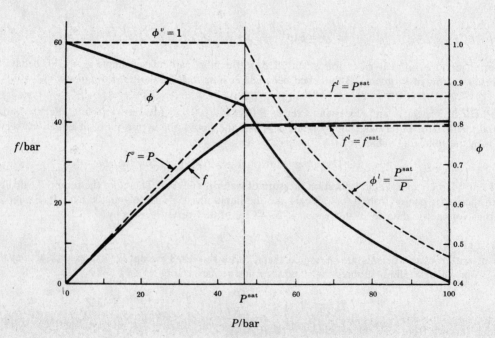

Fig. 7-5

The third common approximation rests on the assumption that $f^l = f^{sat} = \text{const.}$, for $P \geqq P^{sat}$. This assumption almost always leads to smaller errors than the second, giving in this case a maximum error of 2.5% at 100 bar. According to (7.61), the approximation amounts to assigning a value of unity to the Poynting factor, or, equivalently, of assuming that $V(P - P^{sat})/RT \ll 1$ for all $P > P^{sat}$. A much better assumption is to treat the liquid molar volume as independent of pressure, equal to its saturation value V^{sat} at the temperature of interest. (With respect to the Tait equation, this amounts to setting either or both of the parameters A and B equal to zero.) In this case (7.61) becomes

$$f^l = f^{sat} \exp\left[\frac{V^{sat}(P - P^{sat})}{RT}\right] \qquad (7.62)$$

Equation (7.62) is frequently used for representing the behavior of the fugacity of pure liquids at subcritical temperatures.

7.4   THE IDEAL SOLUTION

The correlation and prediction of the properties of real materials is greatly facilitated when one has a standard or model of "ideal" behavior to which real behavior can be compared. Such models should be simple, and they should conform to the behavior of real materials at least in some limiting condition. Thus, as we have seen, the ideal gas serves as a useful model of the behavior of gases, and it represents real-gas behavior in the limit as P approaches zero.

Another important model is the *ideal solution*, which serves as a standard of behavior for the composition dependence of the properties of mixtures. Motivation for the ideal-solution model is provided by (7.22), an expression for the chemical potential of a species in an ideal-gas mixture. However, the presence in (7.22) of the *ideal-gas* Gibbs energy G_i^{ig} disqualifies that expression as a descriptor of limiting behavior of mixtures comprising *real* substances. One overcomes this deficiency by the simple replacement of G_i^{ig} by G_i, the Gibbs energy of pure i at the T and P of the mixture and in the same physical state (*real* gas, liquid, or solid) as the mixture. Thus we *define* an ideal solution as one for which

$$\boxed{\mu_i^{id} \equiv G_i + RT \ln x_i} \qquad (7.63)$$

for all constituent species at all temperatures, pressures, and compositions.

In (7.63), the superscript "id" on μ_i denotes "ideal solution". We may view (7.22) as a special case of (7.63); hence, ideal-gas mixtures are ideal solutions. However, ideal solutions are not necessarily ideal-gas mixtures. Nevertheless, many features of the ideal solution are superficially similar to those of the ideal-gas mixture, as will now be demonstrated.

The Gibbs energy of an ideal solution is given by (7.14) and (7.63) as

$$G^{id} = \sum x_i G_i + RT \sum x_i \ln x_i \qquad (7.64)$$

All other ideal-solution mixture properties follow from (7.64), by methods illustrated in Example 7.4; e.g.,

$$S^{id} = \sum x_i S_i - R \sum x_i \ln x_i \qquad (7.65)$$

$$V^{id} = \sum x_i V_i \qquad (7.66)$$

$$H^{id} = \sum x_i H_i \qquad (7.67)$$

$$C_P^{id} = \sum x_i (C_P)_i \qquad (7.68)$$

to which correspond the partial properties

$$\bar{S}_i^{id} = S_i - R \ln x_i \qquad (7.69)$$

$$\overline{V}_i^{\,\mathrm{id}} = V_i \qquad (7.70)$$

$$\overline{H}_i^{\,\mathrm{id}} = H_i \qquad (7.71)$$

$$(\overline{C}_P^{\,\mathrm{id}})_i = (C_P)_i \qquad (7.72)$$

Equations (7.65) through (7.72) are nearly exact analogs of the ideal-gas-mixture equations (5) through (12) of Example 7.5. The important difference is that the pure-species properties S_i, V_i, H_i, and $(C_p)_i$ appearing in (7.65) through (7.72) are for *real physical states*. In other words, no assumptions are made in the ideal-solution model about the PVT behavior of the *pure* materials, whereas in the ideal-gas-mixture model the pure materials are described by the ideal-gas equation of state.

EXAMPLE 7.9 Determine the behavior of the fugacity $\hat{f}_i$ and of the fugacity coefficient $\hat{\phi}_i$ for a species in an ideal solution.

We start with (7.25), one of the defining equations for $\hat{f}_i$:

$$d\mu_i \equiv RT \, d(\ln \hat{f}_i) \qquad (\text{constant } T)$$

Integration at constant T and P from the state of pure i (where $\hat{f}_i = f_i$ and $\mu_i = G_i$), to the hypothetical state of i in an ideal solution at arbitrary composition, yields on rearrangement

$$\mu_i^{\mathrm{id}} = G_i + RT \ln \frac{\hat{f}_i^{\mathrm{id}}}{f_i}$$

Comparison of this with (7.63) yields

$$\boxed{\hat{f}_i^{\mathrm{id}} = x_i f_i} \qquad (7.73)$$

Thus the fugacity of species i in an ideal solution is proportional to the mole fraction of the species; the coefficient of proportionality is the fugacity of *pure i* at the T and P of the solution and in the same physical state as the solution. The corresponding expression for the fugacity coefficient is, by (7.30) and (7.73),

$$\hat{\phi}_i^{\mathrm{id}} \equiv \frac{\hat{f}_i^{\mathrm{id}}}{x_i P} = \frac{f_i}{P}$$

or

$$\boxed{\hat{\phi}_i^{\mathrm{id}} = \phi_i} \qquad (7.74)$$

Hence the fugacity coefficient of species i in an ideal solution is equal to the fugacity coefficient of *pure i* in the same physical state at the same T and P.

Either (7.73) or (7.74) can in fact be used as the *definition* of an ideal solution, in which case motivation is again provided by the ideal-gas model. For a species in an ideal-gas mixture,

$$\hat{\phi}_i^{\mathrm{ig}} \equiv \frac{\hat{f}_i^{\mathrm{ig}}}{x_i P} = 1$$

But $\phi_i^{\mathrm{ig}} \equiv f_i^{\mathrm{ig}}/P = 1$, whence

$$\hat{f}_i^{\mathrm{ig}} = x_i f_i^{\mathrm{ig}} \qquad \text{and} \qquad \hat{\phi}_i^{\mathrm{ig}} = \phi_i^{\mathrm{ig}}$$

Equations (7.73) and (7.74) can be viewed as generalizations of these two relations, just as (7.63) is a generalization of the ideal-gas prescription (7.22).

7.5 ACTIVITY COEFFICIENT AND EXCESS PROPERTIES

The ideal-solution model and its implications, particularly the expressions (7.66) and (7.67) for mixture volumes and enthalpies, are readily tested against experiment (see Example 7.11). One finds for nonelectrolyte mixtures that the model suffices for highly concentrated solutions, but that its performance generally deteriorates as compositions become equimolar. Disparities can be particularly large for mixtures of *liquids*. Thus, just as real gases are not ideal gases, real solutions are not ideal solutions.

One can formally describe the behavior of real mixtures through a simple extension of the ideal-solution definition (7.63). Thus we write for species i in a *real* solution

$$\boxed{\mu_i \equiv G_i + RT \ln \gamma_i x_i} \qquad (7.75)$$

where γ_i is the *activity coefficient* of species i. The activity coefficient is a dimensionless function of temperature, pressure, and composition. If the γ_i are identically unity for all T, P, and x, then (7.75) reduces to (7.63); thus, an ideal solution is characterized by $\gamma_i^{id} = 1$, for all i. Since μ_i becomes equal to G_i for $x_i = 1$, we also have by (7.75)

$$\lim_{x_i \to 1} \gamma_i = 1 \qquad \text{(real solution)}$$

EXAMPLE 7.10 Relate activity coefficients to fugacities and fugacity coefficients.
We start with (7.25):

$$d\mu_i \equiv RT \, d(\ln \hat{f}_i) \qquad \text{(constant } T\text{)}$$

Integration at constant T and P from the state of pure i to the state of i in a real solution at arbitrary composition yields on rearrangement

$$\mu_i = G_i + RT \ln \frac{\hat{f}_i}{f_i}$$

Comparison with the defining equation (7.75) for γ_i gives

$$\boxed{\gamma_i = \frac{\hat{f}_i}{x_i f_i}} \qquad (7.76)$$

Since $\hat{\phi}_i \equiv \hat{f}_i / x_i P$ and $\phi_i \equiv f_i / P$, (7.76) can also be written as

$$\gamma_i = \frac{\hat{\phi}_i}{\phi_i} \qquad (7.77)$$

Either (7.76) or (7.77) can serve as the *definition* of the activity coefficient, and in fact (7.76) is often so employed. Consider Fig. 7-6, a schematic plot of $\hat{f}_1$ vs. x_1 for a species in a binary mixture. The solid line represents real behavior; the dashed line, ideal-solution behavior as expressed by (7.73). According to (7.76), the activity coefficient at any composition x_1 is the ratio of the ordinates of the two curves at that composition. This provides the interpretation

$$\gamma_i = \frac{\hat{f}_i}{\hat{f}_i^{id}} \qquad (7.78)$$

In the case of Fig. 7-6, $\hat{f}_1 \gtrless \hat{f}_1^{id}$ and thus $\gamma_1 \gtrless 1$. Note that both $\hat{f}_1$ and $\hat{f}_1^{id}$ approach zero as x_1 approaches zero, and hence in this limit γ_1 is indeterminate. Experiment shows that for nonelectrolyte solutions this limiting value of the activity coefficient (the *activity coefficient at infinite dilution*) is finite and generally nonzero.

Other features of the activity coefficient follow from the fact that $\ln \gamma_i$ is a *partial property*, as we now demonstrate.

Subtraction of (7.63) from (7.75) gives $\mu_i - \mu_i^{id} = RT \ln \gamma_i$. But μ_i is a partial property with respect to G, and μ_i^{id} is a partial property with respect to G^{id}; hence, $\overline{G}_i - \overline{G}_i^{id} = RT \ln \gamma_i$, or

$$\ln \gamma_i = \frac{\overline{G}_i^E}{RT} \qquad (7.79)$$

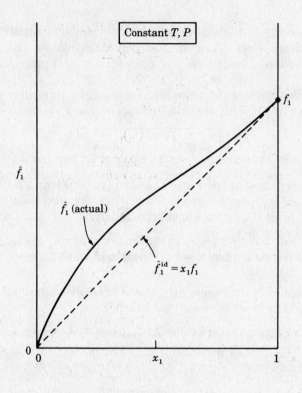

Fig. 7-6

where G^E is the *excess Gibbs energy* of the mixture:

$$G^E \equiv G - G^{\mathrm{id}} \tag{7.80}$$

Thus the logarithm of the activity coefficient is the partial molar property with respect to the dimensionless excess Gibbs energy. This means, in view of (7.1), that

$$\ln \gamma_i = \left(\frac{\partial (nG^E/RT)}{\partial n_i} \right)_{T,P,n_j} \tag{7.81}$$

and, in view of (7.4), that

$$\frac{G^E}{RT} = \sum x_i \ln \gamma_i \tag{7.82}$$

Equation (7.80) serves as inspiration for the introduction of a class of mixture properties called *excess properties*, given the generic symbol M^E and defined as the difference between a real mixture property M and that value M^{id} it would have were it an ideal solution at the same temperature, pressure, and composition:

$$\boxed{M^E \equiv M - M^{\mathrm{id}}} \tag{7.83}$$

Comparison of (7.83) and (4.22) suggests that excess properties and residual properties are closely related; we explore this relationship in Problem 7.6.

Specializing (7.83) to the cases $M = G, S, V, H, C_P$, and invoking the ideal-solution-mixture equations (7.64) through (7.68), we obtain the following explicit expressions for the principal excess properties:

$$G^E = G - \sum x_i G_i - RT \sum x_i \ln x_i \qquad (7.84)$$

$$S^E = S - \sum x_i S_i + R \sum x_i \ln x_i \qquad (7.85)$$

$$V^E = V - \sum x_i V_i \qquad (7.86)$$

$$H^E = H - \sum x_i H_i \qquad (7.87)$$

$$C_P^E = C_P - \sum x_i (C_P)_i \qquad (7.88)$$

The right side of each of these equations contains a pair of leading terms of identical functional form. These combinations of terms are examples of *property changes of mixing*, denoted generically as ΔM and defined by

$$\Delta M \equiv M - \sum x_i M_i \qquad (7.89)$$

In (7.89), as in (7.83), the terms are evaluated at the temperature, pressure, and composition of the mixture; note that M_i is a property of *pure i* at the T and P of the mixture.

EXAMPLE 7.11　Property change of mixing ΔM has a conceptual and experimental basis in a *standard mixing process* which forms a solution at (standard) T and P from appropriate amounts of pure chemical species, each of them also at T and P. Such a process, for mixing of two pure species 1 and 2, is represented in Fig. 7-7. Initially, the system is divided by a partition into two parts, with n_1 moles of pure species 1 on the left and n_2 moles of pure species 2 on the right, both at T and P. Withdrawal of the partition initiates mixing, and in time the system becomes a homogeneous solution of composition $x_1 = n_1/(n_1 + n_2)$. The observable phenomena accompanying the mixing process are expansion (or contraction) of the fluid and heat transfer to (or from) the system. Expansion is accommodated by movement of the piston, and this maintains the pressure at P. Heat transfer is controlled so as to hold the temperature at T.

When mixing is complete, the piston has moved a distance d from its initial position, and from this measured displacement one calculates the total volume change ΔV^t resulting from mixing. Since this volume change is the difference between final and initial volumes, we may write

$$\Delta V^t = V^t - V_1^t - V_2^t = (n_1 + n_2)V - n_1 V_1 - n_2 V_2$$

Division by $n_1 + n_2$ yields the volume change per mole of solution (*volume change of mixing*):

$$\Delta V = V - x_1 V_1 - x_2 V_2 \qquad (1)$$

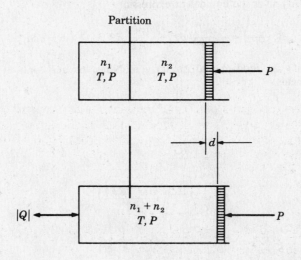

Fig. 7-7

Heat transferred to the system during mixing can also be measured. Now, Example 1.7 shows that for a mechanically reversible, constant-pressure process, the heat transferred equals the enthalpy change. If our process proceeds slowly and if the piston moves with negligible friction, mechanical reversibility is for practical purposes achieved. We can then identify the total heat transfer with the total enthalpy change $\Delta H'$ resulting from mixing; thus, analogous to (1), we have the *heat of mixing*:

$$\Delta H = H - x_1 H_1 - x_2 H_2$$

A key feature of $\Delta V\ (=V^E)$ and $\Delta H\ (=H^E)$ is their direct measurability. Values for other property changes of mixing and excess properties must be inferred through G^E from phase-equilibrium experiments.

For an ideal solution, the left sides of (7.84) through (7.88) are by definition zero, and the pairs of terms to the right of the equals signs represent ΔM^{id} values. Thus

$$\Delta G^{id} = RT \sum x_i \ln x_i \tag{7.90}$$

$$\Delta S^{id} = -R \sum x_i \ln x_i \tag{7.91}$$

$$\Delta V^{id} = 0 \tag{7.92}$$

$$\Delta H^{id} = 0 \tag{7.93}$$

$$\Delta C_P^{id} = 0 \tag{7.94}$$

Note that the entropy change and quantities related to it (e.g., ΔG) are not zero for an ideal solution.

EXAMPLE 7.12 Show how other excess properties follow from knowledge of G^E as a function of T, P, and x. Write (3.35) for the special case of an ideal solution:

$$d(nG^{id}) = -(nS^{id})\,dT + (nV^{id})\,dP + \sum \mu_i^{id}\,dn_i$$

Subtracting this from (3.35) and recalling the definitions of excess properties and of the activity coefficient, we obtain

$$d(nG^E) = -(nS^E)\,dT + (nV^E)\,dP + RT \sum (\ln \gamma_i)\,dn_i \tag{7.95}$$

Similar manipulation of (7.16) gives the equivalent expression

$$d\left(\frac{nG^E}{RT}\right) = -\frac{nH^E}{RT^2}\,dT + \frac{nV^E}{RT}\,dP + \sum (\ln \gamma_i)\,dn_i \tag{7.96}$$

Equations (7.95) and (7.96) are fundamental property relations for nG^E and nG^E/RT; inspection of the coefficients of dT and dP yields

$$S^E = -\left(\frac{\partial G^E}{\partial T}\right)_{P,x} \tag{7.97}$$

$$V^E = \left(\frac{\partial G^E}{\partial P}\right)_{T,x} \tag{7.98}$$

$$H^E = -RT^2\left(\frac{\partial (G^E/RT)}{\partial T}\right)_{P,x} \tag{7.99}$$

and (7.99) leads via (1.9) to

$$C_P^E = -T\left(\frac{\partial^2 G^E}{\partial T^2}\right)_{P,x} \tag{7.100}$$

Again, a *canonical* formulation is "complete"; see Example 7.4.

7.6 REAL BINARY MIXTURES

Peculiarities of mixture behavior are most succintly described through the excess properties. For liquids at normal temperatures, properties are not strongly influenced by pressure; here, the quantities of primary interest are G^E and its temperature derivatives [see particularly (7.99) for H^E]. The simplest liquid mixtures are those consisting of two nonelectrolyte species; these are the only kind we consider here.

Figure 7-8 illustrates the composition dependence of G^E, H^E, and S^E for six binary liquid systems at 50 °C and approximately atmospheric pressure: (a) chloroform (1)/n-heptane (2); (b) acetone (1)/methanol (2); (c) acetone (1)/chloroform (2); (d) ethanol (1)/n-heptane (2); (e) ethanol (1)/chloroform (2); (f) ethanol (1)/water (2). Although the systems exhibit a diversity of behavior, we may note the following common features:

(1) All excess properties become zero as either species approaches purity.

(2) Although G^E versus x_1 is approximately parabolic in shape, both H^E and S^E exhibit individualistic composition dependencies.

(3) When an excess property M^E has a single sign (as does G^E in all six cases), the extreme value of M^E (maximum or minimum) often occurs near to equimolar composition. [H^E of system (f) does not exhibit this feature.]

Feature (1) is a consequence of (7.83): as any x_i approaches unity, M and M^{id} both approach M_i, the corresponding property of pure i. Features (2) and (3) are generalizations based on observation, and admit occasional exceptions. The relatively simple behavior of $G^E(x)$ is most fortunate, because it is the composition derivatives of G^E, as expressed by the activity coefficients, that are important in phase-equilibrium applications. Modeling of $G^E(x)$ is not so formidable a task as is modeling of $G^E(T, x)$ or $H^E(x)$ or $S^E(x)$.

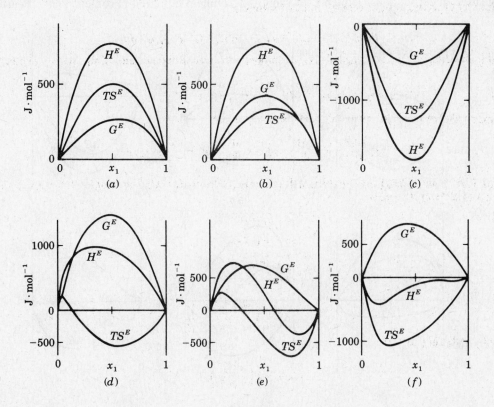

Fig. 7-8

Figure 7-9 displays the property-changes of mixing ΔG, ΔH, and ΔS for the same systems and conditions as in Fig. 7-8. Again one notes diversity of behavior, but again all systems have certain common features:

    I.   Property changes of mixing are zero for pure materials.

    II.   The Gibbs-energy change of mixing ΔG is negative.

    III.   The entropy change of mixing ΔS is positive.

Feature I follows at once from (7.89). Feature II is a consequence of the requirement that the Gibbs energy be a minimum for equilibrium states at specified T and P (see Section 3.6, and Problem 7.7). Feature III *is not* a consequence of the second law of thermodynamics: negative entropy changes of mixing, though unusual, are commonly observed for certain special classes of mixtures. The second law merely forbids negative entropy changes under conditions of *isolation*, whereas ΔS is defined for conditions of constant T and P.

The last figure in this series, Fig. 7-10, shows the logarithms of the activity coefficients for the systems of Figs. 7-8 and 7-9. These are determined from G^E (Fig. 7-8) by application of (7.81). We note the following features of these plots:

    (i)   As x_i approaches unity, $\ln \gamma_i$ approaches zero with zero slope.

    (ii) G^E and $\ln \gamma_i$ generally have the same sign. That is, positive G^E implies activity coefficients greater than unity and negative G^E implies activity coefficients less than unity, at least over most of the composition range.

    (iii)   The activity coefficient at infinite dilution (γ_i^∞) usually represents an extreme (maximum or minimum) value for γ_i.

The limiting value of unity for γ_i, implied by feature (i), has already been inferred from (7.75); the limiting slope of zero is a consequence of the Gibbs/Duhem equation (Problem 7.35). Features (ii)

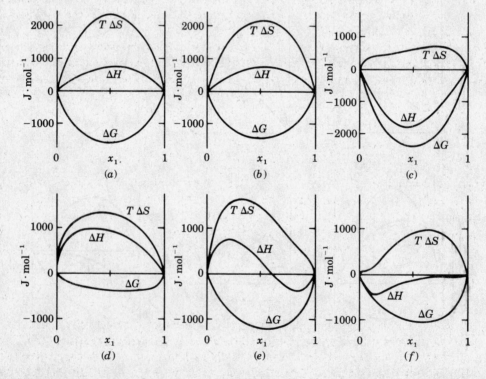

Fig. 7-9

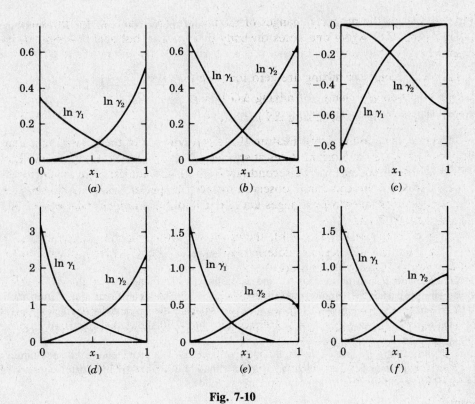

Fig. 7-10

and (iii) follow from the well-behavedness of G^E: for binary mixtures, G^E is most commonly a strictly concave or strictly convex function of composition, and hence both activity coefficients are monotonic functions of mole fraction (Problem 7.10).

Classical thermodynamics provides no analytical formulation for the composition dependence of G^E and the other excess properties. Such formulations must be established either empirically or on the basis of some molecular theory. For binary solutions, equations for G^E are often special cases of

$$\frac{G^E}{x_1 x_2 RT} = B + C(x_1 - x_2) + D(x_1 - x_2)^2 + \cdots \qquad (7.101)$$

$$\frac{x_1 x_2 RT}{G^E} = B' + C'(x_1 - x_2) + D'(x_1 - x_2)^2 + \cdots \qquad (7.102)$$

Here, where the coefficients B, B', C, C', etc. (not to be confused with virial coefficients) are functions of T and P, but not of x. Equations (7.101) and (7.102) may be converted into power series in either x_1 or x_2 by use of the identities $x_1 - x_2 \equiv 2x_1 - 1 \equiv 1 - 2x_2$.

The choice between (7.101) and (7.102), and the degree of truncation, depend on the real behavior to be modeled. The assignment $C = D = \cdots = 0$ or $C' = D' = \cdots = 0$ yields the simplest nontrivial expression for G^E:

$$\frac{G^E}{RT} = Bx_1 x_2 \qquad (7.103)$$

Equation (7.103), the *Porter equation*, is rarely suitable for accurate numerical work. However, it captures many of the essential algebraic features of the composition dependence of G^E, and, if parameter B is allowed to depend on T and P, leads to nontrivial expressions for other excess properties. The Porter equation for G^E, like the van der Waals equation for the PVT equation of state, thus serves primarily for explaining phenomena and for predicting and rationalizing trends.

EXAMPLE 7.13 Determine expressions for H^E, S^E, V^E, and the activity coefficients for a binary mixture described by (7.103).

We begin with the activity coefficients. Since $\ln \gamma_i$ is a partial property with respect to G^E/RT—see (7.81)—

$$\ln \gamma_1 = \left(\frac{\partial(nG^E/RT)}{\partial n_1}\right)_{T,P,n_2} = Bx_2^2 \qquad \ln \gamma_2 = \left(\frac{\partial(nG^E/RT)}{\partial n_2}\right)_{T,P,n_1} = Bx_1^2 \qquad (7.104)$$

The excess enthalpy is found by application of (7.99) to (7.103):

$$H^E = -RT^2\left(\frac{\partial B}{\partial T}\right)_P x_1 x_2 \qquad (7.105)$$

An expression for S^E follows from (7.97):

$$S^E = -R\left[B + T\left(\frac{\partial B}{\partial T}\right)_P\right]x_1 x_2 \qquad (7.106)$$

Finally, (7.98) gives

$$V^E = RT\left(\frac{\partial B}{\partial P}\right)_T x_1 x_2 \qquad (7.107)$$

Note that H^E, S^E, and V^E all depend on x_1 and x_2 in exactly the same way as does G^E; that is, they are described by simple parabolas symmetrical about $x_1 = x_2 = \frac{1}{2}$. This feature of the Porter model is rarely exhibited by real mixtures: see Fig. 7-8. Note also, by (7.104), that the two activity-coefficient curves are mirror images of each other in the line $x_1 = x_2 = \frac{1}{2}$. Reference to Fig. 7-10 shows that this is an idealization seldom obeyed by real mixtures.

Numerical illustration of some of the equations just developed is provided by data for liquid mixtures of benzene (1) and cyclohexane (2), for which G^E is approximately described by the Porter equation. Experimental values for B at pressures near to atmospheric are as follows:

$T/°C$	B
35	0.479
40	0.458
45	0.439

The temperature derivative of B at 40 °C is estimated as

$$\left(\frac{\partial B}{\partial T}\right)_P \approx \frac{0.439 - 0.479}{45 - 35} = -4.00 \times 10^{-3} \text{ K}^{-1}$$

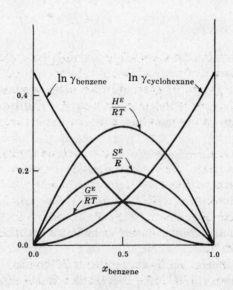

Fig. 7-11

From this and from $B = 0.458$ at 40 °C (= 313.15 K), we obtain from (7.103), (7.105), (7.106), and (7.104) the numerical correlations

$$\frac{G^E}{RT} = 0.458x_1x_2 \qquad \frac{H^E}{RT} = 1.25x_1x_2 \qquad \frac{S^E}{R} = 0.792x_1x_2$$

$$\ln \gamma_1 = 0.458x_2^2 \qquad \ln \gamma_2 = 0.458x_1^2$$

The correlations are plotted against composition in Fig. 7-11; the symmetries about the equimolar line are apparent.

In this numerical example we do not compute V^E, because the pressure dependence of parameter B is not given. In fact, correlations for G^E for liquid mixtures rarely display the pressure dependence of their parameters. At conditions where the excess-property concept is useful, the effect of P on the excess properties is small; when pressure corrections to G^E *are* required, they are effected via (7.98) with independently measured values for V^E.

7.7   PHASE DIAGRAMS FOR BINARY SYSTEMS

Quantitative methods for the calculation of phase equilibria can be formulated from concepts presented in Sections 7.1 through 7.6. First, however, we show how the phase behavior of binary systems may be represented by diagrams.

For a nonreacting PVT system containing two chemical species ($m = 2$), the phase rule (3.88) becomes $F = 4 - \pi$, and the maximum number of independent intensive variables required to specify the thermodynamic state of a stable system is therefore *three*, corresponding to the case of a single equilibrium phase ($\pi = 1$). If the three intensive variables are chosen as P, T, and one of the mole fractions (or a mass fraction), then all equilibrium states of the system can be represented in three-dimensional P-T-composition space. Within this space, the states of *pairs* of phases, coexisting at equilibrium, define *surfaces* ($F = 4 - 2 = 2$); similarly, the states of *three* phases in equilibrium are represented as *space curves* ($F = 4 - 3 = 1$).

Two-dimensional *phase diagrams* are obtained as the intersections of the three-dimensional surfaces and curves with planes of constant pressure or constant temperature. By the former construction, one obtains a *Tx diagram*, having temperature and composition as its coordinates; the second construction yields a *Px diagram*, with the coordinates pressure and composition. Some features of these diagrams are indicated in Figs. 7-12 through 7-20.

Vapor/Liquid Systems

Figure 7-12 is a Tx diagram for the vapor/liquid equilibrium (VLE) of the cyclohexane/toluene system at 101.3 kPa total pressure, obtained from the intersection of the VLE surface with the plane $P = 101.3$ kPa. Figure 7-13 is a Px diagram for the same system at 90 °C, obtained from the intersection of the VLE surface with the plane $T = 90$ °C. The lenslike shape of these figures is typical for systems containing components of similar chemical nature but having dissimilar vapor pressures.

Curve *ABC* in Figs. 7-12 and 7-13 represents the states of saturated-liquid mixtures; it is called the *bubblepoint* curve. Curve *ADC*, the *dewpoint* curve, represents states of saturated vapor. The two curves converge to the pure-component saturation values on the T and P axes at the composition extremes $x = 0$ and $x = 1$. Thus curves *ABC* and *ADC* in Fig. 7-12 intersect the T axes at 110.6 °C and 80.7 °C, which are the boiling points at 101.3 kPa of toluene and cyclohexane. Similarly, curves *ABC* and *ADC* in Fig. 7-13 intersect the P axes at 54.2 kPa and 132.7 kPa, the vapor pressures of toluene and cyclohexane at 90 °C.

The region below *ABC* in Fig. 7-12, and above *ABC* in Fig. 7-13, corresponds to states of subcooled liquid; the region above *ADC* in Fig. 7-12, and below *ADC* in Fig. 7-13, is for superheated vapor. The area between *ABC* and *ADC* in both figures is the vapor/liquid two-phase region.

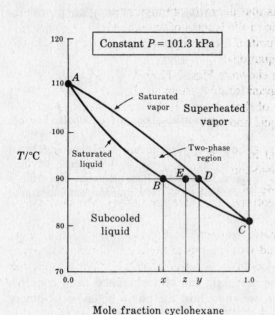

Fig. 7-12

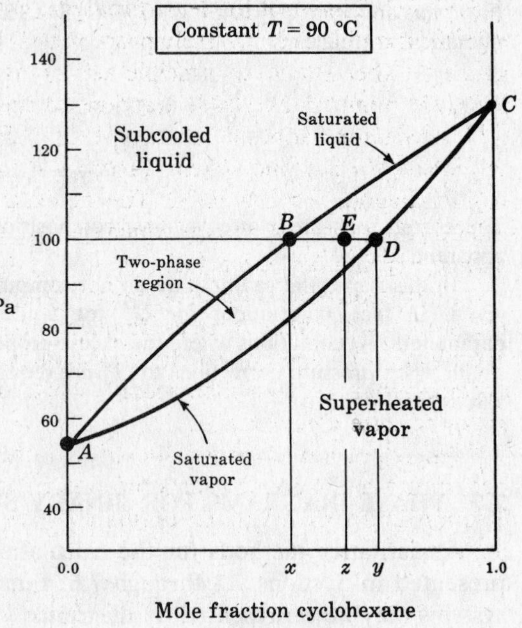

Fig. 7-13

Mixtures whose (T, x) or (P, x) coordinates fall within this area spontaneously split into a liquid and a vapor phase. The equilibrium compositions of the phases formed in such a separation are determined from the phase diagrams by the intersections with the dew- and bubblepoint curves of a horizontal straight line drawn through the point representing the overall state of the mixture; this construction derives from the requirement that T and P of coexisting phases be the same. For example, a mixture with mole fraction z of cyclohexane, when brought to the temperature 90 °C at 101.3 kPa total pressure (point E in Figs. 7-12 and 7-13), forms a liquid containing mole fraction x (point B) and a vapor of mole fraction y (point D). Straight lines such as BD which connect states in phase equilibrium are called *tie lines*.

Tie lines have a useful stoichiometric property, derived as follows. Let n be the total number of moles of a mixture having a mole fraction z_1 of component 1, which separates into n^l moles of liquid with mole fraction x_1 and n^v moles of vapor of mole fraction y_1. A mole-number balance on component 1 gives

$$x_1 n^l + y_1 n^v = z_1 n$$

while an overall mole-number balance yields

$$n^l + n^v = n$$

Solving these equations for the mole-number ratio n^l/n^v, we obtain

$$\frac{n^l}{n^v} = \frac{y_1 - z_1}{z_1 - x_1}$$

But we see from Figs. 7-12 and 7-13 (letting component 1 be cyclohexane) that $y_1 - z_1$ and $z_1 - x_1$ are the lengths of segments ED and BE of the tie line BD. Thus the last equation becomes

$$\frac{n^l}{n^v} = \frac{\overline{ED}}{\overline{BE}}$$

This is a statement of the *lever principle*, which asserts that the ratio of the numbers of moles of the equilibrium phases is inversely proportional to the ratio of the lengths of the corresponding segments of the tie line. The lever principle also holds for mass units if the analogous quantities are substituted for mole numbers and mole fractions in the above equations.

A common variation on the type of VLE behavior shown in Figs. 7-12 and 7-13 is exemplified by the ethanol/n-heptane system. Figure 7-14 is a Tx diagram for this system at 101.3 kPa, and Fig. 7-15 is a Px diagram for $T = 30$ °C. The significant feature of these figures is the occurrence of a state of intermediate composition for which the equilibrium liquid and vapor compositions are identical. Such a state is called an *azeotrope*.

The azeotropic state of a binary system is special in that it possesses only a single degree of freedom, rather than two as required for normal two-component, two-phase equilibrium. Thus specification of any one of the coordinates T, P, or x_1 for a binary azeotrope fixes the other two, provided that the azeotrope actually exists. Binary azeotropes are therefore similar to the saturation states of pure components.

An important characteristic of the azeotropic state, apparent in Figs. 7-14 and 7-15, is the occurrence of a minimum or a maximum on the Tx and Px diagrams at the azeotropic composition. The maximum or minimum occurs on both the bubblepoint and dewpoint curves, and satisfies the appropriate pair of equations:

$$\left(\frac{\partial T}{\partial x_1}\right)_{P,az} = \left(\frac{\partial T}{\partial y_1}\right)_{P,az} = 0$$

or

$$\left(\frac{\partial P}{\partial x_1}\right)_{T,az} = \left(\frac{\partial P}{\partial y_1}\right)_{T,az} = 0$$

Although the ethanol/n-heptane system has a maximum-pressure (minimum-temperature) azeotrope, minimum-pressure (maximum-temperature) azeotropes are common. Azeotropes also may occur in systems containing more than two components.

EXAMPLE 7.14 A 50 mole % liquid mixture of cyclohexane in toluene is confined in a vertical piston-and-cylinder apparatus at 85 °C and 101.3 kPa. The temperature of the mixture is increased to 105 °C by addition of heat. Show on the appropriate phase diagram the physical states assumed by the system during the process.

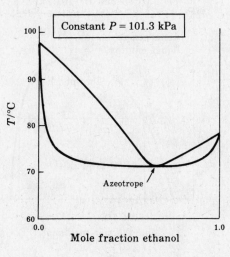

Fig. 7-14

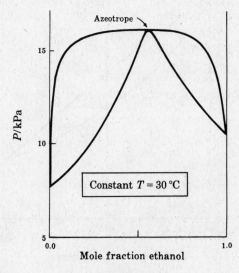

Fig. 7-15

If the piston can be considered frictionless, and if the heating is done slowly enough, then the process occurs at uniform pressure 101.3 kPa and all states assumed by the system are equilibrium states. The appropriate phase diagram is the Tx diagram of Fig. 7-12, redrawn for this example as Fig. 7-16.

At 85 °C and 101.3 kPa the state of the mixture is subcooled liquid, shown as point a on the diagram. The system is closed, and the overall composition remains constant during the process. Therefore the states of the system *as a whole* fall on a vertical line passing through a. When the temperature reaches the bubble point, 90.6 °C (point b), the first bubble of vapor appears. This vapor, represented by point b' with mole fraction $y_{b'} = 0.70$, is richer in cyclohexane than the original mixture. As heating is continued, the amount of vapor increases and the amount of liquid decreases, with the states of the two phases following the paths $b'c'$ and bc, respectively. Finally, at the dew point, 97.1 °C (point c'), the last drop (dew) of liquid with composition $x_c = 0.29$ disappears. The system now contains only vapor of the original 50 mole % composition, and further heating to the final state 105 °C and 101.3 kPa (point d) is through the superheated-vapor region.

Liquid/Liquid Systems

Pairs of components are often incompletely miscible in the liquid state over certain ranges of P, T, and composition. Since the behavior of condensed phases is often quite insensitive to pressure, the important variables are T and composition. A Tx diagram of the liquid/liquid equilibria for the partially miscible system methanol/n-heptane is shown in Fig. 7-17; the data were taken at atmospheric pressure.

The liquid/liquid two-phase region is delineated by the dome-shaped curve, which extends to lower temperatures than are shown in the figure. Mixtures whose (T, x) coordinates fall outside of the dome (but not inside any other two-phase region) exist as single homogeneous liquid phases at 101.3 kPa. A mixture whose (T, x) coordinates are inside the dome cannot exist as a single stable phase, but separates into two liquid phases of different compositions. Thus, as shown in Fig. 7-17, a 50 mole % mixture of methanol in n-heptane, when brought to 40 °C at 101.3 kPa (point E), splits into a n-heptane-rich phase (point B) containing 24 mole % methanol, and a methanol-rich phase (point A) containing 85 mole % methanol. The ratio of the total numbers of moles in the two phases

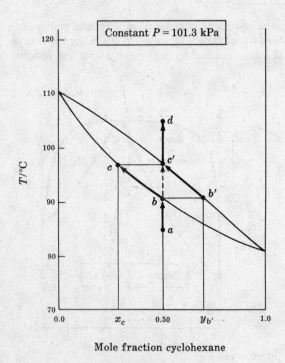

Mole fraction cyclohexane

Mole fraction methanol

Fig. 7-16   **Fig. 7-17**

at these conditions is then given by the lever principle as

$$\frac{n(\text{methanol-rich})}{n(\text{heptane-rich})} = \frac{0.50 - 0.24}{0.85 - 0.50} = 0.743$$

Liquid/liquid Tx diagrams are sometimes called *mutual-solubility curves*, and the compositions of the coexisting phases are commonly reported as *solubilities*. Thus, at 40 °C and 101.3 kPa the solubility of methanol in *n*-heptane is 24 mole % and the solubility of *n*-heptane in methanol is $100 - 85 = 15$ mole %.

Point C on Fig. 7-17 represents the highest temperature (51.2 °C) for which the two liquid phases can coexist in equilibrium at 101.3 kPa pressure. It is called an upper *consolute temperature*, or upper *critical-solution temperature*. Liquid systems may have an upper consolute temperature, a lower consolute temperature, or in some cases both.

Liquid/Solid and Solid/Solid Systems

The greatest variety of binary phase-behavior is exhibited by such systems. The liquid/solid Tx diagram (composition in *mass* fraction copper) for the copper/silver system at 101.3 kPa, shown in Fig. 7-18, is one of the simpler diagrams obtained for such systems. This diagram is typical of binary systems in which the components form *solid solutions* with limited ranges of composition.

Curves AE and EB are *freezing curves* for copper/silver mixtures; they intersect the T axes at 960.5 °C and 1083 °C, the melting points of pure silver and pure copper, respectively. Mixtures whose (T, x) coordinates lie above these curves (but below other equilibrium curves) are stable homogeneous liquid solutions. The area enclosed by $AECA$ is a liquid/solid two-phase region, within which crystals of silver-rich α solid solution with compositions given by curve AC coexist with liquid mixtures with compositions given by AE. Area $BEDB$ is also a liquid/solid two-phase region, within

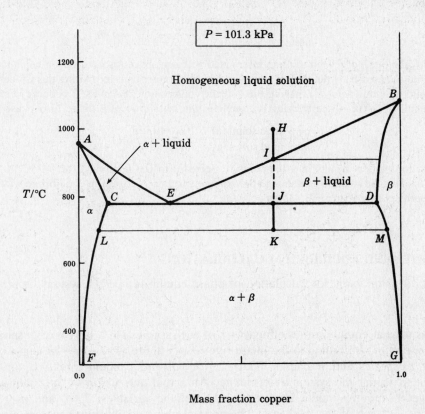

Fig. 7-18

which liquid mixtures with compositions given by curve BE are in equilibrium with crystals of copper-rich β solid solution with compositions given by BD.

The area to the left of ACF is a region of homogeneous α solid solution; similarly, the area to the right of BDG is a region of homogeneous β solid solution. Mixtures with (T, x) coordinates falling in the area $FCEDGF$ separate into α and β solid solutions, with compositions given by the curves CF and DG, respectively.

There is a single temperature at which a liquid solution is in equilibrium with both α and β solid solutions; this temperature is defined by the intersection of curves AE and EB. The point of intersection at E is called the *eutectic* point. For this system, the eutectic occurs at 779 °C and 28.5 mass % copper; at the eutectic, the equilibrium α and β solid solutions contain 8.8 and 92.0 mass % copper, respectively. Curves AC and BD are the *melting curves*, and thus the eutectic temperature is the lowest melting point for systems of this type. Except for mixtures of eutectic composition, melting points and freezing points are in general different for binary systems. Significantly, binary systems can have mixture freezing points and melting points which are considerably lower than those of the constituent pure components. Note that points A and B represent *both* the freezing points and the melting points of the pure constituents.

EXAMPLE 7.15 A 60 mass % mixture of copper in silver is cooled from 1000 °C to 700 °C. Describe the physical states assumed by the mixture.

The path of the cooling process is shown by the vertical line $HIJK$ in Fig. 7-18. The initial state of the mixture (point H) is homogeneous liquid solution. When the temperature reaches approximately 910 °C (point I), crystals of β solid solution containing about 93 mass % copper freeze out. On further cooling, the amount of solid phase increases relative to that of liquid, while the equilibrium liquid and β solid solution both become richer in silver. By the time the eutectic temperature 779 °C is reached, the silver composition in the liquid (point E) has increased to $100 - 28.5 = 71.5$ mass %, and the relative mass of solid solution to liquid solution has become

$$\frac{m(\beta \text{ solid solution})}{m(\text{eutectic solution})} = \frac{0.60 - 0.285}{0.92 - 0.60} = 0.98$$

At exactly 779 °C crystals of α solid solution intermixed with crystals of β solid solution begin to freeze out, and further extraction of heat from the system results in no temperature change. During this isothermal process the remaining liquid solution (point E), having the eutectic composition 28.5 mass % copper, freezes into α and β phases in the ratio $\overline{ED}/\overline{CE}$. Once freezing is complete, the mass ratio of β phase to α phase is

$$\frac{m(\beta \text{ solid solution})}{m(\alpha \text{ solid solution})} = \frac{0.60 - 0.088}{0.92 - 0.60} = 1.6$$

Further extraction of heat results in a temperature decrease to the final value of 700 °C, in which state the system consists of α solid solution containing approximately 6 mass % copper in equilibrium with β solid solution containing approximately 95 mass % copper.

7.8 VAPOR/LIQUID EQUILIBRIUM CALCULATIONS

A thermodynamic basis for calculation of phase equilibria in PVT systems is provided by

$$\hat{f}_i^\alpha = \hat{f}_i^\beta = \cdots = \hat{f}_i^\pi \qquad (i = 1, 2, \ldots, m) \tag{7.52}$$

which asserts that, at equilibrium, the fugacity $\hat{f}_i$ of each species i in a system of m species is constant over all the π phases. According to the phase rule (3.88), there are $2 - \pi + m$ degrees of freedom in a system of m species and π phases. Thus, $2 - \pi + m$ independent intensive variables must be specified just to render the system determinate. An actual *description* of the intensive state of the system, however, requires values for $2 + (m - 1)\pi$ of the variables: T, P, and $m - 1$ independent mole fractions for each of the π phases. The general phase-equilibrium problem may therefore be posed as follows: Given values for $2 - \pi + m$ of the $2 + (m - 1)\pi$ intensive variables, what values (if any) of the remaining $m(\pi - 1)$ variables cause the equilibrium criterion (7.52) to be satisfied?

VLE Equations. Raoult's Law

For equilibrium between a vapor phase and a single liquid phase, (7.52) gives

$$\hat{f}_i^v = \hat{f}_i^l \qquad (i = 1, 2, \ldots, m) \tag{7.108}$$

Equation (7.108) as it stands is too abstract for application to practical problems, because none of the variables T, P, x_i (liquid mole fractions) or y_i (vapor mole fractions) appears explicitly. Useful equilibrium equations result when the fugacities $\hat{f}_i$ are eliminated in favor of the corresponding fugacity coefficients $\hat{\phi}_i$ or activity coefficients γ_i. There are in principle four choices, but only two find general use. The first is the "phi/phi" (or "equation-of-state") formalism, in which fugacity coefficients are employed for both phases. From (7.30) and (7.108), we obtain

$$\boxed{y_i \hat{\phi}_i^v = x_i \hat{\phi}_i^l \qquad (i = 1, 2, \ldots, m)} \tag{7.109}$$

Fugacity coefficients are determined from an equation of state, which—see (7.32) and (7.50)—must apply from zero pressure to the pressure, and for the phase, of interest. Effective use of (7.109) therefore depends on the availability of a single EOS, valid for vapor and liquid mixtures over wide ranges of pressure. Such equations are invariably explicit in pressure, and computations involve considerable iteration. The electronic computer is therefore an essential tool for solving VLE problems by the phi/phi approach. We do not here consider further this method for the calculation of VLE.

The second, or "gamma/phi," approach is the one usually taken at low-to-moderate pressures. Here one uses (7.30) for the vapor phase and (7.76) for the liquid phase, along with (7.108), to obtain

$$\boxed{y_i \hat{\phi}_i P = x_i \gamma_i f_i \qquad (i = 1, 2, \ldots, m)} \tag{7.110}$$

Superscripts have been suppressed in (7.110), as it is understood that $\hat{\phi}_i$ refers to the vapor phase, and γ_i and f_i to the liquid.

To apply (7.110), we require an EOS for the vapor phase and an expression for G^E for the liquid; from the former we find $\hat{\phi}_i$ by (7.32) or (7.50); from the latter, γ_i by (7.81). Also required is the fugacity f_i of pure liquid i. As illustrated in Example 7.8, this is evaluated from the vapor-phase EOS and from liquid-phase volumetric data for pure i. Specifically, we invoke the following identity for $f_i(P)$, the fugacity of pure liquid i at pressure P:

$$f_i(P) \equiv P_i^{\text{sat}} \cdot \frac{f_i^{\text{sat}}}{P_i^{\text{sat}}} \cdot \frac{f_i(P)}{f_i^{\text{sat}}}$$

The first ratio on the right side is just the fugacity coefficient ϕ_i^{sat} of pure liquid or vapor i at the liquid/vapor saturation pressure P_i^{sat}; it is given by (7.41). The second ratio is the Poynting factor for pure liquid i, evaluated for system pressure P relative to reference pressure P_i^{sat}; it is given by (7.61). Substitution of the resulting expression for $f_i(P)$ in (7.110) yields

$$\boxed{y_i \hat{\phi}_i P = x_i \gamma_i P_i^{\text{sat}} \exp \left[\int_0^{P_i^{\text{sat}}} (Z_i - 1) \frac{dP}{P} + \frac{1}{RT} \int_{P_i^{\text{sat}}}^{P} V_i \, dP \right] \qquad (i = 1, 2, \ldots, m)} \tag{7.111}$$

where Z_i is the vapor-phase compressibility factor and V_i is the liquid-phase molar volume.

Equation (7.111) rests on the tacit assumption that each pure liquid i *has a vapor pressure* P_i^{sat}; i.e., that system temperature T is below the liquid/vapor critical temperature T_{ci} for each species i. The great advantage of (7.111) is that it displays explicitly some of the "practical" intensive variables. Moreover, it is written in a form permitting the introduction of clearly stated approximations. A list of common approximations is as follows.

(1)   The liquid-phase activity coefficients γ_i are independent of pressure.

(2)   The liquid-phase molar volumes V_i are independent of pressure, and are equal to the molar volumes V_i^{sat} of the saturated liquids:

$$\frac{1}{RT} \int_{P_i^{\text{sat}}}^{P} V_i \, dP = \frac{V_i^{\text{sat}}}{RT} (P - P_i^{\text{sat}})$$

(3)   The liquid-phase fugacities f_i are independent of pressure, and are equal to their saturation values. The underlying assumption by (7.61) is

$$\frac{1}{RT} \int_{P_i^{\text{sat}}}^{P} V_i \, dP \approx 0$$

(4)   The vapor phase is an ideal solution, so that all $\hat{\phi}_i = \phi_i$.

(5)   The vapor phase is a mixture of ideal gases; thus all $\phi_i = 1$, $Z_i = 1$, and the first integral in (7.111) vanishes.

(6)   The liquid phase is an ideal solution, so that all $\gamma_i = 1$.

Depending on the chemical nature of the system, and on the temperature and pressure levels, one or more of the above approximations may be applicable to a given VLE problem. Approximation (1) is invariably invoked [either explicitly or as an implication of (6)]; at conditions where the gamma/phi approach is useful, the effect of P on γ_i is less significant than are other complicating factors. Approximations (3) and (5) are often suitable when pressures are low, and approximation (2) usually applies up to moderate pressure levels. Approximation (4) is most likely to be valid for systems containing species of similar chemical nature. Approximation (6) is the most drastic of the six, and is rarely suitable for quantitative work: most mixtures of real liquids simply do not behave as ideal solutions. Approximations (3), (5), and (6) conjointly lead to

$$\boxed{y_i P = x_i P_i^{\text{sat}} \qquad (i = 1, 2, \ldots, m)} \qquad (7.112)$$

and (1), (3), and (5) yield

$$\boxed{y_i P = x_i \gamma_i P_i^{\text{sat}} \qquad (i = 1, 2, \ldots, m)} \qquad (7.113)$$

Equation (7.112) is *Raoult's law*; it is the simplest possible equation for VLE, and thus represents a standard against which the behavior of real systems can be compared (see the following subsection). Equation (7.113) is the simplest *realistic* equation for VLE; unlike Raoult's law, it can account for azeotropy, a common feature of VLE in real systems.

EXAMPLE 7.16   VLE in the benzene/toluene system is well represented by Raoult's law at low and moderate pressures. Construct the Px diagram at 90 °C and the Tx diagram at 101.3 kPa (1 atm) total pressure for this system. Vapor-pressure data for benzene (1) and toluene (2) are presented in Table 7-2.

Table 7-2

$T/°C$	$P_1^{\text{sat}}/\text{kPa}$	$P_2^{\text{sat}}/\text{kPa}$	$T/°C$	$P_1^{\text{sat}}/\text{kPa}$	$P_2^{\text{sat}}/\text{kPa}$
80.1	101.3	38.9	98	170.5	69.8
84	114.1	44.5	100	180.1	74.2
88	128.5	50.8	104	200.4	83.6
90	136.1	54.2	108	222.5	94.0
94	152.6	61.6	110.6	237.8	101.3

Application of Raoult's law (7.112) gives

$$y_1 P = x_1 P_1^{sat} \qquad y_2 P = x_2 P_2^{sat} \tag{7.114}$$

The mole fractions in each phase sum to unity; the vapor-phase mole fractions may therefore be eliminated by addition of the two equations, giving

$$P = P_2^{sat} + x_1(P_1^{sat} - P_2^{sat}) \tag{7.115}$$

Substitution of (7.115) in the first (7.114) and solution for y_1 yields

$$y_1 = \frac{x_1 P_1^{sat}}{P_2^{sat} + x_1(P_1^{sat} - P_2^{sat})} \tag{7.116}$$

Equations (7.115) and (7.116) are equations for the bubblepoint and dewpoint curves, respectively, written in forms convenient for construction of the Px diagram. Since the temperature is fixed for this type of phase diagram, the pure-component vapor pressures are constant over the entire composition range. Thus the bubblepoint pressure P and the corresponding equilibrium value of y_1 at the dew point are given directly as functions of x_1.

Pressure, not temperature, is the fixed coordinate for a Tx diagram, and solution of (7.115) for the equilibrium temperatures at specified x_1 is inconvenient. The bubblepoint curve is most readily constructed by solving (7.115) for the bubblepoint compositions x_1 at representative values of T between the saturation temperatures of the pure components:

$$x_1 = \frac{P - P_2^{sat}}{P_1^{sat} - P_2^{sat}} \tag{7.117}$$

The corresponding equilibrium compositions on the dewpoint curve are obtained by substitution of (7.117) in the first (7.114) and solution for y_1:

$$y_1 = \left(\frac{P_1^{sat}}{P}\right)\left(\frac{P - P_2^{sat}}{P_1^{sat} - P_1^{sat}}\right) \tag{7.118}$$

We now illustrate the application of the above equations for the benzene (1)/toluene (2) system. For the Px diagram, we calculate the P and equilibrium y_1 for $x_1 = 0.20$ at 90 °C. For 90 °C, we are given

$$P_1^{sat} = 136.1 \text{ kPa} \qquad P_2^{sat} = 54.2 \text{ kPa}$$

Then, by (7.115) and (7.116),

$$P = 54.2 + (0.2)(136.1 - 54.2) = 70.6 \text{ kPa} \qquad y_1 = (0.20)\left(\frac{136.1}{70.6}\right) = 0.386$$

Thus, at 90 °C and 70.6 kPa, liquid containing 20 mole % benzene is in equilibrium with vapor containing 38.6 mole % benzene. These compositions are shown as points B and D on Fig. 7-19, which is the complete Px diagram.

For the Tx diagram, we calculate the equilibrium values of x_1 and y_1 for $T = 100$ °C at 101.3 kPa total pressure. At 100 °C, from the data,

$$P_1^{sat} = 180.1 \text{ kPa} \qquad P_2^{sat} = 74.2 \text{ kPa}$$

Equations (7.117) and (7.118) then yield

$$x_1 = \frac{101.3 - 74.2}{180.1 - 74.2} = 0.256 \qquad y_1 = \left(\frac{180.1}{101.3}\right)(0.256) = 0.455$$

These compositions are shown as points B' and D' on the complete Tx diagram of Fig. 7-20.

Deviations from Raoult's Law

A significant result of Example 7.16 is that the bubblepoint curve of a system which obeys Raoult's law is linear on the Px diagram (Fig. 7-19). Similarly, the *partial pressures* $y_1 P$ and $y_2 P$ of the two components are proportional to x_1 and x_2, respectively, as shown by the two dashed lines in Fig. 7-19. These linear relationships resulting from Raoult's law provide a convenient basis for classification of VLE behavior. A system for which the bubblepoint and partial-pressure curves lie *above* the Raoult's-law lines, as in Fig. 7-21, is said to exhibit *positive deviations* from Raoult's law.

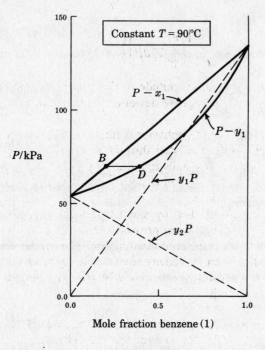

Fig. 7-19

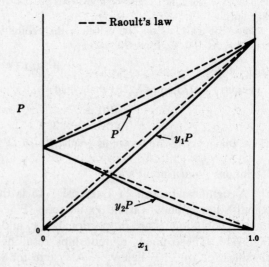

Fig. 7-20

Similarly, if these curves fall *below* the Raoult's-law lines, as in Fig. 7-22, the system shows *negative deviations* from Raoult's law.

The types of effects which promote deviations from Raoult's law can be determined by comparison of a rigorous expression for partial pressure with the partial-pressure equation of Raoult's law. Designating the partial pressure $y_i P$ by the symbol P_i, we can write Raoult's law (*7.112*) as

$$P_i(\text{RL}) = x_i P_i^{\text{sat}}$$

where (RL) signifies Raoult's law. Combination of this equation with the exact expression (*7.111*)

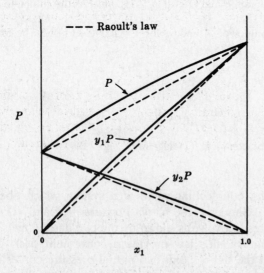

Fig. 7-21

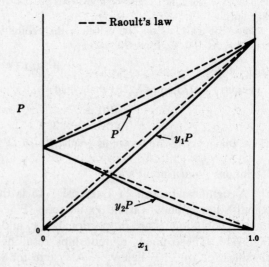

Fig. 7-22

then gives, on rearrangement,

$$\frac{P_i - P_i(\text{RL})}{P_i(\text{RL})} = \left\{ \frac{\gamma_i}{\hat{\phi}_i} \exp\left[\int_0^{P_i^{\text{sat}}} (Z_i - 1) \frac{dP}{P} + \frac{1}{RT} \int_{P_i^{\text{sat}}}^{P} V_i \, dP \right] \right\} - 1 \qquad (7.119)$$

The sign of the deviation $P_i - P_i(\text{RL})$ is thus determined by the magnitude of the term in braces; if this term is greater than unity, positive deviations result, while negative deviations obtain if the term is less than unity.

It is clear from (7.119) that both liquid-phase and vapor-phase behavior contribute to deviations from Raoult's law, the former through γ_i and the integral of V_i, the latter through $\hat{\phi}_i$ and the integral containing Z_i. At low pressures, however, the dominant effect is usually that due to γ_i. Thus, systems with liquid-phase activity coefficients greater than unity generally show positive deviations, while systems for which γ_i is less than unity exhibit negative deviations.

Significant deviations from Raoult's law are often manifested by azeotrope formation. For systems in which the pure components have significantly different vapor pressures (or boiling points), the occurrence of azeotropes generally implies very large (for maximum-pressure azeotropes) or very small (for minimum-pressure azeotropes) values of the liquid-phase activity coefficients. Systems with similar pure-component vapor pressures, however, can exhibit azeotropes with only moderately nonideal liquid-solution behavior. This is illustrated in Example 7.17.

EXAMPLE 7.17   Construct the Px diagram for the cyclohexane (1)/benzene (2) system at 40 °C. Use (7.113) and the activity coefficients derived in Example 7.13. At 40 °C, $P_1^{\text{sat}} = 24.6$ kPa and $P_2^{\text{sat}} = 24.4$ kPa.

Application of (7.113) gives

$$y_1 P = x_1 \gamma_1 P_1^{\text{sat}} \qquad y_2 P = x_2 \gamma_2 P_2^{\text{sat}} \qquad (7.120)$$

Addition of these equations yields

$$P = \gamma_2 P_2^{\text{sat}} + x_1 (\gamma_1 P_1^{\text{sat}} - \gamma_2 P_2^{\text{sat}}) \qquad (7.121)$$

Substitution for the activity coefficients from (7.104) yields

$$P = P_2^{\text{sat}} \exp(B x_1^2) + x_1 \{ P_1^{\text{sat}} \exp[B(1 - x_1)^2] - P_2^{\text{sat}} \exp(B x_1^2) \} \qquad (7.122)$$

Equation (7.122) is the equation of the bubblepoint curve. Corresponding equilibrium points on the dewpoint curve are found by substituting (7.122) into the first (7.120) and solving for y_1, noting that $\gamma_1 = \exp[B(1 - x_1)^2]$.

The Px diagram, calculated from these equations with the given vapor pressures and with $B = 0.458$, is shown on Fig. 7-23. It is assumed in these calculations that B is independent of pressure. Because the benzene and cyclohexane vapor pressures are nearly identical, the effect of the activity coefficients is to produce a maximum-pressure azeotrope; if the liquid solutions were ideal, the bubblepoint curve would follow Raoult's law, as indicated by the dashed line.

K-Values and Flash Calculations

The primary application of phase-equilibrium relationships is in the design of separation processes which depend on the tendency of given chemical species to distribute themselves preferentially in one or another equilibrium phase. A convenient measure of this tendency with respect to VLE is the equilibrium ratio $K_i \equiv y_i/x_i$, which can be expressed as a function of thermodynamic variables via (7.110):

$$K_i = \frac{\gamma_i f_i}{\hat{\phi}_i P} \qquad (7.123)$$

The use of K_i to represent the ratio y_i/x_i is so common that this quantity is usually called an equilibrium *K-value*; another name is the vapor/liquid *distribution coefficient*. Although the introduction of K-values adds nothing to our thermodynamic knowledge of VLE, a K-value serves as a measure of the "lightness" of a component; i.e., of its tendency to concentrate in the vapor phase. If K_i is greater than unity, component i concentrates in the vapor phase; if K_i is less than unity,

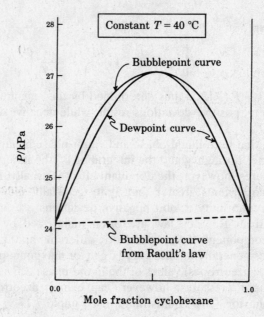

Fig. 7-23

component i concentrates in the liquid phase, and is regarded as a "heavy" component. Furthermore, the use of K_i provides for computational convenience in material-balance calculations, allowing formal elimination of one of the sets of variables x_i or y_i in favor of the other. However, K_i as a *thermodynamic* variable is related by (*7.123*) to γ_i, which is a function of T, P, and liquid composition, and to $\hat{\phi}_i$, which is a function of T, P, and vapor composition; it is therefore a complex function of T, P, the x_i, and the y_i. For this reason, no generally satisfactory direct correlation of K_i with these variables has ever been developed. The simplest behavior displayed by K-values is for systems that follow Raoult's law; for such systems (*7.123*) becomes $K_i(\text{RL}) = P_i^{\text{sat}}/P$. Even in this case, K_i is a strong function of T and P.

Duhem's theorem (Problem 3.18) asserts that the equilibrium state of a closed PVT system formed from specified initial amounts of prescribed chemical species is completely determined by any *two* properties of the system, provided that these two properties are independently variable at equilibrium. Temperature and pressure qualify as such properties for any system of more than one component. According to Duhem's theorem, then, one can in principle calculate the compositions of equilibrium phases at a specified T and P if one knows the overall mole fractions $z_1, z_2, \ldots, z_m$ of the m components. This type of computation when done for a VLE problem is called a *flash calculation*. The term derives from the fact that a liquid mixture at a pressure above its bubblepoint pressure will "flash" or partially evaporate if the pressure is lowered to a value between the bubblepoint and dewpoint pressures. Such a process can be carried out continuously if a liquid is throttled through an orifice into a tank maintained at an appropriate pressure. The liquid and vapor phases formed in the flash tank are equilibrium phases existing at a particular T and P.

The x_i and y_i values which result from a flash calculation must of course satisfy the equilibrium criterion as expressed by (*7.123*); in addition, they must satisfy certain material-balance requirements, derived as follows. At the specified T and P, one mole of mixture, of composition $z_1, z_2, \ldots, z_m$, will separate into L moles of liquid, of composition $x_1, x_2, \ldots, x_m$, and V moles of vapor, of composition $y_1, y_2, \ldots, y_m$. An overall mole balance requires that

$$1 = L + V$$

and the component mole balances are

$$z_i = x_i L + y_i V \qquad (i = 1, 2, \ldots, m)$$

Elimination of V between the two equations yields

$$z_i = x_i L + y_i(1 - L) \qquad (i = 1, 2, \ldots, m)$$

Now substituting $y_i = K_i x_i$ and solving for x_i, we obtain

$$x_i = \frac{z_i}{L + K_i(1 - L)} \qquad (i = 1, 2, \ldots, m) \tag{7.124}$$

But the x_i must sum to unity; hence

$$\sum \frac{z_i}{L + K_i(1 - L)} = 1 \tag{7.125}$$

Analogous equations result if y_i and V, rather than x_i and L, are retained in the derivation:

$$y_i = \frac{K_i z_i}{(1 - V) + K_i V} \tag{7.126}$$

$$\sum \frac{K_i z_i}{(1 - V) + K_i V} = 1 \tag{7.127}$$

Solution of (7.124) and (7.125) for L and the x_i, or of (7.126) and (7.127) for V and the y_i, constitutes the material-balance portion of a flash calculation. The thermodynamic requirement is expressed by (7.123), and it is here that the difficulty lies. The K-values used in (7.124) through (7.127) should be those which satisfy (7.123). However, γ_i is a function of the x_i, and $\hat{\phi}_i$ is a function of the y_i. Thus K-values depend on the values of x_i and y_i to be determined, and iterative calculations are necessary. K-values are first estimated on any reasonable basis (perhaps the assumption of Raoult's law), and the x_i or the y_i are determined from the appropriate material-balance equations. (Either set of values determines the other via the definition of the K_i.) The two sets of values, together with the specified T and P, allow (if sufficient data are available) the calculation of K-values by (7.123). With these new K-values the calculations are repeated, and the process is continued to convergence.

The quantities L and V must, as molar fractions of liquid and vapor, satisfy the inequalities

$$0 \leqq L \leqq 1 \qquad 0 \leqq V \leqq 1$$

These inequalities will in fact be satisfied (assuming the use of correct K-values) only if the specified T and P are within the range for which a two-phase system can exist. It is therefore important to establish this first, before one makes a flash calculation.

The limiting case $L = 0$, $V = 1$, represents the dew point, for which (7.125) becomes

$$\sum \frac{z_i}{K_i} = 1 \tag{7.128}$$

where the $z_i \, (\equiv y_i)$ represent the composition of the saturated vapor. The temperature and pressure at the dew point for a vapor of fixed composition are not independently variable, and specification of one automatically fixed the other. Thus the dewpoint temperature (or pressure) at a specified P (or T) is the value for which the equilibrium K-values satisfy (7.128). The corresponding equilibrium liquid mole fractions are found as $x_i = y_i/K_i$. As with a flash calculation, the equilibrium values of x_i and $y_i \, (\equiv z_i)$ are those consistent with (7.123). Since the K_i are in general functions of composition (both liquid and vapor) as well as of T and P, the calculation of equilibrium values of either x_i or y_i usually requires an iterative procedure.

At the bubble point, $L = 1$, $V = 0$, and (7.127) becomes

$$\sum K_i z_i = 1 \tag{7.129}$$

where the $z_i \, (\equiv x_i)$ give the composition of the saturated liquid. The bubblepoint conditions are determined by calculations similar to those for the dew point. Again, an iterative procedure is generally necessary to establish values of P (or T), $x_i \, (\equiv z_i)$, and y_i that satisfy (7.123).

EXAMPLE 7.18 A hydrocarbon mixture contains methane (1), ethane (2), and propane (3). Its overall composition is $z_1 = 0.10$, $z_2 = 0.20$, and $z_3 = 0.70$. At 10 °C and 15 bar, equilibrium K-values are, approximately, $K_1 = 9.1$, $K_2 = 1.7$, and $K_3 = 0.50$. Determine for these conditions the fraction of the system that is liquid and the compositions of the equilibrium liquid and vapor phases.

The "flash calculation" will employ (7.124) and (7.125); (7.126) and (7.127) could equally well be used. Direct solution of (7.125) for L is impractical for anything but a binary mixture; calculations are done instead by iterative numerical techniques. For this example we solve the equation by trial (Table 7-3). A value of L is assumed, and the x_i are computed from (7.124). If the calculated x_i sum to unity, then the assumed value of L is consistent with the given K-values and feed composition; if not, then a new value of L is chosen and the calculations are repeated until $\Sigma x_i = 1$. Notice that (7.124) implies

$$\sum K_i x_i = \frac{1 - L \sum x_i}{1 - L}$$

so that $\Sigma K_i x_i = 1$ automatically follows from $\Sigma x_i = 1$.

The K-values for this example were estimated from a correlation in which the phase mole fractions x_i and y_i do not appear as variables. Were this a "real" flash calculation—where, for example, the K-values are represented by (7.123)—the K-values would generally depend upon the x_i and y_i. In this case, a multiply iterative solution would be required, in which computed equilibrium compositions are compared against those assumed for evaluation of the K_i, until satisfactory agreement is obtained.

Table 7-3

Species	z_i	K_i	x_i ($L = 0.70$)	x_i ($L = 0.80$)	x_i ($L = 0.763$)	$y_i = K_i x_i$ ($L = 0.763$)
Methane	0.10	9.1	0.029	0.038	0.034	0.312
Ethane	0.20	1.7	0.165	0.175	0.172	0.291
Propane	0.70	0.5	0.824	0.778	0.794	0.397
			$\Sigma x_i = 1.018$	$\Sigma x_i = 0.991$	$\Sigma x_i = 1.000$	$\Sigma y_i = 1.000$

7.9 CHEMICAL-REACTION STOICHIOMETRY. PROPERTY CHANGES OF REACTION

Up to this point, interphase transport of matter has been the only mechanism considered as causing changes in the chemical composition of a phase in a closed system. But composition changes may also occur as a result of depletion or creation of chemical species due to *chemical reaction*. The calculation of property changes associated with chemical reactions and of equilibrium conditions in reacting systems are the subjects of the remaining sections of this chapter.

Stoichiometric Numbers. Reaction Coordinates

Most scientists and engineers are familiar with the usual "chemical" notation for single chemical reactions. Thus, the combination of gaseous hydrogen, $H_2(g)$, with gaseous oxygen, $O_2(g)$, to form liquid water, $H_2O(l)$, is written as

$$H_2(g) + \tfrac{1}{2}O_2(g) \rightarrow H_2O(l)$$

The use of chemical notation serves adequately as an aid in material-balance or thermodynamic calculations for simple systems in which only a few chemical reactions occur. However, the systematic development of the methods of chemical-reaction thermodynamics, and the application of the results to complex reacting systems, is facilitated by use of an equivalent but slightly more abstract "algebraic" notation. We find it convenient to use both notations; the algebraic notation is described briefly below.

A single chemical reaction in a system containing m chemical species is written algebraically as

$$0 = \sum_i \nu_i X_i(p) \qquad (i = 1, 2, \ldots, m) \tag{7.130}$$

where $X_i(p)$ represents the chemical formula for species i and includes a designation (p) of the physical state of i: solid, $(p)=(s)$; liquid, $(p)=(l)$; gas, $(p)=(g)$. The ν_i are *stoichiometric numbers* determined from the "chemical" notation for a reaction. By convention, ν_i is negative for a *reactant* and positive for a *product*. If a particular species k in a system does not participate in the reaction (if it is "inert"), then $\nu_k = 0$. For the reaction $H_2(g) + \frac{1}{2}O_2(g) \rightarrow H_2O(l)$, (7.130) becomes

$$0 = \nu_1 X_1(g) + \nu_2 X_2(g) + \nu_3 X_3(l)$$

where
$$X_1(g) \equiv H_2(g) \qquad \nu_1 = -1$$
$$X_2(g) \equiv O_2(g) \qquad \nu_2 = -\tfrac{1}{2}$$
$$X_3(l) \equiv H_2O(l) \qquad \nu_3 = 1$$

Chemical reactions can be "combined" in the algebraic sense. Consider q chemical reactions for which the stoichiometric numbers of the species $X_i(p)$ are $\nu_{i,j}$, where $j = 1, 2, \ldots, q$ identifies the particular reaction. We write (7.130) for each reaction:

$$0 = \sum_i \nu_{i,1} X_i(p)$$

$$0 = \sum_i \nu_{i,2} X_i(p) \tag{7.131}$$

$$\cdots\cdots\cdots\cdots$$

$$0 = \sum_i \nu_{i,q} X_i(p)$$

where each set of species $\{i\}$ is specific to the particular reaction and need not include all species of the system. Summation of these q equations gives on rearrangement

$$0 = \sum_i \left(\sum_j \nu_{i,j} \right) X_i(p) \tag{7.132}$$

If on comparing (7.132) with (7.130) we find that for each species i

$$\nu_i = \sum_j \nu_{i,j} \tag{7.133}$$

then the single chemical reaction (7.130) may formally be considered the "sum" of the q reactions (7.131).

EXAMPLE 7.19   Consider the water-gas synthesis reaction

$$C(s) + H_2O(g) \rightarrow CO(g) + H_2(g)$$

In the algebraic form (7.130), this reaction is written as

$$0 = \nu_1 X_1(s) + \nu_2 X_2(g) + \nu_3 X_3(g) + \nu_4 X_4(g) \tag{1}$$

where
$$X_1(s) \equiv C(s) \qquad \nu_1 = -1$$
$$X_2(g) \equiv H_2O(g) \qquad \nu_2 = -1$$
$$X_3(g) \equiv CO(g) \qquad \nu_3 = 1 \tag{2}$$
$$X_4(g) \equiv H_2(g) \qquad \nu_4 = 1$$

We wish to show that reaction (1) may be considered the sum of the two reactions

$$C(s) + \tfrac{1}{2}O_2(g) \rightarrow CO(g)$$
$$H_2O(g) \rightarrow H_2(g) + \tfrac{1}{2}O_2(g)$$

In algebraic notation these become

$$0 = \nu_{1,1} X_1(s) + \nu_{5,1} X_5(g) + \nu_{3,1} X_3(g) \tag{3}$$

$$0 = \nu_{2,2} X_2(g) + \nu_{4,2} X_4(g) + \nu_{5,2} X_5(g) \tag{4}$$

where

$$
\begin{aligned}
X_1(s) &\equiv C(s) & \nu_{1,1} &= -1 \\
X_2(g) &\equiv H_2O(g) & & & \nu_{2,2} &= -1 \\
X_3(g) &\equiv CO(g) & \nu_{3,1} &= 1 \\
X_4(g) &\equiv H_2(g) & & & \nu_{4,2} &= 1 \\
X_5(g) &\equiv O_2(g) & \nu_{5,1} &= -\tfrac{1}{2} & \nu_{5,2} &= \tfrac{1}{2}
\end{aligned}
$$

It is now necessary to show that (7.133) is satisfied for each stoichiometric coefficient ν_i of reaction (1).

$$
\begin{aligned}
\nu_1 &= \nu_{1,1} + \nu_{1,2} = -1 + 0 = -1 \\
\nu_2 &= \nu_{2,1} + \nu_{2,2} = 0 - 1 = -1 \\
\nu_3 &= \nu_{3,1} + \nu_{3,2} = 1 + 0 = 1 \\
\nu_4 &= \nu_{4,1} + \nu_{4,2} = 0 + 1 = 1 \\
\nu_5 &= \nu_{5,1} + \nu_{5,2} = -\tfrac{1}{2} + \tfrac{1}{2} = 0
\end{aligned}
$$

Since these results are in agreement with (2), reaction (1) is indeed the sum of reactions (3) and (4).

Given a system containing many chemical species, one could postulate the occurrence of a great number of chemical reactions. Similarly, a single known overall reaction could conceivably be considered, in the *algebraic* sense, the net result of many combinations of simpler reactions. Application of thermodynamic methods to prediction of equilibrium states of such systems requires a set of *independent* chemical reactions consistent with the particular species assumed present in a system. The number r of a set of independent reactions is determined as follows:

(1)   Write reaction equations for the formation of each chemical species from its elements (*formation reactions*).

(2)   Combine these equations so as to eliminate from the set of equations all elements not considered to be present in the system.

The resulting r equations, or any r equations obtained as independent linear combinations of them, represent the desired independent reactions.

EXAMPLE 7.20 Determine a set of independent chemical reactions in a system containing $C(s)$, $H_2O(g)$, $CO(g)$, $CO_2(g)$, and $H_2(g)$.

We can write three formation reactions:

$$H_2(g) + \tfrac{1}{2}O_2(g) \rightarrow H_2O(g) \tag{1}$$

$$C(s) + \tfrac{1}{2}O_2(g) \rightarrow CO(g) \tag{2}$$

$$C(s) + O_2(g) \rightarrow CO_2(g) \tag{3}$$

The element $O_2(g)$ is assumed not present in the system, and it must be eliminated from (1), (2), and (3). Subtraction of (1) from (2) gives

$$C(s) + H_2O(g) \rightarrow CO(g) + H_2(g) \tag{4}$$

Similarly, multiplication of (1) by 2 and subtraction of the result from (3) yields

$$C(s) + 2H_2O(g) \rightarrow CO_2(g) + 2H_2(g) \tag{5}$$

Thus there are *two* possible independent reactions ($r = 2$) for the given system. The two independent reactions need not necessarily be taken as (4) and (5); multiplication of (2) by 2 and subtraction of (3) from the result gives

$$C(s) + CO_2(g) \rightarrow 2CO(g) \tag{6}$$

Either (4) or (5) in combination with (6) would also constitute a pair of independent reactions.

If $CO_2(g)$ were assumed *not present* in the system, then the formation equation (3) would not be written. Subtraction of (1) from (2) to eliminate $O_2(g)$ would then yield the single independent ($r = 1$) reaction (4), which is the water-gas reaction of Example 7.19.

In a closed system undergoing chemical reaction, the changes in the numbers of moles of the chemical species present are not independent of one another. For the general case of r independent reactions, they are related through the stoichiometric numbers to r independent variables called *reaction coordinates* ε_j. In the remainder of this section we restrict treatment to the case of a single reaction, for which the reaction coordinate is ε. We represent the reaction as

$$0 = \nu_1 X_1 + \nu_2 X_2 + \nu_3 X_3 + \nu_4 X_4$$

As the reaction proceeds, changes in the numbers of moles of the species X_1, X_2, X_3, and X_4 are directly related to the stoichiometric numbers ν_1, ν_2, ν_3, and ν_4 according to

$$\frac{dn_2}{dn_1} = \frac{\nu_2}{\nu_1} \quad \text{or} \quad \frac{dn_2}{\nu_2} = \frac{dn_1}{\nu_1}$$

$$\frac{dn_3}{dn_1} = \frac{\nu_3}{\nu_1} \quad \text{or} \quad \frac{dn_3}{\nu_3} = \frac{dn_1}{\nu_1}$$

$$\frac{dn_4}{dn_1} = \frac{\nu_4}{\nu_1} \quad \text{or} \quad \frac{dn_4}{\nu_4} = \frac{dn_1}{\nu_1}$$

The above relations may be combined to give

$$\frac{dn_1}{\nu_1} = \frac{dn_2}{\nu_2} = \frac{dn_3}{\nu_3} = \frac{dn_4}{\nu_4} \equiv d\varepsilon$$

Thus the general relation between a mole number and the reaction coordinate is by definition,

$$\boxed{dn_i \equiv \nu_i\, d\varepsilon} \qquad (i = 1, 2, \ldots, m) \tag{7.134}$$

In (7.134) the ν_i are considered dimensionless, whence ε is in moles. The definition is *completed* by the specification that ε vanish for some particular condition of the system, usually the initial unreacted state. Designating the mole numbers for this initial state as $n_{i,0}$, we have

$$\varepsilon = 0 \quad \text{for} \quad n_i = n_{i,0} \qquad (i = 1, 2, \ldots, m)$$

Integration of (7.134) for each species i then gives

$$\int_{n_{i,0}}^{n_i} dn_i = n_i - n_{i,0} = \nu_i \int_0^\varepsilon d\varepsilon = \nu_i \varepsilon$$

from which

$$\boxed{n_i = n_{i,0} + \nu_i \varepsilon} \qquad (i = 1, 2, \ldots, m) \tag{7.135}$$

Summation of (7.135) over the m species gives

$$\boxed{n = n_0 + \nu \varepsilon} \tag{7.136}$$

where
$$n = \sum n_i \qquad n_0 = \sum n_{i,0} \qquad \nu = \sum \nu_i$$

Division of (7.135) by (7.136) gives the mole fraction z_i of species i in the system:

$$z_i \equiv \frac{n_i}{n} = \frac{n_{i,0} + \nu_i \varepsilon}{n_0 + \nu \varepsilon}$$

or

$$z_i = \frac{z_{i,0} + \nu_i \xi}{1 + \nu \xi} \qquad (i = 1, 2, \ldots, m) \tag{7.137}$$

where $z_{i,0} \equiv n_{i,0}/n_0$, and $\xi \equiv \varepsilon/n_0$ is a dimensionless reaction coordinate.

EXAMPLE 7.21 In a bomb calorimeter 0.02 mol of propane (C_3H_8) is completely oxidized to carbon dioxide (CO_2) and water vapor (H_2O) on burning with 0.6 mol of air. What is the final composition of the system? Take air to contain 21 mole % oxygen (O_2) and 79 mole % nitrogen (N_2).

The oxidation reaction is written in conventional chemical notation as

$$C_3H_8(g) + 5O_2(g) \rightarrow 3CO_2(g) + 4H_2O(g) \tag{1}$$

However, the system contains *five* species—C_3H_8 (1), O_2 (2), N_2 (3), CO_2 (4), and H_2O (5)—and thus reaction (1) becomes in algebraic notation

$$0 = \sum_{i=1}^{5} \nu_i X_i(p)$$

with

$$\begin{aligned}
X_1(g) &\equiv C_3H_8(g) & \nu_1 &= -1 \\
X_2(g) &\equiv O_2(g) & \nu_2 &= -5 \\
X_3(g) &\equiv N_2(g) & \nu_3 &= 0 \\
X_4(g) &\equiv CO_2(g) & \nu_4 &= 3 \\
X_5(g) &\equiv H_2O(g) & \nu_5 &= 4 \\
\hline
& & \nu = \sum \nu_i &= 1
\end{aligned}$$

The initial mole numbers are given as $n_{1,0} = 0.02$ mol, $n_{2,0} = (0.21)(0.6) = 0.126$ mol, $n_{3,0} = (0.79)(0.6) = 0.474$ mol, $n_{4,0} = n_{5,0} = 0$; and the final number of moles of propane is zero ($n_1 = 0$). The corresponding value of the reaction coordinate is found by solution of (7.135) for ε with $i = 1$:

$$\varepsilon = \frac{n_1 - n_{1,0}}{\nu_1} = \frac{0 - 0.02}{-1} = 0.02 \text{ mol}$$

The remaining final mole numbers n_i ($i = 2, 3, 4, 5$) can now be found from (7.135). The results are summarized in Table 7-4 as mole numbers n_i and mole fractions y_i.

Table 7-4

Species	Initial		Final	
	$n_{i,0}$/mol	$y_{i,0}$	n_i/mol	y_i
C_3H_8	0.02	0.0323	0.0	0.0
O_2	0.126	0.2032	0.026	0.0406
N_2	0.474	0.7645	0.474	0.7406
CO_2	0.0	0.0	0.06	0.0938
H_2O	0.0	0.0	0.08	0.1250
	$n_0 = \Sigma\, n_{i,0} = 0.62$		$n = \Sigma\, n_i = 0.64$	

Property Changes of Reaction

For each phase of a π-phase, m-component system (3.35) provides an expression for the total differential of the Gibbs energy expressed as a function of its canonical variables, and it is rewritten here as

$$d(nG)^p = -(nS)^p \, dT^p + (nV)^p \, dP^p + \sum \mu_i^p \, dn_i^p \tag{7.138}$$

where $\mu_i^p \ (\equiv \overline{G}_i^p)$ is the chemical potential of species i in phase p. If the system is in phase equilibrium, then the temperature, pressure, and [see (3.87)] chemical potential of each species are uniform throughout all phases, and we may sum the π equations represented by (7.138) to get an equation for the differential of the total Gibbs energy G^t for the entire system of π phases:

$$dG^t = -S^t \, dT + V^t \, dP + \sum \mu_i \, dn_i \qquad (7.139)$$

Although this is a very general equation, its usual application is to reacting systems for which any particular species appears in just one phase. In this event, the identification of a species also identifies its phase. If changes in the mole numbers n_i occur as the result of a single chemical reaction in a closed system, then by (7.134) each dn_i may be eliminated in favor of the product $\nu_i \, d\varepsilon$, and (7.139) becomes

$$dG^t = -S^t \, dT + V^t \, dP + \Delta G^{rx} \, d\varepsilon \qquad (7.140)$$

where

$$\Delta G^{rx} \equiv \sum \nu_i \mu_i \qquad (7.141)$$

The property ΔG^{rx}, the *Gibbs energy change of reaction*, is evidently equal to $(\partial G^t/\partial \varepsilon)_{T,P}$, and it therefore provides a measure of the variation of the total system property G^t at constant T and P with the extent of reaction as represented by the reaction coordinate ε. Moreover, the units of ΔG^{rx} must be the units of G^t divided by the units of ε. Thus, when G^t is expressed in joules and ε is expressed in moles, ΔG^{rx} has the units $J \cdot mol^{-1}$ and is therefore regarded as an intensive property of the system for a given reaction with a specific set of (dimensionless) stoichiometric numbers.

As a matter of convenience we write ΔG^{rx} as a sum of two terms, the first of which refers to the chemical species in *standard states*. The basis for this is (7.25), which yields on integration between the standard state and the actual state of i in solution

$$\mu_i = \mu_i^\circ + RT \ln (\hat{f}_i/f_i^\circ) \qquad (7.142)$$

The standard states are denoted by the superscript (°); they are states of the *pure* chemical species, at the temperature of the system but at *fixed* pressure. Substitution of (7.142) into (7.141) gives

$$\Delta G^{rx} = \Delta G^\circ + RT \ln \prod (\hat{f}_i/f_i^\circ)^{\nu_i} \qquad (7.143)$$

where

$$\Delta G^\circ \equiv \sum \nu_i \mu_i^\circ \qquad (7.144)$$

and $\prod$ signifies the product over all species i. The function ΔG° is the *standard* (-state) *Gibbs energy change of reaction*; it depends on temperature only. Other *standard* (-state) *property changes of reaction* are defined similarly:

$$\boxed{\Delta M^\circ = \sum \nu_i M_i^\circ} \qquad (7.145)$$

and are related to ΔG° by equations analogous to property relations already derived. Thus we have a Gibbs/Helmholtz equation (Problem 3.9) which gives the *standard enthalpy change of reaction* ΔH°:

$$\boxed{\Delta H^\circ = -RT^2 \frac{d(\Delta G^\circ/RT)}{dT}} \qquad (7.146)$$

This quantity is often called a *standard heat of reaction*. In addition, the *standard heat capacity change of reaction* ΔC_P° and the *standard entropy change of reaction* ΔS° are given by:

$$\boxed{\Delta C_P^\circ = \frac{d \, \Delta H^\circ}{dT}} \qquad (7.147)$$

$$\Delta S° = \frac{\Delta H° - \Delta G°}{T} \tag{7.148}$$

Ordinary derivatives are used in (7.146) and (7.147) because the standard property changes of reaction are functions of temperature only.

Two important stoichiometric properties of the $\Delta M°$ follow directly from the definition (7.145). First, if the stoichiometric coefficients for a given reaction are all multiplied by the same factor α, then the value of $\Delta M°$ for the new reaction is just α times that for the original reaction. For example, the value of $\Delta M°$ for the reaction

$$2H_2O(g) \rightarrow 2H_2(g) + O_2(g)$$

is -2 times the value of $\Delta M°$ for the reaction

$$H_2(g) + \tfrac{1}{2}O_2(g) \rightarrow H_2O(g)$$

Second, there is an additivity property. If we can consider a particular overall chemical reaction to be the algebraic sum of q subsidiary reactions, then the stoichiometric numbers of the overall and subsidiary reactions are related by (7.133). Substitution of this equation into (7.145) yields

$$\Delta M° = \sum_j \Delta M_j° \qquad \text{where} \qquad \Delta M_j° \equiv \sum_i \nu_{i,j} M_i° \tag{7.149}$$

Thus the value of $\Delta M°$ for an overall reaction is the sum of the values $\Delta M_j°$ for any set of subsidiary reactions that sum to the overall reaction.

EXAMPLE 7.22 Derive equations for the calculation of $\Delta H°$ and $\Delta G°$ at any temperature T from given values $\Delta H_0°$ and $\Delta G_0°$ at some reference temperature T_0. Assume that $\Delta C_P°$ is known as a function of T.

Integration of (7.147) from T_0 to T gives the desired expression for $\Delta H°$:

$$\Delta H° = \Delta H_0° + \int_{T_0}^{T} \Delta C_P° \, dT \tag{7.150}$$

Similarly, integration of (7.146) from T_0 to T yields

$$\frac{\Delta G°}{RT} = \frac{\Delta G_0°}{RT_0} - \int_{T_0}^{T} \frac{\Delta H°}{RT^2} \, dT$$

Substituting (7.150) for $\Delta H°$ and integrating, we obtain

$$\frac{\Delta G°}{RT} = \frac{\Delta G_0° - \Delta H_0°}{RT_0} + \frac{\Delta H_0°}{RT} + \frac{1}{RT} \int_{T_0}^{T} \Delta C_P° \, dT - \frac{1}{R} \int_{T_0}^{T} \frac{\Delta C_P°}{T} \, dT \tag{7.151}$$

Compilation of $\Delta M°$ values as a function of T for all reactions of practical interest is neither feasible nor necessary. Equations (7.150) and (7.151) show that $\Delta H°$ and $\Delta G°$ for *any* T can be calculated from values of $\Delta H°$ and $\Delta G°$ at a single reference temperature T_0, provided one knows $\Delta C_P°$ over the temperature range from T_0 to T. Equation (7.149) shows further that it is unnecessary to tabulate $\Delta H°$ and $\Delta G°$, even at T_0, for *every* reaction of interest; one needs no more than entries for a representative set of subsidiary reactions which can be combined algebraically to yield other reactions. Such a set is provided by the *formation reactions* (see Example 7.20).

Standard property changes of formation are designated $\Delta M_f°$. The standard state for a gas is taken as the hypothetical ideal-gas state of the pure gas at either 1 bar (newer convention) or 1 atm (older convention), and the standard state for a liquid or solid is taken as the real state of the pure liquid or solid at either 1 bar or 1 atm. We give the standard-state pressure the symbol $P°$. For a pure ideal gas, the fugacity is equal to the pressure, and thus

$$f_i° = P° \qquad (gases) \qquad (7.152a)$$

For liquids or solids, $f_i°$ is the fugacity of the pure liquid or solid at temperature T and at pressure $P°$:

$$f_i° = f_i(T, P°) \qquad (liquids, solids) \qquad (7.152b)$$

For standard formation reactions, modern compilations of data are for the temperature $T = 298.15$ K (25 °C).

The temperature dependence of $\Delta H°$ and $\Delta G°$ is established as follows. According to (7.145), $\Delta C_P°$ is given by

$$\Delta C_P° = \sum \nu_i (C_P°)_i \qquad (7.153)$$

For most purposes the temperature variation of the $(C_P°)_i$ may be described by [cf. (4.13)]

$$\frac{(C_P°)_i}{R} = A_i + B_i T + C_i T^2 + \frac{D_i}{T^2} \qquad (7.154)$$

where A_i, B_i, C_i, and D_i are constants specific to species i. Substitution in (7.153) yields

$$\frac{\Delta C_P°}{R} = \Delta A + (\Delta B)T + (\Delta C)T^2 + \frac{\Delta D}{T^2} \qquad (7.155)$$

where $\Delta A \equiv \sum \nu_i A_i, \dots, \Delta D \equiv \sum \nu_i D_i$. Combination of (7.155) with (7.150) and (7.151) then gives, on integration and rearrangement,

$$\frac{\Delta H°}{RT_0} = \frac{\Delta H_0°}{RT_0} + (\Delta A)(\tau - 1) + \frac{1}{2}(\Delta B)T_0(\tau^2 - 1) + \frac{1}{3}(\Delta C)T_0^2(\tau^3 - 1) + \frac{\Delta D}{T_0^2}\left(\frac{\tau - 1}{\tau}\right) \qquad (7.156)$$

$$\frac{\Delta G°}{RT_0} = \frac{\Delta H_0°}{RT_0} + \left(\frac{\Delta G_0° - \Delta H_0°}{RT_0}\right)\tau + (\Delta A)(\tau - 1 - \tau \ln \tau) - \frac{1}{2}(\Delta B)T_0(\tau - 1)^2$$

$$\qquad - \frac{1}{6}(\Delta C)T_0^2(\tau + 2)(\tau - 1)^2 - \frac{1}{2}\frac{\Delta D}{T_0^2}\frac{(\tau - 1)^2}{\tau} \qquad (7.157)$$

Here, $\tau \equiv T/T_0$ is a dimensionless absolute temperature.

EXAMPLE 7.23 Calculate $\Delta H°$ and $\Delta G°$ at 1000 °C for the water-gas-synthesis reaction

$$C(s) + H_2O(g) \rightarrow CO(g) + H_2(g) \qquad (1)$$

Standard formation data at 25 °C and for $P° = 1$ bar are: for $H_2O(g)$, $\Delta H_f° = -241.818$ kJ $\cdot$ mol^{-1}, $\Delta G_f° = -228.572$ kJ $\cdot$ mol^{-1}; for $CO(g)$, $\Delta H_f° = -110.525$ kJ $\cdot$ mol^{-1}, $\Delta G_f° = -137.169$ kJ $\cdot$ mol^{-1}. Constants in the heat-capacity equation (7.154) for the pure species in their standard states are $C = 0$ and

Species	A	$B/10^{-3}$ K^{-1}	$D/10^5$ K^2
C(s)	1.771	0.771	-0.867
$H_2O(g)$	3.470	1.450	0.121
$CO(g)$	3.376	0.557	-0.031
$H_2(g)$	3.249	0.422	0.083

As is shown in Example 7.19, reaction (1) can be considered the sum of the two reactions

$$C(s) + \tfrac{1}{2}O_2(g) \rightarrow CO(g) \qquad (2)$$

$$H_2O(g) \rightarrow H_2(g) + \tfrac{1}{2}O_2(g) \qquad (3)$$

But (2) is the formation reaction for $CO(g)$, and (3) is the reverse of the formation reaction for $H_2O(g)$. Thus, for reaction (1) at 298.15 K,

$$\Delta H_0^\circ = \Delta H_f^\circ[CO(g)] - \Delta H_f^\circ[H_2O(g)] = -110.525 - (-241.818) = 131.293 \text{ kJ} \cdot \text{mol}^{-1}$$

$$\Delta G_0^\circ = \Delta G_f^\circ[CO(g)] - \Delta G_f^\circ[H_2O(g)] = -137.169 - (-228.572) = 91.403 \text{ kJ} \cdot \text{mol}^{-1}$$

The stoichiometric numbers in (1) and the tabulated data give

$$\Delta A = (-1)(1.771) + (-1)(3.470) + (1)(3.376) + (1)(3.249) = 1.384$$

Similarly, $\Delta B = -1.242 \times 10^{-3} \text{ K}^{-1}$, $\Delta C = 0.0 \text{ K}^{-2}$, $\Delta D = 0.798 \times 10^5 \text{ K}^2$.

We can now calculate ΔH° and ΔG°. For $T = 1273.15$ K and $T_0 = 298.15$ K, the dimensionless temperature is $\tau = 1273.15/298.15 = 4.270$. Substitution of numerical values into (7.156) and (7.157) gives $\Delta H^\circ/RT_0 = 54.988$ and $\Delta G^\circ/RT_0 = -18.945$, whence

$$\Delta H^\circ = (0.008\,314)(298.15)(54.988) = 136.310 \text{ kJ} \cdot \text{mol}^{-1}$$

$$\Delta G^\circ = (0.008\,314)(298.15)(-18.945) = -46.960 \text{ kJ} \cdot \text{mol}^{-1}$$

Comparing the values just determined to ΔH_0° and ΔG_0°, we see that the effect of a 975 K increase in temperature is to increase ΔH° by a modest 5 kJ $\cdot$ mol^{-1}, whereas the same change in T results in a decrease in ΔG° of about 138 kJ $\cdot$ mol^{-1}. This relative insensitivity of ΔH° and extreme sensitivity of ΔG° to changes in temperature is not uncommon in chemical reactions.

7.10 CHEMICAL-REACTION-EQUILIBRIUM CALCULATIONS

A mixture of chemical species, when brought in contact with a catalyst in a well-stirred vessel, reacts to a greater or lesser extent to form new chemical species at the expense of all or some of the original compounds present in the system. After the passage of sufficient time (perhaps forever, if the catalyst is poor), the composition of the system will attain a true steady-state value, different in general from the original composition. This section deals with the application of thermodynamic methods to the prediction of such states of *chemical equilibrium*. The equilibrium states, while possible in the thermodynamic sense, may or may not be realizable on a realistic time scale in the plant or the laboratory, and questions relating to reaction *rates* cannot be answered by thermodynamics. Nonetheless, knowledge of the equilibrium states of a reacting system is important, because the *equilibrium* conversion of reactant to product species (as measured by, e.g., the reaction coordinate ε) is the *maximum* conversion possible under specified conditions of T and P.

A Criterion for Chemical Equilibrium

The condition for equilibrium with respect to variations in mole numbers in a closed, heterogeneous system is given by

$$(dG^t)_{T,P} = 0 \tag{3.80}$$

If the changes in mole numbers occur as the result of a single chemical reaction, then (7.140) becomes $(dG^t)_{T,P} = \Delta G^{rx} \, d\varepsilon$, and by (3.80) we have $\Delta G^{rx} \, d\varepsilon = 0$. Since this last equation must hold for arbitrary $d\varepsilon$, it follows that

$$\Delta G^{rx} = 0 \tag{7.158}$$

is a criterion for chemical equilibrium in a system in which there occurs a single chemical reaction. This result is readily generalized to systems in which there occur r independent chemical reactions:

$$\Delta G_j^{rx} = 0 \qquad (j = 1, 2, \ldots, r) \tag{7.159}$$

where ΔG_j^{rx} is the Gibbs energy change of reaction for the jth independent reaction.

Since $\Delta G_j^{rx} = \Sigma_i \nu_{i,j} \mu_i$ by definition (7.141), and since each μ_i is in general a function of T, P, and the mole fractions of the various species, (7.159) constitutes r implicit relationships among the intensive variables at chemical equilibrium. Attainment of a state of complete equilibrium in a multiphase reacting system requires in addition that the condition (3.87) for phase equilibrium be satisfied. The generalization of the phase rule (3.88) to include the additional r constraints of (7.159) is obtained by a simple extension of the arguments of Example 3.14; the result is

$$F = 2 - \pi + m - r \qquad (7.160)$$

Thus the effect of the occurrence of chemical reactions is to *decrease* the number of degrees of freedom of a PVT system at equilibrium. As noted in Problem 3.18, however, Duhem's theorem remains unchanged.

EXAMPLE 7.24 Determine the number of degrees of freedom at equilibrium of a chemically reactive system containing solid sulfur S and the three gases O_2, SO_2, and SO_3.

There are four chemical species $(m = 4)$ and two phases $(\pi = 2)$ in the system. The number of independent reactions is *two* $(r = 2)$, because the elements $S(s)$ and $O_2(g)$, which appear in the formation reactions

$$S(s) + O_2(g) \rightarrow SO_2(g) \qquad S(s) + \tfrac{3}{2}O_2(g) \rightarrow SO_3(g)$$

are both assumed present at equilibrium. The extended phase rule (7.160) then gives $F = 2 - 2 + 4 - 2 = 2$.

Thus there are *two* degrees of freedom for the system, and specification of any two of the intensive variables P, T, and the equilibrium mole fractions of the gaseous species (the mole fraction of sulfur in the solid phase is always unity) fixes the equilibrium state of the system.

Practical application of the chemical equilibrium criterion (7.158) is facilitated if the thermodynamic variables are displayed more explicitly. This is done by use of relationships presented in Section 7.9. Substitution of (7.143) into (7.158) gives, on rearrangement,

$$\prod (\hat{f}_i/f_i^\circ)^{\nu_i} = \exp\left(\frac{-\Delta G^\circ}{RT}\right)$$

or

$$\prod (\hat{f}_i/f_i^\circ)^{\nu_i} = K \qquad (7.161)$$

where

$$K \equiv \exp\left(\frac{-\Delta G^\circ}{RT}\right) \qquad (7.162)$$

Quantity K, which depends on T only, is called a chemical *equilibrium "constant"*. Numerical values of ΔG°, and hence of K, can be determined at any temperature from tabulated formation data and heat-capacity values, as illustrated in Example 7.23. By contrast, expressions for $\hat{f}_i$ must in general be found from an equation of state or from a G^E expression.

Applications to Gas-Phase Reactions

As the case of a single gas-phase reaction is of special importance, it will be treated in some detail. By $(7.152a)$, the standard-state fugacity f_i° of a gaseous species is just the standard-state pressure P°. Thus (7.161) becomes

$$\prod \hat{f}_i^{\nu_i} = (P^\circ)^\nu K$$

where $\nu = \Sigma \nu_i$ is the total stoichiometric number for the reaction. Eliminating $\hat{f}_i$ in favor of the fugacity coefficient $\hat{\phi}_i$ via (7.30), with the notation y_i for a gas-phase mole fraction, we obtain on rearrangement

$$\prod (y_i \hat{\phi}_i)^{\nu_i} = (P/P^\circ)^{-\nu} K \tag{7.163}$$

Involving as it does the *ratio* P/P°, (7.163) accommodates any choice of standard-state pressure P° and any choice of units for pressure. Many texts omit P° from (7.163); in such versions a particular value of P° is *presumed* (1 bar or 1 atm), and the corresponding unit for P is then *implied*.

The usual application of (7.163) is to the determination of equilibrium conditions for a gas mixture of known initial composition. This amounts to an application of Duhem's theorem, which requires that two independently variable equilibrium conditions be specified in order to make the problem determinate. In the present case, the component mole fractions y_i are eliminated in favor of their initial values $y_{i,0}$ and the dimensionless reaction coordinate ξ by use of (7.137) (written in terms of y in place of z); (7.163) then becomes a single equation relating the three variables T, P, and ξ. Specification of any two of them, in conformance with Duhem's theorem, allows determination of the remaining one. However, the solution of the problem may be complicated because of the dependence of $\hat{\phi}_i$ on T, P, and ξ.

Ideal-Gas Reactions

The simplest case of (7.163) obtains for a reacting mixture of ideal gases, for which $\hat{\phi}_i = 1$ for each species:

$$\prod y_i^{\nu_i} = (P/P^\circ)^{-\nu} K \tag{7.164}$$

Since (7.164) is in the form $f(\xi) = g(P) K(T)$, the solution for one variable in terms of the other two is straightforward.

EXAMPLE 7.25   Ethanol (C_2H_5OH) can be manufactured by the vapor-phase hydration of ethylene (C_2H_4) according to the reaction

$$C_2H_4(g) + H_2O(g) \rightarrow C_2H_5OH(g)$$

The feed to a reactor in which the above reaction takes place is a gas mixture containing 25 mole % ethylene and 75 mole % steam. Estimate the product composition if the reaction occurs at 125 °C and 1.5 bar. A value for ΔG° ($P^\circ = 1$ bar), calculated by the method of Example 7.23, is 4.570 kJ $\cdot$ mol^{-1}.

The best estimate we can make of the product composition is the *equilibrium* composition at the given T and P. Since the pressure is low, we assume ideal-gas behavior for the mixture of reacting gases; therefore (7.164) is applicable. For C_2H_4 (1), H_2O (2), and C_2H_5OH (3), the stoichiometric coefficients are $\nu_1 = -1$, $\nu_2 = -1$, and $\nu_3 = 1$. Then $\nu = \Sigma \nu_i = -1$, and the material-balance equation (7.137) gives

$$y_1 = \frac{y_{1,0} + \nu_1 \xi}{1 + \nu \xi} = \frac{0.25 - \xi}{1 - \xi}$$

$$y_2 = \frac{y_{2,0} + \nu_2 \xi}{1 + \nu \xi} = \frac{0.75 - \xi}{1 - \xi} \tag{1}$$

$$y_3 = \frac{y_{3,0} + \nu_3 \xi}{1 + \nu \xi} = \frac{\xi}{1 - \xi}$$

The equilibrium constant K is calculated from the definition (7.162):

$$K = \exp \left(\frac{-4.570}{0.008\,314 \times 398.15} \right) = 0.251$$

The equilibrium equation (7.164) then becomes

$$\left(\frac{0.25 - \xi}{1 - \xi} \right)^{-1} \left(\frac{0.75 - \xi}{1 - \xi} \right)^{-1} \left(\frac{\xi}{1 - \xi} \right)^{+1} = \left(\frac{1.5}{1} \right)^{+1} (0.251)$$

or

$$\frac{\xi(1 - \xi)}{(0.25 - \xi)(0.75 - \xi)} = 0.377$$

Solution for ξ yields $\xi = 0.0543$. Substitution of this value in (1) gives the equilibrium composition of the product:

$$y_1 = 0.207 \qquad y_2 = 0.736 \qquad y_3 = 0.057$$

By definition, the reaction coordinate ξ (or ε) measures the extent of reaction or the "yield" of a reaction. Thus, an increase of ξ (or ε) for a given reaction indicates a "shift to the right," or the conversion of more reactant species to product species. For the particular case of an ideal-gas reaction, one can make some important generalizations regarding the effects of thermodynamic properties and of process variables on the equilibrium value of ξ (or ε); these generalizations, although strictly valid only for the ideal-gas case, often may be applied to systems of real gases as well.

We start with (7.164), written in the equivalent form

$$K_y = (P/P^\circ)^{-\nu}K \tag{7.165}$$

where

$$K_y \equiv \prod y_i^{\nu_i} \tag{7.166}$$

According to (7.165), K_y is a function of T and P only, because K is a function of T only. Alternatively, by (7.166) and (7.137) in y, K_y can be considered a function of the single variable ξ for fixed values of the initial mole fractions $y_{i,0}$. It can be shown moreover (the proof is lengthy and we shall not reproduce it here) that

$$\frac{d\xi}{dK_y} > 0 \tag{7.167}$$

and (7.167) lets us infer trends in ξ from trends in K_y.

The first generalization has to do with the relative magnitude of ξ. According to the definition (7.162) of K, a large negative value of ΔG° gives a large value of K, and hence, by (7.165), a large value of K_y. This implies, through (7.167), a large equilibrium ξ relative to the equilibrium ξ for a reaction of identical stoichiometry but smaller negative ΔG°. Conversely, a reaction with a large positive ΔG° yields a small equilibrium ξ relative to that for a similar reaction with smaller positive ΔG°. The sign and magnitude of ΔG° thus offer clues as to the yield one can expect from a given reaction; large negative values of ΔG° imply high equilibrium conversions to the product species, whereas large positive values of ΔG° imply small equilibrium conversions to the product species.

Two other generalizations concern the effects of changes in temperature and pressure on the equilibrium value of ξ. The temperature derivative of ξ is given by

$$\left(\frac{\partial \xi}{\partial T}\right)_P = \left(\frac{d\xi}{dK_y}\right)\left(\frac{\partial K_y}{\partial T}\right)_P$$

But, from (7.165),

$$\left(\frac{\partial K_y}{\partial T}\right)_P = (P/P^\circ)^{-\nu}\frac{dK}{dT}$$

and, from (7.162)

$$\frac{dK}{dT} = K\frac{d(-\Delta G^\circ/RT)}{dT}$$

Combining this with (7.146), we get

$$\frac{dK}{dT} = \frac{K\,\Delta H^\circ}{RT^2}$$

The equation for $(\partial \xi/\partial T)_P$ then becomes

$$\left(\frac{\partial \xi}{\partial T}\right)_P = \frac{d\xi}{dK_y}\left[(P/P^\circ)^{-\nu}\frac{K\,\Delta H^\circ}{RT^2}\right]$$

or, using (7.165),

$$\left(\frac{\partial \xi}{\partial T}\right)_P = \left[\frac{K_y}{RT^2}\frac{d\xi}{dK_y}\right]\Delta H° \tag{7.168}$$

Similarly, we find for the pressure derivative of ξ

$$\left(\frac{\partial \xi}{\partial P}\right)_T = \left[\frac{K_y}{P}\frac{d\xi}{dK_y}\right](-\nu) \tag{7.169}$$

The bracketed terms in (7.168) and (7.169) are always positive, and therefore the signs of $(\partial \xi/\partial T)_P$ and $(\partial \xi/\partial P)_T$ are determined solely by the signs of $\Delta H°$ and $\nu \ (= \Sigma \nu_i)$, respectively. We may therefore conclude the following:

(1) If, at a given temperature, $\Delta H°$ is positive—i.e., if the standard reaction is *endothermic*—then an increase in T at constant P causes an increase in the equilibrium ξ. If $\Delta H°$ is negative—i.e., if the standard reaction is *exothermic*—then an increase in T at constant P causes a decrease in the equilibrium ξ.

(2) If the total stoichiometric coefficient ν is *negative*, then an increase in P at constant T results in an increase in the equilibrium ξ. If ν is *positive*, then an increase in P at constant T causes a decrease in the equilibrium ξ.

EXAMPLE 7.26   State the probable effects of increasing P and T on the yield of ammonia (NH_3) from the reaction

$$\tfrac{1}{2}N_2(g) + \tfrac{3}{2}H_2(g) \rightarrow NH_3(g) \tag{1}$$

For this reaction, $\Delta H° = -46.220 \text{ kJ} \cdot \text{mol}^{-1}$ at 298 K.

In the absence of data on which to base a rigorous calculation, we assume that the reacting system is a mixture of ideal gases. The total stoichiometric coefficient is

$$\nu = \sum \nu_i = -\tfrac{1}{2} - \tfrac{3}{2} + 1 = -1$$

According to (7.169), then, an increase in pressure tends to *increase* the production of ammonia by reaction (1). The standard reaction is exothermic at 298 K and therefore, by (7.168), increasing T at this temperature level causes a *decrease* in the conversion to ammonia. Actually, $\Delta H°$ for reaction (1) is negative for all temperatures of practical interest; so this last conclusion holds generally for the ideal-gas ammonia-synthesis reaction.

Multireaction Equilibria

So far we have considered the prediction of equilibrium states only for systems in which a single independent chemical reaction occurs. The approach adopted in this special case was to develop an expression for $(dG')_{T,P}$ as a function of the composition variables and to set it equal to zero, in accord with the equilibrium condition (3.80). This method is readily generalized to apply when there are r independent chemical reactions. However, one is then faced with r equations which require simultaneous solution, and hand calculations can be tedious. Moreover, these equations are not well suited to solution by automatic machine computation. The modern procedure for solving multireaction equilibrium problems is therefore based on the alternative and equivalent method noted in Section 3.6. This method requires an expression for G' in terms of the composition variables, and the equilibrium composition is then determined directly as the set of mole fractions which *minimizes G'* at a given T and P, as required by (3.79). Solution of the equations developed is again most tedious by hand calculation, but solution by automatic machine computation is facilitated. In the present treatment, we consider only the calculation of equilibrium compositions for gas-phase reactions at specified T, P, and initial compositions.

The total Gibbs energy G' of the system is given by

$$G' = \sum_i n_i \overline{G}_i = \sum_i n_i \mu_i$$

This is the function to be minimized with respect to the mole numbers n_i at constant T and P, subject to constraints imposed by material balances for a closed system. The μ_i are given by (7.142) and (7.152a) as

$$\mu_i = \mu_i^\circ + RT \ln (\hat{f}_i/P^\circ)$$

and $\hat{f}_i$ is related to $\hat{\phi}_i$ by (7.30):

$$\hat{f}_i = \hat{\phi}_i P y_i$$

The standard states are those adopted previously for chemical-reaction calculations, and μ_i° is taken as $(\Delta G_f^\circ)_i$, the Gibbs energy change for the standard formation reaction of species i. Combination of the above equations then gives

$$G^t = \sum_i n_i (\Delta G_f^\circ)_i + nRT \ln (P/P^\circ) + RT \sum_i n_i \ln n_i - RT \sum_i n_i \ln n + RT \sum_i n_i \ln \hat{\phi}_i$$
$$(7.170)$$

where $n \equiv \Sigma_i \, n_i$ and where the y_i have been temporarily replaced by the n_i/n.

We must now find the set $\{n_i\}$ which minimizes the function (7.170) at constant T and P, subject to the constraints of the material balances. The standard solution to this type of problem is through the method of Lagrange's undetermined multipliers. This requires that the constraint equations be incorporated in the expression for G^t. The material-balance equations describe the conservation of the numbers of moles of the chemical elements composing the system. Let A_k ($k = 1, 2, \ldots, w$) be the number of moles of the kth element, as determined by the initial constitution of the system. Let a_{ik} be the number of atoms of the kth element present in each molecule of chemical species i. Then, for each element k, $\Sigma_i \, n_i a_{ik} = A_k$, or

$$A_k - \sum_i n_i a_{ik} = 0 \qquad (k = 1, 2, \ldots, w) \qquad (7.171)$$

The kth equation (7.171) is multiplied by an undetermined constant λ_k, and the expression is then summed over all k, giving

$$\sum_k \lambda_k \left(A_k - \sum_i n_i a_{ik} \right) = 0$$

Since this quantity is zero, it may be added to the right-hand side of (7.170):

$$G^t = \sum_i n_i (\Delta G_f^\circ)_i + nRT \ln (P/P^\circ) + RT \sum_i n_i \ln n_i - RT \sum_i n_i \ln n$$
$$+ RT \sum_i n_i \ln \hat{\phi}_i + \sum_k \lambda_k \left(A_k - \sum_i n_i a_{ik} \right) \qquad (7.172)$$

The final step is to differentiate (7.172) with respect to each mole number n_i (bearing in mind that $n = \Sigma \, n_i$) to form the derivatives $(\partial G^t/\partial n_i)_{T,P,n_j}$, which are then set equal to zero. This leads to

$$(\Delta G_f^\circ)_i + RT \ln (P/P^\circ) + RT \ln y_i + RT \ln \hat{\phi}_i - \sum_k \lambda_k a_{ik} = 0 \qquad (i = 1, 2, \ldots, m)$$
$$(7.173)$$

The m equations (7.173), plus the w equations (7.171) with ny_i replacing n_i, plus the normalizing condition $\Sigma_i \, y_i = 1$, suffice to determine the $m + w + 1$ unknowns y_i, λ_k, and n. However, (7.173) was derived on the presumption that all the $\hat{\phi}_i$ are known. If the phase is an ideal gas, then each $\hat{\phi}_i$ is indeed known, and is unity, but for real gases each $\hat{\phi}_i$ is a function of the various y_i, which are to be determined. Thus an iterative procedure is indicated, which is initiated by setting the $\hat{\phi}_i$ equal to unity. Solution of the equations then provides a preliminary set of y_i. For low pressures or high temperatures this result is usually quite adequate. Where it is not, an equation of state is used together with the calculated y_i to give a new and more nearly correct set of $\hat{\phi}_i$ for use in (7.173), and a new set of y_i is determined. The process is repeated until successive iterations produce no significant change in the y_i. All calculations are done by computer, including calculation of the $\hat{\phi}_i$ by equations such as (7.58).

It is important to note that in the minimization procedure the question of what chemical reactions are involved never enters directly into any of the equations. However, the choice of a set of species is entirely equivalent to the choice of a set of independent reactions among the species. In any event, a set of species or a set of independent reactions must be assumed, and different assumptions in general produce different results.

EXAMPLE 7.27 Calculate at 1000 K and 1 atm the equilibrium composition of a system containing the gaseous species CH_4, H_2O, CO, CO_2, and H_2, if in the initial unreacted state there are present 2 mol of CH_4 and 3 mol of H_2O. Values of ΔG_f° ($P^\circ = 1$ atm) at 1000 K are (in $kJ \cdot mol^{-1}$): CH_4, 19.300; H_2O, −192.720; CO, −200.715; CO_2, −396.110; H_2, 0.

The material-balance data are presented in Table 7-5. At 1 atm and 1000 K the assumption of ideal gases should be fully justified, and the $\ln \hat{\phi}_i$ terms of (7.173) can be omitted. In addition, $\ln (P/P^\circ)$ is zero, so that the five equations resulting from (7.173) are (with $RT = 8.314 \, kJ \cdot mol^{-1}$)

$$
\begin{aligned}
\text{CH}_4: &\quad 19.300 + 8.314 \ln y_{CH_4} - \lambda_C - 4\lambda_H = 0 \\
\text{H}_2\text{O}: &\quad -192.720 + 8.314 \ln y_{H_2O} - 2\lambda_H - \lambda_O = 0 \\
\text{CO}: &\quad -200.715 + 8.314 \ln y_{CO} - \lambda_C - \lambda_O = 0 \\
\text{CO}_2: &\quad -396.110 + 8.314 \ln y_{CO_2} - \lambda_C - 2\lambda_O = 0 \\
\text{H}_2: &\quad 8.314 \ln y_{H_2} - 2\lambda_H = 0
\end{aligned}
\tag{1}
$$

Replacing n_i by $y_i n$ in (7.171), we get the three material balances:

$$
\begin{aligned}
\text{C:} &\quad y_{CH_4} + y_{CO} + y_{CO_2} = 2/n \\
\text{H:} &\quad 4y_{CH_4} + 2y_{H_2O} + 2y_{H_2} = 14/n \\
\text{O:} &\quad y_{H_2O} + y_{CO} + 2y_{CO_2} = 3/n
\end{aligned}
\tag{2}
$$

In addition, we have

$$
y_{CH_4} + y_{H_2O} + y_{CO} + y_{CO_2} + y_{H_2} = 1
\tag{3}
$$

Solution of these nine equations yields the results:

$$
\begin{aligned}
y_{CH_4} &= 0.0199 \\
y_{H_2O} &= 0.0995 \qquad & n &= 8.656 \text{ mol} \\
y_{CO} &= 0.1753 \qquad & \lambda_C &= -6.626 \, kJ \cdot mol^{-1} \\
y_{CO_2} &= 0.0359 \qquad & \lambda_O &= -208.680 \, kJ \cdot mol^{-1} \\
y_{H_2} &= 0.6694 \qquad & \lambda_H &= -1.671 \, kJ \cdot mol^{-1} \\
\overline{\Sigma \, y_i} &= 1.0000
\end{aligned}
$$

Table 7-5

Species i	Element k		
	Carbon	Oxygen	Hydrogen
	$A_k \equiv$ number of moles of k in system		
	$A_C = 2$	$A_O = 3$	$A_H = 14$
	$a_{ik} \equiv$ number of atoms of k per molecule of i		
CH_4	$a_{CH_4,C} = 1$	$a_{CH_4,O} = 0$	$a_{CH_4,H} = 4$
H_2O	$a_{H_2O,C} = 0$	$a_{H_2O,O} = 1$	$a_{H_2O,H} = 2$
CO	$a_{CO,C} = 1$	$a_{CO,O} = 1$	$a_{CO,H} = 0$
CO_2	$a_{CO_2,C} = 1$	$a_{CO_2,O} = 2$	$a_{CO_2,H} = 0$
H_2	$a_{H_2,C} = 0$	$a_{H_2,O} = 0$	$a_{H_2,H} = 2$

Solved Problems

SOLUTION THERMODYNAMICS (Sections 7.1 through 7.6)

7.1 Apply the Gibbs/Duhem equation (7.11) to a binary solution at constant T and P, and give graphical interpretations of the results.

For a binary system at constant T and P, (7.11) becomes

$$x_1 \, d\overline{M}_1 + x_2 \, d\overline{M}_2 = 0 \qquad (1)$$

However, the pure-component properties M_i are functions of T and P only, so that (1) may be rewritten as

$$x_1 \, d(\overline{M}_1 - M_1) + x_2 \, d(\overline{M}_2 - M_2) = 0 \qquad (2)$$

Dividing (1) by dx_1 and noting that $x_2 = 1 - x_1$, we obtain

$$x_1 \frac{d\overline{M}_1}{dx_1} = -(1 - x_1) \frac{d\overline{M}_2}{dx_1} \qquad (3)$$

Alternatively, we may divide (2) by dx_1, giving

$$x_1 \frac{d(\overline{M}_1 - M_1)}{dx_1} = -(1 - x_1) \frac{d(\overline{M}_2 - M_2)}{dx_1} \qquad (4)$$

Equations (3) and (4) relate the composition derivatives of the $\overline{M}_i$ (or the $\overline{M}_i - M_i$) at constant T and P. For example, the slopes of the two representative ($\overline{M}_i - M_i$) curves shown in Fig. 7-24 are of opposite sign at every value of x_1, as required by (4).

The composition extremes, $x_1 = 0$ and $x_1 = 1$, are of special interest. If $d\overline{M}_1/dx_1$ is finite at $x_1 = 0$; then, from (3) and (4),

$$\frac{d\overline{M}_2}{dx_1} = \frac{d(\overline{M}_2 - M_2)}{dx_1} = 0 \qquad (x_1 = 0)$$

Similarly, if $d\overline{M}_2/dx_2$ is finite at $x_2 = 0$ (i.e., if $d\overline{M}_2/dx_1$ is finite at $x_1 = 1$), then

$$\frac{d\overline{M}_1}{dx_1} = \frac{d(\overline{M}_1 - M_1)}{dx_1} = 0 \qquad (x_1 = 1)$$

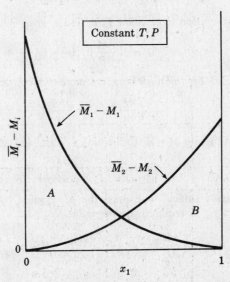

Fig. 7-24

The limiting derivatives $d\overline{M}_i/dx_i$ at $x_i = 0$ *are* finite for nonelectrolyte solutions, and the last two equations require that the $(\overline{M}_i - M_i)$ curves become tangent to the composition axis at $x_i = 1$. The two curves in Fig. 7-24 conform to this requirement.

A further important property of the $(\overline{M}_i - M_i)$ plot follows from (2). Consider the property change of mixing ΔM defined by (7.89):

$$\Delta M = M - \sum x_i M_i = x_1(\overline{M}_1 - M_1) + x_2(\overline{M}_2 - M_2)$$

The total differential of ΔM is

$$
\begin{aligned}
d(\Delta M) &= x_1 \, d(\overline{M}_1 - M_1) + (\overline{M}_1 - M_1) \, dx_1 + x_2 \, d(\overline{M}_2 - M_2) + (\overline{M}_2 - M_2) \, dx_2 \\
&= [(\overline{M}_1 - M_1) - (\overline{M}_2 - M_2)] \, dx_1
\end{aligned}
$$

where the second equality follows from $dx_2 = -dx_1$ and (2). Now $\Delta M = 0$ at $x_1 = 0$ and at $x_1 = 1$; as a result, we obtain on integrating the last equation from $x_1 = 0$ to $x_1 = 1$:

$$\int_0^1 [(\overline{M}_1 - M_1) - (\overline{M}_2 - M_2)] \, dx_1 = 0 \tag{5}$$

Condition (5) requires that the net area under a plot of $[(\overline{M}_1 - M_1) - (\overline{M}_2 - M_2)]$ versus x_1 be zero; in terms of Fig. 7-24, areas A and B must be numerically equal.

Equations (3) through (5) are examples of *thermodynamic consistency requirements* for the partial molar properties $\overline{M}_i$. Such requirements constitute necessary, but not sufficient, conditions for the validity of thermodynamic data.

7.2 The conventional definition (7.1) of a partial property $\overline{M}_i$ presumes that T, P, and composition are natural independent variables. This is indeed the case with experimental data, or with a volume-explicit equation of state. With a pressure-explicit EOS, however, temperature, molar density, and composition are the favored independent variables, and partial properties $\tilde{M}_i$ are defined as

$$\tilde{M}_i \equiv \left[\frac{\partial(nM)}{\partial n_i} \right]_{T,\rho,n_j} \tag{1}$$

Show how $\overline{M}_i$ and $\tilde{M}_i$ are related.

We start with (7.2):

$$d(nM) = n\left(\frac{\partial M}{\partial T}\right)_{P,x} dT + n\left(\frac{\partial M}{\partial P}\right)_{T,x} dP + \sum \overline{M}_i \, dn_i$$

Division by dn_i and restriction to constant T, ρ, and n_j gives immediately

$$\tilde{M}_i = n\left(\frac{\partial M}{\partial P}\right)_{T,x} \left(\frac{\partial P}{\partial n_i}\right)_{T,\rho,n_j} + \overline{M}_i \tag{2}$$

From the EOS, $P = Z\rho RT = (nZ)(\rho RT/n)$, the mole-number derivative of P is found as

$$\left(\frac{\partial P}{\partial n_i}\right)_{T,\rho,n_j} = \frac{\rho RT}{n}(\tilde{Z}_i - Z) \tag{3}$$

where $\tilde{Z}_i$ is given by (1) with $M = Z$. Combining (2) and (3) and solving for $\overline{M}_i$ gives us

$$\overline{M}_i = \tilde{M}_i - \rho RT\left(\frac{\partial M}{\partial P}\right)_{T,x}(\tilde{Z}_i - Z) \tag{4}$$

Equation (4) usually suffices to express the $\overline{M}_i$ implied by a P-explicit EOS. When an *explicit* expression for $(\partial M/\partial P)_{T,x}$ is required, the formula

$$\left(\frac{\partial M}{\partial P}\right)_{T,x} = \frac{1}{RT}\left(\frac{\partial M}{\partial \rho}\right)_{T,x}\left[Z + \rho\left(\frac{\partial Z}{\partial \rho}\right)_{T,x}\right]^{-1}$$

may be used.

7.3   Prove that all δ_{jk} vanish [see (5.32)] for an ideal solution of real gases.

Substitution of coefficients from (5.23) allows the virial equation (5.21) to be written in the form

$$Z = 1 + \frac{BP}{RT} + \frac{(C - B^2)}{(RT)^2} P^2 + \cdots$$

An expression for $\ln \phi$ then follows from application of (7.36):

$$\ln \phi = \int_0^P (Z - 1) \frac{dP}{P} \qquad \text{(constant } T, x)$$

$$= \frac{BP}{RT} + (\ \ )P^2 + (\ \ )P^3 + \cdots \tag{1}$$

where the coefficients represented by parentheses are functions of temperature and composition only. For a pure component i, (1) gives

$$\ln \phi_i = \frac{B_{ii}P}{RT} + (\ \ )P^2 + (\ \ )P^3 + \cdots \tag{2}$$

An expression for the fugacity coefficient of component i in solution is obtained from (1) as indicated in Example 7.7. The extension of (7.58) of that example is

$$\ln \hat{\phi}_i = \frac{B_{ii}P}{RT} + \left[\frac{1}{2RT} \sum_j \sum_k x_j x_k (2\delta_{ji} - \delta_{jk}) \right] P + \{\ \} P^2 + \{\ \} P^3 + \cdots \tag{3}$$

where the coefficients represented by braces are functions of temperature and composition only. By (7.77),

$$\ln \gamma_i = \ln \hat{\phi}_i - \ln \phi_i \tag{7.77}$$

Substitution of (2) and (3) into this equation gives

$$\ln \gamma_i = \left[\frac{1}{2RT} \sum_j \sum_k x_j x_k (2\delta_{ji} - \delta_{jk}) \right] P + [\ ] P^2 + [\ ] P^3 + \cdots \tag{4}$$

where, again, the coefficients represented by brackets are functions of temperature and composition only.

For an ideal solution, $\gamma_i = 1$ for every component i, at all temperatures, pressures, and compositions. Thus the coefficient in brackets for each term of the expansion (4) must be identically zero; in particular,

$$\sum_j \sum_k x_j x_k (2\delta_{ji} - \delta_{jk}) = 0 \qquad (i = 1, 2, \ldots, m) \tag{5}$$

The only way (5) can hold for arbitrary composition is for δ_{jk} (and hence δ_{ji}) to be zero for all possible pairs j, k.

7.4   Derive (7.50), the appropriate expression for $\ln \hat{\phi}_i$ given a pressure-explicit equation of state.

We have

$$\ln \hat{\phi}_i = \overline{(\ln \phi)_i} \tag{7.46}$$

and, for a P-explicit EOS,

$$\ln \phi = \int_0^\rho (Z - 1) \frac{d\rho}{\rho} + Z - 1 - \ln Z \qquad \text{(constant } T, x) \tag{7.48}$$

Now, the partial property $\widetilde{(\ln \phi)}_i$ is easily found from its definition: multiply (7.48) by n and then differentiate partially with respect to n_i, holding T and ρ constant. The result is

$$\widetilde{(\ln \phi)}_i = \int_0^\rho (\tilde{Z}_i - 1) \frac{d\rho}{\rho} + \left(\frac{Z - 1}{Z} \right) \tilde{Z}_i - \ln Z \tag{1}$$

By (4) of Problem 7.2 we have, with $M = \ln \phi$,

$$\overline{(\ln \phi)_i} = (\widetilde{\ln \phi})_i - \rho RT \left(\frac{\partial \ln \phi}{\partial P} \right)_{T,x} (\tilde{Z}_i - Z) \tag{2}$$

Finally, differentiation of (7.36) yields

$$\left(\frac{\partial \ln \phi}{\partial P} \right)_{T,x} = \frac{1}{P} (Z - 1) = \frac{1}{\rho RT} \left(\frac{Z-1}{Z} \right) \tag{3}$$

Combination of (7.46) with (1), (2), and (3) gives us, on simplification,

$$\ln \hat{\phi}_i = \int_0^\rho (\tilde{Z}_i - 1) \frac{d\rho}{\rho} + Z - 1 - \ln Z \qquad (\text{constant } T, x)$$

which is (7.50).

7.5   A cleaning solution is to be manufactured from equal masses of acetone (a) and dichloro-methane (d), both at 298 K. If these components are mixed adiabatically at 1 bar, with negligible stirring work, what is the temperature of the cleaning solution formed?
Data: Heat capacities of the pure components at 298 K and 1 bar:

$$(C_P)_a = 2.173 \text{ kJ} \cdot \text{kg}^{-1} \cdot \text{K}^{-1} \qquad (C_P)_d = 1.193 \text{ kJ} \cdot \text{kg}^{-1} \cdot \text{K}^{-1}$$

Heats of mixing for the equal-mass solution at 1 bar:

T/K	$\Delta H / \text{kJ} \cdot \text{kg}^{-1}$
293	12.468
298	12.380
303	12.292

For purposes of calculation we consider the process to consist of two steps, as indicated in Fig. 7-25. Since the process may be considered mechanically reversible, application of the first law to the overall process gives simply (see Example 1.7):

$$\Delta H^t_{13} = \Delta H^t_{12} + \Delta H^t_{23} = 0 \tag{1}$$

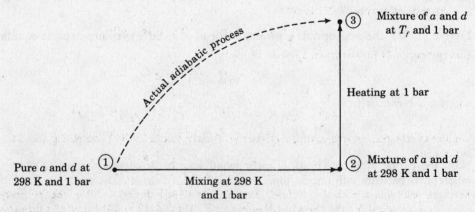

Fig. 7-25

Step 1-2 is just the standard mixing process of Example 7.11; thus,

$$\Delta H'_{12} = (m_a + m_d)\, \Delta H \tag{2}$$

where ΔH is the heat of mixing per unit mass of solution.

The enthalpy change for step 2-3 is given by

$$\Delta H'_{23} = (m_a + m_d) \int_{298}^{T_f} C_P\, dT = (m_a + m_d) C_P (T_f - 298) \tag{3}$$

where C_P is the specific heat capacity of the mixture, assumed independent of T for the small temperature range considered here. This heat capacity is found by application of (7.89):

$$C_P = x_a (C_P)_a + x_d (C_P)_d + \Delta C_P \tag{4}$$

The heat-capacity change of mixing, ΔC_P, is related to ΔH in the same way that C_P is related to H:

$$\Delta C_P = \left(\frac{\partial(\Delta H)}{\partial T} \right)_{P,x} \tag{5}$$

Combination of (3), (4), and (5) yields

$$\Delta H'_{23} = \left[m_a (C_P)_a + m_d (C_P)_d + (m_a + m_d) \left(\frac{\partial(\Delta H)}{\partial T} \right)_{P,x} \right] (T_f - 298) \tag{6}$$

Substitution of (2) and (6) into (1) and solution for T_f gives

$$T_f = 298 - \frac{\Delta H}{x_a (C_P)_a + x_d (C_P)_d + (\partial(\Delta H)/\partial T)_{P,x}} \tag{7}$$

All the quantities in (7), except for the derivative, are given. The derivative can be approximated from the data for ΔH at 293 and 303 K:

$$\left(\frac{\partial(\Delta H)}{\partial T} \right)_{P,x} \approx \frac{-12.292 - (-12.468)}{303 - 293} = 0.0176 \text{ kJ} \cdot \text{kg}^{-1} \cdot \text{K}^{-1}$$

Substituting numerical values into (7), we get the required answer:

$$T_f = 298 - \frac{-12.380}{(0.5)(2.173) + (0.5)(1.193) + 0.0176} = 305.3 \text{ K}$$

The adiabatic mixing process considered here is seen to cause a temperature rise of 7.3 K. We note that the sign of the temperature change is determined by the sign of ΔH. In the present case ΔH is negative (*exothermic* mixing). Had ΔH been positive (*endothermic* mixing), the temperature would have decreased. We note further that the temperature change is independent of the amount of 50% solution formed.

7.6 Develop an expression relating excess property M^E and residual property M^R.

By the definitions (7.83) and (4.22),

$$M^E \equiv M - M^{id} \qquad M^R \equiv M - M^{ig}$$

Elimination of M between these equations gives

$$M^E = M^R - (M^{id} - M^{ig}) \tag{1}$$

But, from (7.89),

$$M^{id} = \sum x_i M_i + \Delta M^{id} \qquad M^{ig} = \sum x_i M_i^{ig} + \Delta M^{ig} \tag{2}$$

where the ΔM are property changes of mixing. Substituting (2) in (1) and noting that $M_i - M_i^{ig} = M_i^R$, we obtain

$$M^E = M^R - \sum x_i M_i^R - (\Delta M^{id} - \Delta M^{ig}) \tag{3}$$

Ideal-gas mixtures *are* ideal solutions; according to (7.90) through (7.94), the property changes of mixing are identical for the two classes of mixtures. Thus, $\Delta M^{ig} = \Delta M^{id}$, and (3) becomes

$$M^E = M^R - \sum x_i M_i^R \tag{4}$$

Equation (*4*) is the desired result. It is used for determination of excess properties from a volume-explicit equation of state, and as an aid for explaining the qualitative behavior of excess properties of liquid mixtures.

7.7   Prove that the Gibbs energy change of mixing ΔG is *negative* for a homogeneous mixture.

The governing relation here is (*3.79*), which asserts that the total Gibbs energy G' assumes a minimum value for equilibrium states at specified T and P. For the present application, we consider the standard mixing process (Example 7.11), which by definition occurs at constant T and P. The total Gibbs energy for the unmixed species is

$$G'(\text{unmixed}) = \sum n_i G_i \qquad (1)$$

After mixing, G' for the solution is, by (*7.89*),

$$G'(\text{mixed}) = \sum n_i G_1 + n \, \Delta G \qquad (2)$$

If a homogeneous solution is in fact obtained on mixing, then our equilibrium criterion requires that

$$G'(\text{mixed}) < G'(\text{unmixed}) \qquad (3)$$

It follows from (*1*), (*2*), and (*3*) that $\Delta G < 0$.

7.8   The composition dependence of activity coefficients in real binary systems is only infrequently described by the symmetrical equations (*7.104*). One can generate more flexible expressions for $\ln \gamma_i$ by retaining more terms in the dimensionless G^E equations (*7.101*) and (*7.102*). Derive the two-parameter *Margules* and *van Laar* expressions for $\ln \gamma_i$ by letting $D = \cdots = 0$ and $D' = \cdots = 0$ in (*7.101*) and (*7.102*), respectively. Give graphical comparisons of the equations for representative values of the parameters.

Solution of (*7.101*), with $D = \cdots = 0$, for G^E/RT yields

$$\frac{G^E}{RT} = Bx_1 x_2 + Cx_1 x_2 (x_1 - x_2)$$

from which

$$\frac{nG^E}{RT} = \frac{Bn_1 n_2}{n} + \frac{Cn_1 n_2 (n_1 - n_2)}{n^2}$$

Thus, by (*7.81*),

$$\ln \gamma_1 = \left[\frac{\partial (nG^E/RT)}{\partial n_1} \right]_{T,P,n_2} = \frac{Bn_2}{n} - \frac{Bn_1 n_2}{n^2} + \frac{C(2n_1 - n_2)n_2}{n^2} - \frac{2C(n_1 - n_2)n_1 n_2}{n^3}$$

which becomes, in terms of mole fractions,

$$\ln \gamma_1 = x_2^2 [B - C(4x_2 - 3)]$$

The corresponding equation for $\ln \gamma_2$ is found analogously, and is

$$\ln \gamma_2 = x_1^2 [B + C(4x_1 - 3)]$$

The last two equations are the two-parameter Margules equations for a binary mixture. They are commonly put in a slightly different form through use of the substitutions $A_{12} \equiv B - C$, $A_{21} \equiv B + C$:

$$\ln \gamma_1 = A_{12} x_2^2 \left[1 + 2x_1 \left(\frac{A_{21}}{A_{12}} - 1 \right) \right]$$

$$\ln \gamma_2 = A_{21} x_1^2 \left[1 + 2x_2 \left(\frac{A_{12}}{A_{21}} - 1 \right) \right] \qquad (1)$$

The van Laar equations are found from (*7.102*) in a similar manner. Solution of (*7.102*), with $D' = \cdots = 0$, for G^E/RT yields

$$\frac{G^E}{RT} = \frac{x_1 x_2}{B' + C'(x_1 - x_2)}$$

from which we obtain

$$\frac{nG^E}{RT} = \frac{n_1 n_2}{nB' + C'(n_1 - n_2)}$$

and

$$\ln \gamma_1 = \left[\frac{\partial(nG^E/RT)}{\partial n_1} \right]_{T,P,n_2} = \frac{(B' - C')n_2^2}{[nB' + C'(n_1 - n_2)]^2} = \frac{(B' - C')x_2^2}{[B' - C'(2x_2 - 1)]^2}$$

Similarly, we find

$$\ln \gamma_2 = \frac{(B' + C')x_1^2}{[B' + C'(2x_1 - 1)]^2}$$

These are the van Laar equations. Making the substitutions

$$A'_{12} \equiv \frac{1}{B' - C'} \qquad A'_{21} \equiv \frac{1}{B' + C'}$$

we can write them as

$$\ln \gamma_1 = \frac{A'_{12} x_2^2}{\left[1 + \left(\dfrac{A'_{12}}{A'_{21}} - 1 \right) x_1 \right]^2}$$

$$\ln \gamma_2 = \frac{A'_{21} x_1^2}{\left[1 + \left(\dfrac{A'_{21}}{A'_{12}} - 1 \right) x_2 \right]^2}$$

(2)

The parameters A_{12} and A_{21} (or A'_{12} and A'_{21}) are respectively equal to $\ln \gamma_1^\infty$ and $\ln \gamma_2^\infty$, the logarithms of the infinite-dilution values of the activity coefficients. Thus one basis for comparison of (1) and (2) is provided if we set

$$A_{12} = A'_{12} \qquad \text{and} \qquad A_{21} = A'_{21}$$

equivalently, we can set

$$A_{12} = A'_{12} \qquad \text{and} \qquad A_{21}/A_{12} = A'_{21}/A'_{12}$$

Using the second pair of equations, we illustrate the behavior of the Margules and van Laar equations by specifying

$$A_{12} = A'_{12} = 0.5 \qquad \text{(fixed)}$$
$$A_{21}/A_{12} = A'_{21}/A'_{12} = 1.0, 1.5, 2.0, 2.5 \qquad \text{(four cases)}$$

In each case the Margules and van Laar equations yield the same limiting values of the activity coefficients at $x_1 = 0$ and $x_1 = 1$, but intermediate values depend on the form of the equation. Results for the Margules equations (1) are shown by Fig. 7-26, and for the van Laar equations (2) by Fig. 7-27. When $A_{21}/A_{12} = A'_{21}/A'_{12} = 1$, the two sets of equations (1) and (2) become identical, and in fact reduce to the symmetrical equations (7.104), with $B = 0.5$. However, as the ratio $A_{21}/A_{12} = A'_{21}/A'_{12}$ increases, differences between the sets of equations become apparent. In particular, the $\ln \gamma_2$ curves given by the van Laar equation approach their limiting values at $x_1 = 1$ with steeper slopes than do the corresponding curves given by the Margules equation. On the other hand, the $\ln \gamma_1$ curves given by the Margules equation show a much stronger dependence on the value of $A_{21}/A_{12} = A'_{21}/A'_{12}$ than do the corresponding curves determined from the van Laar equation. In fact, as seen from Fig. 7-26 the $\ln \gamma_1$ curves given by the Margules equation exhibit a maximum (with a corresponding minimum in the $\ln \gamma_2$ curve) for values of $A_{21}/A_{12} > 2$.

The use of the Margules and van Laar equations is in the fitting of experimental data for activity coefficients, and the different characteristics of the two equations provide for the accommodation of a considerable range of behavior.

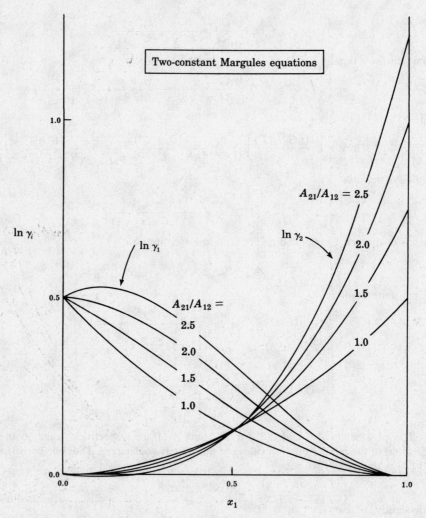

Fig. 7-26

7.9 Examine the T and P dependence of parameter B in (7.103) for solutions for which (a) $H^E = 0$, (b) $V^E = 0$, (c) $H^E = V^E = 0$.

(a) From (7.105),

$$\left(\frac{\partial B}{\partial T}\right)_P = 0 \quad \text{or} \quad B = B(P)$$

(b) From (7.107),

$$\left(\frac{\partial B}{\partial P}\right)_T = 0 \quad \text{or} \quad B = B(T)$$

(c) From (a) and (b), $B = \text{const}$.

7.10 Prove that if G^E for a binary mixture is a strictly concave or strictly convex function of composition, then $\ln \gamma_1$ and $\ln \gamma_2$ vary monotonically with composition.

Since $\ln \gamma_i$ is the partial property corresponding to G^E/RT, we have by (7.9) and (7.10)

$$RT \ln \gamma_1 = G^E + (1 - x_1)\frac{dG^E}{dx_1} \qquad RT \ln \gamma_2 = G^E - x_1\frac{dG^E}{dx_1}$$

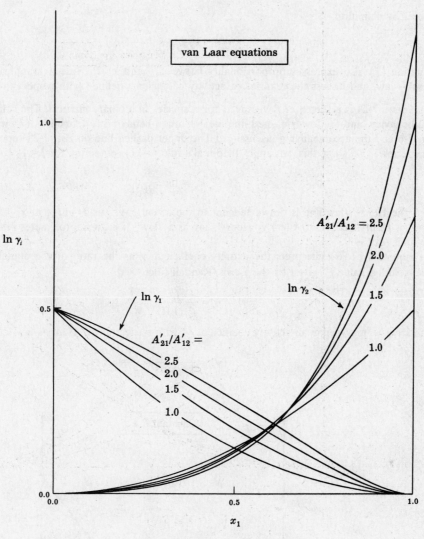

Fig. 7-27

Differentiation of these expressions with respect to x_1 gives

$$RT \frac{d(\ln \gamma_1)}{dx_1} = (1 - x_1) \frac{d^2 G^E}{dx_1^2} \qquad RT \frac{d(\ln \gamma_2)}{dx_1} = -x_1 \frac{d^2 G^E}{dx_1^2}$$

Thus if $d^2 G^E / dx_1^2$ has a single sign, then so do the composition derivatives of $\ln \gamma_1$ and $\ln \gamma_2$. This theorem is a particular consequence of the general results

$$\frac{d\overline{M}_1}{dx_1} = (1 - x_1) \frac{d^2 M}{dx_1^2} \qquad \frac{d\overline{M}_2}{dx_1} = -x_1 \frac{d^2 M}{dx_1^2}$$

7.11 The statement of ideal-solution behavior expressed by (7.73) has as its essential feature a simple proportionality between $\hat{f}_i$ and x_i:

$$\hat{f}_i^{\text{id}}(\text{LR}) = x_i f_i \tag{1}$$

Equation (1) is known as the *Lewis/Randall* (LR) *rule*; it is a "rule" (i.e., it approximates real behavior) only in the limit as $x_i \to 1$. An alternative definition of ideal-solution behavior is provided by *Henry's law* (HL):

$$\hat{f}_i^{\text{id}}(\text{HL}) = x_i k_i \tag{2}$$

where, by definition,

$$k_i \equiv \lim_{x_i \to 0} \frac{\hat{f}_i}{x_i} \qquad (3)$$

Equation (2) also expresses proportionality between $\hat{f}_i$ and x_i. Provide a graphical interpretation of Henry's law, and discuss the features of activity coefficients defined with respect to (2).

Figure 7-28 is a sketch of $\hat{f}_1$ versus x_1 for a species in a binary mixture. The solid curve represents real behavior, and the lower dashed line ideal-solution behavior as defined by (1) with $i = 1$. (See also Fig. 7-6 and the accompanying discussion.) The upper dashed line on Fig. 7-28 represents Henry's law for species 1. To show this, we apply l'Hôpital's rule to (3), obtaining for $i = 1$

$$k_1 \equiv \lim_{x_1 \to 0} \frac{d\hat{f}_1}{dx_1}$$

Thus the Henry's-law line is drawn tangent to the actual $\hat{f}_1$-versus-x_1 curve at $x_1 = 0$. The intercept at $x_1 = 1$ is Henry's constant. Clearly, Henry's law is a "law" (i.e., it approximates real behavior) only in the limit as $x_i \to 0$.

Equation (7.78) interprets the activity coefficient γ_i as the ratio of the actual fugacity $\hat{f}_i$ to the ideal-solution value $\hat{f}_i^{id}$ given by the Lewis/Randall rule:

$$\gamma_i = \frac{\hat{f}_i}{\hat{f}_i^{id}(\text{LR})} = \frac{\hat{f}_i}{x_i f_i} \qquad (4)$$

Similarly, we may define an activity coefficient γ_i^* with respect to the Henry's-law ideal-solution model:

$$\gamma_i^* = \frac{\hat{f}_i}{\hat{f}_i^{id}(\text{HL})} = \frac{\hat{f}_i}{x_i k_i} \qquad (5)$$

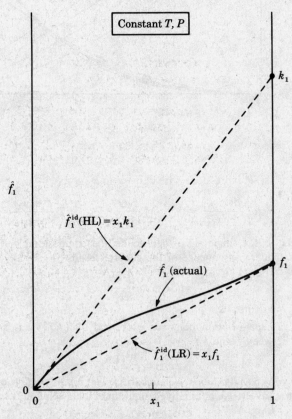

Fig. 7-28

We saw in Section 7.5 that for species i in a real solution,

$$\lim_{x_i \to 1} \gamma_i = 1$$

Analogous to this result, we have, by (5) and (3),

$$\lim_{x_i \to 0} \gamma_i^* = 1$$

Lewis/Randall activity coefficients and Henry's-law activity coefficients are related. By (4) and (3),

$$\gamma_i^\infty = \frac{k_i}{f_i} \qquad (6)$$

where γ_i^∞ is the LR activity coefficient at infinite dilution:

$$\gamma_i^\infty \equiv \lim_{x_i \to 0} \gamma_i$$

Eliminating the ratio $\hat{f}_i/x_i$ between (4) and (5), we obtain $\gamma_i^* = (f_i/k_i)\gamma_i$, which combines with (6) to give

$$\gamma_i^* = \frac{\gamma_i}{\gamma_i^\infty} \qquad (7)$$

The important concept here is that of *convertibility*: (6) shows how to convert f_i to k_i; (7) shows how to convert γ_i to γ_i^*. In both cases, the conversion factor is γ_i^∞.

PHASE EQUILIBRIA   (Sections 7.7 and 7.8)

7.12 Discuss the liquid/solid Tx diagrams shown in Fig. 7-29.

(a) Figure 7-29(a) is for a system in which the components form a continuous series of homogeneous *solid solutions*. The properties of such diagrams are analogous to those of vapor/liquid Tx diagrams of the type shown in Fig. 7-12. Thus, mixtures with Tx coordinates falling above the freezing curve ACD are homogeneous liquid solutions; similarly, states below the melting curve ABD are solid solutions. States of coexisting liquid and solid solutions are connected as shown by horizontal tie lines. Examples of pairs of compounds having this type of liquid/solid phase diagram are nitrogen/carbon monoxide, silver chloride/sodium chloride, and copper/nickel.

(b) Figure 7-29(b) represents a system which exhibits a eutectic, but in which the pure components are completely insoluble in the solid phase. This type of diagram is a limiting case of the kind shown in Fig. 7-18, for which partial solubility obtains in the solid phase. Thus, in Fig. 7-29(b), the regions $AECA$ and $BEDB$ are two-phase liquid/solid regions. In $AECA$, liquid mixtures with compositions given by the curve AE are in equilibrium with crystals of pure solid 2, while in $BEDB$ liquid mixtures with compositions given by BE are in equilibrium with crystals of pure solid 1. Below the melting line CED, pure solid 1 is in equilibrium with pure solid 2. *Complete* solid/solid immiscibility is an idealization which never strictly obtains for real systems; however, some systems approach this behavior.

(c) Figure 7-29(c) depicts a system for which a third chemical species, represented by the vertical line $DFEK$, is formed in the solid phase and for which all species are completely insoluble as solids. If the composition coordinate is mole fraction, then the chemical formula of the compound can be directly determined from the value of x_1 for the line $DFEK$. In the present case the compound is found at mole fraction 0.5, so that the chemical formula is MN (or a multiple of MN, say M_lN_l), where M and N represent species 1 and 2.

Figure 7-29(c) can be considered a combination of two diagrams of the type of Fig. 7-29(b) placed side by side, with the properties of the two halves of the diagram (to the left and to the right of $DFEK$) separately similar to those of Fig. 7-29(b). Thus, points C and G represent eutectic mixtures of N (component 2) with MN, and M (component 1) with MN, respectively. Because of the formation of the intermediate compound, pure solid M and N cannot coexist as equilibrium phases; within the area $BCEKLB$, pure solid N is in equilibrium with solid MN, while pure solid M coexists with solid MN in the area $EFGHJKE$. As with Fig. 7-29(b), Fig. 7-29(c) represents a type of limiting behavior which is only approximated by real systems.

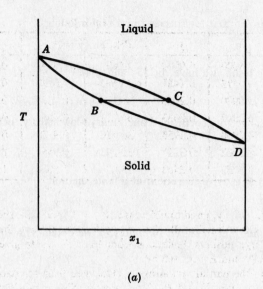

(a)

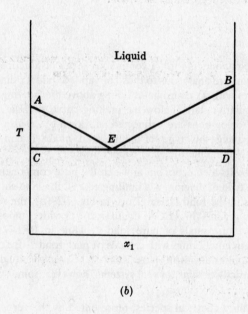

(b)

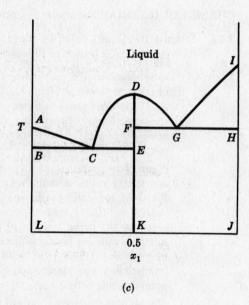

(c)

Fig. 7-29

7.13 Table 7-6 presents experimental VLE data for the isopropanol (1)/benzene (2) system at 45 °C [Brown, I., *Aust. J. Chem.*, **9**, 364 (1956)]. (*a*) Plot the dew- and bubblepoint curves versus mole fraction of isopropanol. Calculate and plot versus x_1 on the same diagram the vapor-phase partial pressures P_1 and P_2. Compare these curves with the bubblepoint and partial-pressure curves given by Raoult's law. (*b*) Derive values of $\ln \gamma_1$ and $\ln \gamma_2$ from the data and plot them against x_1. Plot on the same graph the G^E/x_1x_2RT curve, and show for comparison the G^E/x_1x_2RT curve one would obtain from the two-parameter Margules equation (Problem 7.8) if the constants were determined from experimental activity coefficients at infinite dilution.

Table 7-6

x_1	y_1	P/kPa	x_1	y_1	P/kPa
0.0000	0.0000	29.829	0.5504	0.3692	35.319
0.0472	0.1467	33.633	0.6198	0.3951	34.577
0.0980	0.2066	35.214	0.7096	0.4378	33.023
0.2047	0.2663	36.271	0.8073	0.5107	30.282
0.2960	0.2953	36.450	0.9120	0.6658	25.235
0.3862	0.3211	36.292	0.9655	0.8252	21.305
0.4753	0.3463	35.928	1.0000	1.0000	18.138

(a) The dewpoint $(P - y_1)$ and bubblepoint $(P - x_1)$ curves are shown on Fig. 7-30; the corresponding bubblepoint curve for Raoult's law is given as the straight line $P - x_1(\text{RL})$. This system exhibits sufficiently large positive deviations from Raoult's law to give a maximum-pressure azeotrope at about $x_1 = 0.29$ and $P = 36.465$ kPa.

Values of the partial pressures are calculated from the data by the definition $P_i = y_i P$, and are shown as the curves P_1 and P_2. The corresponding curves for Raoult's law are the straight lines P_1 (RL) and P_2 (RL).

(b) For these low pressures, the approximate equation (7.113) is entirely adequate for description of the VLE of the system. Solving for γ_i and taking logarithms, we obtain

$$\ln \gamma_i = \ln \left(\frac{y_i P}{x_i P_i^{\text{sat}}} \right) \tag{1}$$

Equation (1) is used to reduce the data to $\ln \gamma_1$ and $\ln \gamma_2$ values for the liquid phase. The value of $G^E/x_1 x_2 RT$ corresponding to each of these pairs of $\ln \gamma_i$ values is calculated from

$$\frac{G^E}{x_1 x_2 RT} = \frac{\ln \gamma_1}{x_2} + \frac{\ln \gamma_2}{x_1} \tag{2}$$

where (2) is obtained from (7.82).

We apply these equations to the data for $x_1 = 0.4753$. The pure-component vapor pressures are equal to the total pressures at $x_1 = 0.0$ and $x_1 = 1.0$; thus $P_1^{\text{sat}} = 18.138$ kPa and $P_2^{\text{sat}} = 29.829$ kPa. Equation (1) then gives

$$\ln \gamma_1 = \ln \left(\frac{y_1 P}{x_1 P_1^{\text{sat}}} \right) = \ln \left(\frac{0.3463 \times 35.928}{0.4753 \times 18.138} \right) = 0.3669$$

$$\ln \gamma_2 = 0.4059$$

Substitution of these values into (2) then gives the corresponding value of $G^E/x_1 x_2 RT$:

$$\frac{G^E}{x_1 x_2 RT} = \frac{0.3669}{1 - 0.4753} + \frac{0.4059}{0.4753} = 1.553$$

The complete curves for $\ln \gamma_1$, $\ln \gamma_2$, and $G^E/x_1 x_2 RT$ are plotted on Fig. 7-31. It can be shown that the limiting values of $G^E/x_1 x_2 RT$ as $x_1 \to 0$ and $x_1 \to 1$ are $\ln \gamma_1^\infty$ and $\ln \gamma_2^\infty$, respectively. Thus we find by extrapolation that $\ln \gamma_1^\infty = 2.18$ and $\ln \gamma_2^\infty = 1.44$. The $\ln \gamma_i$ curves, which are more difficult to extrapolate, are drawn so as to yield these intercepts.

According to the results of Problem 7.8, the Margules equation gives

$$G^E/x_1 x_2 RT = A_{12} + (A_{21} - A_{12})x_1$$

where A_{12} and A_{21} are the intercepts with the $G^E/x_1 x_2 RT$ axes at $x_1 = 0$ and $x_1 = 1$. If these intercepts are made equal to the experimental values of $\ln \gamma_1^\infty$ and $\ln \gamma_2^\infty$, respectively, then the dashed straight line shown on Fig. 7-31 results. Clearly, the two-parameter Margules equation is incapable of accurately representing the liquid-phase activity coefficients for this system; for even if other values were chosen for A_{12} and A_{21}, a straight line would always result for the $G^E/x_1 x_2 RT$ curve.

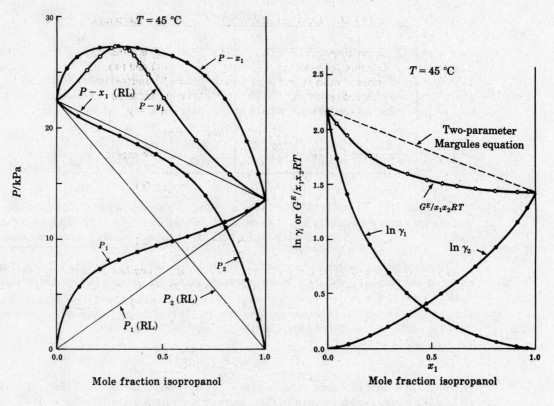

Fig. 7-30   Fig. 7-31

7.14 Consider the case of binary VLE at low pressures. (a) Determine an expression for the *relative volatility* α_{12} of species 1 to species 2; by definition,

$$\alpha_{12} \equiv \frac{K_1}{K_2} = \frac{y_1/x_1}{y_2/x_2} \tag{1}$$

where the K_i are the equilibrium K-values. (b) Use the result of part (a) to develop a procedure for checking for azeotrope formation in a binary system.

(a)   For low-pressure VLE, we assume the validity of the approximate equation (7.113):

$$\frac{y_i}{x_i} = \frac{\gamma_i P_i^{\text{sat}}}{P} \qquad (i = 1, 2)$$

Thus, by (1),

$$\alpha_{12} = \frac{\gamma_1 P_1^{\text{sat}}}{\gamma_2 P_2^{\text{sat}}} \tag{2}$$

Consider binary *isothermal* VLE. Since the P_i^{sat} depend only on T, variations in α_{12} result solely from the composition dependence of the activity coefficients. Normally, the ratio γ_1/γ_2 varies monotonically with x_1, and thus by (2) extreme values of α_{12} occur at the composition limits $x_1 = 0$ and $x_1 = 1$. At these limits, $\gamma_1 = \gamma_1^{\infty}$ and $\gamma_2 = 1$ (for $x_1 = 0$), and $\gamma_1 = 1$ and $\gamma_2 = \gamma_2^{\infty}$ (for $x_1 = 1$). Thus, by (2),

$$\alpha_{12}(x_1 = 0) = \frac{\gamma_1^{\infty} P_1^{\text{sat}}}{P_2^{\text{sat}}} \tag{3}$$

and

$$\alpha_{12}(x_1 = 1) = \frac{P_1^{\text{sat}}}{\gamma_2^{\infty} P_2^{\text{sat}}} \tag{4}$$

Equations (3) and (4) also apply for isobaric VLE, but here the temperatures are different at $x_1 = 0$ and at $x_1 = 1$.

(b) For "normal" (nonazeotropic) VLE, the relative volatility α_{12} is either greater than unity or less than unity over the entire composition range. According to (3) and (4), such behavior is expected when the vapor-pressure ratio is sufficiently large (or small) and/or when the γ_i^∞ do not differ greatly from unity. *Azeotropy* occurs when α_{12} becomes unity for some composition $0 \leq x_1 \leq 1$. To see this, recall that an azeotrope is defined by the condition $x_i = y_i$ $(i = 1, 2)$, whence by (1)

$$\alpha_{12} = 1 \qquad \text{(azeotrope)} \tag{5}$$

Sufficient conditions for (5) follow from (3) and (4). Thus, we expect azeotropy if either

$$\frac{\gamma_1^\infty P_1^{sat}}{P_2^{sat}} \geq 1 \qquad \text{and} \qquad \frac{P_1^{sat}}{\gamma_2^\infty P_2^{sat}} \leq 1 \tag{6}$$

or

$$\frac{\gamma_1^\infty P_1^{sat}}{P_2^{sat}} \leq 1 \qquad \text{and} \qquad \frac{P_1^{sat}}{\gamma_2^\infty P_2^{sat}} \geq 1 \tag{7}$$

The tests suggested by (6) and (7) are easily applied if one has available estimates for the activity coefficients at infinite dilution. Consider the cyclohexane (1)/benzene (2) system treated in Examples 7.13 and 7.17. At 40 °C, $P_1^{sat} = 24.6$ kPa and $P_2^{sat} = 24.4$ kPa. From Example 7.13,

$$\gamma_1^\infty = \gamma_2^\infty = \exp 0.458 = 1.58$$

and thus by (3) and (4)

$$\alpha_{12}(x_1 = 0) = \frac{(1.58)(24.6)}{24.4} = 1.59 > 1$$

$$\alpha_{12}(x_1 = 1) = \frac{24.6}{(1.58)(24.4)} = 0.638 < 1$$

Hence α_{12} passes through unity for some composition $0 < x_1 < 1$, and (as demonstrated by direct calculation in Example 7.17) an azeotrope does occur. Note that the present procedure does not give the azeotrope *coordinates*; it merely tells us whether an azeotrope exists. Note also that azeotropy is practically inevitable for systems like cyclohexane/benzene, where the pure-species vapor pressures are nearly equal; in any case where the vapor pressures are similar, only modest liquid-phase nonidealities are required to satisfy the inequalities of (6) or (7).

7.15 Develop equations for binary VLE when species 1 is a supercritical and species 2 a subcritical species. Specialize the equations to the case of low-pressure VLE with species 1 very sparingly soluble in the liquid phase.

Species 2 is treated by the gamma/phi approach as developed in Section 7.8. Thus by (7.110) we write

$$y_2 \hat{\phi}_2 P = x_2 \gamma_2 f_2 \tag{1}$$

Equation (7.110) is awkward to apply for species 1, because by assumption $T > T_{c1}$; at such conditions species 1 cannot exist as a pure liquid and therefore the liquid-phase fugacity f_1 is a hypothetical quantity. It is convenient in this case to apply the Henry's-law concepts of Problem 7.11 to species 1. By (5) of Problem 7.11, we write for species 1 in the liquid phase $\hat{f}_1^l = x_1 \gamma_1^* k_1$. Substituting this expression into (7.108) and employing (7.30) for $\hat{f}_1^v$, we obtain

$$y_1 \hat{\phi}_1 P = x_1 \gamma_1^* k_1 \tag{2}$$

Equation (2) is the Henry's-law analog of (1). Quantity k_1 is Henry's constant and γ_1^* is the activity coefficient of species 1 defined with respect to the Henry's-law statement of ideal-solution behavior: see the discussion of Problem 7.11. Equations (1) and (2) together constitute a rigorous formulation of the VLE problem posed. Equation (2) is subject to simplification and approximation along the lines discussed with respect to (7.110) in Section 7.8.

If the pressure is sufficiently low, then $\hat{\phi}_1$ and $\hat{\phi}_2$ are approximately unity. Moreover, for low pressure, $f_2 \approx P_2^{\text{sat}}(T)$ and $k_1 \approx k_1(T)$. (Generally, k_1 depends on P as well as on T.) Under these conditions, (1) and (2) become

$$y_2 P = x_2 \gamma_2 P_2^{\text{sat}} \tag{3}$$

$$\gamma_1 P = x_1 \gamma_1^* k_1 \tag{4}$$

If the liquid phase is highly concentrated in species 2, then $\gamma_2 \approx 1$ and $\gamma_1^* \approx 1$. These equations *do not* presume that the liquid phase is an ideal solution; they merely reflect the limiting behavior of the activity coefficients defined with respect to the Lewis/Randall and Henry's-law conventions, respectively (see Problem 7.11). For conditions of high dilution of species 1 in the liquid phase, then, (3) and (4) further simplify to

$$y_2 P = x_2 P_2^{\text{sat}} \tag{5}$$

$$y_1 P = x_1 k_1 \tag{6}$$

Equation (5) is just Raoult'a law (7.112), applied to species 2; equation (6) is the historical statement of Henry's law.

The liquid-phase mole fraction x_1 of supercritical species 1 is often called the *solubility* of gas 1 in liquid 2. Combination of (5) and (6) and solution for x_1 gives

$$x_1 = \frac{P - P_2^{\text{sat}}}{k_1 - P_2^{\text{sat}}} \tag{7}$$

According to (7), the solubility of a gas in a liquid increases with increasing pressure P; also, a large Henry's-law constant k_1 implies a small gas solubility, and conversely. Henry's constants, unlike the pure-species vapor pressures P_i^{sat}, require *mixture* data for their evaluation; (6) and (7) are the simplest bases for obtaining estimates of k_1 from gas-solubility data.

7.16 Derive equations for description of the VLE behavior of a binary system in which the components are completely immiscible in the liquid phase. Base the derivation on the approximate VLE relationship (7.113).

Solution of the problem is expedited by consideration of the Tx diagram of the system, shown in Fig. 7-32. Comparison with Fig. 7-29(b) shows that the present system is the vapor/liquid analog of liquid-solid systems in which the solid phases are immiscible, and for which eutectic behavior is observed. Temperatures T_2 and T_1 are the boiling points at pressure P of pure 2 and 1, respectively. There are two distinct VLE regions. Within the area $AECA$ (region I) vapor mixtures with compositions given by the curve AE are in equilibrium with pure liquid 2; within the area $BEDB$ (region II) vapor mixtures with compositions given by BE are in equilibrium with pure liquid 1. At the temperature T^{az}, vapor of composition y_1^{az} is in equilibrium with *two* liquid phases containing pure 2 and pure 1, respectively. The exceptional point E represents a *heterogeneous azeotrope*, the vapor/liquid analog of a liquid/solid eutectic.

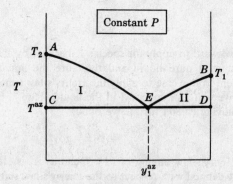

Mole fraction component 1

Fig. 7-32

For VLE in region I, (7.113) written for component 2 is $y_2P = x_2\gamma_2P_2^{\text{sat}}$. But $x_2 = 1.0$, and therefore $\gamma_2 = 1.0$, in region I. Thus we obtain for region I

$$y_2(\text{I}) = \frac{P_2^{\text{sat}}}{P} \qquad y_1(\text{I}) = 1 - \frac{P_2^{\text{sat}}}{P} \tag{1}$$

For VLE in region II, (7.113) written for component 1 is $y_1P = x_1\gamma_1P_1^{\text{sat}}$. But $x_1 = 1.0$ and $\gamma_1 = 1.0$, and we obtain

$$y_1(\text{II}) = \frac{P_1^{\text{sat}}}{P} \qquad y_2(\text{II}) = 1 - \frac{P_1^{\text{sat}}}{P} \tag{2}$$

Either of the equations (1) represents the dewpoint curve AE; similarly, either of the equations (2) represents the dewpoint curve BE. These equations are the basis for the familiar statement that the liquid components of such systems "exert their own vapor pressures." Thus, the partial pressure $P_2(\equiv y_2P)$ of 2 in region I is given by (1) as $P_2(\text{I}) = P_2^{\text{sat}}$. Similarly, the partial pressure $P_1(\equiv y_1P)$ of 1 in region II is given by (2) as $P_1(\text{II}) = P_1^{\text{sat}}$.

The two dewpoint curves converge at E, and therefore at this point $y_1(\text{I}) = y_1(\text{II})$, or $y_2(\text{I}) = y_2(\text{II})$. Combination of (1) and (2) thus gives, on rearrangement, the following implicit equation for T^{az}:

$$P = P_1^{\text{sat}} + P_2^{\text{sat}} \tag{3}$$

The azeotropic composition y_1^{az} is found by substitution of (3) into (1) or (2):

$$y_1^{\text{az}} = \frac{P_1^{\text{sat}}}{P_1^{\text{sat}} + P_2^{\text{sat}}} \tag{4}$$

7.17 (a) Derive equations for determination of equilibrium conditions in two-component, liquid/liquid systems. (b) Determine the equation of the solubility curve of a binary liquid system for which G^E is given by (7.103).

(a) We designate the two liquid phases by the superscripts α and β. Equation (7.52), written for the two components 1 and 2, gives $\hat{f}_1^\alpha = \hat{f}_1^\beta$ and $\hat{f}_2^\alpha = \hat{f}_2^\beta$. Elimination of the $\hat{f}_i$ in favor of activity coefficients by use of (7.76) gives

$$x_1^\alpha\gamma_1^\alpha = x_1^\beta\gamma_1^\beta \tag{1}$$

$$x_2^\alpha\gamma_2^\alpha = x_2^\beta\gamma_2^\beta \tag{2}$$

where the pure-species fugacities f_1 and f_2 cancel in the two equations. The γ_i are functions of T, P, and composition; thus, for fixed T and P, (1) and (2) constitute two equations in the two unknowns x_1^α and x_1^β (or x_2^α and x_2^β). The solubility curve is the curve determined by the sets of x_1^α, x_1^β, and T (or x_2^α, x_2^β, and T) that satisfy both (1) and (2) for a given pressure P.

(b) Equations (1) and (2) may be written in alternate forms which are more convenient for application:

$$\ln\left(\frac{x_1^\alpha}{x_1^\beta}\right) = \ln\gamma_1^\beta - \ln\gamma_1^\alpha \tag{3}$$

$$\ln\left(\frac{1 - x_1^\alpha}{1 - x_1^\beta}\right) = \ln\gamma_2^\beta - \ln\gamma_2^\alpha \tag{4}$$

The expressions for $\ln\gamma_i$ consistent with (7.103) are given by (7.104). When applied to components 1 and 2 in the α and β phases, they yield

$$\ln\gamma_1^\beta - \ln\gamma_1^\alpha = B(x_1^\beta - x_1^\alpha)[(x_1^\beta - 1) + (x_1^\alpha - 1)] \tag{5}$$

$$\ln\gamma_2^\beta - \ln\gamma_2^\alpha = B(x_1^\beta - x_1^\alpha)(x_1^\alpha + x_1^\beta) \tag{6}$$

Substitution of (5) in (3), and (6) in (4), and elimination of B from the resulting equations, gives

$$(x_1^\alpha + x_1^\beta)\ln\left(\frac{x_1^\alpha}{x_1^\beta}\right) = [(1 - x_1^\beta) + (1 - x_1^\alpha)]\ln\left(\frac{1 - x_1^\beta}{1 - x_1^\alpha}\right) \tag{7}$$

Equation (7) is satisfied by all pairs (x_1^α, x_1^β) for which

$$x_1^\beta = 1 - x_1^\alpha \tag{8}$$

This requires that the solubility curve be symmetrical about $x_1 = 1/2$. The equation of the curve is found by combination of (3), (5), and (8); the result is

$$\ln\left(\frac{x_1^\alpha}{1 - x_1^\alpha}\right) = B(2x_1^\alpha - 1) \tag{9}$$

Numerical values of the solubilities x_1^α and x_1^β for specified T and P can be found from (8) and (9) if B is given as a function of T and P.

7.18 For the completely miscible binary system benzene (1)/chloroform (2) the VLE K-values at 343 K and 1 bar are $K_1 = 0.719$, $K_2 = 1.31$. Calculate the equilibrium compositions of the liquid and vapor phases and the fraction of the total system which is liquid for a total system composition of (a) 40 mole % benzene, and (b) 50 mole % benzene.

The mole fractions of components 1 and 2 must sum to unity in each phase:

$$x_1 + x_2 = 1 \qquad y_1 + y_2 = 1$$

Formal elimination of y_1 and y_2 in favor of x_1 and x_2 by the definition of K_i reduces second equation to $K_1 x_1 + K_2 x_2 = 1$, which combines with the first equation to give

$$x_1 = \frac{K_2 - 1}{K_2 - K_1} \tag{1}$$

or, since $y_1 = K_1 x_1$,

$$y_1 = K_1 \frac{K_2 - 1}{K_2 - K_1} \tag{2}$$

For specified T and P, there is only one possible set of equilibrium compositions, and hence one set of K-values, in a two-phase, two-component system. Thus the equilibrium compositions (if two phases actually exist) are independent of the overall system composition for given values of T and P. This is reflected by the fact that neither (1) nor (2) contains the overall compositions z_1 and z_2. Thus we obtain both for part (a) and for part (b):

$$x_1 = \frac{1.31 - 1}{1.31 - 0.719} = 0.525 \qquad x_2 = 1 - x_1 = 0.475$$
$$y_1 = (0.719)(0.525) = 0.377 \qquad y_2 = 1 - y_1 = 0.623$$

Although L (the fraction of the system that is liquid) could be determined by application of the general equation (7.125), it is more simply calculated for a binary system by direct material balance. Thus, on the basis of a total system of one mole, we can write $x_1 L + y_1(1 - L) = z_1$. Substitution of expressions (1) and (2) for x_1 and y_1 and solution for L then gives

$$L = \frac{z_1}{K_2 - 1} + \frac{z_1 - K_1}{1 - K_1} \tag{3}$$

Using the data we obtain for (a):

$$L = \frac{0.40}{1.31 - 1} + \frac{0.40 - 0.719}{1 - 0.719} = 0.155$$

and for (b):

$$L = 0.834$$

Thus, although the equilibrium compositions resulting from a flash calculation for a binary system are independent of the feed composition, the relative amounts of liquid and vapor formed are not.

CHEMICAL-REACTION CALCULATIONS (Sections 7.9 and 7.10)

7.19 The single independent chemical reaction $0 = \Sigma \, \nu_i X_i(p)$ is to be carried out in a steady-flow reactor. All chemical species present in either the feed or the product streams are in their standard states at the same temperature T. Derive an expression for the rate of heat transfer $\dot{Q}$ to the chemical reactor.

Equation (6.16) is applicable. Neglecting potential- and kinetic-energy terms, and assuming that no shaft work is done, we obtain

$$\dot{Q} = \dot{n}_P H_P - \dot{n}_F H_F \tag{1}$$

where $\dot{n}_P$ is the total molar product rate, $\dot{n}_F$ is the total molar feed rate, H_P is the molar enthalpy of the products, and H_F is the molar enthalpy of the feed.

All chemical species are in their standard states, and therefore

$$H_P = \sum (x_i)_P H_i^\circ \tag{2}$$

$$H_F = \sum (x_i)_F H_i^\circ \tag{3}$$

where $(x_i)_P$ is the overall mole fraction of species i in the product, $(x_i)_F$ is the overall mole fraction of i in the feed, and H_i° is the standard-state enthalpy of i. Equations (2) and (3) are valid when each species i enters and leaves the reactor in its standard state as a *pure component*. However, they also hold for ideal-gas mixtures, because the pure ideal-gas state at standard pressure P° *is* a standard state, and because there is neither a pressure effect on the enthalpy nor an enthalpy change of mixing for ideal gases.

Combination of (1), (2), and (3) gives

$$\dot{Q} = \sum [(\dot{n}_i)_P - (\dot{n}_i)_F] H_i^\circ \tag{4}$$

where $(\dot{n}_i)_P$ and $(\dot{n}_i)_F$ are the molar product and feed rates, respectively, of component i. These rates are not independent, however, but are related through the reaction stoichiometry to the rate of conversion, as measured by the time derivative $\dot{\varepsilon}$ of the reaction coordinate ε. The required relationship is obtained from (7.135) on formal substitution of $(\dot{n}_i)_P$ for n_i, $(\dot{n}_i)_F$ for $n_{i,0}$, and $\dot{\varepsilon}$ for ε:

$$(\dot{n}_i)_P = (\dot{n}_i)_F + \nu_i \dot{\varepsilon} \tag{5}$$

Substitution of (5) in (4) gives

$$\dot{Q} = \sum \nu_i \dot{\varepsilon} H_i^\circ = \left(\sum \nu_i H_i^\circ \right) \dot{\varepsilon}$$

or

$$\dot{Q} = \Delta H^\circ \dot{\varepsilon} \tag{6}$$

since, by (7.145), $\Delta H^\circ = \sum \nu_i H_i^\circ$. Equation (6) asserts that $\dot{Q}$ for the prescribed process is proportional to the standard enthalpy change of reaction. This expression is the basis for many isothermal "heat of reaction" calculations in chemical-reaction engineering.

7.20 Hydrogen gas flowing at $1 \, \text{mol} \cdot \text{min}^{-1}$ is completely oxidized to water vapor by burning with the theoretical amount of air in an adiabatic reactor. The hydrogen and air enter the reactor as ideal gases at 1 bar and 300 K, and the products leave the reactor as an ideal-gas mixture at 1 bar and at some higher temperature. Estimate this product temperature. For the reaction

$$H_2(g) + \tfrac{1}{2}O_2(g) \rightarrow H_2O(g)$$

$\Delta H^\circ = -241.818 \, \text{kJ} \cdot \text{mol}^{-1}$ at 300 K. Molar heat capacities of $N_2(g)$ and $H_2O(g)$ are given by

$$C_P^\circ/R = 3.352 + (0.513 \times 10^{-3})(T/\text{K}) \qquad \text{and} \qquad C_P^\circ/R = 3.608 + (1.288 \times 10^{-3})(T/\text{K})$$

respectively. Take air to contain 21 mole % O_2 and 79 mole % N_2.

Neglecting potential- and kinetic-energy changes, assuming that $\dot{W}_s = 0$, and noting that $\dot{Q} = 0$ (adiabatic reaction), we may write (6.16) as

$$\dot{n}_p H_p(T) - \dot{n}_f H_f(300) = 0$$

where $\dot{n}_f$ and $\dot{n}_p$ are molar feed and product rates, $H_f(300)$ is the molar enthalpy of the feed at 300 K, and $H_p(T)$ is the molar enthalpy of the products at the unknown temperature T. Addition and subtraction of the term $\dot{n}_p H_p(300)$ from the left-hand side of this equation gives

$$\dot{n}_p[H_p(T) - H_p(300)] + [\dot{n}_p H_p(300) - \dot{n}_f H_f(300)] = 0$$

But, by Problem 7.19, the second term in brackets is just equal to $\dot{\varepsilon}\,\Delta H°(300)$. Moreover, the first term is given by

$$\dot{n}_p[H_p(T) - H_p(300)] = \left[\sum \dot{n}_i(\langle C_P° \rangle_T)_i\right]_p (T - 300)$$

Thus the equation to be solved for the final temperature T is

$$\left[\sum \dot{n}_i(\langle C_P° \rangle_T)_i\right]_p (T - 300) + \dot{\varepsilon}\,\Delta H°(300) = 0 \qquad (1)$$

The quantities $(\dot{n}_i)_p$ and $\dot{\varepsilon}$ must now be evaluated. Letting $H^2 \equiv (1)$, $O_2 \equiv (2)$, $H_2O \equiv (3)$, and $N_2 \equiv (4)$, we have from the statement of the problem

$$(\dot{n}_1)_f = 1\ \text{mol} \cdot \text{min}^{-1} \qquad (\dot{n}_2)_f = 0.5\ \text{mol} \cdot \text{min}^{-1} \qquad (\dot{n}_3)_f = 0\ \text{mol} \cdot \text{min}^{-1}$$

$$(\dot{n}_4)_f = (0.5)(79/21) = 1.881\ \text{mol} \cdot \text{min}^{-1}$$

The stoichiometric numbers of the species are $\nu_1 = -1$, $\nu_2 = -\frac{1}{2}$, $\nu_3 = 1$, and $\nu_4 = 0$. Since the hydrogen is completely oxidized, $(\dot{n}_1)_p = 0$ and

$$\dot{\varepsilon} = \frac{(\dot{n}_1)_p - (\dot{n}_1)_f}{\nu_1} = \frac{0 - 1}{-1} = 1\ \text{mol} \cdot \text{min}^{-1}$$

[see (5) of Problem 7.19]. The molar product rates of the other species are then found to be

$$(\dot{n}_2)_p = 0\ \text{mol} \cdot \text{min}^{-1} \qquad (\dot{n}_3)_p = 1\ \text{mol} \cdot \text{min}^{-1} \qquad (\dot{n}_4)_p = 1.881\ \text{mol} \cdot \text{min}^{-1}$$

Substitution of expressions for the mean heat capacities and of numerical values for $\Delta H°(300)$, $(\dot{n}_3)_p$, and $(\dot{n}_4)_p$, in (1) then gives

$$(0.008\,314)[(1)(3.608 + 1.288 \times 10^{-3} T_{\text{am}}) + (1.881)(3.352 + 0.513 \times 10^{-3} T_{\text{am}})](T - 300)$$
$$+ (1)(-241.818) = 0$$

or
$$T = \frac{241.818}{(0.008\,314)(9.913 + 2.253 \times 10^{-3} T_{\text{am}})} + 300 \qquad (2)$$

Solution of (2) for T, which appears in T_{am} as well as on the left-hand side of the equation, is most easily carried out by iteration (instead of by the quadratic formula). One obtains $T = 2522\ \text{K}$.

7.21   Estimate the equilibrium composition at 400 K and 1 atm of a gaseous mixture containing the three isomers n-pentane (1), isopentane (2), and neopentane (3). Standard formation data at 400 K are: $(\Delta G_f°)_1 = 40.195\ \text{kJ} \cdot \text{mol}^{-1}$, $(\Delta G_f°)_2 = 34.415\ \text{kJ} \cdot \text{mol}^{-1}$, $(\Delta G_f°)_3 = 37.640\ \text{kJ} \cdot \text{mol}^{-1}$.

We first determine the independent chemical reactions. The three formation reactions are

$$5C(s) + 6H_2(g) \rightarrow n\text{-}C_5H_{12}(g) \qquad (1)$$

$$5C(s) + 6H_2(g) \rightarrow \text{iso-}C_5H_{12}(g) \qquad (2)$$

$$5C(s) + 6H_2(g) \rightarrow \text{neo-}C_5H_{12}(g) \qquad (3)$$

Elimination of $C(s)$ and $H_2(g)$ by subtraction of (1) from (2) and from (3) gives $r = 2$ *independent* reactions:

$$n\text{-}C_5H_{12}(g) \rightarrow \text{iso-}C_5H_{12}(g) \qquad (4)$$

$$n\text{-}C_5H_{12}(g) \rightarrow \text{neo-}C_5H_{12}(g) \qquad (5)$$

Reaction equilibrium calculations for the case of more than one independent chemical reaction can occasionally be done quite simply by straightforward extensions of the method based on (7.161). At

1 atm pressure, we assume ideal-gas behavior, and the generalization of the ideal-gas equilibrium equation (7.164) becomes

$$\prod_i y_i^{\nu_{i,j}} = (P/P^\circ)^{-(\Sigma\,\nu)_j} K_j \qquad (j = 1, 2, \ldots, r) \tag{6}$$

Here, $\nu_{i,j}$ is the stoichiometric number of i in the jth reaction, and $(\Sigma\,\nu)_j \equiv \Sigma_i\,\nu_{i,j}$. Designating reaction (4) by $j = 1$ and reaction (5) by $j = 2$, the stoichiometric numbers for the present problem are $\nu_{1,1} = -1$, $\nu_{2,1} = 1$, $\nu_{1,2} = -1$, and $\nu_{3,2} = 1$. Thus $(\Sigma\,\nu)_1 = 0$ and $(\Sigma\,\nu)_2 = 0$, and (6) becomes

$$y_2 = y_1 K_1 \qquad\qquad y_3 = y_1 K_2 \tag{7}$$

where
$$K_1 = \exp\left(-\Delta G_1^\circ / RT\right) \qquad K_2 = \exp\left(-\Delta G_2^\circ / RT\right) \tag{8}$$

There are only two independent mole fractions; therefore, for specified T, (7) constitutes two equations in two unknowns. The equilibrium mole fractions are found by combination of (7) with the equation $y_1 + y_2 + y_3 = 1$:

$$\begin{aligned}
y_1 &= 1/(1 + K_1 + K_2) \\
y_2 &= K_1/(1 + K_1 + K_2) \\
y_3 &= K_2/(1 + K_1 + K_2)
\end{aligned} \tag{9}$$

Numerical values of ΔG_1° and ΔG_2° are found from the given standard formation values; thus,

$$\Delta G_1^\circ = 34.415 - 40.195 = -5.780 \text{ kJ} \cdot \text{mol}^{-1} \qquad \Delta G_2^\circ = 37.640 - 40.195 = -2.555 \text{ kJ} \cdot \text{mol}^{-1}$$

and, from (8),

$$K_1 = \exp\left[\frac{+5.780}{(0.008\,314)(400)}\right] = 5.686 \qquad K_2 = \exp\left[\frac{+2.555}{(0.008\,314)(400)}\right] = 2.156$$

Then, from (9), $y_1 = 0.113$, $y_2 = 0.643$, $y_3 = 0.244$.

7.22   One method for the production of hydrogen cyanide (HCN) is the gas-phase nitrogenation of acetylene (C_2H_2) according to the reaction

$$N_2(g) + C_2H_2(g) \rightarrow 2HCN(g) \tag{1}$$

The feed to a reactor consists of gaseous N_2 and C_2H_2 in their stoichiometric proportions; the reaction temperature is controlled at 573 K. Estimate the maximum mole fraction of HCN in the product stream if the reactor pressure is (a) 1 bar and (b) 200 bar. At 573 K, ΔG° ($P^\circ = 1$ bar) for the reaction is 30.100 kJ $\cdot$ mol^{-1}. Physical properties for HCN are: $T_c = 456.7$ K, $P_c = 49.6$ bar, and acentric factor $\omega = 0.4$.

(a)   The maximum yield of HCN is that corresponding to equilibrium conversion. At 1 bar, the gas mixtures may be assumed ideal, and (7.164) therefore applies:

$$\prod_i y_i^{\nu_i} = (P/P^\circ)^{-\nu} K \tag{7.164}$$

Letting $N_2 \equiv (1)$, $C_2H_2 \equiv (2)$, and $HCN \equiv (3)$, we see from (1) that $\nu_1 = -1$, $\nu_2 = -1$, and $\nu_3 = 2$. Thus $\nu = \Sigma\,\nu_i = 0$. The equilibrium constant K is found from (7.162) and the given value for ΔG°:

$$K = \exp\left[\frac{-30.100}{(0.008\,314)(573)}\right] = 1.803 \times 10^{-3}$$

By the statement of the problem, the initial mole fractions are $y_{1,0} = 0.5$, $y_{2,0} = 0.5$, and $y_{3,0} = 0.0$. Thus the material-balance equation (7.137) gives

$$y_1 = 0.5 - \xi \qquad y_2 = 0.5 - \xi \qquad y_3 = 2\xi \tag{2}$$

Substitution of numerical values in (7.164) now yields

$$\frac{\xi^2}{(0.5 - \xi)^2} = 4.508 \times 10^{-4}$$

from which $\xi = 0.0104$. The equilibrium mole fractions of the product stream are then found from (2):

$$y_1 = 0.4896 \qquad y_2 = 0.4896 \qquad y_3 = 0.0208$$

The maximum mole fraction of HCN at 573 K and 1 bar is thus 0.0208.

(b) At 200 bar, the assumption of ideal-gas behavior is certain to be invalid. We assume instead that the gas mixtures are *ideal solutions*, for which (7.74) gives $\hat{\phi}_i = \phi_i$. The general equilibrium equation (7.163) for gas-phase reactions then simplifies to

$$\prod y_i^{\nu_i} = \left(\prod \phi_i^{-\nu_i} \right) (P/P^\circ)^{-\nu} K \qquad (3)$$

Calculation of the equilibrium composition therefore requires numerical values for the pure-species fugacity coefficients ϕ_i. Here we use (7.55) and the generalized correlation of Section 5.4, which combine to give (with subscript i suppressed for clarity)

$$\ln \phi = (B^0 + \omega B^1) \frac{P_r}{T_r} \qquad (4)$$

Quantities P_r and T_r are reduced pressure and reduced temperature; B^0 and B^1 are given as functions of T_r by (5.73). Physical constants for N_2 and C_2H_2 are obtained from Appendix C; those for HCN are given in the problem statement. Table 7-7 summarizes the calculation of the ϕ's. Substitution of (2) and the numerical values of ν, the ν_i, and the ϕ_i in (3) gives, on rearrangement,

$$\frac{\xi^2}{(0.5 - \xi)^2} = 1.438 \times 10^{-3}$$

from which $\xi = 0.0183$. The equilibrium mole fractions are found from (2):

$$y_1 = 0.4817 \qquad y_2 = 0.4817 \qquad y_3 = 0.0366$$

The effect of increasing the reaction pressure from 1 bar to 200 bar is to nearly double the mole fraction of HCN. This effect results solely from the vapor-phase nonidealities, and not from the reaction stoichiometry, because $\nu = 0$ and, by (7.169), *no* pressure effect on conversion obtains for the ideal-gas reaction.

Table 7-7

Compound	T/K	P/bar	ω	T_r	P_r	B^0	B^1	ϕ
N_2 (1)	126.2	33.9	0.040	4.54	5.90	+0.046	+0.139	1.069
C_2H_2 (2)	308.3	61.4	0.184	1.86	3.26	−0.073	+0.126	0.916
HCN (3)	456.7	49.6	0.4	1.25	4.03	−0.212	+0.072	0.554

7.23 Derive an expression for determination of the equilibrium states of a system in which there occurs the single (independent) reaction

$$0 = \nu_1 X_1(s1) + \nu_2 X_2(s2) + \nu_3 X_3(g)$$

The designations $s1$ and $s2$ indicate that the compounds X_1 and X_2 are present as two, distinct, pure solid phases.

All three species are present as pure components in their phases, and therefore (7.161) becomes

$$(f_1/f_1^\circ)^{\nu_1} (f_2/f_2^\circ)^{\nu_2} (\phi_3 P/P^\circ)^{\nu_3} = K \qquad (1)$$

in which the relationship $f_3 = \phi_3 P$ has been used. Expressions for the fugacity ratios of the solids are obtained by application of (7.42):

$$f_i/f_i^\circ = \exp\left[\frac{1}{RT} \int_{P^\circ}^P V_i \, dP \right] \qquad (i = 1, 2) \qquad (2)$$

where V_i is the molar volume of the pure solid, and the integration is carried out at the system temperature. Combination of (1) with (2) then gives

$$(\phi_3 P/P^\circ)^{\nu_3} \exp\left[\frac{1}{RT}\int_{P^\circ}^{P}(\nu_1 V_1 + \nu_2 V_2)\,dP\right] = K \qquad (3)$$

For the type of system under consideration, there are three phases ($\pi = 3$), three components ($m = 3$), and one independent chemical reaction ($r = 1$). The extended phase rule (7.160) then gives $F = 1$. Thus there is only a single degree of freedom for the system. This is in accord with (3), which shows that the equilibrium pressure is determined on specification of the equilibrium temperature. The compound present in the gas phase (species 3) is invariably a *product* species ($\nu_3 > 0$) for this type of reaction, and the equilibrium pressures defined by (3) are called *decomposition pressures* (see Problem 7.57).

7.24 Consider a gas-phase reaction with total stoichiometric number ν. Show how to convert ΔG° and K from standard-state pressure P_1° to standard-state pressure P_2°.

For a gas-phase reaction, the standard state for species i is the ideal-gas state of pure i at temperature T and pressure P°. Thus, by the definition (7.144),

$$\Delta G^\circ(T, P^\circ) = \sum \nu_i G_i^{ig}(T, P^\circ)$$

But—see the equation following (7.21)—

$$G_i^{ig}(T, P^\circ) = \Gamma_i(T) + RT\ln P^\circ$$

Combining these equations, we obtain

$$\Delta G^\circ(T, P^\circ) = \sum \nu_i \Gamma_i(T) + \nu RT\ln P^\circ \qquad (1)$$

Specializing (1) to standard pressures P_1° and P_2° and subtracting gives, on rearrangement,

$$\Delta G^\circ(T, P_2^\circ) = \Delta G^\circ(T, P_1^\circ) + \nu RT\ln(P_2^\circ/P_1^\circ) \qquad (2)$$

Equation (2) is the required conversion formula for ΔG°. The corresponding formula for K follows from (2) and the definition (7.162):

$$K(T, P_2^\circ) = (P_1^\circ/P_2^\circ)^\nu K(T, P_1^\circ) \qquad (3)$$

Equations (2) and (3) show that numerical values of ΔG° and K are *independent of the standard pressure* for gas-phase reactions for which $\nu = 0$.

Supplementary Problems

SOLUTION THERMODYNAMICS   (Sections 7.1 through 7.6)

7.25 (a) Show that the "partial molar mass" of a component in solution is equal to the molar mass of the component. (b) Write (7.8) for the components in a ternary solution.

Ans.   (b) $\overline{M}_1 = M - x_2(\partial M/\partial x_2)_{T,P,x_3} - x_3(\partial M/\partial x_3)_{T,P,x_2}$

$\overline{M}_2 = M - x_1(\partial M/\partial x_1)_{T,P,x_3} - x_3(\partial M/\partial x_3)_{T,P,x_1}$

$\overline{M}_3 = M - x_1(\partial M/\partial x_1)_{T,P,x_2} - x_2(\partial M/\partial x_2)_{T,P,x_1}$

7.26 At 303 K and 1 bar the volumetric data for liquid mixtures of benzene (1) and cyclohexane (2) are represented by the simple quadratic expression $V = 109.4 - 16.8x_1 - 2.64x_1^2$, where x_1 is the mole fraction of benzene and V has units $cm^3 \cdot mol^{-1}$. Find expressions for $\overline{V}_1$, $\overline{V}_2$, and ΔV at 303 K and 1 bar.

Ans. $\overline{V}_1 = 92.6 - 5.28x_1 + 2.64x_1^2$ $\overline{V}_2 = 109.4 + 2.64x_1^2$ $\Delta V = 2.64x_1 x_2$

7.27 The following pair of equations has been suggested for representation of partial-molar-volume data for simple binary systems at constant T and P:

$$\overline{V}_1 - V_1 = a + (b - a)x_1 - bx_1^2 \qquad \overline{V}_2 - V_2 = a + (b - a)x_2 - bx_2^2$$

Here a and b are functions of T and P only, and V_1 and V_2 are the molar volumes of the pure components. Are these equations thermodynamically sound? (*Hint*: See Problem 7.1.)

Ans. No. Although the equations give $\overline{V}_1 = V_1$ and $\overline{V}_2 = V_2$ for $x_1 = 1$ and $x_2 = 1$, and in addition satisfy the area test for thermodynamic consistency [(5) of Problem 7.1], they *do not* satisfy the slope test [(3) or (4) of Problem 7.1]. Values generated from such equations would be thermodynamically inconsistent, and therefore incorrect.

7.28 Data for properties of components in solution are sometimes reported as "apparent molar properties." For a binary system of constituents 1 and 2, the apparent molar property $\mathscr{M}_1$ of component 1 is defined by

$$\mathscr{M}_1 = \frac{M - x_2 M_2}{x_1}$$

where x is mole fraction, M is the molar property of the mixture, and M_2 is the molar property of pure 2 at the solution T and P. (*a*) Derive equations for determination of the partial molar properties $\overline{M}_1$ and $\overline{M}_2$ from knowledge of $\mathscr{M}_1$ as a function of x_1 at constant T and P. The equations should include only the quantities x_1, M_2, $\mathscr{M}_1$, and $d\mathscr{M}_1/dx_1$. (*b*) Find $\overline{M}_1$ and $\overline{M}_2$ at infinite dilution.

Ans. (*a*) $\quad \overline{M}_1 = \mathscr{M}_1 + x_1(1 - x_1)\dfrac{d\mathscr{M}_1}{dx_1} \qquad \overline{M}_2 = M_2 - x_1^2 \dfrac{d\mathscr{M}_1}{dx_1}$

(*b*) For $x_1 = 0$, $\overline{M}_1^\infty = \mathscr{M}_1^\infty$; for $x_1 = 1$, $\overline{M}_2^\infty = M_2 - \left(\dfrac{d\mathscr{M}_1}{dx_1}\right)_{x_1=1}$

7.29 A ternary gas mixture contains 20 mole % A, 35 mole % B, and 45 mole % C. At pressure 60 bar and temperature 350 K, the fugacity coefficients of components A, B, and C in this mixture are 0.7, 0.6, and 0.9, respectively. What is the fugacity of the mixture? *Ans.* 44.6 bar

7.30 Estimate the fugacity of liquid $CHClF_2$ at 255.4 K and 13.79 MPa. [*Hint*: See (7.54) and Example 7.8.]

Data: (i) Molar mass $= 0.0865 \text{ kg} \cdot \text{mol}^{-1}$

(ii) $P^{\text{sat}} = 0.2674$ MPa at 255.4 K

(iii) $Z = 0.932$ for the saturated vapor at 255.4 K

(iv) At 255.4 K:

P/MPa	$V/\text{cm}^3 \cdot \text{g}^{-1}$	P/MPa	$V/\text{cm}^3 \cdot \text{g}^{-1}$
0.069	347.9	6.90	0.48
0.276	0.743	10.35	0.35
3.45	0.612	13.79	0.22

Ans. $f \approx 0.336$ MPa

PHASE EQUILIBRIA (Sections 7.7 and 7.8)

7.31 Develop (7.33) through (7.37), for the fugacity of a mixture, and (7.38) through (7.42), for the fugacity of pure species i. (*Hint*: Follow Section 7.3.)

7.32 Determine an expression for $\ln \phi$ of a gas mixture described by the van der Waals equation of state (5.42), and an expression for $\ln \hat{\phi}_i$ for species i in a van der Waals mixture. (See Problem 7.4.)

Ans. $\ln \phi = \dfrac{b\rho}{1-b\rho} - \dfrac{2a\rho}{RT} - \ln(1-b\rho)Z$ $\ln \hat{\phi}_i = \dfrac{\overline{b}_i\rho}{1-b\rho} - \dfrac{(a+\overline{a}_i)\rho}{RT} - \ln(1-b\rho)Z$

where $\overline{a}_i$ and $\overline{b}_i$ are defined through (7.1).

7.33   Prove:

$$\left(\frac{\partial \ln \gamma_i}{\partial T}\right)_{P,x} = -\frac{\overline{H}_i^E}{RT^2} \qquad \left(\frac{\partial \ln \gamma_i}{\partial P}\right)_{T,x} = \frac{\overline{V}_i^E}{RT}$$

[*Hint*:   Use (7.81), (7.99), and (7.98).]

7.34   The *Wilson equation* for G^E of a binary mixture is

$$\frac{G^E}{RT} = -x_1 \ln(x_1 + x_2\Lambda_{12}) - x_2 \ln(x_2 + x_1\Lambda_{21})$$

where parameters Λ_{12} and Λ_{21} depend on T only. Find the activity coefficients implied by the Wilson equation.

Ans.
$$\ln \gamma_1 = -\ln(x_1 + x_2\Lambda_{12}) + x_2\left(\frac{\Lambda_{12}}{x_1 + x_2\Lambda_{12}} - \frac{\Lambda_{21}}{x_2 + x_1\Lambda_{21}}\right)$$

$$\ln \gamma_2 = -\ln(x_2 + x_1\Lambda_{21}) + x_1\left(\frac{\Lambda_{21}}{x_2 + x_1\Lambda_{21}} - \frac{\Lambda_{12}}{x_1 + x_2\Lambda_{12}}\right)$$

7.35   Develop the following form of the Gibbs/Duhem equation (7.11) for the property $M = G^E/RT$ of a binary mixture:

$$x_1 \frac{d(\ln \gamma_1)}{dx_1} + x_2 \frac{d(\ln \gamma_2)}{dx_1} = -\frac{H^E}{RT^2}\left(\frac{dT}{dx_1}\right) + \frac{V^E}{RT}\left(\frac{dP}{dx_1}\right)$$

7.36   The *activity* $\hat{a}_i$ of species i in solution is defined as $\hat{a}_i \equiv \hat{f}_i/f_i$. Prove that $\ln \hat{a}_i$ is a partial property with respect to $\Delta G/RT$.

7.37   For a binary mixture, show that $\overline{M}_1^E = \overline{M}_2^E$ at compositions for which M^E is a maximum or a minimum.

7.38   Prove: For the entropy change of mixing ΔS to be negative, the excess enthalpy H^E and the excess entropy S^E must both be negative. (*Hint*: $\Delta G < 0$ and $\Delta S^{id} > 0$.)

7.39   The equilibrium total pressure P for VLE in a binary system described by (7.113) is

$$P = x_1\gamma_1 P_1^{sat} + x_2\gamma_2 P_2^{sat}$$

If some or all of the assumptions upon which (7.113) is based are waived, then a similar but more rigorous equation may be written for P:

$$P = \frac{x_1\gamma_1 P_1^{sat}}{\Phi_1} + \frac{x_2\gamma_2 P_2^{sat}}{\Phi_2}$$

where quantities Φ_1 and Φ_2 depend in general on T, P, and composition. If the vapor phase is described by the two-term virial equation in pressure (5.33), and if liquid-phase molar volumes are independent of pressure, show that

$$\Phi_1 = \exp\left[\frac{(B_{11} - V_1^{sat})(P - P_1^{sat}) + y_2^2\,\delta_{12}P}{RT}\right] \qquad \Phi_2 = \exp\left[\frac{(B_{22} - V_2^{sat})(P - P_2^{sat}) + y_1^2\,\delta_{12}P}{RT}\right]$$

[*Hint*:   Write (7.59) in terms of the gas-phase y_i, and use together with (7.111).]

7.40   Components 1 and 2 are essentially insoluble in the liquid phase. Estimate the dewpoint temperatures and the compositions of the first drops of liquid formed when vapor mixtures of 1 and 2 containing (*a*) 75 mol % component 1, and (*b*) 25 mol % component 1, are cooled at 101.3 kPa total pressure. (*Hint*:   See Problem 7.16.) Vapor-pressure data for the pure components are displayed in Table 7-8.

Table 7-8

$T/°C$	P_1^{sat}/kPa	P_2^{sat}/kPa
85	23.53	57.82
90	33.56	70.13
95	39.58	84.53
100	46.59	101.32
105	54.61	120.78
110	63.97	143.27
115	74.31	169.01
120	86.25	198.50
125.6	101.32	239.53

Ans.   (*a*) 115.7 °C, liquid is pure 1   (*b*) 92.1 °C, liquid is pure 2

7.41   (*a*) Derive expressions for the bubblepoint pressure P_b and the dewpoint pressure P_d of a system which obeys Raoult's law. (*b*) Assuming that Raoult's law applies, determine P_b and P_d at 110 °C for an overall system composition of 45 mole % *n*-octane (1), 10 mole % 2,5-dimethylhexane (2), and 45 mole % 2,2,4-trimethylpentane (3). The pure-component vapor pressures at 110 °C are: $P_1^{sat} = 64.24$ kPa, $P_2^{sat} = 104.06$ kPa, and $P_3^{sat} = 137.8$ kPa.
Ans.   (*a*) $P_b = \Sigma\, z_i P_i^{sat}$, $P_d = [\Sigma\, (z_i/P_i^{sat})]^{-1}$ (*b*) $P_b = 101.3$ kPa, $P_d = 89.06$ kPa

7.42   For binary isothermal VLE described by (*7.113*), show that the slope of the bubblepoint curve is

$$\frac{dP}{dx_1} = (\gamma_1 P_1^{sat} - \gamma_2 P_2^{sat})\left[x_1\,\frac{d(\ln \gamma_1)}{dx_1} + 1\right]$$

[*Hint*: Use the Gibbs/Duhem equation as presented in Problem 7.35, noting that pressure independence of the activity coefficients requires that $V^E/RT = 0$ (see Problem 7.33).]

7.43   Consider binary isothermal VLE described by (*7.113*). Prove that if an azeotrope exists, its composition is given by

$$x_1^{az} = y_1^{az} = \frac{1}{2}\left(1 + \frac{1}{B}\,\ln\frac{P_1^{sat}}{P_2^{sat}}\right)$$

Thus the azeotrope pressure is

$$P^{az} = P_2^{sat}\exp\left[\frac{B}{4}\left(1 + \frac{1}{B}\,\ln\frac{P_1^{sat}}{P_2^{sat}}\right)\right]^2$$

7.44   Justify the following rule of thumb: For an equimolar liquid mixture, the vapor composition of a binary system in subcritical isothermal VLE is approximately

$$y_1 = \frac{P_1^{sat}}{P_1^{sat} + P_2^{sat}} \qquad \left(x_1 = x_2 = \frac{1}{2}\right)$$

Note that this result is trivially true if Raoult's law applies. Show why it might be approximately true for nonideal systems.

7.45   Table 7-9 shows experimental data for of VLE states for 2-propanone (1)/*n*-hexane (2) at two temperatures. (*a*) Assuming the validity of (*7.113*), evaluate G^E/RT for the liquid phase ($x_1 = 0.300$) at each temperature. (*b*) What value of H^E for the liquid phase ($x_1 = 0.300$) at $T = 323.15$ K is consistent with the results of part (*a*)?
Ans.   (*a*) 0.329 (318.15 K), 0.321 (328.15 K)   (*b*) ≈ 686 J $\cdot$ mol^{-1}

Table 7-9

	$T = 318.15$ K	$T = 328.15$ K
x_1	0.300	0.300
y_1	0.562	0.543
P/kPa	81.7	113.4
P_1^{sat}/kPa	68.39	97.52
P_2^{sat}/kPa	45.13	64.39

7.46 Sulfur hexafluoride vapor at moderate pressures is used as a dielectric in large outdoor circuit breakers. Since H_2O (1) and SF_6 (2) are immiscible as liquids, the moisture content of the SF_6 vapor must be low enough that a liquid phase of pure water does not condense when the dew point of the system is reached in cold weather. If the lowest temperature expected is 0 °C, what is the maximum mole fraction of H_2O allowable in the vapor SF_6? At 0 °C, $P_1^{sat} = 0.61$ kPa, $P_2^{sat} = 1273.9$ kPa.   *Ans.* $y_1 = 4.8 \times 10^{-4}$

7.47 A quick estimate is needed for the solubility x_1 of supercritical hydrogen (1) in liquid toluene (2) at 190 °C and 10 bar. Only a single data point is available at conditions near to these; viz.,

$$T = 461.9 \text{ K} \qquad P = 20.3 \text{ bar} \qquad x_1 = 0.0113 \qquad y_1 = 0.6674$$

Using only these data, determine the required solubility. (*Hint*: See Problem 7.15.)
Ans. $x_1 \approx 0.0027$

7.48 A system comprising methane (1) and a light oil (2) at 200 K and 30 bar has a vapor phase containing 95 mole % methane and a liquid phase containing oil and dissolved methane. The fugacity of methane in the liquid phase is described by Henry's law, with $k_1 = 200$ bar. The second virial coefficient of methane is $105 \text{ cm}^3 \cdot \text{mol}^{-1}$. Estimate the mole fraction of methane in the liquid phase. (*Hint*: See Problem 7.15.)   *Ans.* $x_1 \approx 0.118$

7.49 The relative volatility α_{12} (see Problem 7.14) for the carbon disulfide (1)/acetone (2) system at 35.17 °C is described by the empirical equation

$$\alpha_{12} = \frac{1 + 7.1875x_2 - 2.4064x_2^2}{1 + 1.4685x_1 + 2.5891x_1^2}$$

At 35.17 °C, the vapor pressures of the pure species are approximately $P_1^{sat} = 0.698$ bar and $P_2^{sat} = 0.469$ bar. Estimate γ_1^∞, γ_2^∞, and the azeotropic composition at 35.17 °C.
Ans. $\gamma_1^\infty \approx 3.84$ $\gamma_2^\infty \approx 7.53$ $x_1^{az} = y_1^{az} \approx 0.667$

7.50 The excess Gibbs energy for liquid mixtures of chloroform (1) and ethanol (2) at 55 °C is well represented by the two-parameter Margules equation

$$\frac{G^E}{RT} = (1.42x_1 + 0.59x_2)x_1x_2$$

Pure-species vapor pressures at 55 °C are $P_1^{sat} = 82.37$ kPa and $P_2^{sat} = 37.31$ kPa. Does this system form a VLE azeotrope at 55 °C?   *Ans.*   Yes; see Problem 7.14.

7.51 Consider a binary liquid mixture described by the Porter equation (*7.103*). (*a*) Infer from Problem 7.17 that a single homogeneous liquid phase obtains over the entire composition range if $B < 2$. (*b*) If $B = 2.5$, determine the equilibrium compositions x_1^α and x_1^β of the coexisting liquid phases.
Ans.   (*b*) $x_1^\alpha = 1 - x_1^\beta = 0.1448$

7.52 Liquids 2 and 3 are for practical purposes immiscible in one another. Liquid 1 is soluble in both liquid 2 and liquid 3. One mole of liquid 1, one mole of liquid 2, and one mole of liquid 3 are shaken together to form an equilibrium mixture of two liquid phases: an α-phase containing species 1 and 2, and a β-phase containing species 1 and 3. What are the mole fractions of species 1 in the α and β phases? At the temperature of the experiment, the excess Gibbs energies of the phases are given by

$$\frac{(G^E)^\alpha}{RT} = 0.4 x_1^\alpha x_2^\alpha \qquad \frac{(G^E)^\beta}{RT} = 0.8 x_1^\beta x_3^\beta$$

Ans. $x_1^\alpha = 0.3711$ $x_1^\beta = 0.2907$

CHEMICAL-REACTION CALCULATIONS   (Sections 7.9 and 7.10)

7.53 In compilations of standard formation data, one also often finds entries for *standard heats of combustion* ΔH_c° for standard combustion reactions. A standard combustion reaction for a given chemical species is the reaction between one mole of that species and oxygen to form specified products, with all the reactant and product species present in their standard states. For compounds containing only carbon, hydrogen, and oxygen, the products are customarily taken as $CO_2(g)$ and $H_2O(l)$. Thus, the standard combustion reaction for gaseous *n*-butane is

$$n\text{-}C_4H_{10}(g) + \tfrac{13}{2}O_2(g) \rightarrow 4CO_2(g) + 5H_2O(l)$$

where the gaseous species are present as pure ideal gases, and water is present as a pure liquid, all at the standard pressure P°.

The standard heats of combustion ($P^\circ = 1$ atm) for benzene $[C_6H_6(g)]$, hydrogen, and cyclohexane $[C_6H_{12}(g)]$ at 25 °C are -3303.7 kJ $\cdot$ mol^{-1}, -286.0 kJ $\cdot$ mol^{-1}, and -3955.6 kJ $\cdot$ mol^{-1}, respectively. Using these data, calculate ΔH° at 25 °C for the reaction

$$C_6H_6(g) + 3H_2(g) \rightarrow C_6H_{12}(g)$$

Ans. -206.1 kJ $\cdot$ mol^{-1}

7.54 The reaction of Example 7.21 is carried out at the standard pressure $P^\circ = 1$ atm and at the constant temperature 150 °C. How much heat is transferred from the gases to the surroundings? Assume ideal-gas mixtures. At 150 °C, in units of kJ $\cdot$ mol^{-1}, $\Delta H_f^\circ[C_3H_8(g)] = -93.07$, $\Delta H_f^\circ[CO_2(g)] = -388.81$, and $\Delta H_f^\circ[H_2O(g)] = -237.74$.   *Ans.*   40.49 kJ

7.55 Determine the numbers of degrees of freedom of the following systems:

  (a) A chemically reactive mixture of $ZnO(s)$, $ZnSO_4(s)$, $SO_2(g)$, $SO_3(g)$, and $O_2(g)$. The two solid phases are completely immiscible.

  (b) A reactive gas mixture containing k isomeric hydrocarbons.

  (c) A system in which the reaction $H_2(g) + \tfrac{1}{2}O_2(g) \rightarrow H_2O(l)$ takes place.

  (d) Same as part (c), except that air (21 mole % O_2, 79 mole % N_2) is the oxidant.

  (e) Same as part (c), only "real" air (21.0 mole % O_2, 78.1 mole % N_2, 0.9 mole % Ar) is the oxidant.

Ans. (a) 2   (b) 2   (c) 2   (d) 3   (e) 3 (The Ar-to-N_2 ratio is constant.)

7.56 (a) A compound X is known to polymerize to the compound X_l in the gas phase according to the reaction $lX(g) \rightarrow X_l(g)$. For constant $l > 1$, and assuming ideal-gas behavior, show that the extent of equilibrium polymerization increases with increasing pressure at constant T. (b) The following experimental data were recorded for the equilibrium mole fractions of monomer X in two gas-phase monomer/polymer mixtures:

$T/°C$	P/kPa	y_X
100	101.3	0.807
100	152.0	0.750

What is the value of l for the polymerization reaction of part (a)?   *Ans.* (b) $l = 2$

7.57 Limestone ($CaCO_3$) decomposes upon heating to yield quicklime (CaO) and carbon dioxide (CO_2). Determine the temperature at which limestone exerts a decomposition pressure of 1 atm (see Problem

7.23). Standard formation data at 25 °C ($P° = 1$ atm) and constants for the heat-capacity equation are given below.

$$C_P°/kJ \cdot mol^{-1} \cdot K^{-1} = a + b \times (T/K) + d \times (T/K)^{-2}$$

Compound	$10^3 a$	$10^6 b$	$10^{-2} d$	$\Delta H_f°/kJ \cdot mol^{-1}$	$\Delta G_f°/kJ \cdot mol^{-1}$
$CaCO_3(s)$	104.59	21.94	−25.96	0.682	−1129.515
$CaO(s)$	48.86	4.52	−6.53	−635.975	−604.574
$CO_2(g)$	44.17	9.04	−8.54	−393.777	−394.648

Ans.   890 °C

7.58   Ethanol is manufactured by the hydrogenation of acetaldehyde according to the reaction

$$CH_3CHO(g) + H_2(g) \rightarrow C_2H_5OH(g)$$

The feed gas to the reactor contains 30 mole % CH_3CHO, 40 mole % H_2, 20 mole % N_2, and 10 mole % H_2O. If the reaction comes to equilibrium at 650 K and 2 bar, and if no other reaction occurs, what is the composition of the product gas from the reactor? Assume ideal gases and, for the given reaction, $\Delta G° = -4.027$ kJ $\cdot$ mol^{-1} at 650 K ($P° = 1$ bar).
Ans. $y_{CH_3CHO} = 0.163$ $y_{H_2} = 0.283$ $y_{C_2H_5OH} = 0.195$ $y_{N_2} = 0.239$ $y_{H_2O} = 0.120$

7.59   The relative concentrations of the pollutants NO and NO_2 in air are governed by the reaction

$$NO + \tfrac{1}{2}O_2 \rightarrow NO_2$$

For air containing 21 mole % O_2 at 25 °C and 1 atm (1.013 bar), what is the concentration of NO in parts per million, if the total concentration of both nitrogen oxides is 5 ppm? For the given reaction at 25 °C, $\Delta G° = -35.240$ kJ $\cdot$ mol^{-1} ($P° = 1$ bar).   *Ans.* $y_{NO} = 7.3 \times 10^{-6}$ ppm

7.60   Suppose that experiment provides a set of equilibrium constants K as a function of T. It is conventional to plot such data as $\ln K$ versus $1/T$. (a) Show that the standard heat of reaction $\Delta H°$ can be found from the slope of such a plot. (b) Show that such a plot yields a *straight line* if $\Delta C_P° = 0$.   *Ans.*   (a) $-\Delta H°/R = d(\ln K)/d(1/T)$

Chapter 8

Real Flows: Work and Entropy Analyses

The purpose of this final chapter is to develop methods of thermodynamic analysis which yield the energy efficiencies of steady-flow processes and which show how much energy is dissipated or wasted as the result of irreversibilities in the individual parts of a process. The methods, based on a simple combination of the first and second laws of thermodynamics, are applied to a variety of steady-flow processes to demonstrate the practicality and power of thermodynamic analysis.

8.1 IDEAL WORK

We consider steady-flow processes for which the energy equation is

$$\Delta[\dot{m}(H + \tfrac{1}{2}u^2 + zg)]_{\substack{\text{flowing}\\\text{streams}}} = \sum \dot{Q} - \dot{W}_s \qquad (6.16)$$

Figure 8-1 represents the essential features of any such process. We presume that the process takes place in surroundings which constitute a heat reservoir at the constant temperature T_0. For such a process, the inequality (6.17) can be replaced by the equation:

$$\Delta(\dot{m}S)_{\substack{\text{flowing}\\\text{streams}}} - \frac{\sum \dot{Q}}{T_0} \equiv \dot{S}_{\text{total}}$$

where the positive quantity $\dot{S}_{\text{total}}$ is the total rate of *entropy generation* as a result of the process. Solving this equation for $\sum \dot{Q}$ gives

$$\boxed{\sum \dot{Q} = T_0\, \Delta(\dot{m}S)_{\substack{\text{flowing}\\\text{streams}}} - T_0 \dot{S}_{\text{total}}}$$

Since $\dot{S}_{\text{total}}$ is usually unknown for real processes, this equation is quantitatively useful only for the special case of a *completely reversible* process, where it is zero. In this event

$$\sum \dot{Q} = T_0\, \Delta(\dot{m}S)_{\substack{\text{flowing}\\\text{streams}}}$$

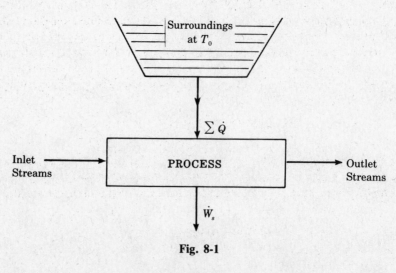

Fig. 8-1

Substitution into (6.16) then gives

$$\dot{W}_s(\text{rev}) = T_0 \, \Delta(\dot{m}S)_{\substack{\text{flowing}\\\text{streams}}} - \Delta[\dot{m}(H + \tfrac{1}{2}u^2 + zg)]_{\substack{\text{flowing}\\\text{streams}}}$$

In this equation $\dot{W}_s(\text{rev})$ is the work associated with a *completely reversible* process that causes the change of state implied by the changes in entropy, enthalpy, velocity, and elevation indicated by the terms to the right of the equals sign. When these same property changes result from a real, irreversible process, then this equation gives the work that *would be associated with a completely reversible process that accomplishes exactly the same change of state.* This stipulation of *complete* reversibility implies not only reversibility within the process, but also reversibility of heat transfer between the system and its surroundings. Such a completely reversible process is taken as the *ideal* against which to measure the efficiencies of real processes that accomplish the same change of state. In order to indicate this explicitly, we replace $\dot{W}_s(\text{rev})$ by $\dot{W}_{\text{ideal}}$ and rewrite this fundamental equation as:

$$\boxed{\dot{W}_{\text{ideal}} = T_0 \, \Delta(\dot{m}S)_{\substack{\text{flowing}\\\text{streams}}} - \Delta[\dot{m}(H + \tfrac{1}{2}u^2 + zg)]_{\substack{\text{flowing}\\\text{streams}}}} \qquad (8.1)$$

For the common special case of a single stream flowing through the control volume, (8.1) becomes

$$\dot{W}_{\text{ideal}} = \dot{m}[T_0 \, \Delta S - \Delta(H + \tfrac{1}{2}u^2 + zg)] \qquad (8.2)$$

Division by $\dot{m}$ puts (8.2) on a unit-mass basis:

$$W_{\text{ideal}} = T_0 \, \Delta S - \Delta H - \tfrac{1}{2} \, \Delta u^2 - g \, \Delta z \qquad (8.3)$$

In most applications, the kinetic- and potential-energy terms are negligible compared with the others; in this event, (8.1) is written

$$\dot{W}_{\text{ideal}} = T_0 \, \Delta(\dot{m}S)_{\substack{\text{flowing}\\\text{streams}}} - \Delta(\dot{m}H)_{\substack{\text{flowing}\\\text{streams}}} \qquad (8.4)$$

For the special case of a single stream flowing through the control volume, this becomes

$$\dot{W}_{\text{ideal}} = \dot{m}(T_0 \, \Delta S - \Delta H) \qquad (8.5)$$

Division of (8.5) by $\dot{m}$ yields

$$W_{\text{ideal}} = T_0 \, \Delta S - \Delta H \qquad (8.6)$$

Equations (8.1) through (8.6) give the power or work for completely reversible processes associated with given property changes in the flowing streams. When the same property changes occur in an actual process, the actual power $\dot{W}_s$ (or work W_s) is given by an energy balance, and we can compare the actual quantity with the ideal power or work. When $\dot{W}_{\text{ideal}}$ (or W_{ideal}) is positive, it is the *maximum power or work obtainable* from given changes in the properties of the flowing streams, and is larger than $\dot{W}_s$. In this case we define a *thermodynamic efficiency* η_t as the ratio of the actual power to the ideal power:

$$\eta_t(\text{work produced}) \equiv \frac{\dot{W}_s}{\dot{W}_{\text{ideal}}} \qquad (8.7)$$

When $\dot{W}_{\text{ideal}}$ (or W_{ideal}) is negative, $|\dot{W}_{\text{ideal}}|$ (or $|W_{\text{ideal}}|$) is the *minimum power (or work) required* to bring about the given changes in the properties of the flowing streams and is smaller than $|\dot{W}_s|$. In this case the thermodynamic efficiency is defined as the ratio of the ideal power to the actual power:

$$\eta_t(\text{work required}) \equiv \frac{\dot{W}_{\text{ideal}}}{\dot{W}_s} \qquad (8.8)$$

EXAMPLE 8.1   Problem 6.20 asks for the minimum power requirement to cool an air stream from 38 °C to 15 °C when the surroundings are at 38 °C. Taking the kinetic- and potential-energy terms to be negligible, we apply (8.5) to the air stream. But (8.5), rewritten in terms of molar properties, coincides with (2) of Problem 6.20.

EXAMPLE 8.2   In Example 6.6 saturated steam at 17.5 bar enters a process and serves as the source of energy for making heat available at a temperature level of 260 °C. The steam is assumed to leave the process as saturated liquid at 20 °C, the temperature of the surroundings (cooling water). The question is: How much heat can be made available at a temperature level of 260 °C for every kilogram of steam passing through the process?

The maximum work obtainable from each kilogram of steam is given by (8.6):

$$W_{ideal} = T_0 \Delta S - \Delta H$$
$$= (20 + 273.15)(0.2963 - 6.3853) - (83.9 - 2794.1) = 925.2 \text{ kJ} \cdot \text{kg}^{-1}$$

This amount of work may be used to operate a reversible heat pump between the surroundings (cooling water) at 20 °C and a heat reservoir at 260 °C. For such a device, (2.3) gives

$$|Q_H| = |W|\frac{T_H}{T_H - T_C} = (925.2)\left(\frac{260 + 273.15}{240}\right) = 2055.3 \text{ kJ} \cdot \text{kg}^{-1}$$

(in exact agreement with Example 6.6).

EXAMPLE 8.3 (a) What is the thermodynamic efficiency η_t of the turbine described in Problem 6.13, if the temperature of the surroundings is $T_0 = 293$ K? (b) Devise a completely reversible process by which the actual change of state of the steam may be accomplished.

(a)   For this work-producing process, η_t is given by (8.7). We need to know both W_s, the actual shaft work of the turbine, and W_{ideal} as given by (8.6) for the actual change of state. Since the turbine of Problem 6.13 is adiabatic and since the kinetic- and potential-energy terms are negligible, the energy equation (6.23) gives

$$W_s = -\Delta H = 620.6 \text{ kJ} \cdot \text{kg}^{-1}$$

Further, $\Delta S = S_2 - S_1 = 8.1511 - 7.2410 = 0.9101 \text{ kJ} \cdot \text{kg}^{-1} \cdot \text{K}^{-1}$. Then, from (8.6),

$$W_{ideal} = (293)(0.9101) - (-620.6) = 887.3 \text{ kJ} \cdot \text{kg}^{-1}$$

and from (8.7),

$$\eta_t = \frac{W_s}{W_{ideal}} = \frac{620.6}{887.3} = 0.699$$

(b)   One such process consists of two steps: (1) a reversible adiabatic expansion of the steam from its initial state to the final pressure 10 kPa (as shown in Problem 6.13, this produces wet steam); (2) reversible transfer of heat from the surroundings at 293 K to the wet steam, so as to evaporate the moisture and produce saturated steam at 10 kPa. These two steps are shown on Fig. 8-2. As shown in Problem 6.13, step (1) produces work in the amount

$$W_1 = -(\Delta H)_S = 910.9 \text{ kJ} \cdot \text{kg}^{-1}$$

The resulting wet steam has an enthalpy at point 2′ of $H_2' = 2294.5 \text{ kJ} \cdot \text{kg}^{-1}$. Step (2) is a constant-pressure, constant-temperature heating process that vaporizes the moisture content of the steam, producing saturated steam at $P_2 = 10$ kPa and $t_2 = 45.83$ °C, with an enthalpy $H_2 = 2584.8 \text{ kJ} \cdot \text{kg}^{-1}$. The heat required is given by the energy equation, $Q_2 = H_2 - H_2'$. Since this heat transfer is accomplished reversibly, we imagine a reversible heat pump operating on the Carnot cycle, taking heat from the surroundings at 293 K and releasing heat to the steam at temperature t_2. The amount of work required by the heat pump is given by (2.3) as

$$|W| = |Q_H|\left(1 - \frac{T_C}{T_H}\right)$$

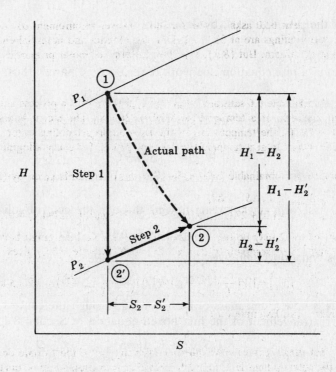

Fig. 8-2

We identify $|W|$ with the work of step (2); because work is required, $|W| = -W_2$. We take $|Q_H|$ to be the heat added to the steam; T_C is T_0, and T_H is T_2. Thus,

$$W_2 = -Q_2\left(1 - \frac{T_0}{T_2}\right) = -Q_2 + T_0\frac{Q_2}{T_2}$$

However,

$$Q_2 = H_2 - H_2' = 2584.8 - 2294.5 = 290.3 \text{ kJ} \cdot \text{kg}^{-1}$$

$$\frac{Q_2}{T_2} = S_2 - S_2' = 8.1511, - 7.2410 = 0.9101 \text{ kJ} \cdot \text{kg}^{-1} \cdot \text{K}^{-1}$$

Therefore, $W_2 = -290.3 + (293)(0.9101) = -23.6 \text{ kJ} \cdot \text{kg}^{-1}$ and

$$W_{\text{ideal}} = W_1 + W_2 = 910.9 - 23.6 = 887.3 \text{ kJ} \cdot \text{kg}^{-1}$$

in agreement with (a) above.

In Problem 6.13 it was shown that the turbine efficiency as based on *isentropic* expansion of the steam is

$$\eta_e = \frac{\Delta H}{(\Delta H)_S} = \frac{W_s}{W_1} \quad (=0.681)$$

whereas that based on *completely reversible* expansion is

$$\eta_t = \frac{W_s}{W_{\text{ideal}}} = \frac{W_s}{W_1 + W_2} = \frac{\eta_e}{1 - |W_2/W_1|} \quad (=0.699)$$

The difference between the two efficiencies is thus made plain.

8.2 LOST WORK

The energy that becomes unavailable for work as the result of irreversibilities in a process is called the *lost work*, and is defined as the difference between the ideal work for a process and the actual work of the process. Thus, by definition,

$$W_{\text{lost}} \equiv W_{\text{ideal}} - W_s \tag{8.9}$$

When rewritten as a rate equation, the work terms become power; thus, the lost power is defined by

$$\dot{W}_{\text{lost}} \equiv \dot{W}_{\text{ideal}} - \dot{W}_s \tag{8.10}$$

With $\dot{W}_{\text{ideal}}$ expressed by (8.1) and $\dot{W}_s$ by (6.16), we find from (8.10) that

$$\boxed{\dot{W}_{\text{lost}} = T_0 \, \Delta(\dot{m}S)_{\substack{\text{flowing} \\ \text{streams}}} - \sum \dot{Q}} \tag{8.11}$$

For the special case of a single stream flowing through the control volume,

$$\dot{W}_{\text{lost}} = \dot{m} T_0 \, \Delta S - \sum \dot{Q} \tag{8.12}$$

Division of this equation by $\dot{m}$ gives

$$W_{\text{lost}} = T_0 \, \Delta S - \sum Q \tag{8.13}$$

where the basis is now a unit mass of fluid flowing through the control volume.

By simple rearrangement of the first boxed equation of Section 8.1, we have:

$$T_0 \dot{S}_{\text{total}} = T_0 \, \Delta(\dot{m}S)_{\substack{\text{flowing} \\ \text{streams}}} - \sum \dot{Q}$$

Since the right-hand sides of this equation and (8.11) are identical, it follows that

$$\boxed{\dot{W}_{\text{lost}} = T_0 \dot{S}_{\text{total}}} \tag{8.14}$$

or, on the basis of a unit mass of fluid,

$$W_{\text{lost}} = T_0 \, \Delta S_{\text{total}} \tag{8.15}$$

Since the second law of thermodynamics requires that $\dot{S}_{\text{total}} \geq 0$ and $\Delta S_{\text{total}} \geq 0$, it follows that $\dot{W}_{\text{lost}} \geq 0$ and $W_{\text{lost}} \geq 0$. When a process is completely reversible, these express equalities and the lost work is zero. For irreversible processes the inequalities hold, and the lost work, i.e., the energy that becomes unavailable for work, is positive. The significance of this result is clear: The greater the irreversibility of a process, the greater the rate of entropy generation and the greater the amount of energy that becomes unavailable for work. Thus every irreversibility carries with it a price.

Many processes consist of a number of steps, and lost-work calculations are then made for each step separately. Summing (8.14) over the steps of a process gives $\sum \dot{W}_{\text{lost}} = T_0 \sum \dot{S}_{\text{total}}$. Dividing (8.14) by this result, we get

$$\frac{\dot{W}_{\text{lost}}}{\sum \dot{W}_{\text{lost}}} = \frac{\dot{S}_{\text{total}}}{\sum \dot{S}_{\text{total}}}$$

Thus an analysis of the lost work, made by calculation of the fraction that each individual lost-work term represents of the total lost work, is the same as an analysis of the rate of entropy generation, made by expressing each individual entropy-generation term as a fraction of the sum of all entropy-generation terms.

An alternative to the lost-work or entropy-generation analysis is a work analysis. For this, we write (8.10) as

$$\sum \dot{W}_{\text{lost}} = \dot{W}_{\text{ideal}} - \dot{W}_s \tag{8.16}$$

For a work-producing process, all of these work quantities are positive and $\dot{W}_{\text{ideal}} > \dot{W}_s$. We therefore write the preceding equation as

$$\dot{W}_{\text{ideal}} = \dot{W}_s + \Sigma \dot{W}_{\text{lost}} \qquad (8.17)$$

A work analysis then expresses each of the individual work terms on the right as a fraction of $\dot{W}_{\text{ideal}}$.

For a work-requiring process, $\dot{W}_s$ and $\dot{W}_{\text{ideal}}$ are negative, $|\dot{W}_s| > |\dot{W}_{\text{ideal}}|$. Equation (8.16) is therefore best written:

$$|\dot{W}_s| = |\dot{W}_{\text{ideal}}| + \Sigma \dot{W}_{\text{lost}} \qquad (8.18)$$

A work analysis here expresses each of the individual work terms on the right as a fraction of $|\dot{W}_s|$. A work analysis cannot be carried out in the case where a process is so inefficient that $\dot{W}_{\text{ideal}}$ is positive, indicating that the process should produce work, but $\dot{W}_s$ is negative, indicating that the process in fact requires work. A lost-work or entropy-generation analysis is always possible.

EXAMPLE 8.4 Perform a work analysis of the turbine/condenser of Problem 6.25.

We have here a two-step process—adiabatic expansion of steam through a turbine, and condensation of the exhaust steam in a condenser. Since the entering cooling water is at 15 °C, take this as the temperature of the surroundings; i.e., $T_0 = 288.15$ K. Label the initial state of the steam by subscript 1, the state of the exhaust steam by subscript 2, and the final state of liquid water by subscript 3. The following data are taken from the steam tables:

$$H_1 = 3411.3 \text{ kJ} \cdot \text{kg}^{-1} \qquad S_1 = 7.3365 \text{ kJ} \cdot \text{kg}^{-1} \cdot \text{K}^{-1}$$
$$H_2 = 2584.8 \text{ kJ} \cdot \text{kg}^{-1} \qquad S_2 = 8.1511 \text{ kJ} \cdot \text{kg}^{-1} \cdot \text{K}^{-1}$$
$$H_3 = 125.7 \text{ kJ} \cdot \text{kg}^{-1} \qquad S_3 = 0.4365 \text{ kJ} \cdot \text{kg}^{-1} \cdot \text{K}^{-1}$$

The overall change resulting from the process is from state 1 to state 3; and for this change of state, (8.6) gives

$$W_{\text{ideal}} = T_0 \Delta S - \Delta H = (288.15)(0.4365 - 7.3365) - (125.7 - 3411.3) = 1297.4 \text{ kJ} \cdot \text{kg}^{-1}$$

The lost-work quantities are given by (8.13):

$$W_{\text{lost}}(\text{turb}) = T_0(S_2 - S_1) - Q(\text{turb}) = (288.15)(8.1511 - 7.3365) - 0 = 234.7 \text{ kJ} \cdot \text{kg}^{-1}$$
$$W_{\text{lost}}(\text{cond}) = T_0(S_3 - S_2) - Q(\text{cond}) = T_0(S_3 - S_2) - (H_3 - H_2)$$
$$= (288.15)(0.4365 - 8.1511) - (125.7 - 2584.8) = 236.1 \text{ kJ} \cdot \text{kg}^{-1}$$

The actual work of the process per kilogram is given by (6.23):

$$W_s = -(H_2 - H_1) = -(2584.8 - 3411.3) = 826.5 \text{ kJ} \cdot \text{kg}^{-1}$$

Since this is a work-producing process, a work analysis is done in accord with (8.17), here written

$$W_{\text{ideal}} = W_s + W_{\text{lost}}(\text{turb}) + W_{\text{lost}}(\text{cond})$$

Results are shown in Table 8-1.

Table 8-1

	$W/\text{kJ} \cdot \text{kg}^{-1}$	% of W_{ideal}
W_s	826.5	63.7
$W_{\text{lost}}(\text{turb})$	234.7	18.1
$W_{\text{lost}}(\text{cond})$	236.1	18.2
TOTAL (W_{ideal})	1297.3	100.0

It is seen that of the work which could theoretically or ideally be obtained for the given change of state, 18.1% is unrealized because of mechanical irreversibilities in the turbine and 18.2% is unrealized because of irreversible heat transfer in the condenser. The actual work represents 63.7% of the ideal, and this is η_t, the thermodynamic efficiency of the process. This efficiency is quite different from the turbine efficiency, which is found to be $\eta_e = 76.1\%$.

Solved Problems

8.1 Compare (8.3) with the mechanical-energy balance (6.25), and combine the result with (8.9) to show the relation of Ψ in (6.25) to the lost work.

The two equations to be compared are

$$W_{\text{ideal}} = T_0 \, \Delta S - \Delta H - \tfrac{1}{2} \Delta u^2 - g \, \Delta z \tag{8.3}$$

$$-W_s = \int_1^2 V \, dP + \tfrac{1}{2} \Delta u^2 + g \, \Delta z + \Psi \tag{6.25}$$

Addition gives, in view of (8.9),

$$W_{\text{lost}} = T_0 \, \Delta S - \Delta H + \int_1^2 V \, dP + \Psi \tag{1}$$

However, the property relation (3.48) requires that $dH = T \, dS + V \, dP$, and integration yields

$$\Delta H = \int_1^2 T \, dS + \int_1^2 V \, dP$$

Combining this with (1), we have

$$W_{\text{lost}} = T_0 \, \Delta S - \int_1^2 T \, dS + \Psi \tag{2}$$

In our formal treatment of ideal work and lost work, we assumed that T_0 was fixed at a constant value in any particular application. However, the selection of a surroundings temperature is inevitably a matter of judgment, and we could have considered the more general case of arbitrary "surroundings" temperatures. In particular we could consider an imaginary surroundings made up of an infinite number of heat reservoirs, one for every temperature traversed by the system. Heat transfer between system and "surroundings" could then always be reversible, and T_0 would be variable and equal to T. The term $T_0 \, \Delta S$ in our equations would become $\int_1^2 T \, dS$, and (2) would reduce to

$$W_{\text{lost}} = \Psi \tag{3}$$

Thus, energy dissipated in fluid friction is identified with lost work taken with respect to the *system* temperature, rather than with respect to a real surroundings temperature. This means that W_{lost} here does not include work lost as the result of direct heat transfer between system and surroundings. We then have a lost-work concept that takes into account only irreversibilities internal to the system.

8.2 Find W_{lost} (per mole) and η_t for the continuous adiabatic compression of helium gas from 1 bar and 293 K to 5 bar. Take $\eta_c = 3/4$, $C_p = (5/2)R$, and $T_0 = T_1 = 293$ K.

For the adiabatic compression, (8.13) and (2.10) give

$$W_{\text{lost}} = T_0 \, \Delta S = C_P T_0 \ln \frac{T_2}{T_1} - R T_0 \ln \frac{P_2}{P_1}$$

which, by use of (7) of Problem 6.14, becomes

$$W_{\text{lost}} = C_P T_0 \ln \left\{ 1 + \eta_c^{-1} \left[\left(\frac{P_2}{P_1} \right)^{(\gamma - 1)/\gamma} - 1 \right] \right\} - R T_0 \ln \frac{P_2}{P_1}$$

$$= \frac{5}{2} (8.314)(293) \ln \left\{ 1 + \frac{4}{3} [5^{2/5} - 1] \right\} - (8.314)(293) \ln 5$$

$$= 894.6 \text{ J} \cdot \text{mol}^{-1}$$

Equation (6) of Problem 6.14 gives the shaft work as

$$W_s(\text{comp}) = -\eta_c^{-1} C_P T_1 \left[\left(\frac{P_2}{P_1} \right)^{(\gamma - 1)/\gamma} - 1 \right]$$

$$= -\left(\frac{4}{3} \right)\left(\frac{5}{2} \right)(8.314)(293)[5^{2/5} - 1] = -7337.7 \text{ J} \cdot \text{mol}^{-1}$$

Therefore, by (8.8) and (8.9),

$$\eta_t = \frac{W_{ideal}}{W_s} = \frac{W_s + W_{lost}}{W_s} = \frac{-7337.7 + 894.6}{-7337.7} = 0.878$$

8.3 The process described in Problem 6.21 was shown not to violate the laws of thermodynamics and therefore to be possible. The minimum work required to generate the input nitrogen is the work of reversible, isothermal compression of nitrogen from 100 to 600 kPa at 21 °C. Compare this minimum work requirement with the lost work of the device, taking $T_0 = 294.15$ K.

The work of isothermal, reversible compression in a flow process is given by (6.25). Neglecting the potential- and kinetic-energy terms and setting $\Psi = 0$ (because the process is reversible), we get simply

$$-W_s = \int_1^2 V \, dP \qquad (1)$$

As in Problem 6.21 we assume nitrogen behaves as an ideal gas, so that

$$W_s = -\int_1^2 \frac{RT}{P} \, dP = -RT \ln \frac{P_2}{P_1} = -(8.314)(294.15)(\ln 6) = -4381.9 \text{ J} \cdot \text{mol}^{-1}$$

The lost work of the device is most easily calculated from (8.15), $W_{lost} = T_0 \, \Delta S_{total}$. Since there is no heat exchange with the surroundings, ΔS_{total} is just the entropy change of the nitrogen flowing through the device, and this was determined in Problem 6.21 to be 28.514 $\text{J} \cdot \text{K}^{-1}$, on the basis of one mole of *each* exit steam. If we change the basis to one mole of *entering* nitrogen, then $\Delta S_{total} = 14.257$ $\text{J} \cdot \text{mol}^{-1} \cdot \text{K}^{-1}$ and

$$W_{lost} = (294.15)(14.257) = 4193.7 \text{ J} \cdot \text{mol}^{-1}$$

Thus, of the 4381.9 $\text{J} \cdot \text{mol}^{-1}$ of work required as a minimum to compress the nitrogen, 4193.7 $\text{J} \cdot \text{mol}^{-1}$ is lost because of the irreversibilities of the device, and only 188.2 $\text{J} \cdot \text{mol}^{-1}$, or 4.3% of the work, is effective in bringing about the separation into two streams at different temperatures. The device is clearly highly inefficient. (Such devices do exist, and are known as *Hilsch tubes* or *Ranque/Hilsch tubes*.)

8.4 Make a work analysis of the process described in Problem 6.15, taking $T_0 = 300$ K.

The overall process results in the compression of methane from the initial conditions $T_1 = 26.7$ °C and $P_1 = 0.69$ MPa to the final conditions $T_3 = 38$ °C and $P_3 = 3.45$ MPa. For this change of state the data of Problem 6.15 give

$$\Delta H = 949.6 - 946.7 = 2.9 \text{ kJ} \cdot \text{kg}^{-1} \qquad \Delta S = 5.264 - 6.071 = -0.807 \text{ kJ} \cdot \text{kg}^{-1} \cdot \text{K}^{-1}$$

By (8.6),

$$W_{ideal} = T_0 \, \Delta S - \Delta H = (300)(-0.807) - 2.9 = -245.0 \text{ kJ} \cdot \text{kg}^{-1}$$

The lost-work terms are calculated from (8.13), $W_{lost} = T_0 \, \Delta S - Q$. For the compressor,

$$W_{lost} = (300)(6.238 - 6.071) - 0 = 50.1 \text{ kJ} \cdot \text{kg}^{-1}$$

where $S = 6.238$ $\text{kJ} \cdot \text{kg}^{-1} \cdot \text{K}^{-1}$ is found in the table of Problem 6.15 by interpolation at 182 °C. For the cooler, assuming the heat removed is discarded to the surroundings, we have

$$Q = H_3 - H_2 = 949.6 - 1324.2 = -374.6 \text{ kJ} \cdot \text{kg}^{-1}$$

whence

$$W_{lost} = (300)(5.264 - 6.238) - (-374.6) = 82.4 \text{ kJ} \cdot \text{kg}^{-1}$$

The work analysis, based on (8.18), is given in Table 8-2. The value of W_s agrees exactly with the value calculated in Problem 6.15.

Table 8-2

	$W/\text{kJ} \cdot \text{kg}^{-1}$	% of $|W_s|$
$|W_{\text{ideal}}|$	245.0	64.9
$W_{\text{lost}}(\text{comp})$	50.1	13.3
$W_{\text{lost}}(\text{cool})$	82.4	21.8
TOTAL $(|W_s|)$	377.5	100.0

8.5   Make a power analysis of the process described in Problem 6.26; take $T_0 = 300$ K.

The ideal power of the process is found from (8.2), which in this case becomes

$$\dot{W}_{\text{ideal}} = \dot{m} T_0 \, \Delta S - \dot{m} \, \Delta H - \tfrac{1}{2} \dot{m} \, \Delta u^2$$

Although the potential-energy term is taken as negligible, the kinetic-energy term must be retained. Since the initial and final states are both saturated steam at 1 bar, both ΔS and ΔH are zero. Thus

$$\dot{W}_{\text{ideal}} = -\tfrac{1}{2} \dot{m} \, \Delta u^2 = -\tfrac{1}{2}(2.5)(600^2 - 0^2) = -450\,000 \text{ W} \qquad \text{or} \qquad -450.00 \text{ kW}$$

We apply (8.12), $\dot{W}_{\text{lost}} = \dot{m} T_0 \, \Delta S - \dot{Q}$, to both the compressor and the nozzle. This requires knowledge of the intermediate state of the steam at the compressor discharge or nozzle inlet. The energy equation for the compressor is (6.18), which here reduces to

$$\dot{m} \, \Delta H = \dot{Q} - \dot{W}_s$$

From the statement of Problem 6.26, $\dot{m} = 2.5$ kg $\cdot$ s^{-1} and $\dot{Q} = -150$ kW; the power input is given by the answer to Problem 6.26 as $\dot{W}_s = -600$ kW. Thus, by the preceding equation,

$$\Delta H \equiv H_2 - H_1 = 180 \text{ kJ} \cdot \text{kg}^{-1}$$

In addition, we have from the steam tables for saturated steam at 1 bar (100 kPa)

$$H_1 = H_3 = 2675.4 \text{ kJ} \cdot \text{kg}^{-1} \qquad S_1 = S_3 = 7.3598 \text{ kJ} \cdot \text{kg}^{-1} \cdot \text{K}^{-1}$$

Therefore $H_2 = 2675.4 + 180.0 = 2855.4$ kJ $\cdot$ kg^{-1}. We find from Appendix E that steam at 3 bar (300 kPa) has this enthalpy at 195.2 °C, at which condition $S_2 = 7.2900$ kJ $\cdot$ kg$^{-1} \cdot$ K^{-1}. The lost power in the compressor is therefore

$$\dot{W}_{\text{lost}} = (2.5)(300)(7.2900 - 7.3598) + 150.0 = 97.65 \text{ kW}$$

and in the nozzle it is

$$\dot{W}_{\text{lost}} = (2.5)(300)(7.3598 - 7.2900) = 52.35 \text{ kW}$$

The power analysis in accord with (8.18) is made in Table 8-3.

Table 8-3

	$\dot{W}/\text{kW}$	% of $|\dot{W}_s|$
$|\dot{W}_{\text{ideal}}|$	450.00	75.0
$\dot{W}_{\text{lost}}(\text{comp})$	97.65	16.3
$\dot{W}_{\text{lost}}(\text{nozzle})$	52.35	8.7
TOTAL $(|\dot{W}_s|)$	600.00	100.0

8.6   A standard refrigeration cycle produces 25 kg $\cdot$ s^{-1} of chilled water at 12 °C (see Fig. 8-3). The refrigerant circulating in the cycle is H_2O. At point 1, saturated steam at 1 kPa enters an adiabatic compressor of 80% efficiency based on isentropic operation, and is compressed to

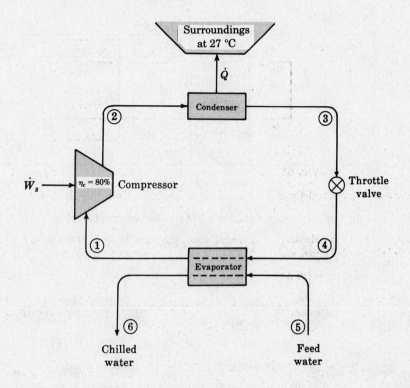

Fig. 8-3

10 kPa at point 2. It then enters a condenser from which it emerges, at point 3, as a saturated liquid at 10 kPa, heat Q having been discharged to the surroundings at 27 °C. The saturated liquid flashes through a valve, which acts as a throttle to reduce the pressure to 1 kPa at point 4. The remaining liquid is vaporized in the evaporator, producing saturated vapor at point 1. The water being chilled also passes through the evaporator, and by heat exchange with the evaporating refrigerant is cooled from its feed temperature, 27 °C, to 12 °C. Determine (a) the power requirement of the compressor and (b) the rate of heat rejection to the surroundings.

First we find the thermodynamic properties at the various points indicated in Fig. 8-3. Data for points 1, 3, 5, and 6 are found directly from the steam tables, and are entered in Table 8-4; calculations are required only for points 2 and 4. Consider first the compressor, for which (see Fig. 6.11),

$$-(H_2 - H_1) = W_s = \frac{W_s(\text{rev})}{\eta_c} = \frac{-(H_2' - H_1)}{\eta_c}$$

From the steam tables, $H_2' = 2914.7 \text{ kJ} \cdot \text{kg}^{-1}$; and H_1 and η_c are known. Therefore, $H_2 = 3014.8 \text{ kJ} \cdot \text{kg}^{-1}$. From the steam tables for superheated vapor at 10 kPa,

$$t_2 = 268.85 \text{ °C} \quad \text{and} \quad S_2 = 9.1692 \text{ kJ} \cdot \text{kg}^{-1} \cdot \text{K}^{-1}$$

For the throttle valve, expansion from 10 to 1 kPa occurs at constant enthalpy, and

$$H_4 = H_3 = 191.8 = H_4^l + x_4 \Delta H_4^{lv} = 29.3 + 2485.1 x_4$$

Solving, $x_4 = 0.0654$. Finally,

$$S_4 = S_4^l + x_4 \Delta S_4^{lv} = 0.1060 + (0.0654)(8.8707) = 0.6861 \text{ kJ} \cdot \text{kg}^{-1} \cdot \text{K}^{-1}$$

and Table 8-4 is complete.

Table 8-4

Point	State	$t/°C$	P/kPa	$H/kJ \cdot kg^{-1}$	$S/kJ \cdot kg^{-1} \cdot K^{-1}$
1	sat. vapor	6.98	1	2514.4	8.9767
2	vapor	268.9	10	3014.8	9.1692
3	sat. liquid	45.83	10	191.8	0.6493
4	wet vapor	6.98	1	191.8	0.6861
5	liquid	27.0	101.33	113.1	0.3949
6	liquid	12.0	101.33	50.4	0.1805

(a) Application of the energy equation (6.16) to the evaporator gives

$$(\dot{m}_1 H_1 - \dot{m}_4 H_4) + (\dot{m}_6 H_6 - \dot{m}_5 H_5) = 0$$

or, since $\dot{m}_4 = \dot{m}_1$ and $\dot{m}_5 = \dot{m}_6$,

$$\dot{m}_1(H_1 - H_4) + \dot{m}_6(H_6 - H_5) = 0$$

With $\dot{m}_6 = 25 \text{ kg} \cdot \text{s}^{-1}$ and numerical values of enthalpy from Table 8-4, we find $\dot{m}_1 = 0.6749 \text{ kg} \cdot \text{s}^{-1}$ as the flow rate of the refrigerant. The power requirement of the compressor is then

$$-\dot{W}_s = -\dot{m}_1 W_s = \dot{m}_1(H_2 - H_1) = 337.7 \text{ kW}$$

(b) Application of the energy equation (6.18) to the condenser yields

$$\dot{Q} = \dot{m}_1(H_3 - H_2) = (0.6749)(191.8 - 3014.8) = -1905.2 \text{ kW}$$

8.7 Make (a) an entropy-generation analysis, and (b) a power analysis, of the refrigeration process of Problem 8.6 ($T_0 = 300.15 \text{ K}$).

(a) The rate of entropy generation for each part of the process is (see Section 8.1)

$$\dot{S}_{\text{total}} = \Delta(\dot{m}S)_{\substack{\text{flowing} \\ \text{streams}}} - \frac{\dot{Q}}{T_0}$$

(i) For the compressor, since $\dot{Q} = 0$:

$$\dot{S}_{\text{total}} = \dot{m}_1(S_2 - S_1) = (0.6749)(9.1692 - 8.9767) = 0.1299 \text{ kW} \cdot \text{K}^{-1}$$

(ii) For the condenser:

$$\dot{S}_{\text{total}} = \dot{m}_2(S_3 - S_2) - \frac{\dot{Q}}{T_0}$$

$$= (0.6749)(0.6493 - 9.1692) - \frac{-1905.2}{300.15} = 0.5974 \text{ kW} \cdot \text{K}^{-1}$$

(iii) For the throttle valve, with $\dot{Q} = 0$:

$$\dot{S}_{\text{total}} = \dot{m}_3(S_4 - S_3) = (0.6749)(0.6861 - 0.6493) = 0.0248 \text{ kW} \cdot \text{K}^{-1}$$

(iv) For the evaporator, $\dot{Q} = 0$ because $\dot{Q}$ is always heat transfer to the *surroundings*, and

$$\dot{S}_{\text{total}} = \dot{m}_4(S_1 - S_4) + \dot{m}_6(S_6 - S_5)$$
$$= (0.6749)(8.9767 - 0.6861) + (25)(0.1805 - 0.3949) = 0.2353 \text{ kW} \cdot \text{K}^{-1}$$

The analysis is presented in Table 8-5.

Table 8-5

	$\dot{S}/\text{kW} \cdot \text{K}^{-1}$	% of $\Sigma \dot{S}_{\text{total}}$
$\dot{S}_{\text{total}}(\text{comp})$	0.1299	13.2
$\dot{S}_{\text{total}}(\text{cond})$	0.5974	60.5
$\dot{S}_{\text{total}}(\text{throttle})$	0.0248	2.5
$\dot{S}_{\text{total}}(\text{evap})$	0.2353	23.8
$\Sigma \dot{S}_{\text{total}}$	0.9874	100.0

(b) Since the process requires power, the appropriate equation upon which to base the analysis is

$$|\dot{W}_s| = |\dot{W}_{\text{ideal}}| + \Sigma \dot{W}_{\text{lost}} \qquad (8.18)$$

The ideal power requirement for chilling water from 27 to 12 °C is given by (8.5) as

$$\dot{W}_{\text{ideal}} = \dot{m}_6[T_0(S_6 - S_5) - (H_6 - H_5)]$$
$$= (25)[(300.15)(0.1805 - 0.3949) - (50.4 - 113.1)] = -41.3 \text{ kW}$$

and the lost-power terms are obtained from Table 8-5 via (8.14). Table 8-6 displays the analysis, which reveals the thermodynamic efficiency η_t of the process to be 12.2%. The mechanical irreversibilities in the compressor result in the loss (or waste) of 11.6% of the actual power of the process. The irreversibilities in both the condenser and evaporator resulting from heat transfer across finite temperature differences cause large lost-power terms; in fact, these terms together amount to 74% of the actual power. Surprisingly, the mechanical irreversibilities of the throttling process produce a very small lost-power term.

Table 8-6

	$\dot{W}/\text{kW}$	% of $|\dot{W}_s|$
$|\dot{W}_{\text{ideal}}|$	41.3	12.2
$\dot{W}_{\text{lost}}(\text{comp})$	39.0	11.6
$\dot{W}_{\text{lost}}(\text{cond})$	179.3	53.1
$\dot{W}_{\text{lost}}(\text{throttle})$	7.5	2.2
$\dot{W}_{\text{lost}}(\text{evap})$	70.6	20.9
TOTAL $(|\dot{W}_s|)$	337.7	100.0

8.8 A nuclear-power-plant cycle is shown in Fig. 8-4. Heat $\dot{q}$ flows from the nuclear reactor to the boiler, to produce saturated steam at point 1 at a pressure of 7200 kPa. This steam flows to an adiabatic turbine which exhausts at 10 kPa at point 2. The turbine efficiency is $\eta_e = 70\%$ compared with isentropic operation. The exhaust steam enters a condenser, from which it emerges at point 3 as liquid water subcooled to 40 °C. Heat $\dot{Q}$ is discarded from the condenser to the surroundings at 20 °C. The liquid from the condenser is fed to an adiabatic pump ($\eta_c = 80\%$) which raises its pressure at point 4 to that of the boiler, 7200 kPa. The plant has the rated capacity 750 MW. Determine (a) the steam rate $\dot{m}$ and (b) the rate of heat transfer $\dot{Q}$ for rated capacity.

We begin by collecting in Table 8-7 the property values of the circulating steam at the various points of the cycle. At point 1 there is saturated steam at 7200 kPa, and its properties are entered directly from the steam tables. The steam at point 2 is wet, and to find its properties we first consider an isentropic expansion of the steam from point 1 to point 2' (see Fig. 6-9):

$$S_2' = S_1 = 5.8020 \text{ kJ} \cdot \text{kg}^{-1} \cdot \text{K}^{-1}$$

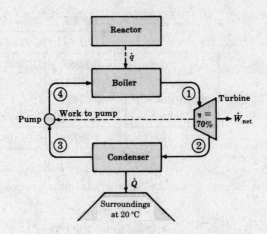

Fig. 8-4

Steam at 10 kPa with this entropy is wet; therefore

$$S_2' = S_2^l + x_2' \Delta S_2^{lv} = 0.6493 + 7.5018 x_2'$$

from which the quality is $x_2' = 0.6869$. The enthalpy is then

$$H_2' = H_2^l + x_2' \Delta H_2^{lv} = 191.8 + (0.6869)(2393.0) = 1835.5 \text{ kJ} \cdot \text{kg}^{-1}$$

whence $(\Delta H)_S = 1835.5 - 2770.9 = -935.4 \text{ kJ} \cdot \text{kg}^{-1}$. Then

$$\begin{aligned} H_2 &= H_1 + \Delta H = H_1 + \eta_e (\Delta H)_S \\ &= 2770.9 + (0.70)(-935.4) = 2116.1 \text{ kJ} \cdot \text{kg}^{-1} \end{aligned}$$

The actual quality is given by

$$H_2 = 2116.1 = 191.8 + 2393.0 x_2 \qquad \text{or} \qquad x_2 = 0.8041$$

and the entropy is calculated from this result as

$$S_2 = 0.6493 + (0.8041)(7.5018) = 6.6818 \text{ kJ} \cdot \text{kg}^{-1} \cdot \text{K}^{-1}$$

Table 8-7

Point	State	$t/°C$	P/kPa	$H/\text{kJ} \cdot \text{kg}^{-1}$	$S/\text{kJ} \cdot \text{kg}^{-1} \cdot \text{K}^{-1}$
1	sat. vapor	287.7	7200	2770.9	5.8020
2	wet vapor	45.8	10	2116.1	6.6818
3	liquid	40.0	10	167.5	0.5721
4	liquid	40.6	7200	176.6	0.5778

At point 3, we have a subcooled liquid at 40 °C and 10 kPa. At 40 °C the pressure of *saturated* liquid is about 7.4 kPa. Since the effect of pressure on liquid properties is small, we take the properties at point 3 to be those of saturated liquid at 40 °C.

At point 4, the discharge from the pump is compressed (or subcooled) liquid at 7200 kPa. The results of Problem 6.22 provide the enthalpy rise across the pump as $9.1 \text{ kJ} \cdot \text{kg}^{-1}$ and the temperature rise as 0.64 K. Thus,

$$H_4 = H_3 + 9.1 = 176.6 \text{ kJ} \cdot \text{kg}^{-1} \qquad \text{and} \qquad t_4 = 40.64 \text{ °C}$$

The entropy change is found from (2) of Example 3.12:

$$\Delta S = C_P \ln \frac{T_4}{T_3} - \beta V_1 (P_4 - P_3)$$

Using the values of C_P, β, and V_1 given in Problem 6.22, we get

$$\Delta S = 4.178 \ln \frac{273.15 + 40.64}{273.15 + 40.00} - (385 \times 10^{-6})(1.008 \times 10^{-3})(7190) = 0.0057 \text{ kJ} \cdot \text{kg}^{-1} \cdot \text{K}^{-1}$$

Thus, $S_4 = S_3 + 0.0057 = 0.5721 + 0.0057 = 0.5778 \text{ kJ} \cdot \text{kg}^{-1} \cdot \text{K}^{-1}$, and this completes Table 8-7.

(a) The power at rated capacity is given as $\dot{W}_{\text{net}} = 750\,000 \text{ kJ} \cdot \text{s}^{-1}$. Since

$$W_{\text{net}} = W_s + W_{\text{pump}} = -(H_2 - H_1) - (H_4 - H_3)$$
$$= 654.8 - 9.1 = 645.7 \text{ kJ} \cdot \text{kg}^{-1}$$

the steam rate is given by

$$\dot{m} = \frac{\dot{W}_{\text{net}}}{W_{\text{net}}} = \frac{750\,000}{645.7} = 1161.5 \text{ kg} \cdot \text{s}^{-1}$$

(b) $\dot{Q} = \dot{m}(H_3 - H_2) = (1161.5)(167.5 - 2116.1) = -2.263 \times 10^6 \text{ kW}$ or -2263 MW

8.9   Make (a) an entropy-generation analysis, and (b) a power analysis, of the plant of Problem 8.8. Treat the reactor as a heat reservoir at 320 °C (593.15 K).

Heretofore, applications have been confined to processes of the type illustrated in Fig. 8-1, for which the rate of entropy generation is

$$\dot{S}_{\text{total}} = \Delta(\dot{m}S)_{\substack{\text{flowing} \\ \text{streams}}} - \frac{\dot{Q}}{T_0}$$

However, in the process considered here, a heat reservoir in addition to the surroundings is also present, and an additional term is needed to account for its entropy change. Thus we write

$$\dot{S}_{\text{total}} = \Delta(\dot{m}S)_{\substack{\text{flowing} \\ \text{streams}}} + \frac{\dot{q}}{T} - \frac{\dot{Q}}{T_0} \tag{1}$$

where $\dot{q}/T$ is the rate of entropy change of the reactor (part of the system). Quantity $\dot{q}$ refers to the reactor and hence is negative, because heat flow is *from* the reactor to the boiler. It is *internal* heat transfer and is distinct from $\dot{Q}$, which is *always* heat transfer between system and surroundings.

(a) Apply (1) to each part of the process.
   (i) For the turbine, both $\dot{Q}$ and $\dot{q}$ are zero, whence

$$\dot{S}_{\text{total}} = \dot{m}(S_2 - S_1) = (1161.5)(6.6818 - 5.8020) = 1021.9 \text{ kW} \cdot \text{K}^{-1}$$

   (ii) For the condenser, with $\dot{q} = 0$, and $T_0 = 293.15 \text{ K}$,

$$\dot{S}_{\text{total}} = \dot{m}(S_3 - S_2) - \frac{\dot{Q}}{T_0}$$

$$= (1161.5)(0.5721 - 6.6818) - \frac{-2.263 \times 10^6}{293.15} = 623.2 \text{ kW} \cdot \text{K}^{-1}$$

   (iii) For the pump, since $\dot{Q} = 0$ and $\dot{q} = 0$,

$$\dot{S}_{\text{total}} = \dot{m}(S_4 - S_3) = (1161.5)(0.5778 - 0.5721) = 6.6 \text{ kW} \cdot \text{K}^{-1}$$

   (iv) Treating the boiler and reactor as a single part of the process for which $\dot{Q} = 0$, we have

$$\dot{S}_{\text{total}} = \dot{m}(S_1 - S_4) + \frac{\dot{q}}{T} = \dot{m}\left(S_1 - S_4 - \frac{H_1 - H_4}{T}\right)$$

$$= (1161.5)\left(5.8020 - 0.5778 - \frac{2770.9 - 176.6}{593.15}\right) = 988.3 \text{ kW} \cdot \text{K}^{-1}$$

Table 8-8 displays the analysis.

Table 8-8

	$\dot{S}/\text{kW} \cdot \text{K}^{-1}$	% of $\Sigma \dot{S}_{total}$
$\dot{S}_{total}(\text{turb})$	1021.9	38.7
$\dot{S}_{total}(\text{cond})$	623.2	23.6
$\dot{S}_{total}(\text{pump})$	6.6	0.3
$\dot{S}_{total}(\text{reac/boil})$	988.3	37.4
$\Sigma \dot{S}_{total}$	2640.0	100.0

(b)   With $\dot{W}_s$ replaced by $\dot{W}_{net}$, (8.17) becomes

$$\dot{W}_{ideal} = \dot{W}_{net} + \Sigma \dot{W}_{lost}$$

The simplest means for calculation of $\dot{W}_{ideal}$ is via the second Carnot equation (2.3):

$$\dot{W}_{ideal} = |\dot{q}|\left(1 - \frac{T_0}{T_{reac}}\right)$$

Here $|\dot{q}| = \dot{m}(H_1 - H_4) = 3013$ MW, so that

$$\dot{W}_{ideal} = (3013)\left(1 - \frac{293.15}{593.15}\right) = 1524 \text{ MW}$$

The terms $\dot{W}_{lost}$ are derived from Table 8-8 via (8.14); the complete analysis is shown in Table 8-9.

Table 8-9

	$\dot{W}/\text{MW}$	% of $\dot{W}_{ideal}$
$\dot{W}_{net}$	750	49.2
$\dot{W}_{lost}(\text{turb})$	299.6	19.7
$\dot{W}_{lost}(\text{cond})$	182.7	12.0
$\dot{W}_{lost}(\text{pump})$	1.9	0.1
$\dot{W}_{lost}(\text{reac/boil})$	289.7	19.0
TOTAL ($\dot{W}_{ideal}$)	1523.9	100.0

The thermodynamic efficiency of this plant, $\eta_t = 49.2\%$, is very high in comparison with most commercial processes. It is quite different from the thermal efficiency, defined in (2.4) as

$$\eta = \frac{\dot{W}_{net}}{|\dot{q}|} = 750/3013 = 0.249$$

This figure means that of the heat transferred from the reactor only 24.9% appears as work, and 75.1% is discarded as heat to the surroundings. However, even if the plant were perfect ($\Sigma \dot{W}_{lost} = 0$), the thermal efficiency would be

$$\eta' = \frac{\dot{W}_{ideal}}{|\dot{q}|} = 1524/3013 = 0.506$$

and about half of $\dot{q}$ would still appear as heat discarded to the surroundings.

8.10   Fresh water is produced from seawater by the process of Fig. 8-5. Seawater at 25 °C and containing 3.45 mass % dissolved salts enters at point 1, and first flows through a heat exchanger, where its temperature is raised. It then goes to an evaporator where 50% of its water content is vaporized at a pressure of 1 bar, producing a more concentrated brine of (it will be shown) 6.67 mass % dissolved salts, which flows out of the evaporator at point 3. If the liquid in the evaporator is well mixed, its concentration of salts must be 6.67 mass % (except

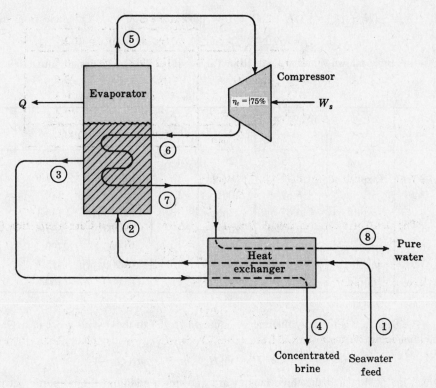

Fig. 8-5

immediately adjacent to the seawater inlet). Because of the boiling-point elevation this solution boils at 100.7 °C at 1 bar, producing a slightly superheated vapor, which leaves the evaporator at point 5 and flows to a compressor, which raises the pressure to 1.6 bar. The compressor operates adiabatically and is 75% efficient compared with an isentropic process. The compressed vapor enters coils in the evaporator where it condenses at 1.6 bar (saturation temperature, 113.3 °C), providing heat for the evaporation process. Design conditions call for saturated liquid water to leave the coils at point 7. This condensate and the hot brine solution from the evaporator are used in the heat exchanger to preheat the incoming seawater. Heat losses to the surroundings are considered negligible from the compressor and heat exchanger, but a heat loss of 1.7 kJ per kg of pure water produced is estimated for the evaporator. Make a work analysis of the process.

We take as a *basis* for working the problem a unit mass (1 kg) of pure water produced at point 8 in steady operation of the process. Since this represents half of the water in the entering seawater, the mass of seawater entering is

$$m_1 = \frac{\text{mass of water entering}}{\text{fraction water in seawater}} = \frac{2(1 \text{ kg})}{1.0000 - 0.0345} = 2.0715 \text{ kg}$$

The mass of salt entering is therefore $2.0715 - 2.0000 = 0.0715$ kg; the amount of concentrated brine produced at point 4 is $2.0715 - 1.0000 = 1.0715$ kg, with a concentration of

$$\frac{0.0715}{1.0715} \times 100 = 6.67\% \text{ dissolved salts}$$

We first calculate W_s for the compressor (Section 6.4) with data from the steam tables:

$$W_s = \frac{-(\Delta H)_S}{\eta_c} = \frac{-(H_6' - H_5)}{\eta_c} = \frac{-(2762.2 - 2677.6)}{0.75} = -112.8 \text{ kJ} \cdot \text{kg}^{-1}$$

Since $W_s = -(H_6 - H_5) = -(H_6 - 2677.6)$, we have $H_6 = 2790.4 \text{ kJ} \cdot \text{kg}^{-1}$, and from the steam tables

$$t_6 = 159.2 \text{ °C} \quad \text{and} \quad S_6 = 7.4319 \text{ kJ} \cdot \text{kg}^{-1} \cdot \text{K}^{-1}$$

These and other known values are entered in Table 8-10; additional values are found below.

Table 8-10

Point	State	Mass % salts	$\dfrac{t}{°C}$	$\dfrac{P}{\text{bar}}$	$\dfrac{m}{\text{kg}}$	$\dfrac{H}{\text{kJ} \cdot \text{kg}^{-1}}$	$\dfrac{S}{\text{kJ} \cdot \text{kg}^{-1} \cdot \text{K}^{-1}}$
1	liquid	3.45	25	1	2.0715	99.6	0.3342
2	liquid	3.45	93.9	1	2.0715	376.6	1.1689
3	sat. liq.	6.67	100.7	1	1.0715	388.1	1.2010
4	liquid	6.67	38.4	1	1.0715	146.1	0.4618
5	vapor	0.0	100.7	1	1.0000	2677.6	7.3655
6	vapor	0.0	159.2	1.6	1.0000	2790.4	7.4319
7	sat. liq.	0.0	113.3	1.6	1.0000	475.4	1.4550
8	liquid	0.0	38.4	1.6	1.0000	160.8	0.5504

The next step is to apply the energy equation (6.16) to the *overall* process, neglecting kinetic- and potential-energy terms:

$$m_4 H_4 + m_8 H_8 - m_1 H_1 = m_8 (Q - W_s) \tag{1}$$

Values for Q, W_s, and the three masses are known; in addition, the properties at point 1 are fixed, because the conditions are fully specified. Available data for seawater include $H_1 = 99.6 \text{ kJ} \cdot \text{kg}^{-1}$. Thus (1) becomes

$$(1.0715)(H_4) + (1.0000)(H_8) - (2.0715)(99.6) = 1.0000[-1.7 - (-112.8)] \tag{2}$$

This equation contains two unknowns, H_4 and H_8, and we must make an additional specification, if we are to determine both of them. To this end, we assume that the heat exchanger is so designed that $t_4 = t_8$. In this case, the selection of a single temperature allows values for both H_4 and H_8 to be found from available data. The procedure is to select a trial value of $t \equiv t_4 = t_8$ and to substitute the corresponding enthalpies into (2). If the equation is satisfied, then the correct t was used. The temperature so found by this trial procedure is 38.4 °C, and properties for points 4 and 8 at this temperature are listed in Table 8-10. Only data for point 2 are lacking. The enthalpy at point 2 is found by application of the energy equation (6.16) to an appropriate portion of the process—the heat exchanger. Since $m_4 = m_3$, $m_8 = m_7$, and $m_2 = m_1$, (6.16) becomes

$$m_4 (H_4 - H_3) + m_1 (H_2 - H_1) + m_8 (H_8 - H_7) = 0$$

The only unknown is H_2, and substitution of numerical values leads to $H_2 = 376.6 \text{ kJ} \cdot \text{kg}^{-1}$. Seawater has this enthalpy at 93.9 °C. Table 8-10 is now complete, and the work analysis can be made.

The ideal work is given by (8.4), integrated over the time interval required for production of 1 kg of output. Taking $T_0 = 298.15 \text{ K}$, we obtain

$$\begin{aligned} W_{\text{ideal}} &= (298.15)[(1.0715)(0.4618) + 0.5504 - (2.0715)(0.3342)] \\ &\quad - [(1.0715)(146.1) + 160.8 - (2.0715)(99.6)] \\ &= -5.90 \text{ kJ} \end{aligned}$$

The lost work for the same time interval follows from (8.11): $W_{\text{lost}} = T_0 \, \Delta(mS) - Q$. Application of this equation to the three parts of the process, the heat exchanger, the evaporator, and the compressor, gives the following results.

Heat exchanger $(Q = 0)$

$$\begin{aligned} W_{\text{lost}} &= T_0 [m_4 (S_4 - S_3) + m_1 (S_2 - S_1) + m_8 (S_8 - S_7)] \\ &= (298.15)[(1.0715)(0.4618 - 1.2010) + (2.0715)(1.1689 - 0.3342) + (0.5504 - 1.4550)] \\ &= 9.67 \text{ kJ} \end{aligned}$$

Evaporator ($Q = 1.7$ kJ)

$$W_{\text{lost}} = T_0[m_5S_5 + m_3S_3 - m_2S_2 + m_6(S_7 - S_6)] - (-1.7)$$
$$= (298.15)[7.3655 + (1.0715)(1.2010) - (2.0715)(1.1689) + (1.4550 - 7.4319)] + 1.7$$
$$= 77.46 \text{ kJ}$$

Compressor ($Q = 0$)

$$W_{\text{lost}} = T_0[m_6(S_6 - S_5)] = (298.15)[7.4319 - 7.3655] = 19.80 \text{ kJ}$$

The work analysis by (*8.18*) is presented in Table 8-11. The value of W_s is in essential agreement with the value calculated earlier.

Table 8-11

	W/kJ	% of $|W_s|$
$|W_{\text{ideal}}|$	5.90	5.2
W_{lost}(exch)	9.67	8.6
W_{lost}(evap)	77.46	68.7
W_{lost}(comp)	19.80	17.5
TOTAL ($|W_s|$)	112.83	100.0

8.11 The following reactions are carried out as steady-flow processes:

$$H_2(g) + \tfrac{1}{2}O_2(g) \rightarrow H_2O(l) \tag{1}$$
$$H_2(g) + \tfrac{1}{2}O_2(g) \rightarrow H_2O(g) \tag{2}$$

(*a*) What is the maximum work that can be obtained from (*1*) at 1 bar and 300 K? For (*1*) at 300 K, $\Delta H°(1) = -285.830$ kJ and $\Delta G°(1) = -237.129$ kJ. (*b*) What is the maximum work that can be obtained from (*2*) with pure H_2 and *air* (21 mole % O_2 and 79 mole % N_2) at 1 bar and 300 K as reactants and a mixture of H_2O and N_2 gases at 1 bar and 500 K as products? For (*2*) at 300 K, $\Delta H°(2) = -241.818$ kJ and $\Delta G°(2) = -228.572$ kJ. (*c*) If H_2 is burned completely with the theoretical amount of air, both initially at 1 bar and 300 K, in an adiabatic reactor at constant pressure, what is the maximum work that can be obtained from the flue gases if they are brought to 500 K at a constant pressure of 1 bar? Where is the irreversibility in this process? What has increased in entropy, and by how much? (*d*) A simple power-plant cycle operates as follows. The flue gases of (*c*) are cooled to 500 K at the constant pressure 1 bar by the transfer of heat to a boiler used to generate superheated steam at 35 bar and 823.15 K (550 °C). A turbine ($\eta_e = 80\%$) expands the steam from the boiler to 0.1 bar. A condenser at 0.1 bar delivers saturated-liquid water to a pump ($\eta_c = 80\%$) that returns the water to the boiler. Make a work analysis of the cycle. For N_2 the heat capacity is given as a function of Kelvin temperature by

$$\frac{C_P°}{R} = 3.352 + (0.513 \times 10^{-3})(T/K) \tag{3}$$

and for $H_2O(g)$, by

$$\frac{C_P°}{R} = 3.608 + (1.288 \times 10^{-3})(T/K) \tag{4}$$

For all parts of this problem take $T_0 = 300$ K.

We assume throughout that all gases at 1 bar are ideal, and we take as a basis the reaction of 1 mol of H_2. According to (*7.148*), the standard entropy changes of reactions (*1*) and (*2*) at 300 K are:

$$\Delta S°(1) = \frac{-285.830 + 237.129}{300} = -0.162\,337 \text{ kJ} \cdot \text{K}^{-1}$$

$$\Delta S°(2) = \frac{-241.818 + 228.572}{300} = -0.044\,153 \text{ kJ} \cdot \text{K}^{-1}$$

(a) Since the gases are at 1 bar and are assumed ideal, the reactants and products are in their standard states at 300 K. The maximum work is then given by (8.6) as

$$W_{ideal} = T_0 \, \Delta S°(1) - \Delta H°(1) = (300)(-0.162\,337) - (-285.830) = +237.129 \text{ kJ}$$

This value is the negative of $\Delta G°$ because T_0, the temperature of the surroundings, has been taken equal to T, the reaction temperature; W_{ideal} would be obtained as the electrical work of a reversible H_2/O_2 fuel cell operating at 300 K. Such a cell would discard heat in the amount $Q = \Delta H°(1) + W_{ideal} = -48.701$ kJ.

(b) The reaction of one mole of H_2 with half a mole of atmospheric O_2 requires $0.5/0.21 = 2.381$ mol of air made up of 0.5 mol of O_2 and 1.881 mol of N_2. As the reaction in question is not the *standard* reaction, the property changes ΔH and ΔS for the overall process are calculated from the scheme of Fig. 8-6. All states are considered ideal-gas states at 1 bar. We now determine ΔH and ΔS for the various steps of this scheme.

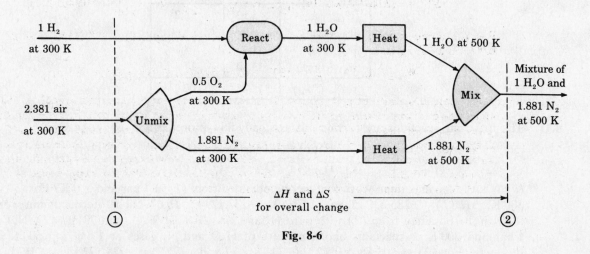

Fig. 8-6

For the unmixing of air: Since an ideal-gas mixture is a special case of an ideal solution, the applicable equations are (7.93) and (7.91):

$$\Delta H^{ig} = 0 \qquad \text{and} \qquad \Delta S^{ig} = -R \sum x_i \ln x_i$$

These equations are for a *mixing* process; for the reverse process, we merely change sign:

$$\Delta H^{unmix} = 0 \qquad \Delta S^{unmix} = R \sum x_i \ln x_i$$

Substitution of numerical values in the last equation gives

$$\Delta S^{unmix} = (0.008\,314)\left(\frac{0.5}{2.381} \ln \frac{0.5}{2.381} + \frac{1.881}{2.381} \ln \frac{1.881}{2.381} \right) = -0.004\,273 \text{ kJ} \cdot \text{mol}^{-1} \cdot \text{K}^{-1}$$

This is the value for 1 mol of mixture (air); the value for 2.381 mol is

$$\Delta S^{unmix} = (2.381)(-0.004\,273) = -0.010\,174 \text{ kJ} \cdot \text{K}^{-1}$$

For the reaction: Since the reactants and products are in their standard states at 300 K, we have simply

$$\Delta H^{react} = \Delta H°(2) = -241.818 \text{ kJ} \qquad \Delta S^{react} = \Delta S°(2) = -0.044\,153 \text{ kJ} \cdot \text{K}^{-1}$$

For the heating steps: The property changes per mole are found from (3) or (4) and the methods of Section 4.7:

$$\Delta H^{heat} = \langle C_P^{ig} \rangle_T (500 - 300) = R(A + BT_{am})(500 - 300)$$

$$\Delta S^{heat} = \langle C_P^{ig} \rangle_{\ln T} \ln \frac{500}{300} = R(A + BT_{lm}) \ln \frac{500}{300}$$

Thus, for 1 mol $H_2O(g)$,

$$\Delta H^{\text{heat}} = (0.008\,314)[3.608 + (1.288 \times 10^{-3})(400)](200) = 6.856 \text{ kJ}$$

$$\Delta S^{\text{heat}} = (0.008\,314)[3.608 + (1.288 \times 10^{-3})(391.52)] \ln \frac{500}{300} = 0.017\,465 \text{ kJ} \cdot \text{K}^{-1}$$

and, for 1.881 mol N_2,

$$\Delta H^{\text{heat}} = (1.881)(0.008\,314)[3.352 + (0.513 \times 10^{-3})(400)](200) = 11.126 \text{ kJ}$$

$$\Delta S^{\text{heat}} = (1.881)(0.008\,314)[3.352 + (0.513 \times 10^{-3})(391.52)] \ln \frac{500}{300} = 0.028\,382 \text{ kJ} \cdot \text{K}^{-1}$$

For mixing the product streams: By the above-cited *mixing* formulas, $\Delta H^{\text{mix}} = 0$ and

$$\Delta S^{\text{mix}} = -(2.881)(0.008\,314)\left(\frac{1}{2.881}\ln\frac{1}{2.881} + \frac{1.881}{2.881}\ln\frac{1.881}{2.881}\right) = 0.015\,465 \text{ kJ} \cdot \text{K}^{-1}$$

For the entire process we sum the above values, obtaining $\Delta H = -223.836$ kJ and $\Delta S = 0.006\,985$ kJ $\cdot$ K^{-1}; hence,

$$W_{\text{ideal}} = T_0\,\Delta S - \Delta H = (300)(0.006\,985) - (-223.836) = 225.932 \text{ kJ}$$

(*c*)   The process involves two steps (Fig. 8-7): first, adiabatic combustion to produce flue gases at a high temperature T, and second, reversible cooling to extract the maximum work from the hot flue gases as they are cooled to 500 K. Throughout both steps the pressure is constant at 1 bar. Our first task is to determine T, the temperature resulting from complete combustion of the reactants in an adiabatic reactor. Exactly this problem was solved as Problem 7.20 and we simply make use of the answer: $T = 2522$ K. This value is needed to allow calculation of the entropy change of step $2 \rightarrow 3$. Since this step is a simple cooling process, we have

$$\Delta S_{23} = \left[\sum n_i(\langle C_P^{\text{ig}}\rangle_{\ln T})_i\right] \ln \frac{500}{2522}$$

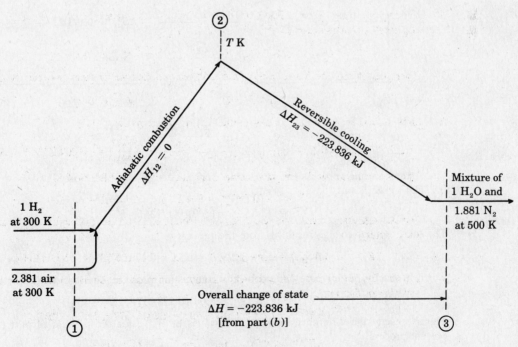

Fig. 8-7

Combining the mole numbers with the heat-capacity expressions gives the value of the sum as

$$(0.008\,314)(9.913 + 2.253 \times 10^{-3} T_{lm}) \quad kJ \cdot K^{-1}$$

With $T_{lm} = 1249.54$ K, we have

$$\Delta S_{23} = 0.105\,82 \ln \frac{500}{2522} = -0.171\,24 \; kJ \cdot K^{-1}$$

Thus, for the cooling process, application of (8.6) gives

$$W_{ideal} = (300)(-0.171\,24) - (-223.836) = 172.464 \; kJ$$

The difference between the process considered here and that of part (b), for which $W_{ideal} = 225.932$ kJ is that the reaction itself is taken to be reversible in part (b). Here an adiabatic combustion process takes place, and such processes are inherently irreversible. In Fig. 8-8 we examine the entropy changes resulting from steps $1 \rightarrow 2$ and $2 \rightarrow 3$. The value $\Delta S_{13} = 0.006\,985 \; kJ \cdot K^{-1}$ comes from part (b), where exactly the same overall change of state was considered. Since $\Delta S_{12} + \Delta S_{23} = \Delta S_{13}$,

$$\Delta S_{12} = \Delta S_{13} - \Delta S_{23} = 0.006\,985 - (-0.171\,24) = 0.178\,23 \; kJ \cdot K^{-1}$$

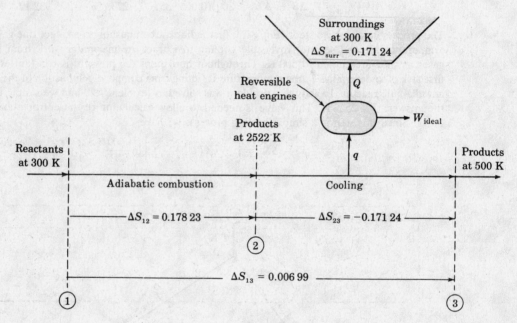

Fig. 8-8

Since the cooling process is reversible, $\Delta S_{23} + \Delta S_{surr} = 0$; therefore, as shown, $\Delta S_{surr} = 0.171\,24 \; kJ \cdot K^{-1}$, and

$$Q_{surr} = T_0 \Delta S_{surr} = (300)(0.17124) = 51.372 \; kJ$$

The total entropy change resulting from the process is

$$\Delta S_{total} = \Delta S_{13} + \Delta S_{surr} = \Delta S_{12} = 0.178\,23 \; kJ \cdot K^{-1}$$

which is exactly the increase caused by the combustion process. The lost work of the process is

$$W_{lost} = T_0 \Delta S_{total} = (300)(0.178\,23) = 53.468 \; kJ$$

This is exactly the difference between the values of W_{ideal} for the process of part (b) and the present process.

(d)   The process is shown schematically in Fig. 8-9. It is a power-plant cycle little different from that of Problem 8-8, and property values at the various points are determined just as in that problem—see

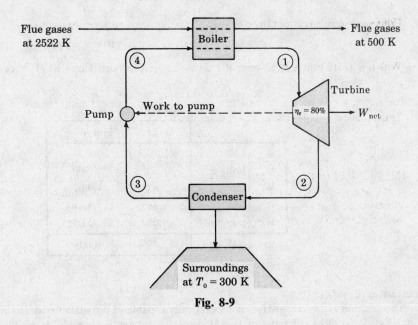

Fig. 8-9

Table 8-12. From these values we get the total shaft work of the turbine as

$$W_s = -(H_2 - H_1) = -(2563.2 - 3563.4) = 1000.2 \text{ kJ} \cdot \text{kg}^{-1}$$

and the pump work as

$$W_{\text{pump}} = -(H_4 - H_3) = -(196.2 - 191.8) = -4.4 \text{ kJ} \cdot \text{kg}^{-1}$$

The net work of the cycle is therefore $W_{\text{net}} = 1000.2 - 4.4 = 995.8 \text{ kJ} \cdot \text{kg}^{-1}$. Finally, the heat transferred to the surroundings is $Q = H_3 - H_2 = 191.8 - 2563.2 = -2371.4 \text{ kJ} \cdot \text{kg}^{-1}$.

Table 8-12

Point	State	$t/°C$	P/bar	$H/\text{kJ} \cdot \text{kg}^{-1}$	$S/\text{kJ} \cdot \text{kg}^{-1} \cdot \text{K}^{-1}$
1	vapor	550.0	35	3563.4	7.2993
2	wet vapor	45.8	0.1	2563.2	8.0834
3	sat. liquid	45.8	0.1	191.8	0.6493
4	liquid	46.1	35	196.2	0.6520

On the basis of 1 mol of H_2 burned to form the flue gases, the energy equation for the boiler is

$$\Delta H_{\text{flue gases}} + m_{\text{steam}}(H_1 - H_4) = 0$$

The value of $\Delta H_{\text{flue gases}}$ is just ΔH_{23} of part (c); thus,

$$-223.836 + m_{\text{steam}}(3563.4 - 196.2) = 0 \qquad \text{or} \qquad m_{\text{steam}} = 0.066\,48 \text{ kg}$$

On this basis,

$$W_{\text{net}} = (995.8)(0.066\,48) = 66.201 \text{ kJ}$$
$$Q = (-2371.4)(0.066\,48) = -157.651 \text{ kJ}$$

The lost-work terms, given by $W_{\text{lost}} = T_0 \Delta S - Q$, have the following values [$\Delta S_{\text{flue gases}} = \Delta S_{23}$ of part (c)]:

$$\textbf{\textit{boiler}} \qquad W_{\text{lost}} = (300)[-0.171\,24 + (7.2993 - 0.6520)(0.066\,48)] + 0 = 81.202 \text{ kJ}$$
$$\textbf{\textit{turbine}} \qquad W_{\text{lost}} = (300)(8.0834 - 7.2993)(0.066\,48) + 0 = 15.638 \text{ kJ}$$
$$\textbf{\textit{condenser}} \qquad W_{\text{lost}} = (300)(0.6493 - 8.0834)(0.066\,48) - (-157.651) = 9.385 \text{ kJ}$$
$$\textbf{\textit{pump}} \qquad W_{\text{lost}} = (300)(0.6520 - 0.6493)(0.066\,48) + 0 = 0.054 \text{ kJ}$$

The work analysis is obtained from

$$W_{ideal} = W_{net} + \Sigma W_{lost}$$

which is (8.17) integrated over the time required to burn 1 mol of H_2. See Table 8-13.

Table 8-13

	W/kJ	% of W_{ideal}
W_{net}	66.201	38.38
W_{lost}(boil)	81.202	47.08
W_{lost}(turb)	15.638	9.07
W_{lost}(cond)	9.385	5.44
W_{lost}(pump)	0.054	0.03
TOTAL (W_{ideal})	172.480	100.0

Summary of (a)–(d)

The reaction $H_2 + \frac{1}{2}O_2 \rightarrow H_2O$ is theoretically capable of providing work in the amount determined in part (a), 237.129 kJ per mole of H_2. However, if oxygen is supplied by air and the products of the reaction are discarded at 500 K, then the maximum possible work is reduced to that of part (b), 225.932 kJ per mole of H_2. If, in addition, an actual combustion process produces the reaction, the maximum work is that of part (c), 172.464 kJ per mole of H_2. If, beyond that, one superimposes the irreversibilities of a practical power plant, then the work actually obtained is the net work of the process, calculated in part (d) as 66.201 kJ per mole of H_2. Thus, a conventional power plant produces only $(66.201/237.129)(100) = 27.9\%$ of the work theoretically available from the reaction.

Supplementary Problems

8.12 Determine η_t for the turbine of Problem 6.25. Take $T_0 = 290$ K.   *Ans.* 0.784

8.13 Determine η_t for the compressor of Problem 6.39. Take $T_0 = 295$ K.   *Ans.* 0.8706

8.14 Find η_t for the process of Problem 6.40. Take $T_0 = 290$ K.   *Ans.* 0.8484

8.15 Rework Problem 6.42, making use of (8.2).

8.16 What is the lost power for the process described in Example 6.5? Take $T_0 = 290$ K.   *Ans.* 191.2 kW

8.17 Refrigeration at a temperature level of 85 K is required for a certain process. A cycle using helium gas has been proposed to operate as follows. Helium at 100 kPa is compressed adiabatically to 500 kPa, water-cooled to 290 K and sent to a heat exchanger, where it is cooled by returning helium. From there it goes to an adiabatic expander which delivers work to be used to help drive the compressor. The helium then enters the refrigerator, where it absorbs enough heat to raise its temperature to 80 K. It returns to the compressor by way of the heat exchanger.

Helium may be considered an ideal gas, with $C_P = (5/2)R$. If the efficiencies of the compressor and expander are 80 percent and if the minimum temperature difference in the exchanger is 5 K, at what rate must the helium be circulated to provide refrigeration at the rate of 2 kW? What is the net power requirement of the process? Sketch the cycle, and show the temperatures at the various points. What is the coefficient of performance of the cycle? How does it compare with the Carnot coefficient of performance? Make a power analysis of the process, taking $T_0 = 290$ K.

Ans. $|\dot{W}_{net}| = 21.17$ kW; $\omega = 0.0945$, $\omega^* = 0.415$;   see Table 8-14.

Table 8-14

	% of $|\dot{W}_{net}|$
$|\dot{W}_{ideal}|$	22.8
$\dot{W}_{lost}(\text{comp})$	11.3
$\dot{W}_{lost}(\text{cool})$	35.3
$\dot{W}_{lost}(\text{exch})$	4.4
$\dot{W}_{lost}(\text{exp})$	16.6
$\dot{W}_{lost}(\text{refrig})$	9.6
TOTAL ($|\dot{W}_{net}|$)	100.0

Review Questions for Chapters 5 through 8

For each of the following statements indicate whether it is true or false.

_____1. The compressibility factor Z is always less than or equal to unity.

_____2. For any real gas at constant temperature, as the pressure approaches zero the residual volume V^R approaches zero.

_____3. The virial coefficients B, C, etc., of a gaseous mixture are functions of temperature and composition only.

_____4. The residual enthalpy and residual entropy of a real gas approach zero as the pressure approaches zero.

_____5. Three-parameter corresponding-states correlations are more useful than two-parameter correlations because they work for any compound whatever.

_____6. The inversion curve of a real fluid defines the states for which the Joule/Thomson coefficient is zero.

_____7. The second virial coefficient B of a binary gas mixture is in general calculable from values of B for the pure gases.

_____8. The critical properties T_c, P_c, and Z_c are constants for a given compound.

_____9. All real fluids become simple fluids in the limit as the pressure approaches zero.

_____10. The Redlich/Kwong equation is superior to the van der Waals equation because its mixing rules are exact.

_____11. A closed system is one of constant volume.

_____12. A steady-flow process is one for which the velocities of all streams may be assumed negligible.

_____13. Gravitational potential-energy terms may be ignored in the steady-flow energy equation if all streams entering and leaving the control volume are at the same elevation.

_____14. Frictional effects are difficult to incorporate explicitly in the energy equations because such effects constitute violations of the second law of thermodynamics.

_____15. In an adiabatic flow process, the entropy of the fluid must increase as the result of any irreversibilities within the system.

_____16. The temperature of a gas undergoing a continuous throttling process may either increase or decrease across the throttling device, depending on conditions.

_____17. The Mach number **M** is negative for a subsonic flow.

_____18. When an ideal gas is compressed adiabatically in a flow process and is then cooled to the initial temperature, the heat removed in the cooler is equal to the work done by the compressor. (Assume potential- and kinetic-energy effects are negligible.)

_____19. A total property M^t of a homogeneous mixture is always equal to $\Sigma n_i M_i$, where n_i is the number of moles of species i and M_i is the corresponding molar property of pure i.

_____20. As $x_i \to 1$ the partial molar volume $\bar{V}_i$ of a component in solution becomes equal to V_i, the molar volume of pure i at the T and P of the solution.

_____21. In the limit as $P \to 0$, the ratio f/P for a gas goes to infinity, where f is the fugacity.

_____22. The fugacity coefficient ϕ has units of pressure.

_____23. The residual Gibbs energy G^R is related to ϕ by $G^R = RT \ln \phi$.

_____24. For equilibrium among contacting phases, the fugacity of a given component must be the same in all phases.

_____25. For an ideal solution at constant T and P, the fugacity of a component in solution is proportional to its mole fraction.

_____26. The Gibbs-energy-change of mixing ΔG is necessarily negative.

_____27. The heat of mixing to form a given binary solution at constant T and P increases with increasing temperature if the total heat capacity of the solution formed is greater than the total heat capacity of the pure constituents that are mixed.

_____28. The entropy-change of mixing at constant T and P to form a binary solution from pure constituents is equal to the heat of mixing at the same conditions, divided by the absolute temperature.

_____29. The use of activity coefficients based on the Lewis/Randall rule is necessarily more realistic than the use of activity coefficients based on Henry's law.

_____30. A mixture of ideal gases is an ideal solution.

_____31. All property-changes of mixing are zero for an ideal solution.

_____32. All excess properties are zero for an ideal solution.

_____33. The activity coefficient is zero for a component in an ideal solution.

_____34. The bubblepoint curve of a binary VLE system represents the states of saturated vapor mixtures.

_____35. At a binary azeotrope the dewpoint and bubblepoint curves become tangent to one another.

_____36. The number of degrees of freedom for an azeotropic state in a two-component VLE system is 1.

_____37. The freezing curves and melting curves are in general different in a binary liquid/solid system.

_____38. Liquid-phase activity coefficients are generally less than zero for systems which exhibit negative deviations from Raoult's law.

_____39. The VLE K-values for systems described by Raoult's law are true constants, independent of T, P, and composition.

_____40. There are two independent reactions for a chemically reactive system containing air (21 mole % O_2, 79 mole % N_2), $S(s)$, $SO_2(g)$, and $SO_3(g)$.

_____41. The reaction coordinate ε is a quantity which characterizes only the *equilibrium* conversion of a given chemical reaction.

_____42. ΔG^{rx} for a reaction is zero if all substances participating in the reaction are in their standard states.

_____43. $\Delta S°$ for a reaction is zero at a temperature for which $\Delta H° = \Delta G°$.

_____44. The equilibrium constant K for a chemical reaction is independent of pressure.

_____45. If the standard Gibbs energy change of a reaction is zero, the reaction is thermo-dynamically impossible.

_____46. The chemical equilibrium constant K increases with increasing T, provided that the standard enthalpy change of reaction $\Delta H°$ is positive.

_____47. There are two degrees of freedom in a chemically reactive system containing the gaseous species N_2, H_2, and NH_3.

_____48. At constant temperature, an increase in pressure will cause an increase in the yield of methanol (CH_3OH) from the ideal-gas reaction

$$CO(g) + 2H_2(g) \rightarrow CH_3OH(g)$$

_____49. W_{ideal} is the same for all steady-flow processes that produce the same change in state, provided that the temperature of the surroundings is the same.

_____50. Lost work is a quantity devised to account for exceptions to the first law of thermo-dynamics.

Ans.

1	2	3	4	5	6	7	8	9	10	11	12	13	14	15
F	F	T	T	F	T	F	T	F	F	F	F	T	F	T

16	17	18	19	20	21	22	23	24	25	26	27	28	29	30
T	F	T	F	T	F	F	T	T	T	T	T	F	F	T

31	32	33	34	35	36	37	38	39	40	41	42	43	44	45
F	T	F	F	T	T	T	F	F	T	F	F	T	T	F

46	47	48	49	50
T	F	T	T	F

Appendix A

Conversion Factors

For conciseness, the various units for each quantity given below are referred to a single basic or derived SI unit. Conversions between other pairs of units for a given quantity are made by employing the usual rules for manipulation of units.

EXAMPLE   Convert ft^3 to gal.
From the volume entries we have $1 \text{ m}^3 = 35.3147 \text{ ft}^3 = 264.172 \text{ gal}$, from which

$$1 \text{ ft}^3 = \frac{264.172}{35.3147} = 7.48051 \text{ gal}$$

Quantity	Conversion	Quantity	Conversion
Length	$1 \text{ m} = 100 \text{ cm}$ $= 3.28084 \text{ ft}$ $= 39.3701 \text{ in}$	Energy	$1 \text{ kJ} = 10^3 \text{ kg} \cdot \text{m}^2 \cdot \text{s}^{-2}$ $= 10^3 \text{ N} \cdot \text{m}$ $= 10^3 \text{ W} \cdot \text{s}$ $= 10^{10} \text{ dyne} \cdot \text{cm}$ $= 10^{10} \text{ erg}$ $= 10^4 \text{ cm}^3 \cdot \text{bar}$ $= 239.006 \text{ cal}$ $= 9869.23 \text{ cm}^3 \cdot \text{atm}$ $= 5.12197 \text{ psia} \cdot \text{ft}^3$ $= 737.562 \text{ ft} \cdot \text{lb}_f$ $= 0.947\,831 \text{ Btu}$
Mass	$1 \text{ kg} = 10^3 \text{ g}$ $= 2.20462 \text{ lb}_m$		
Force	$1 \text{ N} = 1 \text{ kg} \cdot \text{m} \cdot \text{s}^{-2}$ $= 10^5 \text{ dyne}$ $= 0.224809 \text{ lb}_f$		
Pressure	$1 \text{ bar} = 10^5 \text{ kg} \cdot \text{m}^{-1} \cdot \text{s}^{-2}$ $= 10^5 \text{ N} \cdot \text{m}^{-2}$ $= 100 \text{ kPa}$ $= 10^6 \text{ dyne} \cdot \text{cm}^{-2}$ $= 0.986923 \text{ atm}$ $= 14.5038 \text{ psia}$ $= 750.061 \text{ mmHg}$ $= 750.061 \text{ torr}$	Power	$1 \text{ kW} = 10^3 \text{ kg} \cdot \text{m}^2 \cdot \text{s}^{-3}$ $= 10^3 \text{ W}$ $= 10^3 \text{ J} \cdot \text{s}^{-1}$ $= 10^3 \text{ V} \cdot \text{A}$ $= 239.006 \text{ cal} \cdot \text{s}^{-1}$ $= 737.562 \text{ ft} \cdot \text{lb}_f \cdot \text{s}^{-1}$ $= 56.8699 \text{ Btu} \cdot \text{min}^{-1}$ $= 1.34102 \text{ hp}$
Volume	$1 \text{ m}^3 = 10^6 \text{ cm}^3$ $= 10^3 \text{ L}$ $= 35.3147 \text{ ft}^3$ $= 264.172 \text{ gal}$		
Density	$1 \text{ kg} \cdot \text{m}^{-3} = 10^{-3} \text{ g} \cdot \text{cm}^{-3}$ $= 1 \text{ g} \cdot \text{L}^{-1}$ $= 0.062\,427\,8 \text{ lb}_m \cdot \text{ft}^{-3}$ $= 0.008\,345\,40 \text{ lb}_m \cdot \text{gal}^{-1}$		

Note: atm $\equiv$ standard atmosphere
cal $\equiv$ thermochemical calorie
Btu $\equiv$ International Steam Table Btu
L $\equiv$ liter

Appendix B

Values of the Universal Gas Constant

$$R = 8.314 \, \text{J} \cdot \text{mol}^{-1} \cdot \text{K}^{-1} = 8.314 \, \text{kJ} \cdot \text{kmol}^{-1} \cdot \text{K}^{-1} = 8.314 \, \text{m}^3 \cdot \text{Pa} \cdot \text{mol}^{-1} \cdot \text{K}^{-1}$$
$$= 0.008\,314 \, \text{kJ} \cdot \text{mol}^{-1} \cdot \text{K}^{-1} = 0.083\,14 \, \text{L} \cdot \text{bar} \cdot \text{mol}^{-1} \cdot \text{K}^{-1}$$
$$= 83.14 \, \text{cm}^3 \cdot \text{bar} \cdot \text{mol}^{-1} \cdot \text{K}^{-1} = 8314 \, \text{cm}^3 \cdot \text{kPa} \cdot \text{mol}^{-1} \cdot \text{K}^{-1}$$
$$= 82.06 \, \text{cm}^3 \cdot \text{atm} \cdot \text{mol}^{-1} \cdot \text{K}^{-1} = 0.082\,06 \, \text{L} \cdot \text{atm} \cdot \text{mol}^{-1} \cdot \text{K}^{-1}$$
$$= 62\,356 \, \text{cm}^3 \cdot \text{torr} \cdot \text{mol}^{-1} \cdot \text{K}^{-1} = 62.356 \, \text{L} \cdot \text{torr} \cdot \text{mol}^{-1} \cdot \text{K}^{-1}$$
$$= 1.987 \, \text{cal} \cdot \text{mol}^{-1} \cdot \text{K}^{-1} = 1.986 \, \text{Btu} \cdot \text{lb mol}^{-1} \cdot \text{R}^{-1}$$
$$= 0.7302 \, \text{ft}^3 \cdot \text{atm} \cdot \text{lb mol}^{-1} \cdot \text{R}^{-1} = 10.73 \, \text{ft}^3 \cdot \text{psia} \cdot \text{lb mol}^{-1} \cdot \text{R}^{-1}$$
$$= 1545 \, \text{ft} \cdot \text{lb}_f \cdot \text{lb mol}^{-1} \cdot \text{R}^{-1}$$

Appendix C

Critical Constants and Acentric Factor

Compound	T_c/K	P_c/bar	$V_c/\text{cm}^3 \cdot \text{mol}^{-1}$	Z_c	ω
Argon	150.8	48.7	74.9	0.291	0.0
Xenon	289.7	58.4	118.0	0.286	0.0
Methane	190.6	46.0	99.0	0.285	0.008
Oxygen	154.6	50.5	73.4	0.288	0.021
Nitrogen	126.2	33.9	89.5	0.290	0.040
Carbon monoxide	132.9	35.0	93.1	0.295	0.049
Ethylene	282.4	50.4	129.0	0.276	0.085
Hydrogen sulfide	373.2	89.4	98.5	0.284	0.100
Propane	369.8	42.5	203.0	0.281	0.152
Acetylene	308.3	61.4	113.0	0.271	0.184
Cyclohexane	553.4	40.7	308.0	0.273	0.213
Benzene	562.1	48.9	259.0	0.271	0.212
Carbon dioxide	304.2	73.8	94.0	0.274	0.225
Ammonia	405.6	112.8	72.5	0.242	0.250
n-Pentane	469.6	33.7	304.0	0.262	0.251
n-Hexane	507.4	29.7	370.0	0.260	0.296
Acetone	508.1	47.0	209.0	0.232	0.309
Water	647.3	221.2	57.1	0.229	0.344
n-Heptane	540.2	27.4	432.0	0.263	0.351
n-Octane	568.8	24.8	492.0	0.259	0.394
Methanol	512.6	81.0	118.0	0.224	0.559
Ethanol	516.2	63.8	167.0	0.248	0.635

Appendix D

Saturated Steam (SI Units)

V = SPECIFIC VOLUME $(cm^3 \cdot g^{-1}$ or $L \cdot kg^{-1})$ H = SPECIFIC ENTHALPY $(kJ \cdot kg^{-1})$

U = SPECIFIC INTERNAL ENERGY $(kJ \cdot kg^{-1})$ S = SPECIFIC ENTROPY $(kJ \cdot kg^{-1} \cdot K^{-1})$

TEMPERATURE		ABS. PRESS.	SPECIFIC VOLUME V			INTERNAL ENERGY U			ENTHALPY H			ENTROPY S		
°C	K	kPa	SAT. LIQUID	EVAP.	SAT. VAPOR	SAT. LIQUID	EVAP.	SAT. VAPOR	SAT. LIQUID	EVAP.	SAT. VAPOR	SAT. LIQUID	EVAP.	SAT. VAPOR
0	273.15	0.611	1.000	206300	206300	-0.04	2375.7	2375.6	-0.04	2501.7	2501.6	0.0000	9.1578	9.1578
0.01	273.16	0.611	1.000	206200	206200	0.0	2375.6	2375.6	0.00	2501.6	2501.6	0.0	9.1575	9.1575
1	274.15	0.657	1.000	192600	192600	4.17	2372.7	2376.9	4.17	2499.2	2503.4	0.0153	9.1158	9.1311
2	275.15	0.705	1.000	179900	179900	8.39	2369.9	2378.3	8.39	2496.8	2505.2	0.0306	9.0741	9.1047
3	276.15	0.757	1.000	168200	168200	12.60	2367.1	2379.7	12.60	2494.5	2507.1	0.0459	9.0326	9.0785
4	277.15	0.813	1.000	157300	157300	16.80	2364.3	2381.1	16.80	2492.1	2508.9	0.0611	8.9915	9.0526
5	278.15	0.872	1.000	147200	147200	21.01	2361.4	2382.4	21.01	2489.7	2510.7	0.0762	8.9507	9.0269
6	279.15	0.935	1.000	137800	137800	25.21	2358.6	2383.8	25.21	2487.4	2512.6	0.0913	8.9102	9.0014
7	280.15	1.001	1.000	129100	129100	29.41	2355.8	2385.2	29.41	2485.0	2514.4	0.1063	8.8699	8.9762
8	281.15	1.072	1.000	121000	121000	33.60	2353.0	2386.6	33.60	2482.6	2516.2	0.1213	8.8300	8.9513
9	282.15	1.147	1.000	113400	113400	37.80	2350.1	2387.9	37.80	2480.3	2518.1	0.1362	8.7903	8.9265
10	283.15	1.227	1.000	106400	106400	41.99	2347.3	2389.3	41.99	2477.9	2519.9	0.1510	8.7510	8.9020
11	284.15	1.312	1.000	99910	99910	46.18	2344.5	2390.7	46.19	2475.5	2521.7	0.1658	8.7119	8.8776
12	285.15	1.401	1.000	93830	93840	50.38	2341.7	2392.1	50.38	2473.2	2523.6	0.1805	8.6731	8.8536
13	286.15	1.497	1.001	88180	88180	54.56	2338.9	2393.4	54.57	2470.8	2525.4	0.1952	8.6345	8.8297
14	287.15	1.597	1.001	82900	82900	58.75	2336.1	2394.8	58.75	2468.5	2527.2	0.2098	8.5963	8.8060
15	288.15	1.704	1.001	77980	77980	62.94	2333.2	2396.2	62.94	2466.1	2529.1	0.2243	8.5582	8.7826
16	289.15	1.817	1.001	73380	73380	67.12	2330.4	2397.6	67.13	2463.8	2530.9	0.2388	8.5205	8.7593
17	290.15	1.936	1.001	69090	69090	71.31	2327.6	2398.9	71.31	2461.4	2532.7	0.2533	8.4830	8.7363
18	291.15	2.062	1.001	65090	65090	75.49	2324.8	2400.3	75.50	2459.0	2534.5	0.2677	8.4458	8.7135
19	292.15	2.196	1.002	61340	61340	79.68	2322.0	2401.7	79.68	2456.7	2536.4	0.2820	8.4088	8.6908
20	293.15	2.337	1.002	57840	57840	83.86	2319.2	2403.0	83.86	2454.3	2538.2	0.2963	8.3721	8.6684
21	294.15	2.485	1.002	54560	54560	88.04	2316.4	2404.4	88.04	2452.0	2540.0	0.3105	8.3356	8.6462
22	295.15	2.642	1.002	51490	51490	92.22	2313.6	2405.8	92.23	2449.6	2541.8	0.3247	8.2994	8.6241
23	296.15	2.808	1.002	48620	48620	96.40	2310.7	2407.1	96.41	2447.2	2543.6	0.3389	8.2634	8.6023
24	297.15	2.982	1.003	45920	45930	100.6	2307.9	2408.5	100.6	2444.9	2545.5	0.3530	8.2277	8.5806
25	298.15	3.166	1.003	43400	43400	104.8	2305.1	2409.9	104.8	2442.5	2547.3	0.3670	8.1922	8.5592
26	299.15	3.360	1.003	41030	41030	108.9	2302.3	2411.2	108.9	2440.2	2549.1	0.3810	8.1569	8.5379
27	300.15	3.564	1.003	38810	38810	113.1	2299.5	2412.6	113.1	2437.8	2550.9	0.3949	8.1218	8.5168
28	301.15	3.778	1.004	36730	36730	117.3	2296.7	2414.0	117.3	2435.4	2552.7	0.4088	8.0870	8.4959
29	302.15	4.004	1.004	34770	34770	121.5	2293.8	2415.3	121.5	2433.1	2554.5	0.4227	8.0524	8.4751
30	303.15	4.241	1.004	32930	32930	125.7	2291.0	2416.7	125.7	2430.7	2556.4	0.4365	8.0180	8.4546
31	304.15	4.491	1.005	31200	31200	129.8	2288.2	2418.0	129.8	2428.3	2558.2	0.4503	7.9839	8.4342
32	305.15	4.753	1.005	29570	29570	134.0	2285.4	2419.4	134.0	2425.9	2560.0	0.4640	7.9500	8.4140
33	306.15	5.029	1.005	28040	28040	138.2	2282.6	2420.8	138.2	2423.6	2561.8	0.4777	7.9163	8.3939
34	307.15	5.318	1.006	26600	26600	142.4	2279.7	2422.1	142.4	2421.2	2563.6	0.4913	7.8828	8.3740
35	308.15	5.622	1.006	25240	25240	146.6	2276.9	2423.5	146.6	2418.8	2565.4	0.5049	7.8495	8.3543
36	309.15	5.940	1.006	23970	23970	150.7	2274.1	2424.8	150.7	2416.4	2567.2	0.5184	7.8164	8.3348
37	310.15	6.274	1.007	22760	22760	154.9	2271.3	2426.2	154.9	2414.1	2569.0	0.5319	7.7835	8.3154
38	311.15	6.624	1.007	21630	21630	159.1	2268.4	2427.5	159.1	2411.7	2570.8	0.5453	7.7509	8.2962
39	312.15	6.991	1.007	20560	20560	163.3	2265.6	2428.9	163.3	2409.3	2572.6	0.5588	7.7184	8.2772
40	313.15	7.375	1.008	19550	19550	167.4	2262.8	2430.2	167.5	2406.9	2574.4	0.5721	7.6861	8.2583
41	314.15	7.777	1.008	18590	18590	171.6	2259.9	2431.6	171.6	2404.5	2576.2	0.5854	7.6541	8.2395
42	315.15	8.198	1.009	17690	17690	175.8	2257.1	2432.9	175.8	2402.1	2577.9	0.5987	7.6222	8.2209
43	316.15	8.639	1.009	16840	16840	180.0	2254.3	2434.2	180.0	2399.7	2579.7	0.6120	7.5905	8.2025
44	317.15	9.100	1.009	16040	16040	184.2	2251.4	2435.6	184.2	2397.3	2581.5	0.6252	7.5590	8.1842
45	318.15	9.582	1.010	15280	15280	188.3	2248.6	2436.9	188.4	2394.9	2583.3	0.6383	7.5277	8.1661
46	319.15	10.09	1.010	14560	14560	192.5	2245.7	2438.3	192.5	2392.5	2585.1	0.6514	7.4966	8.1481
47	320.15	10.61	1.011	13880	13880	196.7	2242.9	2439.6	196.7	2390.1	2586.9	0.6645	7.4657	8.1302
48	321.15	11.16	1.011	13230	13230	200.9	2240.0	2440.9	200.9	2387.7	2588.6	0.6776	7.4350	8.1125
49	322.15	11.74	1.012	12620	12620	205.1	2237.2	2442.3	205.1	2385.3	2590.4	0.6906	7.4044	8.0950
50	323.15	12.34	1.012	12040	12050	209.2	2234.3	2443.6	209.3	2382.9	2592.2	0.7035	7.3741	8.0776
51	324.15	12.96	1.013	11500	11500	213.4	2231.5	2444.9	213.4	2380.5	2593.9	0.7164	7.3439	8.0603
52	325.15	13.61	1.013	10980	10980	217.6	2228.6	2446.2	217.6	2378.1	2595.7	0.7293	7.3138	8.0432
53	326.15	14.29	1.014	10490	10490	221.8	2225.8	2447.6	221.8	2375.7	2597.5	0.7422	7.2840	8.0262
54	327.15	15.00	1.014	10020	10020	226.0	2222.9	2448.9	226.0	2373.2	2599.2	0.7550	7.2543	8.0093
55	328.15	15.74	1.015	9577.9	9578.9	230.2	2220.0	2450.2	230.2	2370.8	2601.0	0.7677	7.2248	7.9925
56	329.15	16.51	1.015	9157.7	9158.7	234.3	2217.2	2451.5	234.4	2368.4	2602.7	0.7804	7.1955	7.9759
57	330.15	17.31	1.016	8758.7	8759.8	238.5	2214.3	2452.8	238.5	2365.9	2604.5	0.7931	7.1663	7.9595
58	331.15	18.15	1.016	8379.8	8380.8	242.7	2211.4	2454.1	242.7	2363.5	2606.2	0.8058	7.1373	7.9431
59	332.15	19.02	1.017	8019.7	8020.8	246.9	2208.6	2455.4	246.9	2361.1	2608.0	0.8184	7.1085	7.9269
60	333.15	19.92	1.017	7677.5	7678.5	251.1	2205.7	2456.8	251.1	2358.6	2609.7	0.8310	7.0798	7.9108
61	334.15	20.86	1.018	7352.1	7353.2	255.3	2202.8	2458.1	255.3	2356.2	2611.4	0.8435	7.0513	7.8948
62	335.15	21.84	1.018	7042.7	7043.7	259.4	2199.9	2459.4	259.5	2353.7	2613.2	0.8560	7.0230	7.8790
63	336.15	22.86	1.019	6748.2	6749.3	263.6	2197.0	2460.7	263.6	2351.3	2614.9	0.8685	6.9948	7.8633
64	337.15	23.91	1.019	6468.0	6469.0	267.8	2194.1	2462.0	267.8	2348.8	2616.6	0.8809	6.9667	7.8477
65	338.15	25.01	1.020	6201.3	6202.3	272.0	2191.2	2463.2	272.0	2346.3	2618.4	0.8933	6.9388	7.8322
66	339.15	26.15	1.020	5947.2	5948.2	276.2	2188.3	2464.5	276.2	2343.9	2620.1	0.9057	6.9111	7.8168
67	340.15	27.33	1.021	5705.2	5706.2	280.4	2185.4	2465.8	280.4	2341.4	2621.8	0.9180	6.8835	7.8015
68	341.15	28.56	1.022	5474.6	5475.6	284.6	2182.5	2467.1	284.6	2338.9	2623.5	0.9303	6.8561	7.7864
69	342.15	29.84	1.022	5254.8	5255.8	288.8	2179.6	2468.4	288.8	2336.4	2625.2	0.9426	6.8288	7.7714
70	343.15	31.16	1.023	5045.2	5046.3	292.9	2176.7	2469.7	293.0	2334.0	2626.9	0.9548	6.8017	7.7565
71	344.15	32.53	1.023	4845.4	4846.4	297.1	2173.8	2470.9	297.2	2331.5	2628.6	0.9670	6.7747	7.7417
72	345.15	33.96	1.024	4654.1	4655.1	301.3	2170.9	2472.2	301.4	2329.0	2630.3	0.9792	6.7478	7.7270
73	346.15	35.43	1.025	4472.7	4473.7	305.5	2168.0	2473.5	305.5	2326.5	2632.0	0.9913	6.7211	7.7124
74	347.15	36.96	1.025	4299.0	4300.0	309.7	2165.1	2474.8	309.7	2324.0	2633.7	1.0034	6.6945	7.6979

°C	K	ABS. PRESS. kPa	SPECIFIC VOLUME V			INTERNAL ENERGY U			ENTHALPY H			ENTROPY S		
			SAT. LIQUID	EVAP.	SAT. VAPOR	SAT. LIQUID	EVAP.	SAT. VAPOR	SAT. LIQUID	EVAP.	SAT. VAPOR	SAT. LIQUID	EVAP.	SAT. VAPOR
75	348.15	38.55	1.026	4133.1	4134.1	313.9	2162.1	2476.0	313.9	2321.5	2635.4	1.0154	6.6681	7.6835
76	349.15	40.19	1.027	3974.6	3975.7	318.1	2159.2	2477.3	318.1	2318.9	2637.1	1.0275	6.6418	7.6693
77	350.15	41.89	1.027	3823.3	3824.3	322.3	2156.3	2478.5	322.3	2316.4	2638.7	1.0395	6.6156	7.6551
78	351.15	43.65	1.028	3678.6	3679.6	326.5	2153.3	2479.8	326.5	2313.9	2640.4	1.0514	6.5896	7.6410
79	352.15	45.47	1.029	3540.3	3541.3	330.7	2150.4	2481.1	330.7	2311.4	2642.1	1.0634	6.5637	7.6271
80	353.15	47.36	1.029	3408.1	3409.1	334.9	2147.4	2482.3	334.9	2308.8	2643.8	1.0753	6.5380	7.6132
81	354.15	49.31	1.030	3281.6	3282.6	339.1	2144.5	2483.5	339.1	2306.3	2645.4	1.0871	6.5123	7.5995
82	355.15	51.33	1.031	3160.6	3161.6	343.3	2141.5	2484.8	343.3	2303.8	2647.1	1.0990	6.4868	7.5858
83	356.15	53.42	1.031	3044.8	3045.8	347.5	2138.6	2486.0	347.5	2301.2	2648.7	1.1108	6.4615	7.5722
84	357.15	55.57	1.032	2933.9	2935.0	351.7	2135.6	2487.3	351.7	2298.6	2650.4	1.1225	6.4362	7.5587
85	358.15	57.80	1.033	2827.8	2828.8	355.9	2132.6	2488.5	355.9	2296.1	2652.0	1.1343	6.4111	7.5454
86	359.15	60.11	1.033	2726.1	2727.2	360.1	2129.7	2489.7	360.1	2293.5	2653.6	1.1460	6.3861	7.5321
87	360.15	62.49	1.034	2628.8	2629.8	364.3	2126.7	2490.9	364.3	2290.9	2655.3	1.1577	6.3612	7.5189
88	361.15	64.95	1.035	2535.4	2536.5	368.5	2123.7	2492.2	368.5	2288.4	2656.9	1.1693	6.3365	7.5058
89	362.15	67.49	1.035	2446.0	2447.0	372.7	2120.7	2493.4	372.7	2285.8	2658.5	1.1809	6.3119	7.4928
90	363.15	70.11	1.036	2360.3	2361.3	376.9	2117.7	2494.6	376.9	2283.2	2660.1	1.1925	6.2873	7.4799
91	364.15	72.81	1.037	2278.0	2279.1	381.1	2114.7	2495.8	381.1	2280.6	2661.7	1.2041	6.2629	7.4670
92	365.15	75.61	1.038	2199.2	2200.2	385.3	2111.7	2497.0	385.4	2278.0	2663.4	1.2156	6.2387	7.4543
93	366.15	78.49	1.038	2123.5	2124.5	389.5	2108.7	2498.2	389.6	2275.4	2665.0	1.2271	6.2145	7.4416
94	367.15	81.46	1.039	2050.9	2051.9	393.7	2105.7	2499.4	393.8	2272.8	2666.6	1.2386	6.1905	7.4291
95	368.15	84.53	1.040	1981.2	1982.2	397.9	2102.7	2500.6	398.0	2270.2	2668.1	1.2501	6.1665	7.4166
96	369.15	87.69	1.041	1914.3	1915.3	402.1	2099.7	2501.8	402.2	2267.5	2669.7	1.2615	6.1427	7.4042
97	370.15	90.94	1.041	1850.0	1851.0	406.3	2096.6	2503.0	406.4	2264.9	2671.3	1.2729	6.1190	7.3919
98	371.15	94.30	1.042	1788.3	1789.3	410.5	2093.6	2504.1	410.6	2262.2	2672.9	1.2842	6.0954	7.3796
99	372.15	97.76	1.043	1729.0	1730.0	414.7	2090.6	2505.3	414.8	2259.6	2674.4	1.2956	6.0719	7.3675
100	373.15	101.33	1.044	1672.0	1673.0	419.0	2087.5	2506.5	419.1	2256.9	2676.0	1.3069	6.0485	7.3554
102	375.15	108.78	1.045	1564.5	1565.5	427.4	2081.4	2508.8	427.5	2251.6	2679.1	1.3294	6.0021	7.3315
104	377.15	116.68	1.047	1465.1	1466.2	435.8	2075.3	2511.1	435.9	2246.3	2682.2	1.3518	5.9560	7.3078
106	379.15	125.04	1.049	1373.1	1374.2	444.3	2069.2	2513.4	444.4	2240.9	2685.3	1.3742	5.9104	7.2845
108	381.15	133.90	1.050	1287.9	1288.9	452.7	2063.0	2515.7	452.9	2235.4	2688.3	1.3964	5.8651	7.2615
110	383.15	143.27	1.052	1208.9	1209.9	461.2	2056.8	2518.0	461.3	2230.0	2691.3	1.4185	5.8203	7.2388
112	385.15	153.16	1.054	1135.6	1136.6	469.6	2050.6	2520.2	469.8	2224.5	2694.3	1.4405	5.7758	7.2164
114	387.15	163.62	1.055	1067.5	1068.5	478.1	2044.3	2522.4	478.3	2219.0	2697.2	1.4624	5.7318	7.1942
116	389.15	174.65	1.057	1004.2	1005.2	486.6	2038.1	2524.6	486.7	2213.4	2700.2	1.4842	5.6881	7.1723
118	391.15	186.28	1.059	945.3	946.3	495.0	2031.8	2526.8	495.2	2207.9	2703.1	1.5060	5.6447	7.1507
120	393.15	198.54	1.061	890.5	891.5	503.5	2025.1	2529.0	503.7	2202.2	2706.0	1.5276	5.6017	7.1293
122	395.15	211.45	1.062	839.4	840.5	512.0	2019.1	2531.1	512.2	2196.6	2708.8	1.5491	5.5590	7.1082
124	397.15	225.04	1.064	791.8	792.8	520.5	2012.7	2533.2	520.7	2190.9	2711.6	1.5706	5.5167	7.0873
126	399.15	239.33	1.066	747.3	748.4	529.0	2006.3	2535.3	529.2	2185.2	2714.4	1.5919	5.4747	7.0666
128	401.15	254.35	1.068	705.8	706.9	537.5	1999.9	2537.4	537.8	2179.4	2717.2	1.6132	5.4330	7.0462
130	403.15	270.13	1.070	667.1	668.1	546.0	1993.4	2539.4	546.3	2173.6	2719.9	1.6344	5.3917	7.0261
132	405.15	286.70	1.072	630.8	631.9	554.5	1986.9	2541.4	554.8	2167.8	2722.6	1.6555	5.3507	7.0061
134	407.15	304.07	1.074	596.9	598.0	563.1	1980.4	2543.4	563.4	2161.9	2725.3	1.6765	5.3099	6.9864
136	409.15	322.29	1.076	565.1	566.2	571.6	1973.8	2545.4	572.0	2155.9	2727.9	1.6974	5.2695	6.9669
138	411.15	341.38	1.078	535.3	536.4	580.2	1967.2	2547.4	580.5	2150.0	2730.5	1.7182	5.2293	6.9475
140	413.15	361.38	1.080	507.4	508.5	588.7	1960.6	2549.3	589.1	2144.0	2733.1	1.7390	5.1894	6.9284
142	415.15	382.31	1.082	481.2	482.3	597.3	1953.9	2551.2	597.7	2137.9	2735.6	1.7597	5.1499	6.9095
144	417.15	404.20	1.084	456.6	457.7	605.9	1947.2	2553.1	606.3	2131.8	2738.1	1.7803	5.1105	6.8908
146	419.15	427.09	1.086	433.5	434.6	614.4	1940.5	2554.9	614.9	2125.7	2740.6	1.8008	5.0715	6.8723
148	421.15	451.01	1.089	411.8	412.9	623.0	1933.7	2556.8	623.5	2119.5	2743.0	1.8213	5.0327	6.8539
150	423.15	476.00	1.091	391.4	392.4	631.6	1926.9	2558.6	632.1	2113.2	2745.4	1.8416	4.9941	6.8358
152	425.15	502.08	1.093	372.1	373.2	640.2	1920.1	2560.3	640.8	2106.9	2747.7	1.8619	4.9558	6.8178
154	427.15	529.29	1.095	354.0	355.1	648.9	1913.2	2562.1	649.4	2100.6	2750.0	1.8822	4.9178	6.8000
156	429.15	557.67	1.098	336.9	338.0	657.5	1906.3	2563.8	658.1	2094.2	2752.3	1.9023	4.8800	6.7823
158	431.15	587.25	1.100	320.8	321.9	666.1	1899.3	2565.5	666.8	2087.7	2754.5	1.9224	4.8424	6.7648
160	433.15	618.06	1.102	305.7	306.8	674.8	1892.3	2567.1	675.5	2081.3	2756.7	1.9425	4.8050	6.7475
162	435.15	650.16	1.105	291.3	292.4	683.5	1885.3	2568.8	684.2	2074.7	2758.9	1.9624	4.7679	6.7303
164	437.15	683.56	1.107	277.8	278.9	692.1	1878.2	2570.4	692.9	2068.1	2761.0	1.9823	4.7309	6.7133
166	439.15	718.31	1.109	265.0	266.1	700.8	1871.1	2571.9	701.6	2061.4	2763.1	2.0022	4.6942	6.6964
168	441.15	754.45	1.112	252.9	254.0	709.5	1863.9	2573.4	710.4	2054.7	2765.1	2.0219	4.6577	6.6796
170	443.15	792.02	1.114	241.4	242.6	718.2	1856.7	2574.9	719.1	2047.9	2767.1	2.0416	4.6214	6.6630
172	445.15	831.06	1.117	230.6	231.7	727.0	1849.5	2576.4	727.9	2041.1	2769.0	2.0613	4.5853	6.6465
174	447.15	871.60	1.120	220.3	221.5	735.7	1842.2	2577.8	736.7	2034.2	2770.9	2.0809	4.5493	6.6302
176	449.15	913.68	1.122	210.6	211.7	744.4	1834.8	2579.3	745.5	2027.3	2772.7	2.1004	4.5136	6.6140
178	451.15	957.36	1.125	201.4	202.5	753.2	1827.4	2580.6	754.3	2020.2	2774.5	2.1199	4.4780	6.5979
180	453.15	1002.7	1.128	192.7	193.8	762.0	1820.0	2581.9	763.1	2013.1	2776.3	2.1393	4.4426	6.5819
182	455.15	1049.6	1.130	184.4	185.5	770.8	1812.5	2583.2	772.0	2006.0	2778.0	2.1587	4.4074	6.5660
184	457.15	1098.3	1.133	176.5	177.6	779.6	1804.9	2584.5	780.8	1998.8	2779.6	2.1780	4.3723	6.5503
186	459.15	1148.8	1.136	169.0	170.2	788.4	1797.3	2585.7	789.7	1991.5	2781.2	2.1972	4.3374	6.5346
188	461.15	1201.0	1.139	161.9	163.1	797.2	1789.7	2586.9	798.6	1984.2	2782.8	2.2164	4.3026	6.5191
190	463.15	1255.1	1.142	155.2	156.3	806.1	1782.0	2588.1	807.5	1976.7	2784.3	2.2356	4.2680	6.5036
192	465.15	1311.1	1.144	148.8	149.9	814.9	1774.2	2589.2	816.5	1969.3	2785.7	2.2547	4.2336	6.4883
194	467.15	1369.0	1.147	142.6	143.8	823.8	1766.4	2590.2	825.4	1961.7	2787.1	2.2738	4.1993	6.4730
196	469.15	1428.9	1.150	136.8	138.0	832.7	1758.6	2591.3	834.4	1954.1	2788.5	2.2928	4.1651	6.4578
198	471.15	1490.9	1.153	131.3	132.4	841.6	1750.6	2592.3	843.3	1946.4	2789.7	2.3117	4.1310	6.4428
200	473.15	1554.9	1.156	126.0	127.2	850.6	1742.6	2593.2	852.4	1938.6	2790.9	2.3307	4.0971	6.4278
202	475.15	1621.0	1.160	121.0	122.1	859.5	1734.6	2594.1	861.4	1930.7	2792.1	2.3495	4.0633	6.4128
204	477.15	1689.3	1.163	116.2	117.3	868.5	1726.5	2595.0	870.5	1922.8	2793.2	2.3684	4.0296	6.3980
206	479.15	1759.8	1.166	111.6	112.8	877.5	1718.3	2595.8	879.5	1914.7	2794.3	2.3872	3.9961	6.3832
208	481.15	1832.6	1.169	107.2	108.4	886.5	1710.1	2596.6	888.6	1906.6	2795.3	2.4059	3.9626	6.3686
210	483.15	1907.7	1.173	103.1	104.2	895.5	1701.8	2597.3	897.7	1898.5	2796.2	2.4247	3.9293	6.3539
212	485.15	1985.2	1.176	99.09	100.26	904.5	1693.5	2598.0	906.9	1890.2	2797.1	2.4434	3.8960	6.3394
214	487.15	2065.1	1.179	95.28	96.46	913.6	1685.1	2598.7	916.0	1881.8	2797.9	2.4620	3.8629	6.3249
216	489.15	2147.5	1.183	91.65	92.83	922.7	1676.6	2599.3	925.2	1873.4	2798.6	2.4806	3.8298	6.3104
218	491.15	2232.4	1.186	88.17	89.36	931.8	1668.0	2599.8	934.4	1864.9	2799.3	2.4992	3.7968	6.2960

TEMPERATURE		ABS. PRESS.	SPECIFIC VOLUME V			INTERNAL ENERGY U			ENTHALPY H			ENTROPY S		
°C	K	kPa	SAT. LIQUID	EVAP.	SAT. VAPOR	SAT. LIQUID	EVAP.	SAT. VAPOR	SAT. LIQUID	EVAP.	SAT. VAPOR	SAT. LIQUID	EVAP.	SAT. VAPOR
220	493.15	2319.8	1.190	84.85	86.04	940.9	1659.4	2600.3	943.7	1856.2	2799.9	2.5178	3.7639	6.2817
222	495.15	2409.9	1.194	81.67	82.86	950.1	1650.7	2600.8	952.9	1847.5	2800.5	2.5363	3.7311	6.2674
224	497.15	2502.7	1.197	78.62	79.82	959.2	1642.0	2601.2	962.2	1838.7	2800.9	2.5548	3.6984	6.2532
226	499.15	2598.2	1.201	75.71	76.91	968.4	1633.1	2601.5	971.5	1829.8	2801.4	2.5733	3.6657	6.2390
228	501.15	2696.5	1.205	72.92	74.12	977.6	1624.2	2601.8	980.9	1820.8	2801.7	2.5917	3.6331	6.2249
230	503.15	2797.6	1.209	70.24	71.45	986.9	1615.2	2602.1	990.3	1811.7	2802.0	2.6102	3.6006	6.2107
232	505.15	2901.6	1.213	67.68	68.89	996.2	1606.1	2602.3	999.7	1802.5	2802.2	2.6286	3.5681	6.1967
234	507.15	3008.6	1.217	65.22	66.43	1005.4	1597.0	2602.5	1009.1	1793.2	2802.3	2.6470	3.5356	6.1826
236	509.15	3118.6	1.221	62.86	64.08	1014.8	1587.7	2602.5	1018.6	1783.8	2802.3	2.6653	3.5033	6.1686
238	511.15	3231.7	1.225	60.60	61.82	1024.1	1578.4	2602.5	1028.1	1774.2	2802.3	2.6837	3.4709	6.1546
240	513.15	3347.8	1.229	58.43	59.65	1033.5	1569.0	2602.5	1037.6	1764.6	2802.2	2.7020	3.4386	6.1406
242	515.15	3467.2	1.233	56.34	57.57	1042.9	1559.5	2602.4	1047.2	1754.9	2802.0	2.7203	3.4063	6.1266
244	517.15	3589.8	1.238	54.34	55.58	1052.3	1549.9	2602.2	1056.8	1745.0	2801.8	2.7386	3.3740	6.1127
246	519.15	3715.7	1.242	52.41	53.66	1061.8	1540.2	2602.0	1066.4	1735.0	2801.4	2.7569	3.3418	6.0987
248	521.15	3844.9	1.247	50.56	51.81	1071.3	1530.5	2601.8	1076.1	1724.9	2801.0	2.7752	3.3096	6.0848
250	523.15	3977.6	1.251	48.79	50.04	1080.8	1520.6	2601.4	1085.8	1714.7	2800.4	2.7935	3.2773	6.0708
252	525.15	4113.7	1.256	47.08	48.33	1090.4	1510.6	2601.0	1095.5	1704.3	2799.8	2.8118	3.2451	6.0569
254	527.15	4253.4	1.261	45.43	46.69	1100.0	1500.5	2600.5	1105.3	1693.8	2799.1	2.8300	3.2129	6.0429
256	529.15	4396.7	1.266	43.85	45.11	1109.6	1490.4	2600.0	1115.2	1683.2	2798.3	2.8483	3.1807	6.0290
258	531.15	4543.7	1.271	42.33	43.60	1119.3	1480.1	2599.3	1125.0	1672.4	2797.4	2.8666	3.1484	6.0150
260	533.15	4694.3	1.276	40.86	42.13	1129.0	1469.7	2598.6	1134.9	1661.5	2796.4	2.8848	3.1161	6.0010
262	535.15	4848.8	1.281	39.44	40.73	1138.7	1459.2	2597.8	1144.9	1650.4	2795.3	2.9031	3.0838	5.9869
264	537.15	5007.1	1.286	38.08	39.37	1148.5	1448.5	2597.0	1154.9	1639.2	2794.1	2.9214	3.0515	5.9729
266	539.15	5169.3	1.291	36.77	38.06	1158.3	1437.8	2596.1	1165.0	1627.8	2792.8	2.9397	3.0191	5.9588
268	541.15	5335.5	1.297	35.51	36.80	1168.2	1426.9	2595.0	1175.1	1616.3	2791.4	2.9580	2.9866	5.9446
270	543.15	5505.8	1.303	34.29	35.59	1178.1	1415.9	2593.9	1185.2	1604.6	2789.9	2.9763	2.9541	5.9304
272	545.15	5680.2	1.308	33.11	34.42	1188.0	1404.7	2592.7	1195.4	1592.8	2788.2	2.9947	2.9215	5.9162
274	547.15	5858.7	1.314	31.97	33.29	1198.0	1393.4	2591.4	1205.7	1580.8	2786.5	3.0131	2.8889	5.9019
276	549.15	6041.5	1.320	30.88	32.20	1208.0	1382.0	2590.1	1216.0	1568.5	2784.6	3.0314	2.8561	5.8876
278	551.15	6228.7	1.326	29.82	31.14	1218.1	1370.4	2588.6	1226.4	1556.2	2782.6	3.0499	2.8233	5.8731
280	553.15	6420.2	1.332	28.79	30.13	1228.3	1358.7	2587.0	1236.8	1543.6	2780.4	3.0683	2.7903	5.8586
282	555.15	6616.1	1.339	27.81	29.14	1238.5	1346.8	2585.3	1247.3	1530.8	2778.1	3.0868	2.7573	5.8440
284	557.15	6816.6	1.345	26.85	28.20	1248.7	1334.8	2583.5	1257.9	1517.8	2775.7	3.1053	2.7241	5.8294
286	559.15	7021.8	1.352	25.93	27.28	1259.0	1322.6	2581.6	1268.5	1504.6	2773.2	3.1238	2.6908	5.8146
288	561.15	7231.5	1.359	25.03	26.39	1269.4	1310.2	2579.6	1279.2	1491.2	2770.5	3.1424	2.6573	5.7997
290	563.15	7446.1	1.366	24.17	25.54	1279.8	1297.7	2577.5	1290.0	1477.6	2767.6	3.1611	2.6237	5.7848
292	565.15	7665.4	1.373	23.33	24.71	1290.3	1284.9	2575.3	1300.9	1463.8	2764.6	3.1798	2.5899	5.7697
294	567.15	7889.7	1.381	22.52	23.90	1300.9	1272.0	2572.9	1311.8	1449.7	2761.5	3.1985	2.5560	5.7545
296	569.15	8118.9	1.388	21.74	23.13	1311.5	1258.9	2570.4	1322.8	1435.4	2758.2	3.2173	2.5218	5.7392
298	571.15	8353.2	1.396	20.98	22.38	1322.2	1245.6	2567.8	1333.9	1420.8	2754.7	3.2362	2.4875	5.7237
300	573.15	8592.7	1.404	20.24	21.65	1333.0	1232.0	2565.0	1345.1	1406.0	2751.0	3.2552	2.4529	5.7081
302	575.15	8837.4	1.412	19.53	20.94	1343.8	1218.3	2562.1	1356.3	1390.9	2747.2	3.2742	2.4182	5.6924
304	577.15	9087.3	1.421	18.84	20.26	1354.8	1204.3	2559.1	1367.7	1375.5	2743.2	3.2933	2.3832	5.6765
306	579.15	9342.7	1.430	18.17	19.60	1365.8	1190.1	2555.9	1379.1	1359.8	2739.0	3.3125	2.3479	5.6604
308	581.15	9603.6	1.439	17.52	18.96	1376.9	1175.6	2552.5	1390.7	1343.9	2734.6	3.3318	2.3124	5.6442
310	583.15	9870.0	1.448	16.89	18.33	1388.1	1161.0	2549.1	1402.4	1327.6	2730.0	3.3512	2.2766	5.6278
312	585.15	10142.1	1.458	16.27	17.73	1399.4	1146.0	2545.4	1414.2	1311.0	2725.2	3.3707	2.2404	5.6111
314	587.15	10420.0	1.468	15.68	17.14	1410.8	1130.8	2541.6	1426.1	1294.1	2720.2	3.3903	2.2040	5.5943
316	589.15	10703.	1.478	15.09	16.57	1422.3	1115.2	2537.5	1438.1	1276.8	2714.9	3.4101	2.1672	5.5772
318	591.15	10993.4	1.488	14.53	16.02	1433.9	1099.4	2533.3	1450.3	1259.1	2709.4	3.4300	2.1300	5.5599
320	593.15	11289.1	1.500	13.98	15.48	1445.7	1083.2	2528.9	1462.6	1241.1	2703.7	3.4500	2.0923	5.5423
322	595.15	11591.0	1.511	13.44	14.96	1457.5	1066.7	2524.3	1475.1	1222.6	2697.6	3.4702	2.0542	5.5244
324	597.15	11899.2	1.523	12.92	14.45	1469.5	1049.9	2519.4	1487.7	1203.6	2691.3	3.4906	2.0156	5.5062
326	599.15	12213.7	1.535	12.41	13.95	1481.7	1032.6	2514.3	1500.4	1184.2	2684.6	3.5111	1.9764	5.4876
328	601.15	12534.8	1.548	11.91	13.46	1494.0	1014.8	2508.8	1513.4	1164.2	2677.6	3.5319	1.9367	5.4685
330	603.15	12862.5	1.561	11.43	12.99	1506.4	996.7	2503.1	1526.5	1143.6	2670.2	3.5528	1.8962	5.4490
332	605.15	13197.0	1.575	10.95	12.53	1519.1	978.0	2497.0	1539.9	1122.5	2662.3	3.5740	1.8550	5.4290
334	607.15	13538.3	1.590	10.49	12.08	1531.9	958.7	2490.6	1553.4	1100.7	2654.1	3.5955	1.8129	5.4084
336	609.15	13886.7	1.606	10.03	11.63	1544.9	938.9	2483.7	1567.2	1078.1	2645.3	3.6172	1.7700	5.3872
338	611.15	14242.3	1.622	9.58	11.20	1558.1	918.4	2476.4	1581.2	1054.8	2636.0	3.6392	1.7261	5.3653
340	613.15	14605.2	1.639	9.14	10.78	1571.5	897.2	2468.7	1595.5	1030.7	2626.2	3.6616	1.6811	5.3427
342	615.15	14975.5	1.657	8.71	10.37	1585.2	875.2	2460.5	1610.0	1005.7	2615.7	3.6844	1.6350	5.3194
344	617.15	15353.5	1.676	8.286	9.962	1599.2	852.5	2451.7	1624.9	979.7	2604.7	3.7075	1.5877	5.2952
346	619.15	15739.3	1.696	7.870	9.566	1613.5	828.9	2442.4	1640.2	952.8	2593.0	3.7311	1.5391	5.2702
348	621.15	16133.1	1.718	7.461	9.178	1628.1	804.5	2432.6	1655.8	924.1	2580.7	3.7553	1.4891	5.2444
350	623.15	16535.1	1.741	7.058	8.799	1643.0	779.2	2422.2	1671.8	895.9	2567.7	3.7801	1.4375	5.2177
352	625.15	16945.5	1.766	6.654	8.420	1659.4	751.5	2410.8	1689.3	864.2	2553.5	3.8071	1.3822	5.1893
354	627.15	17364.4	1.794	6.252	8.045	1676.3	722.4	2398.7	1707.5	830.9	2538.4	3.8349	1.3247	5.1596
356	629.15	17792.2	1.824	5.850	7.674	1693.4	692.2	2385.6	1725.9	796.2	2522.1	3.8629	1.2654	5.1283
358	631.15	18229.0	1.858	5.448	7.306	1710.8	660.5	2371.4	1744.7	759.9	2504.6	3.8915	1.2037	5.0953
360	633.15	18675.1	1.896	5.044	6.940	1728.8	627.1	2355.8	1764.2	721.3	2485.4	3.9210	1.1390	5.0600
361	634.15	18901.7	1.917	4.840	6.757	1738.0	609.5	2347.5	1774.2	701.0	2475.2	3.9362	1.1052	5.0414
362	635.15	19130.7	1.939	4.634	6.573	1747.5	591.2	2338.7	1784.6	679.8	2464.4	3.9518	1.0702	5.0220
363	636.15	19362.1	1.963	4.425	6.388	1757.3	572.1	2329.3	1795.3	657.8	2453.0	3.9679	1.0338	5.0017
364	637.15	19596.1	1.988	4.213	6.201	1767.4	552.0	2319.4	1806.4	634.6	2440.9	3.9846	0.9958	4.9804
365	638.15	19832.6	2.016	3.996	6.012	1778.0	530.8	2308.8	1818.0	610.0	2428.0	4.0021	0.9558	4.9579
366	639.15	20071.6	2.046	3.772	5.819	1789.1	508.2	2297.3	1830.2	583.9	2414.1	4.0205	0.9134	4.9339
367	640.15	20313.2	2.080	3.540	5.621	1801.0	483.8	2284.8	1843.2	555.7	2399.0	4.0401	0.8680	4.9081
368	641.15	20557.5	2.118	3.298	5.416	1813.8	457.3	2271.1	1857.3	525.1	2382.4	4.0613	0.8189	4.8801
369	642.15	20804.4	2.162	3.039	5.201	1827.8	427.9	2255.7	1872.8	491.1	2363.9	4.0846	0.7647	4.8492
370	643.15	21054.0	2.214	2.759	4.973	1843.6	394.5	2238.1	1890.2	452.6	2342.8	4.1108	0.7036	4.8144
371	644.15	21306.4	2.278	2.446	4.723	1862.0	355.3	2217.3	1910.5	407.4	2317.9	4.1414	0.6324	4.7738
372	645.15	21561.6	2.364	2.075	4.439	1884.6	306.6	2191.2	1935.6	351.4	2287.0	4.1794	0.5446	4.7240
373	646.15	21819.7	2.496	1.588	4.084	1916.0	238.9	2154.9	1970.5	273.5	2244.0	4.2325	0.4233	4.6559
374	647.15	22080.5	2.843	0.623	3.466	1983.9	95.7	2079.7	2046.7	109.5	2156.2	4.3493	0.1692	4.5185
374.15	647.30	22120.0	3.170	0.000	3.170	2037.3	0.0	2037.3	2107.4	0.0	2107.4	4.4429	0.0000	4.4429

Appendix E

Superheated Steam (SI Units)

ABS. PRESS., kPa (SAT. TEMP., °C)		SAT. WATER	SAT. STEAM	TEMPERATURE, °C (TEMPERATURE, K)							
				75 (348.15)	100 (373.15)	125 (398.15)	150 (423.15)	175 (448.15)	200 (473.15)	225 (498.15)	250 (523.15)
1 (6.98)	V	1.000	129200	160640	172180	183720	195270	206810	218350	229890	241430
	U	29.334	2385.2	2480.8	2516.4	2552.3	2588.5	2624.9	2661.7	2698.8	2736.3
	H	29.335	2514.4	2641.5	2688.6	2736.0	2783.7	2831.7	2880.1	2928.7	2977.7
	S	0.1060	8.9767	9.3828	9.5136	9.6365	9.7527	9.8629	9.9679	10.0681	10.1641
10 (45.83)	V	1.010	14670	16030	17190	18350	19510	20660	21820	22980	24130
	U	191.822	2438.0	2479.7	2515.6	2551.6	2588.0	2624.5	2661.4	2698.6	2736.1
	H	191.832	2584.8	2640.0	2687.5	2735.2	2783.1	2831.2	2879.6	2928.4	2977.4
	S	0.6493	8.1511	8.3168	8.4486	8.5722	8.6888	8.7994	8.9045	9.0049	9.1010
20 (60.09)	V	1.017	7649.8	8000.0	8584.7	9167.1	9748.0	10320	10900	11480	12060
	U	251.432	2456.9	2478.4	2514.6	2550.9	2587.4	2624.1	2661.0	2698.3	2735.8
	H	251.453	2609.9	2638.4	2686.3	2734.2	2782.3	2830.6	2879.2	2928.0	2977.1
	S	0.8321	7.9094	7.9933	8.1261	8.2504	8.3676	8.4785	8.5839	8.6844	8.7806
30 (69.12)	V	1.022	5229.3	5322.0	5714.4	6104.6	6493.2	6880.8	7267.5	7653.8	8039.7
	U	289.271	2468.6	2477.1	2513.6	2550.2	2586.8	2623.6	2660.7	2698.0	2735.6
	H	289.302	2625.4	2636.8	2685.1	2733.3	2781.6	2830.0	2878.7	2927.6	2976.8
	S	0.9441	7.7695	7.8024	7.9363	8.0614	8.1791	8.2903	8.3960	8.4967	8.5930
40 (75.89)	V	1.027	3993.4	....	4279.2	4573.3	4865.8	5157.2	5447.8	5738.0	6027.7
	U	317.609	2477.1		2512.6	2549.4	2586.2	2623.2	2660.3	2697.7	2735.4
	H	317.650	2636.9		2683.8	2732.3	2780.9	2829.5	2878.2	2927.2	2976.5
	S	1.0261	7.6709		7.8009	7.9268	8.0450	8.1566	8.2624	8.3633	8.4598
50 (81.35)	V	1.030	3240.2	....	3418.1	3654.5	3889.1	4123.0	4356.0	4588.5	4820.5
	U	340.513	2484.0		2511.7	2548.6	2585.6	2622.7	2659.9	2697.4	2735.1
	H	340.564	2646.0		2682.6	2731.4	2780.1	2828.9	2877.7	2926.8	2976.1
	S	1.0912	7.5947		7.6953	7.8219	7.9406	8.0526	8.1587	8.2598	8.3564
75 (91.79)	V	1.037	2216.9	....	2269.8	2429.4	2587.3	2744.2	2900.2	3055.8	3210.9
	U	384.374	2496.7		2509.2	2546.7	2584.2	2621.6	2659.0	2696.7	2734.5
	H	384.451	2663.0		2679.2	2728.9	2778.2	2827.4	2876.6	2925.8	2975.3
	S	1.2131	7.4570		7.5014	7.6300	7.7500	7.8629	7.9697	8.0712	8.1681
100 (99.63)	V	1.043	1693.7	....	1695.5	1816.7	1936.3	2054.7	2172.3	2289.4	2406.1
	U	417.406	2506.1		2506.6	2544.8	2582.7	2620.4	2658.1	2695.9	2733.9
	H	417.511	2675.4		2676.2	2726.5	2776.3	2825.9	2875.4	2924.9	2974.5
	S	1.3027	7.3598		7.3618	7.4923	7.6137	7.7275	7.8349	7.9369	8.0342
101.325 (100.00)	V	1.044	1673.0	....	1673.0	1792.7	1910.7	2027.7	2143.8	2259.3	2374.5
	U	418.959	2506.5		2506.5	2544.7	2582.6	2620.4	2658.1	2695.9	2733.9
	H	419.064	2676.0		2676.0	2726.4	2776.2	2825.8	2875.3	2924.8	2974.5
	S	1.3069	7.3554		7.3554	7.4860	7.6075	7.7213	7.8288	7.9308	8.0280
125 (105.99)	V	1.049	1374.6	....	....	1449.1	1545.6	1641.0	1735.6	1829.6	1923.2
	U	444.224	2513.4			2542.9	2581.2	2619.3	2657.2	2695.2	2733.3
	H	444.356	2685.2			2724.0	2774.4	2824.4	2874.2	2923.9	2973.7
	S	1.3740	7.2847			7.3844	7.5072	7.6219	7.7300	7.8324	7.9300
150 (111.37)	V	1.053	1159.0	....	....	1204.0	1285.2	1365.2	1444.4	1523.0	1601.3
	U	466.968	2519.5			2540.9	2579.7	2618.1	2656.3	2694.4	2732.7
	H	467.126	2693.4			2721.5	2772.5	2822.9	2872.9	2922.9	2972.9
	S	1.4336	7.2234			7.2953	7.4194	7.5352	7.6439	7.7468	7.8447
175 (116.06)	V	1.057	1003.34	....	....	1028.8	1099.1	1168.2	1236.4	1304.1	1371.3
	U	486.815	2524.7			2538.9	2578.2	2616.9	2655.3	2693.7	2732.1
	H	487.000	2700.3			2719.0	2770.5	2821.3	2871.7	2921.9	2972.0
	S	1.4849	7.1716			7.2191	7.3447	7.4614	7.5708	7.6741	7.7724
200 (120.23)	V	1.061	885.44	....	....	897.47	959.54	1020.4	1080.4	1139.8	1198.9
	U	504.489	2529.2			2536.9	2576.6	2615.7	2654.4	2692.9	2731.4
	H	504.701	2706.3			2716.4	2768.5	2819.8	2870.5	2920.9	2971.2
	S	1.5301	7.1268			7.1523	7.2794	7.3971	7.5072	7.6110	7.7096
225 (123.99)	V	1.064	792.97	....	....	795.25	850.97	905.44	959.06	1012.1	1064.7
	U	520.465	2533.2			2534.8	2575.1	2614.5	2653.5	2692.2	2730.8
	H	520.705	2711.6			2713.8	2766.5	2818.2	2869.3	2919.9	2970.4
	S	1.5705	7.0873			7.0928	7.2213	7.3400	7.4508	7.5551	7.6540
250 (127.43)	V	1.068	718.44	....	....	....	764.09	813.47	861.98	909.91	957.41
	U	535.077	2536.8				2573.5	2613.3	2652.5	2691.4	2730.2
	H	535.343	2716.4				2764.5	2816.7	2868.0	2918.9	2969.6
	S	1.6071	7.0520				7.1689	7.2886	7.4001	7.5050	7.6042
275 (130.60)	V	1.071	657.04	....	....	....	693.00	738.21	782.55	826.29	869.61
	U	548.564	2540.0				2571.9	2612.1	2651.6	2690.7	2729.6
	H	548.858	2720.7				2762.5	2815.1	2866.8	2917.9	2968.7
	S	1.6407	7.0201				7.1211	7.2419	7.3541	7.4594	7.5590
300 (133.54)	V	1.073	605.56	....	....	....	633.74	675.49	716.35	756.60	796.44
	U	561.107	2543.0				2570.3	2610.8	2650.6	2689.9	2729.0
	H	561.429	2724.7				2760.4	2813.5	2865.5	2916.9	2967.9
	S	1.6716	6.9909				7.0771	7.1990	7.3119	7.4177	7.5176

ABS. PRESS., kPa (SAT. TEMP., °C)		SAT. WATER	SAT. STEAM	TEMPERATURE, °C (TEMPERATURE, K)							
				300 (573.15)	350 (623.15)	400 (673.15)	450 (723.15)	500 (773.15)	550 (823.15)	600 (873.15)	650 (923.15)
1 (6.98)	V	1.000	129200	264500	287580	310660	333730	356810	379880	402960	426040
	U	29.334	2385.2	2812.3	2889.9	2969.1	3049.9	3132.4	3216.7	3302.6	3390.3
	H	29.335	2514.4	3076.8	3177.5	3279.7	3383.6	3489.2	3596.5	3705.6	3816.4
	S	0.1060	8.9767	10.3450	10.5133	10.6711	10.8200	10.9612	11.0957	11.2243	11.3476
10 (45.83)	V	1.010	14670	26440	28750	31060	33370	35670	37980	40290	42600
	U	191.822	2438.0	2812.2	2889.8	2969.0	3049.8	3132.3	3216.6	3302.6	3390.3
	H	191.832	2584.8	3076.6	3177.3	3279.6	3383.5	3489.1	3596.5	3705.5	3816.3
	S	0.6493	8.1511	9.2820	9.4504	9.6083	9.7572	9.8984	10.0329	10.1616	10.2849
20 (60.09)	V	1.017	7649.8	13210	14370	15520	16680	17830	18990	20140	21300
	U	251.432	2456.9	2812.0	2889.6	2968.9	3049.7	3132.3	3216.5	3302.5	3390.2
	H	251.453	2609.9	3076.4	3177.1	3279.4	3383.4	3489.0	3596.4	3705.4	3816.2
	S	0.8321	7.9094	8.9618	9.1303	9.2882	9.4372	9.5784	9.7130	9.8416	9.9650
30 (69.12)	V	1.022	5229.3	8810.8	9581.2	10350	11120	11890	12660	13430	14190
	U	289.271	2468.6	2811.8	2889.5	2968.7	3049.6	3132.2	3216.5	3302.5	3390.2
	H	289.302	2625.4	3076.1	3176.9	3279.3	3383.3	3488.9	3596.3	3705.4	3816.2
	S	0.9441	7.7695	8.7744	8.9430	9.1010	9.2499	9.3912	9.5257	9.6544	9.7778
40 (75.89)	V	1.027	3993.4	6606.5	7184.6	7762.5	8340.1	8917.6	9494.9	10070	10640
	U	317.609	2477.1	2811.6	2889.4	2968.6	3049.5	3132.1	3216.4	3302.4	3390.1
	H	317.650	2636.9	3075.9	3176.8	3279.1	3383.1	3488.8	3596.2	3705.3	3816.1
	S	1.0261	7.6709	8.6413	8.8100	8.9680	9.1170	9.2583	9.3929	9.5216	9.6450
50 (81.35)	V	1.030	3240.2	5283.9	5746.7	6209.1	6671.4	7133.5	7595.5	8057.4	8519.2
	U	340.513	2484.0	2811.5	2889.2	2968.5	3049.4	3132.0	3216.3	3302.3	3390.1
	H	340.564	2646.0	3075.7	3176.6	3279.0	3383.0	3488.7	3596.1	3705.2	3816.0
	S	1.0912	7.5947	8.5380	8.7068	8.8649	9.0139	9.1552	9.2898	9.4185	9.5419
75 (91.79)	V	1.037	2216.9	3520.5	3829.4	4138.0	4446.4	4754.7	5062.8	5370.9	5678.9
	U	384.374	2496.7	2811.0	2888.9	2968.2	3049.2	3131.8	3216.1	3302.2	3389.9
	H	384.451	2663.0	3075.1	3176.1	3278.6	3382.7	3488.4	3595.8	3705.0	3815.9
	S	1.2131	7.4570	8.3502	8.5191	8.6773	8.8265	8.9678	9.1025	9.2312	9.3546
100 (99.63)	V	1.043	1693.7	2638.7	2870.8	3102.5	3334.0	3565.3	3796.5	4027.7	4258.8
	U	417.406	2506.1	2810.6	2888.6	2968.0	3049.0	3131.6	3216.0	3302.0	3389.8
	H	417.511	2675.4	3074.5	3175.6	3278.2	3382.4	3488.1	3595.6	3704.8	3815.7
	S	1.3027	7.3598	8.2166	8.3858	8.5442	8.6934	8.8348	8.9695	9.0982	9.2217
101.325 (100.00)	V	1.044	1673.0	2604.2	2833.2	3061.9	3290.3	3518.7	3746.9	3975.0	4203.1
	U	418.959	2506.5	2810.6	2888.5	2968.0	3048.9	3131.6	3215.9	3302.0	3389.8
	H	419.064	2676.0	3074.4	3175.6	3278.2	3382.3	3488.1	3595.6	3704.8	3815.7
	S	1.3069	7.3554	8.2105	8.3797	8.5381	8.6873	8.8287	8.9634	9.0922	9.2156
125 (106.99)	V	1.049	1374.6	2109.7	2295.6	2481.2	2666.5	2851.7	3036.8	3221.8	3406.7
	U	444.224	2513.4	2810.2	2888.2	2967.7	3048.7	3131.4	3215.8	3301.9	3389.7
	H	444.356	2685.2	3073.9	3175.2	3277.8	3382.0	3487.9	3595.4	3704.6	3815.5
	S	1.3740	7.2847	8.1129	8.2823	8.4408	8.5901	8.7316	8.8663	8.9951	9.1186
150 (111.37)	V	1.053	1159.0	1757.0	1912.2	2066.9	2221.5	2375.9	2530.2	2684.5	2838.6
	U	466.968	2519.5	2809.7	2887.9	2967.4	3048.5	3131.2	3215.6	3301.7	3389.6
	H	467.126	2693.4	3073.3	3174.7	3277.5	3381.7	3487.6	3595.1	3704.4	3815.3
	S	1.4336	7.2234	8.0280	8.1976	8.3562	8.5056	8.6472	8.7819	8.9108	9.0343
175 (116.06)	V	1.057	1003.34	1505.1	1638.3	1771.1	1903.7	2036.1	2168.4	2300.7	2432.9
	U	486.815	2524.7	2809.3	2887.5	2967.1	3048.3	3131.0	3215.4	3301.6	3389.4
	H	487.000	2700.3	3072.7	3174.2	3277.1	3381.4	3487.3	3594.9	3704.2	3815.1
	S	1.4849	7.1716	7.9561	8.1259	8.2847	8.4341	8.5758	8.7106	8.8394	8.9630
200 (120.23)	V	1.061	885.44	1316.2	1432.8	1549.2	1665.3	1781.2	1897.1	2012.9	2128.6
	U	504.489	2529.2	2808.8	2887.2	2966.9	3048.0	3130.8	3215.3	3301.4	3389.2
	H	504.701	2706.3	3072.1	3173.8	3276.7	3381.1	3487.0	3594.7	3704.0	3815.0
	S	1.5301	7.1268	7.8937	8.0638	8.2226	8.3722	8.5139	8.6487	8.7776	8.9012
225 (123.99)	V	1.064	792.97	1169.2	1273.1	1376.6	1479.9	1583.0	1686.0	1789.0	1891.9
	U	520.465	2533.2	2808.4	2886.9	2966.6	3047.8	3130.6	3215.1	3301.2	3389.1
	H	520.705	2711.6	3071.5	3173.3	3276.3	3380.8	3486.8	3594.4	3703.8	3814.8
	S	1.5705	7.0873	7.8385	8.0088	8.1679	8.3175	8.4593	8.5942	8.7231	8.8467
250 (127.43)	V	1.068	718.44	1051.6	1145.2	1238.5	1331.5	1424.4	1517.2	1609.9	1702.5
	U	535.077	2536.8	2808.0	2886.5	2966.3	3047.6	3130.4	3214.9	3301.1	3389.0
	H	535.343	2716.4	3070.9	3172.8	3275.9	3380.4	3486.5	3594.2	3703.6	3814.6
	S	1.6071	7.0520	7.7891	7.9597	8.1188	8.2686	8.4104	8.5453	8.6743	8.7980
275 (130.60)	V	1.071	657.04	955.45	1040.7	1125.5	1210.2	1294.7	1379.0	1463.3	1547.6
	U	548.564	2540.0	2807.5	2886.2	2966.0	3047.3	3130.2	3214.7	3300.9	3388.8
	H	548.858	2720.7	3070.3	3172.4	3275.5	3380.1	3486.2	3594.0	3703.4	3814.4
	S	1.6407	7.0201	7.7444	7.9151	8.0744	8.2243	8.3661	8.5011	8.6301	8.7538
300 (133.54)	V	1.073	605.56	875.29	953.52	1031.4	1109.0	1186.5	1263.9	1341.2	1418.5
	U	561.107	2543.0	2807.1	2885.8	2965.8	3047.1	3130.0	3214.5	3300.8	3388.7
	H	561.429	2724.7	3069.7	3171.9	3275.2	3379.8	3486.0	3593.7	3703.2	3814.2
	S	1.6716	6.9909	7.7034	7.8744	8.0338	8.1838	8.3257	8.4608	8.5898	8.7135

ABS. PRESS., kPa (SAT. TEMP., °C)		SAT. WATER	SAT. STEAM	TEMPERATURE, °C (TEMPERATURE, K)							
				150 (423.15)	175 (448.15)	200 (473.15)	220 (493.15)	240 (513.15)	260 (533.15)	280 (553.15)	300 (573.15)
325 (136.29)	V	1.076	561.75	583.58	622.41	660.33	690.22	719.81	749.18	778.39	807.47
	U	572.847	2545.7	2568.7	2609.6	2649.6	2681.2	2712.7	2744.0	2775.3	2806.6
	H	573.197	2728.3	2758.4	2811.9	2864.2	2905.6	2946.6	2987.5	3028.2	3069.0
	S	1.7004	6.9640	7.0363	7.1592	7.2729	7.3585	7.4400	7.5181	7.5933	7.6657
350 (138.87)	V	1.079	524.00	540.58	576.90	612.31	640.18	667.75	695.09	722.27	749.33
	U	583.892	2548.2	2567.1	2608.3	2648.6	2680.4	2712.0	2743.4	2774.8	2806.2
	H	584.270	2731.6	2756.3	2810.3	2863.0	2904.5	2945.7	2986.7	3027.6	3068.4
	S	1.7273	6.9392	6.9982	7.1222	7.2366	7.3226	7.4045	7.4828	7.5581	7.6307
375 (141.31)	V	1.081	491.13	503.29	537.46	570.69	596.81	622.62	648.22	673.64	698.94
	U	594.332	2550.6	2565.4	2607.1	2647.7	2679.6	2711.3	2742.8	2774.3	2805.7
	H	594.737	2734.7	2754.1	2808.6	2861.7	2903.4	2944.8	2985.9	3026.9	3067.8
	S	1.7526	6.9160	6.9624	7.0875	7.2027	7.2891	7.3713	7.4499	7.5254	7.5981
400 (143.62)	V	1.084	462.22	470.66	502.93	534.26	558.85	583.14	607.20	631.09	654.85
	U	604.237	2552.7	2563.7	2605.8	2646.7	2678.8	2710.6	2742.2	2773.7	2805.3
	H	604.670	2737.6	2752.0	2807.0	2860.4	2902.3	2943.9	2985.1	3026.2	3067.2
	S	1.7764	6.8943	6.9285	7.0548	7.1708	7.2576	7.3402	7.4190	7.4947	7.5675
425 (145.82)	V	1.086	436.61	441.85	472.47	502.12	525.36	548.30	571.01	593.54	615.95
	U	613.667	2554.8	2562.0	2604.5	2645.7	2678.0	2709.9	2741.6	2773.2	2804.8
	H	614.128	2740.3	2749.8	2805.3	2859.1	2901.2	2942.9	2984.3	3025.5	3066.6
	S	1.7990	6.8739	6.8965	7.0239	7.1407	7.2280	7.3108	7.3899	7.4657	7.5388
450 (147.92)	V	1.088	413.75	416.24	445.38	473.55	495.59	517.33	538.83	560.17	581.37
	U	622.672	2556.7	2560.3	2603.2	2644.7	2677.1	2709.2	2741.0	2772.7	2804.4
	H	623.162	2742.9	2747.7	2803.7	2857.8	2900.2	2942.0	2983.5	3024.8	3066.0
	S	1.8204	6.8547	6.8660	6.9946	7.1121	7.1999	7.2831	7.3624	7.4384	7.5116
475 (149.92)	V	1.091	393.22	393.31	421.14	447.97	468.95	489.62	510.05	530.30	550.43
	U	631.294	2558.5	2558.6	2601.9	2643.7	2676.3	2708.5	2740.4	2772.2	2803.9
	H	631.812	2745.3	2745.5	2802.0	2856.5	2899.1	2941.1	2982.7	3024.1	3065.4
	S	1.8408	6.8365	6.8369	6.9667	7.0850	7.1732	7.2567	7.3363	7.4125	7.4858
500 (151.84)	V	1.093	374.68	....	399.31	424.96	444.97	464.67	484.14	503.43	522.58
	U	639.569	2560.2	....	2600.6	2642.7	2675.5	2707.8	2739.8	2771.7	2803.5
	H	640.116	2747.5	....	2800.3	2855.1	2898.0	2940.1	2981.9	3023.4	3064.8
	S	1.8604	6.8192	....	6.9400	7.0592	7.1478	7.2317	7.3115	7.3879	7.4614
525 (153.69)	V	1.095	357.84	....	379.56	404.13	423.28	442.11	460.70	479.11	497.38
	U	647.528	2561.8	....	2599.3	2641.6	2674.6	2707.1	2739.2	2771.2	2803.0
	H	648.103	2749.7	....	2798.6	2853.8	2896.8	2939.2	2981.1	3022.7	3064.1
	S	1.8790	6.8027	....	6.9145	7.0345	7.1236	7.2078	7.2879	7.3645	7.4381
550 (155.47)	V	1.097	342.48	....	361.60	385.19	403.55	421.59	439.38	457.00	474.48
	U	655.199	2563.3	....	2598.0	2640.6	2673.8	2706.4	2738.6	2770.6	2802.6
	H	655.802	2751.7	....	2796.8	2852.5	2895.7	2938.3	2980.3	3022.0	3063.5
	S	1.8970	6.7870	....	6.8900	7.0108	7.1004	7.1849	7.2653	7.3421	7.4158
575 (157.18)	V	1.099	328.41	....	345.20	367.90	385.54	402.85	419.92	436.81	453.56
	U	662.603	2564.8	....	2596.6	2639.6	2672.9	2705.7	2738.0	2770.1	2802.1
	H	663.235	2753.6	....	2795.1	2851.1	2894.6	2937.3	2979.5	3021.3	3062.9
	S	1.9142	6.7720	....	6.8664	6.9880	7.0781	7.1630	7.2436	7.3206	7.3945
600 (158.84)	V	1.101	315.47	....	330.16	352.04	369.03	385.68	402.08	418.31	434.39
	U	669.762	2566.2	....	2595.3	2638.5	2672.1	2705.0	2737.4	2769.6	2801.6
	H	670.423	2755.5	....	2793.3	2849.7	2893.5	2936.4	2978.7	3020.6	3062.3
	S	1.9308	6.7575	....	6.8437	6.9662	7.0567	7.1419	7.2228	7.3000	7.3740
625 (160.44)	V	1.103	303.54	....	316.31	337.45	353.83	369.87	385.67	401.28	416.75
	U	676.695	2567.5	....	2593.9	2637.5	2671.2	2704.2	2736.8	2769.1	2801.2
	H	677.384	2757.2	....	2791.6	2848.4	2892.3	2935.4	2977.8	3019.9	3061.7
	S	1.9469	6.7437	....	6.8217	6.9451	7.0361	7.1217	7.2028	7.2802	7.3544
650 (161.99)	V	1.105	292.49	....	303.53	323.98	339.80	355.29	370.52	385.56	400.47
	U	683.417	2568.7	....	2592.5	2636.4	2670.3	2703.5	2736.2	2768.5	2800.7
	H	684.135	2758.9	....	2789.8	2847.0	2891.2	2934.4	2977.0	3019.2	3061.0
	S	1.9623	6.7304	....	6.8004	6.9247	7.0162	7.1021	7.1835	7.2611	7.3355
675 (163.49)	V	1.106	282.23	....	291.69	311.51	326.81	341.78	356.49	371.01	385.39
	U	689.943	2570.0	....	2591.1	2635.4	2669.5	2702.8	2735.6	2768.0	2800.3
	H	690.689	2760.5	....	2788.0	2845.6	2890.1	2933.5	2976.2	3018.5	3060.4
	S	1.9773	6.7176	....	6.7798	6.9050	6.9970	7.0833	7.1650	7.2428	7.3173
700 (164.96)	V	1.108	272.68	....	280.69	299.92	314.75	329.23	343.46	357.50	371.39
	U	696.285	2571.1	....	2589.7	2634.3	2668.6	2702.1	2735.0	2767.5	2799.8
	H	697.061	2762.0	....	2786.2	2844.2	2888.9	2932.5	2975.4	3017.7	3059.8
	S	1.9918	6.7052	....	6.7598	6.8859	6.9784	7.0651	7.1470	7.2250	7.2997
725 (166.38)	V	1.110	263.77	....	270.45	289.13	303.51	317.55	331.33	344.92	358.36
	U	702.457	2572.2	....	2588.3	2633.2	2667.7	2701.3	2734.3	2767.0	2799.3
	H	703.261	2763.4	....	2784.4	2842.8	2887.7	2931.6	2974.6	3017.0	3059.1
	S	2.0059	6.6932	....	6.7404	6.8673	6.9604	7.0474	7.1296	7.2078	7.2827

ABS. PRESS., kPa (SAT. TEMP., °C)		SAT. WATER	SAT. STEAM	TEMPERATURE, °C (TEMPERATURE, K)							
				325 (598.15)	350 (623.15)	400 (673.15)	450 (723.15)	500 (773.15)	550 (823.15)	600 (873.15)	650 (923.15)
325 (136.29)	V	1.076	561.75	843.68	879.78	951.73	1023.5	1095.0	1166.5	1237.9	1309.2
	U	572.847	2545.7	2845.9	2885.5	2965.5	3046.9	3129.8	3214.4	3300.6	3388.6
	H	573.197	2728.3	3120.1	3171.4	3274.8	3379.5	3485.7	3593.5	3702.9	3814.1
	S	1.7004	6.9640	7.7530	7.8369	7.9965	8.1465	8.2885	8.4236	8.5527	8.6764
350 (138.87)	V	1.079	524.00	783.01	816.57	883.45	950.11	1016.6	1083.0	1149.3	1215.6
	U	583.892	2548.2	2845.6	2885.1	2965.2	3046.6	3129.6	3214.2	3300.5	3388.4
	H	584.270	2731.6	3119.6	3170.9	3274.4	3379.2	3485.4	3593.3	3702.7	3813.9
	S	1.7273	6.9392	7.7181	7.8022	7.9619	8.1120	8.2540	8.3892	8.5183	8.6421
375 (141.31)	V	1.081	491.13	730.42	761.79	824.28	886.54	948.66	1010.7	1072.6	1134.5
	U	594.332	2550.6	2845.2	2884.8	2964.9	3046.4	3129.4	3214.0	3300.3	3388.3
	H	594.737	2734.7	3119.1	3170.5	3274.0	3378.8	3485.1	3593.0	3702.5	3813.7
	S	1.7526	6.9160	7.6856	7.7698	7.9296	8.0798	8.2219	8.3571	8.4863	8.6101
400 (143.62)	V	1.084	462.22	684.41	713.85	772.50	830.92	889.19	947.35	1005.4	1063.4
	U	604.237	2552.7	2844.8	2884.6	2964.6	3046.2	3129.2	3213.8	3300.2	3388.2
	H	604.670	2737.6	3118.5	3170.0	3273.6	3378.5	3484.9	3592.8	3702.3	3813.5
	S	1.7764	6.8943	7.6552	7.7395	7.8994	8.0497	8.1919	8.3271	8.4563	8.5802
425 (145.82)	V	1.086	436.61	643.81	671.56	726.81	781.84	836.72	891.49	946.17	1000.8
	U	613.667	2554.8	2844.4	2884.1	2964.4	3045.9	3129.0	3213.7	3300.0	3388.0
	H	614.128	2740.3	3118.0	3169.5	3273.3	3378.2	3484.6	3592.5	3702.1	3813.4
	S	1.7990	6.8739	7.6265	7.7109	7.8710	8.0214	8.1636	8.2989	8.4282	8.5520
450 (147.92)	V	1.088	413.75	607.73	633.97	686.20	738.21	790.07	841.83	893.50	945.10
	U	622.672	2556.7	2844.0	2883.8	2964.1	3045.7	3128.8	3213.5	3299.8	3387.9
	H	623.162	2742.9	3117.5	3169.1	3272.9	3377.9	3484.3	3592.3	3701.9	3813.2
	S	1.8204	6.8547	7.5995	7.6840	7.8442	7.9947	8.1370	8.2723	8.4016	8.5255
475 (149.92)	V	1.091	393.22	575.44	600.33	649.87	699.18	748.34	797.40	846.37	895.27
	U	631.294	2558.5	2843.6	2883.4	2963.8	3045.4	3128.6	3213.3	3299.7	3387.7
	H	631.812	2745.3	3116.9	3168.6	3272.5	3377.6	3484.0	3592.1	3701.7	3813.0
	S	1.8408	6.8365	7.5739	7.6585	7.8189	7.9694	8.1118	8.2472	8.3765	8.5004
500 (151.84)	V	1.093	374.68	546.38	570.05	617.16	664.05	710.78	757.41	803.95	850.42
	U	639.569	2560.2	2843.2	2883.1	2963.5	3045.2	3128.4	3213.1	3299.5	3387.6
	H	640.116	2747.5	3116.4	3168.1	3272.1	3377.2	3483.8	3591.8	3701.5	3812.8
	S	1.8604	6.8192	7.5496	7.6343	7.7948	7.9454	8.0879	8.2233	8.3526	8.4766
525 (153.69)	V	1.095	357.84	520.08	542.66	587.58	632.26	676.80	721.23	765.57	809.85
	U	647.528	2561.8	2842.8	2882.7	2963.2	3045.0	3128.2	3213.0	3299.4	3387.5
	H	648.103	2749.7	3115.9	3167.6	3271.7	3376.9	3483.5	3591.6	3701.3	3812.6
	S	1.8790	6.8027	7.5264	7.6112	7.7719	7.9226	8.0651	8.2006	8.3299	8.4539
550 (155.47)	V	1.097	342.48	496.18	517.76	560.68	603.37	645.91	688.34	730.68	772.96
	U	655.199	2563.3	2842.4	2882.4	2963.0	3044.7	3128.0	3212.8	3299.2	3387.3
	H	655.802	2751.7	3115.3	3167.2	3271.3	3376.6	3483.2	3591.4	3701.1	3812.5
	S	1.8970	6.7870	7.5043	7.5892	7.7500	7.9008	8.0433	8.1789	8.3083	8.4323
575 (157.18)	V	1.099	328.41	474.36	495.03	536.12	576.98	617.70	658.30	698.83	739.28
	U	662.603	2564.8	2842.0	2882.1	2962.7	3044.5	3127.8	3212.6	3299.1	3387.2
	H	663.235	2753.6	3114.8	3166.7	3271.0	3376.3	3482.9	3591.1	3700.9	3812.3
	S	1.9142	6.7720	7.4831	7.5681	7.7290	7.8799	8.0226	8.1581	8.2876	8.4116
600 (158.84)	V	1.101	315.47	454.35	474.19	513.61	552.80	591.84	630.78	669.63	708.41
	U	669.762	2566.2	2841.6	2881.7	2962.4	3044.3	3127.6	3212.4	3298.9	3387.1
	H	670.423	2755.5	3114.3	3166.2	3270.6	3376.0	3482.7	3590.9	3700.7	3812.1
	S	1.9308	6.7575	7.4628	7.5479	7.7090	7.8600	8.0027	8.1383	8.2678	8.3919
625 (160.44)	V	1.103	303.54	435.94	455.01	492.89	530.55	568.05	605.45	642.76	680.01
	U	676.695	2567.5	2841.2	2881.4	2962.1	3044.0	3127.4	3212.2	3298.8	3386.9
	H	677.384	2757.2	3113.7	3165.7	3270.2	3375.6	3482.4	3590.7	3700.5	3811.9
	S	1.9469	6.7437	7.4433	7.5285	7.6897	7.8408	7.9836	8.1192	8.2488	8.3729
650 (161.99)	V	1.105	292.49	418.95	437.31	473.78	510.01	546.10	582.07	617.96	653.79
	U	683.417	2568.7	2840.9	2881.0	2961.8	3043.8	3127.2	3212.1	3298.6	3386.8
	H	684.135	2758.9	3113.2	3165.3	3269.8	3375.3	3482.1	3590.4	3700.3	3811.8
	S	1.9623	6.7304	7.4245	7.5099	7.6712	7.8224	7.9652	8.1009	8.2305	8.3546
675 (163.49)	V	1.106	282.23	403.22	420.92	456.07	491.00	525.77	560.43	595.00	629.51
	U	689.943	2570.0	2840.5	2880.7	2961.6	3043.6	3127.0	3211.9	3298.5	3386.7
	H	690.689	2760.5	3112.6	3164.8	3269.4	3375.0	3481.8	3590.2	3700.1	3811.6
	S	1.9773	6.7176	7.4064	7.4919	7.6534	7.8046	7.9475	8.0833	8.2129	8.3371
700 (164.96)	V	1.108	272.68	388.61	405.71	439.64	473.34	506.89	540.33	573.68	606.97
	U	696.285	2571.1	2840.1	2880.3	2961.3	3043.3	3126.8	3211.7	3298.3	3386.5
	H	697.061	2762.0	3112.1	3164.3	3269.0	3374.7	3481.6	3589.9	3699.9	3811.4
	S	1.9918	6.7052	7.3890	7.4745	7.6362	7.7875	7.9305	8.0663	8.1959	8.3201
725 (166.38)	V	1.110	263.77	375.01	391.54	424.33	456.90	489.31	521.61	553.83	585.99
	U	702.457	2572.2	2839.7	2880.0	2961.0	3043.1	3126.6	3211.5	3298.1	3386.4
	H	703.261	2763.4	3111.5	3163.8	3268.7	3374.3	3481.3	3589.7	3699.7	3811.2
	S	2.0059	6.6932	7.3721	7.4578	7.6196	7.7710	7.9140	8.0499	8.1796	8.3038

ABS. PRESS., kPa (SAT. TEMP., °C)		SAT. WATER	SAT. STEAM	175 (448.15)	200 (473.15)	220 (493.15)	240 (513.15)	260 (533.15)	280 (553.15)	300 (573.15)	325 (598.15)
750 (167.76)	V	1.112	255.43	260.88	279.05	293.03	306.65	320.01	333.17	346.19	362.32
	U	708.467	2573.3	2586.9	2632.1	2666.8	2700.6	2733.7	2766.4	2798.9	2839.3
	H	709.301	2764.8	2782.5	2841.4	2886.6	2930.6	2973.7	3016.3	3058.5	3111.0
	S	2.0195	6.6817	6.7215	6.8494	6.9429	7.0303	7.1128	7.1912	7.2662	7.3558
775 (169.10)	V	1.113	247.61	251.93	269.63	283.22	296.45	309.41	322.19	334.81	350.44
	U	714.326	2574.1	2585.4	2631.0	2665.9	2699.8	2733.1	2765.9	2798.4	2838.9
	H	715.189	2766.2	2780.7	2840.0	2885.4	2929.6	2972.9	3015.6	3057.9	3110.5
	S	2.0328	6.6705	6.7031	6.8319	6.9259	7.0137	7.0965	7.1751	7.2502	7.3400
800 (170.41)	V	1.115	240.26	243.53	260.79	274.02	286.88	299.48	311.89	324.14	339.31
	U	720.043	2575.3	2584.0	2629.9	2665.0	2699.1	2732.5	2765.4	2797.9	2838.5
	H	720.935	2767.5	2778.8	2838.6	2884.2	2928.6	2972.1	3014.9	3057.3	3109.9
	S	2.0457	6.6596	6.6851	6.8148	6.9094	6.9976	7.0807	7.1595	7.2348	7.3247
825 (171.69)	V	1.117	233.34	235.64	252.48	265.37	277.90	290.15	302.21	314.12	328.85
	U	725.625	2576.2	2582.5	2628.8	2664.1	2698.4	2731.8	2764.8	2797.5	2838.1
	H	726.547	2768.7	2776.9	2837.1	2883.1	2927.6	2971.2	3014.1	3056.6	3109.4
	S	2.0583	6.6491	6.6675	6.7982	6.8933	6.9819	7.0653	7.1443	7.2197	7.3098
850 (172.94)	V	1.118	226.81	228.21	244.66	257.24	269.44	281.37	293.10	304.68	319.00
	U	731.080	2577.1	2581.1	2627.7	2663.2	2697.6	2731.2	2764.3	2797.0	2837.7
	H	732.031	2769.9	2775.1	2835.7	2881.9	2926.6	2970.4	3013.4	3056.0	3108.8
	S	2.0705	6.6388	6.6504	6.7820	6.8777	6.9666	7.0503	7.1295	7.2051	7.2954
875 (174.16)	V	1.120	220.65	221.20	237.29	249.56	261.46	273.09	284.51	295.79	309.72
	U	736.415	2578.0	2579.6	2626.6	2662.3	2696.8	2730.6	2763.7	2796.5	2837.3
	H	737.394	2771.0	2773.1	2834.2	2880.7	2925.6	2969.5	3012.7	3055.3	3108.3
	S	2.0825	6.6289	6.6336	6.7662	6.8624	6.9518	7.0357	7.1152	7.1909	7.2813
900 (175.36)	V	1.121	214.81	....	230.32	242.31	253.93	265.27	276.40	287.39	300.96
	U	741.635	2578.8	....	2625.5	2661.4	2696.1	2729.9	2763.2	2796.1	2836.9
	H	742.644	2772.1	....	2832.7	2879.5	2924.6	2968.7	3012.0	3054.7	3107.7
	S	2.0941	6.6192	....	6.7508	6.8475	6.9373	7.0215	7.1012	7.1771	7.2676
925 (176.53)	V	1.123	209.28	....	223.73	235.46	246.80	257.87	268.73	279.44	292.66
	U	746.746	2579.6	....	2624.3	2660.5	2695.3	2729.3	2762.6	2795.6	2836.5
	H	747.784	2773.2	....	2831.3	2878.3	2923.6	2967.8	3011.2	3054.1	3107.2
	S	2.1055	6.6097	....	6.7357	6.8329	6.9231	7.0076	7.0875	7.1636	7.2543
950 (177.67)	V	1.124	204.03	....	217.48	228.96	240.05	250.86	261.46	271.91	284.81
	U	751.754	2580.4	....	2623.2	2659.5	2694.6	2728.7	2762.1	2795.1	2836.0
	H	752.822	2774.2	....	2829.8	2877.0	2922.6	2967.0	3010.5	3053.4	3106.6
	S	2.1166	6.6005	....	6.7209	6.8187	6.9093	6.9941	7.0742	7.1505	7.2413
975 (178.79)	V	1.126	199.04	....	211.55	222.79	233.64	244.20	254.56	264.76	277.35
	U	756.663	2581.1	....	2622.0	2658.6	2693.8	2728.0	2761.5	2794.6	2835.6
	H	757.761	2775.2	....	2828.3	2875.8	2921.6	2966.1	3009.7	3052.8	3106.1
	S	2.1275	6.5916	....	6.7064	6.8048	6.8958	6.9809	7.0612	7.1377	7.2286
1000 (179.88)	V	1.127	194.29	....	205.92	216.93	227.55	237.89	248.01	257.98	270.27
	U	761.478	2581.9	....	2620.9	2657.7	2693.0	2727.4	2761.0	2794.2	2835.2
	H	762.605	2776.2	....	2826.8	2874.6	2920.6	2965.2	3009.0	3052.1	3105.5
	S	2.1382	6.5828	....	6.6922	6.7911	6.8825	6.9680	7.0485	7.1251	7.2163
1050 (182.02)	V	1.130	185.45	....	195.45	206.04	216.24	226.15	235.84	245.37	257.12
	U	770.843	2583.3	....	2618.5	2655.8	2691.5	2726.1	2759.9	2793.2	2834.4
	H	772.029	2778.0	....	2823.8	2872.1	2918.5	2963.5	3007.5	3050.8	3104.4
	S	2.1588	6.5659	....	6.6645	6.7647	6.8569	6.9430	7.0240	7.1009	7.1924
1100 (184.07)	V	1.133	177.38	....	185.92	196.14	205.96	215.47	224.77	233.91	245.16
	U	779.878	2584.5	....	2616.2	2653.9	2689.9	2724.7	2758.8	2792.2	2833.6
	H	781.124	2779.7	....	2820.7	2869.6	2916.4	2961.8	3006.0	3049.6	3103.3
	S	2.1786	6.5497	....	6.6379	6.7392	6.8323	6.9190	7.0005	7.0778	7.1695
1150 (186.05)	V	1.136	169.99	....	177.22	187.10	196.56	205.73	214.67	223.44	234.25
	U	788.611	2585.8	....	2613.8	2651.9	2688.3	2723.4	2757.7	2791.3	2832.8
	H	789.917	2781.3	....	2817.6	2867.1	2914.4	2960.0	3004.5	3048.2	3102.2
	S	2.1977	6.5342	....	6.6122	6.7147	6.8086	6.8959	6.9779	7.0556	7.1476
1200 (187.96)	V	1.139	163.20	....	169.23	178.80	187.95	196.79	205.40	213.85	224.24
	U	797.064	2586.9	....	2611.3	2650.0	2686.7	2722.1	2756.5	2790.3	2832.0
	H	798.430	2782.7	....	2814.4	2864.5	2912.2	2958.2	3003.0	3046.9	3101.0
	S	2.2161	6.5194	....	6.5872	6.6909	6.7858	6.8738	6.9562	7.0342	7.1266
1250 (189.81)	V	1.141	156.93	....	161.88	171.17	180.02	188.56	196.88	205.02	215.03
	U	805.259	2588.0	....	2608.9	2648.0	2685.1	2720.8	2755.4	2789.3	2831.1
	H	806.685	2784.1	....	2811.2	2861.9	2910.1	2956.5	3001.5	3045.6	3099.9
	S	2.2338	6.5050	....	6.5630	6.6680	6.7637	6.8523	6.9353	7.0136	7.1064
1300 (191.61)	V	1.144	151.13	....	155.09	164.11	172.70	180.97	189.01	196.87	206.53
	U	813.213	2589.0	....	2606.4	2646.0	2683.5	2719.4	2754.3	2788.4	2830.3
	H	814.700	2785.4	....	2808.0	2859.3	2908.0	2954.7	3000.0	3044.3	3098.8
	S	2.2510	6.4913	....	6.5394	6.6457	6.7424	6.8316	6.9151	6.9938	7.0869

ABS. PRESS., kPa (SAT. TEMP., °C)		SAT. WATER	SAT. STEAM	TEMPERATURE, °C (TEMPERATURE, K)							
				350 (623.15)	375 (648.15)	400 (673.15)	450 (723.15)	500 (773.15)	550 (823.15)	600 (873.15)	650 (923.15)
750 (167.76)	V	1.112	255.43	378.31	394.22	410.05	441.55	472.90	504.15	535.30	566.40
	U	708.467	2573.3	2879.6	2920.1	2960.7	3042.9	3126.3	3211.4	3298.0	3386.2
	H	709.301	2764.8	3163.4	3215.7	3268.3	3374.0	3481.0	3589.5	3699.5	3811.0
	S	2.0195	6.6817	7.4416	7.5240	7.6035	7.7550	7.8981	8.0340	8.1637	8.2880
775 (169.10)	V	1.113	247.61	365.94	381.35	396.69	427.20	457.56	487.81	517.97	548.07
	U	714.326	2574.3	2879.3	2919.8	2960.4	3042.6	3126.1	3211.2	3297.8	3386.1
	H	715.189	2766.2	3162.9	3215.3	3267.9	3373.7	3480.8	3589.2	3699.3	3810.9
	S	2.0328	6.6705	7.4259	7.5084	7.5880	7.7396	7.8827	8.0187	8.1484	8.2727
800 (170.41)	V	1.115	240.26	354.34	369.29	384.16	413.74	443.17	472.49	501.72	530.89
	U	720.043	2575.3	2878.9	2919.5	2960.2	3042.4	3125.9	3211.0	3297.7	3386.0
	H	720.935	2767.5	3162.4	3214.9	3267.5	3373.4	3480.5	3589.0	3699.1	3810.7
	S	2.0457	6.6596	7.4107	7.4932	7.5729	7.7246	7.8678	8.0038	8.1336	8.2579
825 (171.69)	V	1.117	233.34	343.45	357.96	372.39	401.10	429.65	458.10	486.46	514.76
	U	725.625	2576.2	2878.6	2919.1	2959.9	3042.2	3125.7	3210.8	3297.5	3385.8
	H	726.547	2768.7	3161.9	3214.5	3267.1	3373.1	3480.2	3588.8	3698.8	3810.5
	S	2.0583	6.6491	7.3959	7.4786	7.5583	7.7101	7.8533	7.9894	8.1192	8.2436
850 (172.94)	V	1.118	226.81	333.20	347.29	361.31	389.20	416.93	444.56	472.09	499.57
	U	731.080	2577.1	2878.2	2918.8	2959.6	3041.9	3125.5	3210.7	3297.4	3385.7
	H	732.031	2769.9	3161.4	3214.0	3266.7	3372.7	3479.9	3588.5	3698.6	3810.3
	S	2.0705	6.6388	7.3815	7.4643	7.5441	7.6960	7.8393	7.9754	8.1053	8.2296
875 (174.16)	V	1.120	220.65	323.53	337.24	350.87	377.98	404.94	431.79	458.55	485.25
	U	736.415	2578.0	2877.9	2918.5	2959.3	3041.7	3125.3	3210.5	3297.2	3385.6
	H	737.394	2771.0	3161.0	3213.6	3266.3	3372.4	3479.7	3588.3	3698.4	3810.2
	S	2.0825	6.6289	7.3676	7.4504	7.5303	7.6823	7.8257	7.9618	8.0917	8.2161
900 (175.36)	V	1.121	214.81	314.40	327.74	341.01	367.39	393.61	419.73	445.76	471.72
	U	741.635	2578.8	2877.5	2918.2	2959.0	3041.4	3125.1	3210.3	3297.1	3385.4
	H	742.644	2772.1	3160.5	3213.2	3266.0	3372.1	3479.4	3588.1	3698.2	3810.0
	S	2.0941	6.6192	7.3540	7.4370	7.5169	7.6689	7.8124	7.9486	8.0785	8.2030
925 (176.53)	V	1.123	209.28	305.76	318.75	331.68	357.36	382.90	408.32	433.66	458.93
	U	746.746	2579.6	2877.2	2917.9	2958.8	3041.2	3124.9	3210.1	3296.9	3385.3
	H	747.784	2773.2	3160.0	3212.7	3265.6	3371.8	3479.1	3587.8	3698.0	3809.8
	S	2.1055	6.6097	7.3408	7.4238	7.5038	7.6560	7.7995	7.9357	8.0657	8.1902
950 (177.67)	V	1.124	204.03	297.57	310.24	322.84	347.87	372.74	397.51	422.19	446.81
	U	751.754	2580.4	2876.8	2917.6	2958.5	3041.0	3124.7	3209.9	3296.7	3385.1
	H	752.822	2774.2	3159.5	3212.3	3265.2	3371.5	3478.8	3587.6	3697.8	3809.6
	S	2.1166	6.6005	7.3279	7.4110	7.4911	7.6433	7.7869	7.9232	8.0532	8.1777
975 (178.79)	V	1.126	199.04	289.81	302.17	314.45	338.86	363.11	387.26	411.32	435.31
	U	756.663	2581.1	2876.5	2917.3	2958.2	3040.7	3124.5	3209.8	3296.6	3385.0
	H	757.761	2775.2	3159.0	3211.9	3264.8	3371.1	3478.6	3587.3	3697.6	3809.4
	S	2.1275	6.5916	7.3154	7.3986	7.4787	7.6310	7.7747	7.9110	8.0410	8.1656
1000 (179.88)	V	1.127	194.29	282.43	294.50	306.49	330.30	353.96	377.52	400.98	424.38
	U	761.478	2581.9	2876.1	2917.0	2957.9	3040.5	3124.3	3209.6	3296.4	3384.9
	H	762.605	2776.2	3158.5	3211.5	3264.4	3370.8	3478.3	3587.1	3697.4	3809.3
	S	2.1382	6.5828	7.3031	7.3864	7.4665	7.6190	7.7627	7.8991	8.0292	8.1537
1050 (182.02)	V	1.130	185.45	268.74	280.25	291.69	314.41	336.97	359.43	381.79	404.10
	U	770.843	2583.3	2875.4	2916.3	2957.4	3040.0	3123.9	3209.2	3296.1	3384.6
	H	772.029	2778.0	3157.6	3210.6	3263.6	3370.2	3477.7	3586.6	3697.0	3808.9
	S	2.1588	6.5659	7.2795	7.3629	7.4432	7.5958	7.7397	7.8762	8.0063	8.1309
1100 (184.07)	V	1.133	177.38	256.28	267.30	278.24	299.96	321.53	342.98	364.35	385.65
	U	779.878	2584.5	2874.7	2915.7	2956.8	3039.6	3123.5	3208.9	3295.8	3384.3
	H	781.124	2779.7	3156.6	3209.7	3262.9	3369.6	3477.2	3586.2	3696.6	3808.5
	S	2.1786	6.5497	7.2569	7.3405	7.4209	7.5737	7.7177	7.8543	7.9845	8.1092
1150 (186.05)	V	1.136	169.99	244.91	255.47	265.96	286.77	307.42	327.97	348.42	368.81
	U	788.611	2585.8	2874.0	2915.1	2956.2	3039.1	3123.1	3208.5	3295.5	3384.1
	H	789.917	2781.3	3155.6	3208.9	3262.1	3368.9	3476.6	3585.7	3696.2	3808.2
	S	2.1977	6.5342	7.2352	7.3190	7.3995	7.5525	7.6966	7.8333	7.9636	8.0883
1200 (187.96)	V	1.139	163.20	234.49	244.63	254.70	274.68	294.50	314.20	333.82	353.38
	U	797.064	2586.9	2873.3	2914.4	2955.7	3038.6	3122.7	3208.2	3295.2	3383.8
	H	798.430	2782.7	3154.6	3208.0	3261.3	3368.2	3476.1	3585.2	3695.8	3807.8
	S	2.2161	6.5194	7.2144	7.2983	7.3790	7.5323	7.6765	7.8132	7.9436	8.0684
1250 (189.81)	V	1.141	156.93	224.90	234.66	244.35	263.55	282.60	301.54	320.39	339.18
	U	805.259	2588.0	2872.5	2913.8	2955.1	3038.1	3122.3	3207.8	3294.9	3383.5
	H	806.685	2784.1	3153.7	3207.1	3260.5	3367.6	3475.5	3584.7	3695.4	3807.5
	S	2.2338	6.5050	7.1944	7.2785	7.3593	7.5128	7.6571	7.7940	7.9244	8.0493
1300 (191.61)	V	1.144	151.13	216.05	225.46	234.79	253.28	271.62	289.85	307.99	326.07
	U	813.213	2589.0	2871.8	2913.2	2954.5	3037.7	3121.9	3207.5	3294.6	3383.2
	H	814.700	2785.4	3152.7	3206.3	3259.7	3366.9	3475.0	3584.3	3695.0	3807.1
	S	2.2510	6.4913	7.1751	7.2594	7.3404	7.4940	7.6385	7.7754	7.9060	8.0309

TEMPERATURE, °C
(TEMPERATURE, K)

ABS. PRESS., kPa (SAT. TEMP., °C)		SAT. WATER	SAT. STEAM	200 (473.15)	225 (498.15)	250 (523.15)	275 (548.15)	300 (573.15)	325 (598.15)	350 (623.15)	375 (648.15)
1350 (193.35)	V	1.146	145.74	148.79	159.70	169.96	179.79	189.33	198.66	207.85	216.93
	U	820.944	2589.9	2603.9	2653.6	2700.1	2744.4	2787.4	2829.5	2871.1	2912.5
	H	822.491	2786.6	2804.7	2869.2	2929.5	2987.1	3043.0	3097.7	3151.7	3205.4
	S	2.2676	6.4780	6.5165	6.6493	6.7675	6.8750	6.9746	7.0681	7.1566	7.2410
1400 (195.04)	V	1.149	140.72	142.94	153.57	163.55	173.08	182.32	191.35	200.24	209.02
	U	828.465	2590.8	2601.3	2651.7	2698.6	2743.2	2786.4	2828.6	2870.4	2911.9
	H	830.074	2787.8	2801.4	2866.7	2927.6	2985.5	3041.6	3096.5	3150.7	3204.5
	S	2.2837	6.4651	6.4941	6.6285	6.7477	6.8560	6.9561	7.0499	7.1386	7.2233
1450 (196.69)	V	1.151	136.04	137.48	147.86	157.57	166.83	175.79	184.54	193.15	201.65
	U	835.791	2591.6	2598.7	2649.7	2697.1	2742.0	2785.4	2827.8	2869.7	2911.3
	H	837.460	2788.9	2798.1	2864.1	2925.5	2983.9	3040.3	3095.4	3149.7	3203.6
	S	2.2993	6.4526	6.4722	6.6082	6.7286	6.8376	6.9381	7.0322	7.1212	7.2061
1500 (198.29)	V	1.154	131.66	132.38	142.53	151.99	161.00	169.70	178.19	186.53	194.77
	U	842.933	2592.4	2596.1	2647.7	2695.5	2740.8	2784.4	2826.9	2868.9	2910.6
	H	844.663	2789.9	2794.7	2861.5	2923.5	2982.3	3038.9	3094.2	3148.7	3202.8
	S	2.3145	6.4406	6.4508	6.5885	6.7099	6.8196	6.9207	7.0152	7.1044	7.1894
1550 (199.85)	V	1.156	127.55	127.61	137.54	146.77	155.54	164.00	172.25	180.34	188.33
	U	849.901	2593.2	2593.5	2645.8	2694.0	2739.5	2783.4	2826.1	2868.2	2910.0
	H	851.694	2790.8	2791.3	2858.9	2921.5	2980.6	3037.6	3093.1	3147.7	3201.9
	S	2.3292	6.4289	6.4298	6.5692	6.6917	6.8022	6.9038	6.9986	7.0881	7.1733
1600 (201.37)	V	1.159	123.69	....	132.85	141.87	150.42	158.66	166.68	174.54	182.30
	U	856.707	2593.8		2643.7	2692.4	2738.3	2782.4	2825.2	2867.5	2909.3
	H	858.561	2791.7		2856.3	2919.4	2979.0	3036.2	3091.9	3146.7	3201.0
	S	2.3436	6.4175		6.5503	6.6740	6.7852	6.8873	6.9825	7.0723	7.1577
1650 (202.86)	V	1.161	120.05	....	128.45	137.27	145.61	153.64	161.44	169.09	176.63
	U	863.359	2594.5		2641.7	2690.9	2737.1	2781.3	2824.4	2866.7	2908.7
	H	865.275	2792.6		2853.6	2917.4	2977.3	3034.8	3090.8	3145.7	3200.1
	S	2.3576	6.4065		6.5319	6.6567	6.7687	6.8713	6.9669	7.0569	7.1425
1700 (204.31)	V	1.163	116.62	....	124.31	132.94	141.09	148.91	156.51	163.96	171.30
	U	869.866	2595.1		2639.6	2689.3	2735.8	2780.3	2823.5	2866.0	2908.0
	H	871.843	2793.4		2851.0	2915.3	2975.6	3033.5	3089.6	3144.7	3199.2
	S	2.3713	6.3957		6.5138	6.6398	6.7526	6.8557	6.9516	7.0419	7.1277
1750 (205.72)	V	1.166	113.38	....	120.39	128.85	136.82	144.45	151.87	159.12	166.27
	U	876.234	2595.7		2637.6	2687.7	2734.5	2779.3	2822.7	2865.3	2907.4
	H	878.274	2794.1		2848.2	2913.2	2974.0	3032.1	3088.4	3143.7	3198.4
	S	2.3846	6.3853		6.4961	6.6233	6.7368	6.8405	6.9368	7.0273	7.1133
1800 (207.11)	V	1.168	110.32	....	116.69	124.99	132.78	140.24	147.48	154.55	161.51
	U	882.472	2596.3		2635.5	2686.1	2733.3	2778.2	2821.8	2864.5	2906.7
	H	884.574	2794.8		2845.5	2911.0	2972.3	3030.7	3087.3	3142.7	3197.5
	S	2.3976	6.3751		6.4787	6.6071	6.7214	6.8257	6.9223	7.0131	7.0993
1850 (208.47)	V	1.170	107.41	....	113.19	121.33	128.96	136.26	143.33	150.23	157.02
	U	888.585	2596.8		2633.3	2684.4	2732.0	2777.2	2820.9	2863.8	2906.1
	H	890.750	2795.5		2842.8	2908.9	2970.6	3029.3	3086.1	3141.7	3196.6
	S	2.4103	6.3651		6.4616	6.5912	6.7064	6.8112	6.9082	6.9993	7.0856
1900 (209.80)	V	1.172	104.65	....	109.87	117.87	125.35	132.49	139.39	146.14	152.76
	U	894.580	2597.3		2631.2	2682.8	2730.7	2776.2	2820.1	2863.0	2905.4
	H	896.807	2796.1		2840.0	2906.7	2968.8	3027.9	3084.9	3140.7	3195.7
	S	2.4228	6.3554		6.4448	6.5757	6.6917	6.7970	6.8944	6.9857	7.0723
1950 (211.10)	V	1.174	102.031	....	106.72	114.58	121.91	128.90	135.66	142.25	148.72
	U	900.461	2597.7		2629.0	2681.1	2729.4	2775.1	2819.2	2862.3	2904.8
	H	902.752	2796.7		2837.1	2904.6	2967.1	3026.5	3083.7	3139.7	3194.8
	S	2.4349	6.3459		6.4283	6.5604	6.6772	6.7831	6.8809	6.9725	7.0593
2000 (212.37)	V	1.177	99.536	....	103.72	111.45	118.65	125.50	132.11	138.56	144.89
	U	906.236	2598.2		2626.9	2679.5	2728.1	2774.0	2818.3	2861.5	2904.1
	H	908.589	2797.2		2834.3	2902.4	2965.4	3025.0	3082.5	3138.6	3193.9
	S	2.4469	6.3366		6.4120	6.5454	6.6631	6.7696	6.8677	6.9596	7.0466
2100 (214.85)	V	1.181	94.890	....	98.147	105.64	112.59	119.18	125.53	131.70	137.76
	U	917.479	2598.9		2622.4	2676.1	2725.4	2771.9	2816.5	2860.0	2902.8
	H	919.959	2798.2		2828.5	2897.9	2961.9	3022.2	3080.1	3136.6	3192.1
	S	2.4700	6.3187		6.3802	6.5162	6.6356	6.7432	6.8422	6.9347	7.0220
2200 (217.24)	V	1.185	90.652	....	93.067	100.35	107.07	113.43	119.53	125.47	131.28
	U	928.346	2599.6		2617.9	2672.7	2722.7	2769.7	2814.7	2858.5	2901.5
	H	930.953	2799.1		2822.7	2893.4	2958.3	3019.3	3077.7	3134.5	3190.3
	S	2.4922	6.3015		6.3492	6.4879	6.6091	6.7179	6.8177	6.9107	6.9985
2300 (219.55)	V	1.189	86.769	....	88.420	95.513	102.03	108.18	114.06	119.77	125.36
	U	938.866	2600.2		2613.3	2669.2	2720.0	2767.6	2812.9	2857.0	2900.2
	H	941.601	2799.8		2816.7	2888.9	2954.7	3016.4	3075.3	3132.4	3188.5
	S	2.5136	6.2849		6.3190	6.4605	6.5835	6.6935	6.7941	6.8877	6.9759

		ABS. PRESS., kPa (SAT. TEMP., °C)	SAT. WATER	SAT. STEAM	400 (673.15)	425 (698.15)	450 (723.15)	475 (748.15)	500 (773.15)	550 (823.15)	600 (873.15)	650 (923.15)
1350 (193.35)	V		1.146	145.74	225.94	234.88	243.78	252.63	261.46	279.03	296.51	313.93
	U		820.944	2589.9	2953.9	2995.5	3037.2	3079.2	3121.5	3207.1	3294.3	3383.0
	H		822.491	2786.6	3259.0	3312.6	3366.3	3420.2	3474.4	3583.8	3694.5	3806.8
	S		2.2676	6.4780	7.3221	7.4003	7.4759	7.5493	7.6205	7.7576	7.8882	8.0132
1400 (195.04)	V		1.149	140.72	217.72	226.35	234.95	243.50	252.02	268.98	285.85	302.66
	U		828.465	2590.8	2953.4	2994.9	3036.7	3078.7	3121.1	3206.8	3293.9	3382.7
	H		830.074	2787.8	3258.2	3311.8	3365.6	3419.6	3473.9	3583.3	3694.1	3806.4
	S		2.2837	6.4651	7.3045	7.3828	7.4585	7.5319	7.6032	7.7404	7.8710	7.9961
1450 (196.69)	V		1.151	136.04	210.06	218.42	226.72	234.99	243.23	259.62	275.93	292.16
	U		835.791	2591.6	2952.8	2994.4	3036.2	3078.3	3120.7	3206.4	3293.6	3382.4
	H		837.460	2788.9	3257.4	3311.1	3365.0	3419.0	3473.3	3582.9	3693.7	3806.1
	S		2.2993	6.4526	7.2874	7.3658	7.4416	7.5151	7.5865	7.7237	7.8545	7.9796
1500 (198.29)	V		1.154	131.66	202.92	211.01	219.05	227.06	235.03	250.89	266.66	282.37
	U		842.933	2592.4	2952.2	2993.9	3035.8	3077.9	3120.3	3206.0	3293.3	3382.1
	H		844.663	2789.9	3256.6	3310.4	3364.3	3418.4	3472.8	3582.4	3693.3	3805.7
	S		2.3145	6.4406	7.2709	7.3494	7.4253	7.4989	7.5703	7.7077	7.8385	7.9636
1550 (199.85)	V		1.156	127.55	196.24	204.08	211.87	219.63	227.35	242.72	258.00	273.21
	U		849.901	2593.2	2951.7	2993.4	3035.3	3077.4	3119.8	3205.7	3293.0	3381.9
	H		851.694	2790.8	3255.8	3309.7	3363.7	3417.8	3472.2	3581.9	3692.9	3805.3
	S		2.3292	6.4289	7.2550	7.3336	7.4095	7.4832	7.5547	7.6921	7.8230	7.9482
1600 (201.37)	V		1.159	123.69	189.97	197.58	205.15	212.67	220.16	235.06	249.87	264.62
	U		856.707	2593.8	2951.1	2992.9	3034.8	3077.0	3119.4	3205.3	3292.7	3381.6
	H		858.561	2791.7	3255.0	3309.0	3363.0	3417.2	3471.7	3581.4	3692.5	3805.0
	S		2.3436	6.4175	7.2394	7.3182	7.3942	7.4679	7.5395	7.6770	7.8080	7.9333
1650 (202.86)	V		1.161	120.05	184.09	191.48	198.82	206.13	213.40	227.86	242.24	256.55
	U		863.359	2594.5	2950.6	2992.3	3034.3	3076.5	3119.0	3205.0	3292.4	3381.3
	H		865.275	2792.6	3254.2	3308.3	3362.4	3416.7	3471.1	3581.0	3692.1	3804.6
	S		2.3576	6.4065	7.2244	7.3032	7.3794	7.4531	7.5248	7.6624	7.7934	7.9188
1700 (204.31)	V		1.163	116.62	178.55	185.74	192.87	199.97	207.04	221.09	235.06	248.96
	U		869.866	2595.1	2949.9	2991.8	3033.9	3076.1	3118.6	3204.6	3292.1	3381.0
	H		871.843	2793.4	3253.5	3307.6	3361.7	3416.1	3470.6	3580.5	3691.7	3804.3
	S		2.3713	6.3957	7.2098	7.2887	7.3649	7.4388	7.5105	7.6482	7.7793	7.9047
1750 (205.72)	V		1.166	113.38	173.32	180.32	187.26	194.17	201.04	214.71	228.28	241.80
	U		876.234	2595.7	2949.3	2991.3	3033.4	3075.7	3118.2	3204.3	3291.8	3380.8
	H		878.274	2794.1	3252.7	3306.9	3361.1	3415.5	3470.0	3580.0	3691.3	3803.9
	S		2.3846	6.3853	7.1955	7.2746	7.3509	7.4248	7.4965	7.6344	7.7656	7.8910
1800 (207.11)	V		1.168	110.32	168.39	175.20	181.97	188.69	195.38	208.68	221.89	235.03
	U		882.472	2596.3	2948.8	2990.8	3032.9	3075.2	3117.8	3203.9	3291.5	3380.5
	H		884.574	2794.8	3251.9	3306.1	3360.4	3414.9	3469.5	3579.5	3690.9	3803.6
	S		2.3976	6.3751	7.1816	7.2608	7.3372	7.4112	7.4830	7.6209	7.7522	7.8777
1850 (208.47)	V		1.170	107.41	163.73	170.37	176.96	183.50	190.02	202.97	215.84	228.64
	U		888.585	2596.8	2948.2	2990.3	3032.4	3074.8	3117.4	3203.6	3291.1	3380.2
	H		890.750	2795.5	3251.1	3305.4	3359.8	3414.3	3468.9	3579.1	3690.4	3803.2
	S		2.4103	6.3651	7.1681	7.2474	7.3239	7.3980	7.4698	7.6079	7.7392	7.8648
1900 (209.80)	V		1.172	104.65	159.30	165.78	172.21	178.59	184.94	197.57	210.11	222.58
	U		894.580	2597.3	2947.6	2989.7	3031.9	3074.3	3117.0	3203.2	3290.8	3380.0
	H		896.807	2796.1	3250.3	3304.7	3359.1	3413.7	3468.4	3578.6	3690.0	3802.8
	S		2.4228	6.3554	7.1550	7.2344	7.3109	7.3851	7.4570	7.5951	7.7265	7.8522
1950 (211.10)	V		1.174	102.031	155.11	161.43	167.70	173.93	180.13	192.44	204.67	216.83
	U		900.461	2597.7	2947.0	2989.2	3031.5	3073.9	3116.6	3202.9	3290.5	3379.7
	H		902.752	2796.7	3249.5	3304.0	3358.5	3413.1	3467.8	3578.1	3689.6	3802.5
	S		2.4349	6.3459	7.1421	7.2216	7.2983	7.3725	7.4445	7.5827	7.7142	7.8399
2000 (212.37)	V		1.177	99.536	151.13	157.30	163.42	169.51	175.55	187.57	199.50	211.36
	U		906.236	2598.2	2946.4	2988.7	3031.0	3073.5	3116.2	3202.5	3290.2	3379.4
	H		908.589	2797.2	3248.7	3303.3	3357.8	3412.5	3467.3	3577.6	3689.2	3802.1
	S		2.4469	6.3366	7.1296	7.2092	7.2859	7.3602	7.4323	7.5706	7.7022	7.8279
2100 (214.85)	V		1.181	94.890	143.73	149.63	155.48	161.28	167.06	178.53	189.91	201.22
	U		917.479	2598.9	2945.3	2987.6	3030.0	3072.6	3115.3	3201.8	3289.6	3378.9
	H		919.959	2798.2	3247.1	3301.8	3356.5	3411.3	3466.2	3576.7	3688.4	3801.4
	S		2.4700	6.3187	7.1053	7.1851	7.2621	7.3365	7.4087	7.5472	7.6789	7.8048
2200 (217.24)	V		1.185	90.652	137.00	142.65	148.25	153.81	159.34	170.30	181.19	192.00
	U		928.346	2599.6	2944.1	2986.6	3029.1	3071.7	3114.5	3201.1	3289.0	3378.3
	H		930.953	2799.1	3245.5	3300.4	3355.2	3410.1	3465.1	3575.7	3687.6	3800.7
	S		2.4922	6.3015	7.0821	7.1621	7.2393	7.3139	7.3862	7.5249	7.6568	7.7827
2300 (219.55)	V		1.189	86.769	130.85	136.28	141.65	146.99	152.28	162.80	173.22	183.58
	U		938.866	2600.2	2942.9	2985.5	3028.1	3070.8	3113.7	3200.4	3288.5	3377.7
	H		941.601	2799.8	3243.9	3299.0	3353.9	3408.9	3464.0	3574.8	3686.7	3800.0
	S		2.5136	6.2849	7.0598	7.1401	7.2174	7.2922	7.3646	7.5035	7.6355	7.7616

ABS. PRESS., kPa (SAT. TEMP., °C)		SAT. WATER	SAT. STEAM	225 (498.15)	250 (523.15)	275 (548.15)	300 (573.15)	325 (598.15)	350 (623.15)	375 (648.15)	400 (673.15)
2400 (221.78)	V	1.193	83.199	84.149	91.075	97.411	103.36	109.05	114.55	119.93	125.22
	U	949.066	2600.7	2608.6	2665.6	2717.3	2765.4	2811.1	2855.4	2898.8	2941.7
	H	951.929	2800.4	2810.6	2884.2	2951.1	3013.4	3072.8	3130.4	3186.7	3242.3
	S	2.5343	6.2690	6.2894	6.4338	6.5586	6.6699	6.7714	6.8656	6.9542	7.0384
2500 (223.94)	V	1.197	79.905	80.210	86.985	93.154	98.925	104.43	109.75	114.94	120.04
	U	958.969	2601.2	2603.8	2662.0	2714.5	2763.1	2809.3	2853.9	2897.5	2940.6
	H	961.962	2800.9	2804.3	2879.5	2947.4	3010.4	3070.4	3128.2	3184.8	3240.7
	S	2.5543	6.2536	6.2604	6.4077	6.5345	6.6470	6.7494	6.8442	6.9333	7.0178
2600 (226.04)	V	1.201	76.856	....	83.205	89.220	94.830	100.17	105.32	110.33	115.26
	U	968.597	2601.5	....	2658.4	2711.7	2760.9	2807.4	2852.3	2896.1	2939.4
	H	971.720	2801.4	....	2874.7	2943.6	3007.4	3067.9	3126.1	3183.0	3239.0
	S	2.5736	6.2387	....	6.3823	6.5110	6.6249	6.7281	6.8236	6.9131	6.9979
2700 (228.07)	V	1.205	74.025	....	79.698	85.575	91.036	96.218	101.21	106.07	110.83
	U	977.968	2601.8	....	2654.7	2708.8	2758.6	2805.6	2850.7	2894.8	2938.2
	H	981.222	2801.7	....	2869.9	2939.8	3004.4	3065.4	3124.0	3181.2	3237.4
	S	2.5924	6.2244	....	6.3575	6.4882	6.6034	6.7075	6.8036	6.8935	6.9787
2800 (230.05)	V	1.209	71.389	....	76.437	82.187	87.510	92.550	97.395	102.10	106.71
	U	987.100	2602.1	....	2650.9	2705.9	2756.3	2803.7	2849.2	2893.4	2937.0
	H	990.485	2802.0	....	2864.9	2936.0	3001.3	3062.8	3121.9	3179.3	3235.8
	S	2.6106	6.2104	....	6.3331	6.4659	6.5824	6.6875	6.7842	6.8746	6.9601
2900 (231.97)	V	1.213	68.928	....	73.395	79.029	84.226	89.133	93.843	98.414	102.88
	U	996.008	2602.3	....	2647.1	2702.9	2754.0	2801.8	2847.6	2892.0	2935.8
	H	999.524	2802.2	....	2859.9	2932.1	2998.2	3060.3	3119.7	3177.4	3234.1
	S	2.6283	6.1969	....	6.3092	6.4441	6.5621	6.6681	6.7654	6.8563	6.9421
3000 (233.84)	V	1.216	66.626	....	70.551	76.078	81.159	85.943	90.526	94.969	99.310
	U	1004.7	2602.4	....	2643.2	2700.0	2751.6	2799.9	2846.0	2890.7	2934.6
	H	1008.4	2802.3	....	2854.8	2928.2	2995.1	3057.7	3117.5	3175.6	3232.5
	S	2.6455	6.1837	....	6.2857	6.4228	6.5422	6.6491	6.7471	6.8385	6.9246
3100 (235.67)	V	1.220	64.467	....	67.885	73.315	78.287	82.958	87.423	91.745	95.965
	U	1013.2	2602.5	....	2639.2	2697.0	2749.2	2797.9	2844.3	2889.3	2933.4
	H	1017.0	2802.3	....	2849.6	2924.2	2991.9	3055.1	3115.4	3173.7	3230.8
	S	2.6623	6.1709	....	6.2626	6.4019	6.5227	6.6307	6.7294	6.8212	6.9077
3200 (237.45)	V	1.224	62.439	....	65.380	70.721	75.593	80.158	84.513	88.723	92.829
	U	1021.5	2602.5	....	2635.2	2693.9	2746.8	2796.0	2842.7	2887.9	2932.1
	H	1025.4	2802.3	....	2844.4	2920.2	2988.7	3052.5	3113.2	3171.8	3229.2
	S	2.6786	6.1585	....	6.2398	6.3815	6.5037	6.6127	6.7120	6.8043	6.8912
3300 (239.18)	V	1.227	60.529	....	63.021	68.282	73.061	77.526	81.778	85.883	89.883
	U	1029.7	2602.5	....	2631.1	2690.8	2744.4	2794.0	2841.1	2886.5	2930.9
	H	1033.7	2802.3	....	2839.0	2916.1	2985.5	3049.9	3110.9	3169.9	3227.5
	S	2.6945	6.1463	....	6.2173	6.3614	6.4851	6.5951	6.6952	6.7879	6.8752
3400 (240.88)	V	1.231	58.728	....	60.796	65.982	70.675	75.048	79.204	83.210	87.110
	U	1037.6	2602.5	....	2626.9	2687.7	2741.9	2792.0	2839.4	2885.1	2929.7
	H	1041.8	2802.1	....	2833.6	2912.0	2982.2	3047.2	3108.7	3168.0	3225.9
	S	2.7101	6.1344	....	6.1951	6.3416	6.4669	6.5779	6.6787	6.7719	6.8595
3500 (242.54)	V	1.235	57.025	....	58.693	63.812	68.424	72.710	76.776	80.689	84.494
	U	1045.4	2602.4	....	2622.7	2684.5	2739.5	2790.0	2837.8	2883.7	2928.4
	H	1049.8	2802.0	....	2828.1	2907.8	2979.0	3044.5	3106.5	3166.1	3224.2
	S	2.7253	6.1228	....	6.1732	6.3221	6.4491	6.5611	6.6626	6.7563	6.8443
3600 (244.16)	V	1.238	55.415	....	56.702	61.759	66.297	70.501	74.482	78.308	82.024
	U	1053.1	2602.2	....	2618.4	2681.3	2737.0	2788.0	2836.1	2882.3	2927.2
	H	1057.6	2801.7	....	2822.5	2903.6	2975.6	3041.8	3104.2	3164.2	3222.5
	S	2.7401	6.1115	....	6.1514	6.3030	6.4315	6.5446	6.6468	6.7411	6.8294
3700 (245.75)	V	1.242	53.888	....	54.812	59.814	64.282	68.410	72.311	76.055	79.687
	U	1060.6	2602.1	....	2614.0	2678.0	2734.4	2786.0	2834.4	2880.8	2926.0
	H	1065.2	2801.4	....	2816.8	2899.3	2972.3	3039.1	3102.0	3162.2	3220.8
	S	2.7547	6.1004	....	6.1299	6.2841	6.4143	6.5284	6.6314	6.7262	6.8149
3800 (247.31)	V	1.245	52.438	....	53.017	57.968	62.372	66.429	70.254	73.920	77.473
	U	1068.0	2601.9	....	2609.5	2674.7	2731.9	2783.9	2832.7	2879.4	2924.7
	H	1072.7	2801.1	....	2811.0	2895.0	2968.9	3036.4	3099.7	3160.3	3219.1
	S	2.7689	6.0896	....	6.1085	6.2654	6.3973	6.5126	6.6163	6.7117	6.8007
3900 (248.84)	V	1.249	51.061	....	51.308	56.215	60.558	64.547	68.302	71.894	75.372
	U	1075.3	2601.6	....	2605.0	2671.4	2729.3	2781.9	2831.0	2877.9	2923.5
	H	1080.1	2800.8	....	2805.1	2890.6	2965.5	3033.6	3097.4	3158.3	3217.4
	S	2.7828	6.0789	....	6.0872	6.2470	6.3806	6.4970	6.6015	6.6974	6.7868
4000 (250.33)	V	1.252	49.749	....	....	54.546	58.833	62.759	66.446	69.969	73.376
	U	1082.4	2601.3	....	....	2668.0	2726.7	2779.8	2829.3	2876.5	2922.2
	H	1087.4	2800.3	....	....	2886.1	2962.0	3030.8	3095.1	3156.4	3215.7
	S	2.7965	6.0685	....	....	6.2288	6.3642	6.4817	6.5870	6.6834	6.7733

TEMPERATURE, °C
(TEMPERATURE, K)

ABS. PRESS., kPa (SAT. TEMP., °C)		SAT. WATER	SAT. STEAM	425 (698.15)	450 (723.15)	475 (748.15)	500 (773.15)	525 (798.15)	550 (823.15)	600 (873.15)	650 (923.15)
2400 (221.78)	V	1.193	83.199	130.44	135.61	140.73	145.82	150.88	155.91	165.92	175.86
	U	949.066	2600.7	2984.5	3027.1	3069.9	3112.9	3156.1	3199.6	3287.7	3377.2
	H	951.929	2800.4	3297.5	3352.6	3407.7	3462.9	3518.2	3573.8	3685.9	3799.3
	S	2.5343	6.2690	7.1189	7.1964	7.2713	7.3439	7.4144	7.4830	7.6152	7.7414
2500 (223.94)	V	1.197	79.905	125.07	130.04	134.97	139.87	144.74	149.58	159.21	168.76
	U	958.969	2601.2	2983.4	3026.2	3069.0	3112.1	3155.4	3198.9	3287.1	3376.7
	H	961.962	2800.9	3296.1	3351.3	3406.5	3461.7	3517.2	3572.9	3685.1	3798.6
	S	2.5543	6.2536	7.0986	7.1763	7.2513	7.3240	7.3946	7.4633	7.5956	7.7220
2600 (226.04)	V	1.201	76.856	120.11	124.91	129.66	134.38	139.07	143.74	153.01	162.21
	U	968.597	2601.5	2982.3	3025.2	3068.1	3111.2	3154.6	3198.2	3286.5	3376.1
	H	971.720	2801.4	3294.6	3349.9	3405.3	3460.6	3516.2	3571.9	3684.3	3797.9
	S	2.5736	6.2387	7.0789	7.1568	7.2320	7.3048	7.3755	7.4443	7.5768	7.7033
2700 (228.07)	V	1.205	74.025	115.52	120.15	124.74	129.30	133.82	138.33	147.27	156.14
	U	977.968	2601.8	2981.2	3024.2	3067.2	3110.4	3153.8	3197.5	3285.8	3375.6
	H	981.222	2801.7	3293.1	3348.6	3404.0	3459.5	3515.2	3571.0	3683.5	3797.1
	S	2.5924	6.2244	7.0600	7.1381	7.2134	7.2863	7.3571	7.4260	7.5587	7.6853
2800 (230.05)	V	1.209	71.389	111.25	115.74	120.17	124.58	128.95	133.30	141.94	150.50
	U	987.100	2602.1	2980.2	3023.2	3066.3	3109.6	3153.1	3196.8	3285.2	3375.0
	H	990.485	2802.0	3291.7	3347.3	3402.8	3458.4	3514.1	3570.0	3682.6	3796.4
	S	2.6106	6.2104	7.0416	7.1199	7.1954	7.2685	7.3394	7.4084	7.5412	7.6679
2900 (231.97)	V	1.213	68.928	107.28	111.62	115.92	120.18	124.42	128.62	136.97	145.26
	U	996.008	2602.3	2979.1	3022.3	3065.5	3108.8	3152.3	3196.1	3284.6	3374.5
	H	999.524	2802.2	3290.2	3346.0	3401.6	3457.3	3513.1	3569.1	3681.8	3795.7
	S	2.6283	6.1969	7.0239	7.1024	7.1780	7.2512	7.3222	7.3913	7.5243	7.6511
3000 (233.84)	V	1.216	66.626	103.58	107.79	111.95	116.08	120.18	124.26	132.34	140.36
	U	1004.7	2602.4	2978.0	3021.3	3064.6	3107.9	3151.5	3195.4	3284.0	3373.9
	H	1008.4	2803.2	3288.7	3344.6	3400.4	3456.2	3512.1	3568.1	3681.0	3795.0
	S	2.6455	6.1837	7.0067	7.0854	7.1612	7.2345	7.3056	7.3748	7.5079	7.6349
3100 (235.67)	V	1.220	64.467	100.11	104.20	108.24	112.24	116.22	120.17	128.01	135.78
	U	1013.2	2602.5	2976.9	3020.3	3063.7	3107.1	3150.8	3194.7	3283.3	3373.4
	H	1017.0	2802.3	3287.3	3343.3	3399.2	3455.1	3511.0	3567.2	3680.2	3794.3
	S	2.6623	6.1709	6.9900	7.0689	7.1448	7.2183	7.2895	7.3588	7.4920	7.6191
3200 (237.45)	V	1.224	62.439	96.859	100.83	104.76	108.65	112.51	116.34	123.95	131.48
	U	1021.5	2602.5	2975.9	3019.3	3062.8	3106.3	3150.0	3193.9	3282.7	3372.8
	H	1025.4	2802.3	3285.8	3342.0	3398.0	3454.0	3510.0	3566.2	3679.3	3793.6
	S	2.6786	6.1585	6.9738	7.0528	7.1290	7.2026	7.2739	7.3433	7.4767	7.6039
3300 (239.18)	V	1.227	60.529	93.805	97.668	101.49	105.27	109.02	112.74	120.13	127.45
	U	1029.7	2602.5	2974.8	3018.3	3061.9	3105.5	3149.2	3193.2	3282.1	3372.3
	H	1033.7	2802.3	3284.3	3340.6	3396.8	3452.8	3509.0	3565.3	3678.5	3792.9
	S	2.6945	6.1463	6.9580	7.0373	7.1136	7.1873	7.2588	7.3282	7.4618	7.5891
3400 (240.88)	V	1.231	58.728	90.930	94.692	98.408	102.09	105.74	109.36	116.54	123.65
	U	1037.6	2602.5	2973.7	3017.4	3061.0	3104.6	3148.4	3192.5	3281.5	3371.7
	H	1041.8	2802.1	3282.8	3339.3	3395.5	3451.7	3507.9	3564.3	3677.7	3792.1
	S	2.7101	6.1344	6.9426	7.0221	7.0986	7.1724	7.2440	7.3136	7.4473	7.5747
3500 (242.54)	V	1.235	57.025	88.220	91.886	95.505	99.088	102.64	106.17	113.15	120.07
	U	1045.4	2602.4	2972.6	3016.4	3060.1	3103.8	3147.7	3191.8	3280.8	3371.2
	H	1049.8	2802.0	3281.3	3338.0	3394.3	3450.6	3506.9	3563.4	3676.9	3791.4
	S	2.7253	6.1228	6.9277	7.0074	7.0840	7.1580	7.2297	7.2993	7.4332	7.5607
3600 (244.16)	V	1.238	55.415	85.660	89.236	92.764	96.255	99.716	103.15	109.96	116.69
	U	1053.1	2602.2	2971.5	3015.4	3059.2	3103.0	3146.9	3191.1	3280.2	3370.6
	H	1057.6	2801.7	3279.8	3336.6	3393.1	3449.5	3505.9	3562.4	3676.1	3790.7
	S	2.7401	6.1115	6.9131	6.9930	7.0698	7.1439	7.2157	7.2854	7.4195	7.5471
3700 (245.75)	V	1.242	53.888	83.238	86.728	90.171	93.576	96.950	100.30	106.93	113.49
	U	1060.6	2602.1	2970.4	3014.4	3058.2	3102.1	3146.1	3190.4	3279.6	3370.1
	H	1065.2	2801.4	3278.4	3335.3	3391.9	3448.4	3504.9	3561.5	3675.2	3790.0
	S	2.7547	6.1004	6.8989	6.9790	7.0559	7.1302	7.2021	7.2719	7.4061	7.5339
3800 (247.31)	V	1.245	52.438	80.944	84.353	87.714	91.038	94.330	97.596	104.06	110.46
	U	1068.0	2601.9	2969.3	3013.4	3057.3	3101.3	3145.4	3189.6	3279.0	3369.5
	H	1072.7	2801.1	3276.8	3333.9	3390.7	3447.2	3503.8	3560.5	3674.4	3789.3
	S	2.7689	6.0896	6.8849	6.9653	7.0424	7.1168	7.1888	7.2587	7.3931	7.5210
3900 (248.84)	V	1.249	51.061	78.767	82.099	85.383	88.629	91.844	95.033	101.35	107.59
	U	1075.3	2601.6	2968.2	3012.4	3056.4	3100.5	3144.6	3188.9	3278.3	3369.0
	H	1080.1	2800.8	3275.3	3332.6	3389.4	3446.1	3502.8	3559.5	3673.6	3788.6
	S	2.7828	6.0789	6.8713	6.9519	7.0292	7.1037	7.1759	7.2459	7.3804	7.5084
4000 (250.33)	V	1.252	49.749	76.698	79.958	83.169	86.341	89.483	92.598	98.763	104.86
	U	1082.4	2601.3	2967.0	3011.4	3055.5	3099.6	3143.8	3188.2	3277.7	3368.4
	H	1087.4	2800.3	3273.8	3331.2	3388.2	3445.0	3501.7	3558.6	3672.8	3787.9
	S	2.7965	6.0685	6.8581	6.9388	7.0163	7.0909	7.1632	7.2333	7.3680	7.4961

ABS. PRESS., kPa (SAT. TEMP., °C)		SAT. WATER	SAT. STEAM	TEMPERATURE, °C (TEMPERATURE, K)							
				260 (533.15)	275 (548.15)	300 (573.15)	325 (598.15)	350 (623.15)	375 (648.15)	400 (673.15)	425 (698.15)
4100 (251.80)	V	1.256	48.500	50.150	52.955	57.191	61.057	64.680	68.137	71.476	74.730
	U	1089.4	2601.0	2624.6	2664.5	2724.0	2777.7	2827.6	2875.0	2920.9	2965.9
	H	1094.6	2799.9	2830.3	2881.6	2958.5	3028.0	3092.8	3154.4	3214.0	3272.3
	S	2.8099	6.0583	6.1157	6.2107	6.3480	6.4667	6.5727	6.6697	6.7600	6.8450
4200 (253.24)	V	1.259	47.307	48.654	51.438	55.625	59.435	62.998	66.392	69.667	72.856
	U	1096.3	2600.7	2620.4	2661.0	2721.4	2775.6	2825.8	2873.6	2919.7	2964.8
	H	1101.6	2799.4	2824.8	2877.1	2955.0	3025.2	3090.4	3152.4	3212.3	3270.8
	S	2.8231	6.0482	6.0962	6.1929	6.3320	6.4519	6.5587	6.6563	6.7469	6.8323
4300 (254.66)	V	1.262	46.168	47.223	49.988	54.130	57.887	61.393	64.728	67.942	71.069
	U	1103.1	2600.3	2616.2	2657.5	2718.7	2773.4	2824.1	2872.1	2918.4	2963.7
	H	1108.5	2798.9	2819.2	2872.4	2951.4	3022.1	3088.1	3150.4	3210.5	3269.3
	S	2.8360	6.0383	6.0768	6.1752	6.3162	6.4373	6.5450	6.6431	6.7341	6.8198
4400 (256.05)	V	1.266	45.079	45.853	48.601	52.702	56.409	59.861	63.139	66.295	69.363
	U	1109.8	2599.9	2611.8	2653.9	2716.0	2771.3	2822.3	2870.6	2917.1	2962.5
	H	1115.4	2798.3	2813.6	2867.8	2947.8	3019.5	3085.7	3148.4	3208.8	3267.7
	S	2.8487	6.0286	6.0575	6.1577	6.3006	6.4230	6.5315	6.6301	6.7216	6.8076
4500 (257.41)	V	1.269	44.037	44.540	47.273	51.336	54.996	58.396	61.620	64.721	67.732
	U	1116.4	2599.5	2607.4	2650.3	2713.2	2769.1	2820.5	2869.1	2915.8	2961.4
	H	1122.1	2797.9	2807.9	2863.0	2944.2	3016.6	3083.3	3146.4	3207.1	3266.2
	S	2.8612	6.0191	6.0382	6.1403	6.2852	6.4088	6.5182	6.6174	6.7093	6.7955
4600 (258.75)	V	1.272	43.038	43.278	46.000	50.027	53.643	56.994	60.167	63.215	66.172
	U	1122.9	2599.1	2602.9	2646.6	2710.4	2766.9	2818.7	2867.6	2914.5	2960.3
	H	1128.8	2797.0	2802.0	2858.2	2940.5	3013.7	3080.9	3144.4	3205.3	3264.7
	S	2.8735	6.0097	6.0190	6.1230	6.2700	6.3949	6.5050	6.6049	6.6972	6.7838
4700 (260.07)	V	1.276	42.081	....	44.778	48.772	52.346	55.651	58.775	61.773	64.679
	U	1129.3	2598.6	....	2642.9	2707.6	2764.7	2816.9	2866.1	2913.2	2959.1
	H	1135.3	2796.4	....	2853.3	2936.8	3010.7	3078.5	3142.3	3203.6	3263.1
	S	2.8855	6.0004	....	6.1058	6.2549	6.3811	6.4921	6.5926	6.6853	6.7722
4800 (261.37)	V	1.279	41.161	....	43.604	47.569	51.103	54.364	57.441	60.390	63.247
	U	1135.6	2598.1	....	2639.1	2704.8	2762.5	2815.1	2864.6	2911.9	2958.0
	H	1141.8	2795.7	....	2848.4	2933.1	3007.8	3076.1	3140.3	3201.8	3261.6
	S	2.8974	5.9913	....	6.0887	6.2399	6.3675	6.4794	6.5805	6.6736	6.7608
4900 (262.65)	V	1.282	40.278	....	42.475	46.412	49.909	53.128	56.161	59.064	61.874
	U	1141.9	2597.6	....	2635.2	2701.9	2760.2	2813.3	2863.0	2910.6	2956.9
	H	1148.2	2794.9	....	2843.3	2929.3	3004.8	3073.6	3138.2	3200.0	3260.0
	S	2.9091	5.9823	....	6.0717	6.2252	6.3541	6.4669	6.5685	6.6621	6.7496
5000 (263.91)	V	1.286	39.429	....	41.388	45.301	48.762	51.941	54.932	57.791	60.555
	U	1148.0	2597.0	....	2631.3	2699.0	2758.0	2811.5	2861.5	2909.3	2955.7
	H	1154.5	2794.2	....	2838.2	2925.5	3001.8	3071.2	3136.2	3198.3	3258.5
	S	2.9206	5.9735	....	6.0547	6.2105	6.3408	6.4545	6.5568	6.6508	6.7386
5100 (265.15)	V	1.289	38.611	....	40.340	44.231	47.660	50.801	53.750	56.567	59.288
	U	1154.1	2596.5	....	2627.3	2696.1	2755.7	2809.6	2860.0	2908.0	2954.5
	H	1160.7	2793.4	....	2833.1	2921.7	2998.7	3068.7	3134.1	3196.5	3256.9
	S	2.9319	5.9648	....	6.0378	6.1960	6.3277	6.4423	6.5452	6.6396	6.7278
5200 (266.37)	V	1.292	37.824	....	39.330	43.201	46.599	49.703	52.614	55.390	58.070
	U	1160.1	2595.9	....	2623.3	2693.1	2753.4	2807.8	2858.4	2906.7	2953.4
	H	1166.8	2792.6	....	2827.8	2917.8	2995.7	3066.2	3132.0	3194.7	3255.4
	S	2.9431	5.9561	....	6.0210	6.1815	6.3147	6.4302	6.5338	6.6287	6.7172
5300 (267.58)	V	1.296	37.066	....	38.354	42.209	45.577	48.647	51.520	54.257	56.897
	U	1166.1	2595.3	....	2619.2	2690.1	2751.0	2805.9	2856.9	2905.3	2952.2
	H	1172.9	2791.7	....	2822.5	2913.8	2992.6	3063.7	3129.9	3192.9	3253.8
	S	2.9541	5.9476	....	6.0041	6.1672	6.3018	6.4183	6.5225	6.6179	6.7067
5400 (268.76)	V	1.299	36.334	....	37.411	41.251	44.591	47.628	50.466	53.166	55.768
	U	1171.9	2594.6	....	2615.0	2687.1	2748.7	2804.0	2855.3	2904.0	2951.1
	H	1178.8	2790.8	....	2817.0	2909.8	2989.5	3061.2	3127.8	3191.1	3252.2
	S	2.9650	5.9392	....	5.9873	6.1530	6.2891	6.4066	6.5114	6.6072	6.6963
5500 (269.93)	V	1.302	35.628	....	36.499	40.327	43.641	46.647	49.450	52.115	54.679
	U	1177.7	2594.0	....	2610.8	2684.0	2746.3	2802.1	2853.7	2902.7	2949.9
	H	1184.9	2789.9	....	2811.5	2905.8	2986.4	3058.7	3125.7	3189.3	3250.6
	S	2.9757	5.9309	....	5.9705	6.1388	6.2765	6.3949	6.5004	6.5967	6.6862
5600 (271.09)	V	1.306	34.946	....	35.617	39.434	42.724	45.700	48.470	51.100	53.630
	U	1183.5	2593.3	....	2606.5	2680.9	2744.0	2800.2	2852.1	2901.3	2948.7
	H	1190.8	2789.0	....	2805.9	2901.7	2983.2	3056.1	3123.6	3187.5	3249.0
	S	2.9863	5.9227	....	5.9537	6.1248	6.2640	6.3834	6.4896	6.5863	6.6761
5700 (272.22)	V	1.309	34.288	....	34.761	38.571	41.838	44.785	47.525	50.121	52.617
	U	1189.1	2592.6	....	2602.1	2677.8	2741.6	2798.3	2850.5	2899.9	2947.5
	H	1196.6	2788.0	....	2800.2	2897.6	2980.0	3053.5	3121.4	3185.6	3247.5
	S	2.9968	5.9146	....	5.9369	6.1108	6.2516	6.3720	6.4789	6.5761	6.6663

ABS. PRESS., kPa (SAT. TEMP., °C)		SAT. WATER	SAT. STEAM	TEMPERATURE, °C (TEMPERATURE, K)							
				450 (723.15)	475 (748.15)	500 (773.15)	525 (798.15)	550 (823.15)	575 (848.15)	600 (873.15)	650 (923.15)
4100 (251.80)	V	1.256	48.500	77.921	81.062	84.165	87.236	90.281	93.303	96.306	102.26
	U	1089.4	2601.0	3010.4	3054.6	3098.8	3143.0	3187.5	3232.1	3277.1	3367.9
	H	1094.6	2799.9	3329.9	3387.0	3443.9	3500.7	3557.6	3614.7	3671.9	3787.1
	S	2.8099	6.0583	6.9260	7.0037	7.0785	7.1508	7.2210	7.2893	7.3558	7.4842
4200 (253.24)	V	1.259	47.307	75.981	79.056	82.092	85.097	88.075	91.030	93.966	99.787
	U	1096.3	2600.7	3009.4	3053.7	3097.9	3142.3	3186.8	3231.5	3276.5	3367.3
	H	1101.6	2799.4	3328.5	3385.7	3442.7	3499.7	3556.7	3613.8	3671.1	3786.4
	S	2.8231	6.0482	6.9135	6.9913	7.0662	7.1387	7.2090	7.2774	7.3440	7.4724
4300 (254.66)	V	1.262	46.168	74.131	77.143	80.116	83.057	85.971	88.863	91.735	97.428
	U	1103.1	2600.3	3008.4	3052.8	3097.1	3141.5	3186.0	3230.8	3275.8	3366.8
	H	1108.5	2798.9	3327.1	3384.4	3441.6	3498.6	3555.7	3612.9	3670.3	3785.7
	S	2.8360	6.0383	6.9012	6.9792	7.0543	7.1269	7.1973	7.2658	7.3324	7.4610
4400 (256.05)	V	1.266	45.079	72.365	75.317	78.229	81.110	83.963	86.794	89.605	95.177
	U	1109.8	2599.9	3007.4	3051.9	3096.3	3140.7	3185.3	3230.1	3275.2	3366.2
	H	1115.4	2798.3	3325.8	3383.3	3440.5	3497.6	3554.7	3612.0	3669.5	3785.0
	S	2.8487	6.0286	6.8892	6.9674	7.0426	7.1153	7.1858	7.2544	7.3211	7.4498
4500 (257.41)	V	1.269	44.037	70.677	73.572	76.427	79.249	82.044	84.817	87.570	93.025
	U	1116.4	2599.5	3006.3	3050.9	3095.4	3139.9	3184.6	3229.5	3274.6	3365.7
	H	1122.1	2797.7	3324.4	3382.0	3439.3	3496.6	3553.8	3611.1	3668.6	3784.3
	S	2.8612	6.0191	6.8774	6.9558	7.0311	7.1040	7.1746	7.2432	7.3100	7.4388
4600 (258.75)	V	1.272	43.038	69.063	71.903	74.702	77.469	80.209	82.926	85.623	90.967
	U	1122.9	2599.1	3005.3	3050.0	3094.6	3139.2	3183.9	3228.8	3273.9	3365.1
	H	1128.8	2797.0	3323.0	3380.8	3438.2	3495.5	3552.8	3610.2	3667.8	3783.6
	S	2.8735	6.0097	6.8659	6.9444	7.0199	7.0928	7.1636	7.2323	7.2991	7.4281
4700 (260.07)	V	1.276	42.081	67.517	70.304	73.051	75.765	78.452	81.116	83.760	88.997
	U	1129.3	2598.6	3004.3	3049.1	3093.7	3138.4	3183.1	3228.1	3273.3	3364.6
	H	1135.3	2796.4	3321.6	3379.5	3437.1	3494.5	3551.9	3609.3	3667.0	3782.9
	S	2.8855	6.0004	6.8545	6.9332	7.0089	7.0819	7.1527	7.2215	7.2885	7.4176
4800 (261.37)	V	1.279	41.161	66.036	68.773	71.469	74.132	76.768	79.381	81.973	87.109
	U	1135.6	2598.1	3003.3	3048.2	3092.9	3137.6	3182.4	3227.4	3272.7	3364.0
	H	1141.8	2795.7	3320.3	3378.3	3435.9	3493.4	3550.9	3608.5	3666.2	3782.1
	S	2.8974	5.9913	6.8434	6.9223	6.9981	7.0712	7.1422	7.2110	7.2781	7.4072
4900 (262.65)	V	1.282	40.278	64.615	67.303	69.951	72.565	75.152	77.716	80.260	85.298
	U	1141.9	2597.6	3002.3	3047.2	3092.0	3136.8	3181.7	3226.8	3272.0	3363.5
	H	1148.2	2794.9	3318.9	3377.0	3434.8	3492.4	3549.9	3607.6	3665.3	3781.4
	S	2.9091	5.9823	6.8324	6.9115	6.9874	7.0607	7.1318	7.2007	7.2678	7.3971
5000 (263.91)	V	1.286	39.429	63.250	65.893	68.494	71.061	73.602	76.119	78.616	83.559
	U	1148.0	2597.0	3001.2	3046.3	3091.2	3136.0	3181.0	3226.1	3271.4	3362.9
	H	1154.5	2794.2	3317.5	3375.8	3433.7	3491.3	3549.0	3606.7	3664.5	3780.7
	S	2.9206	5.9735	6.8217	6.9009	6.9770	7.0504	7.1215	7.1906	7.2578	7.3872
5100 (265.15)	V	1.289	38.611	61.940	64.537	67.094	69.616	72.112	74.584	77.035	81.888
	U	1154.1	2596.5	3000.2	3045.4	3090.3	3135.3	3180.2	3225.4	3270.8	3362.4
	H	1160.7	2793.4	3316.1	3374.5	3432.5	3490.3	3548.0	3605.8	3663.7	3780.0
	S	2.9319	5.9648	6.8111	6.8905	6.9668	7.0403	7.1115	7.1807	7.2479	7.3775
5200 (266.37)	V	1.292	37.824	60.679	63.234	65.747	68.227	70.679	73.108	75.516	80.282
	U	1160.1	2595.9	2999.2	3044.5	3089.5	3134.5	3179.5	3224.7	3270.2	3361.8
	H	1166.8	2792.6	3314.7	3373.3	3431.4	3489.3	3547.1	3604.9	3662.8	3779.3
	S	2.9431	5.9561	6.8007	6.8803	6.9567	7.0304	7.1017	7.1709	7.2382	7.3679
5300 (267.58)	V	1.296	37.066	59.466	61.980	64.452	66.890	69.300	71.687	74.054	78.736
	U	1166.1	2595.3	2998.2	3043.5	3088.6	3133.7	3178.8	3224.1	3269.5	3361.3
	H	1172.9	2791.7	3313.3	3372.0	3430.2	3488.2	3546.1	3604.0	3662.0	3778.6
	S	2.9541	5.9476	6.7905	6.8703	6.9468	7.0206	7.0920	7.1613	7.2287	7.3585
5400 (268.76)	V	1.299	36.334	58.297	60.772	63.204	65.603	67.973	70.320	72.646	77.248
	U	1171.9	2594.6	2997.1	3042.6	3087.8	3132.9	3178.1	3223.4	3268.9	3360.7
	H	1178.9	2790.8	3311.9	3370.8	3429.1	3487.2	3545.1	3603.1	3661.2	3777.8
	S	2.9650	5.9392	6.7804	6.8604	6.9371	7.0110	7.0825	7.1519	7.2194	7.3493
5500 (269.93)	V	1.302	35.628	57.171	59.608	62.002	64.362	66.694	69.002	71.289	75.814
	U	1177.7	2594.0	2996.1	3041.7	3086.9	3132.1	3177.3	3222.7	3268.3	3360.2
	H	1184.9	2789.9	3310.5	3369.5	3427.9	3486.1	3544.2	3602.2	3660.4	3777.1
	S	2.9757	5.9309	6.7705	6.8507	6.9275	7.0015	7.0731	7.1426	7.2102	7.3402
5600 (271.09)	V	1.306	34.946	56.085	58.486	60.843	63.165	65.460	67.731	69.981	74.431
	U	1183.5	2593.3	2995.0	3040.7	3086.1	3131.3	3176.6	3222.0	3267.6	3359.6
	H	1190.8	2789.0	3309.1	3368.2	3426.8	3485.1	3543.2	3601.3	3659.5	3776.4
	S	2.9863	5.9227	6.7607	6.8411	6.9181	6.9922	7.0639	7.1335	7.2011	7.3313
5700 (272.22)	V	1.309	34.288	55.038	57.403	59.724	62.011	64.270	66.504	68.719	73.096
	U	1189.1	2592.6	2994.0	3039.8	3085.2	3130.5	3175.9	3221.3	3267.0	3359.1
	H	1196.6	2788.0	3307.7	3367.0	3425.6	3484.0	3542.2	3600.4	3658.7	3775.7
	S	2.9968	5.9146	6.7511	6.8316	6.9088	6.9831	7.0549	7.1245	7.1923	7.3226

ABS. PRESS., kPa (SAT. TEMP., °C)		SAT. WATER	SAT. STEAM	TEMPERATURE, °C (TEMPERATURE, K)							
				280 (553.15)	290 (563.15)	300 (573.15)	325 (598.15)	350 (623.15)	375 (648.15)	400 (673.15)	425 (698.15)
5800 (273.35)	V	1.312	33.651	34.756	36.301	37.736	40.982	43.902	46.611	49.176	51.638
	U	1194.7	2591.9	2614.4	2645.7	2674.6	2739.1	2796.3	2848.9	2898.6	2946.4
	H	1202.3	2787.0	2816.0	2856.3	2893.5	2976.8	3051.0	3119.3	3183.8	3245.9
	S	3.0071	5.9066	5.9592	6.0314	6.0969	6.2393	6.3608	6.4683	6.5660	6.6565
5900 (274.46)	V	1.315	33.034	33.953	35.497	36.928	40.154	43.048	45.728	48.262	50.693
	U	1200.3	2591.1	2610.2	2642.1	2671.4	2736.7	2794.4	2847.3	2897.2	2945.2
	H	1208.0	2786.0	2810.5	2851.5	2889.3	2973.6	3048.4	3117.1	3182.0	3244.3
	S	3.0172	5.8986	5.9431	6.0166	6.0830	6.2272	6.3496	6.4578	6.5560	6.6469
6000 (275.55)	V	1.319	32.438	33.173	34.718	36.145	39.353	42.222	44.874	47.379	49.779
	U	1205.8	2590.4	2605.9	2638.4	2668.1	2734.2	2792.4	2845.7	2895.8	2944.0
	H	1213.7	2785.0	2804.9	2846.7	2885.0	2970.4	3045.8	3115.0	3180.1	3242.6
	S	3.0273	5.8908	5.9270	6.0017	6.0692	6.2151	6.3386	6.4475	6.5462	6.6374
6100 (276.63)	V	1.322	31.860	32.415	33.962	35.386	38.577	41.422	44.048	46.524	48.895
	U	1211.2	2589.6	2601.5	2634.6	2664.8	2731.7	2790.4	2844.1	2894.5	2942.8
	H	1219.3	2783.9	2799.3	2841.8	2880.5	2967.1	3043.1	3112.8	3178.3	3241.0
	S	3.0372	5.8830	5.9108	5.9869	6.0555	6.2031	6.3277	6.4373	6.5364	6.6280
6200 (277.70)	V	1.325	31.300	31.679	33.227	34.650	37.825	40.648	43.248	45.697	48.039
	U	1216.6	2588.8	2597.1	2630.8	2661.5	2729.2	2788.5	2842.4	2893.1	2941.6
	H	1224.8	2782.9	2793.5	2836.8	2876.3	2963.8	3040.5	3110.6	3176.4	3239.4
	S	3.0471	5.8753	5.8946	5.9721	6.0418	6.1911	6.3168	6.4272	6.5268	6.6188
6300 (278.75)	V	1.328	30.757	30.962	32.514	33.935	37.097	39.898	42.473	44.895	47.210
	U	1221.9	2588.0	2592.6	2626.9	2658.1	2726.7	2786.5	2840.8	2891.7	2940.4
	H	1230.3	2781.8	2787.6	2831.7	2871.9	2960.4	3037.8	3108.4	3174.5	3237.8
	S	3.0568	5.8677	5.8783	5.9573	6.0281	6.1793	6.3061	6.4172	6.5173	6.6096
6400 (279.79)	V	1.332	30.230	30.265	31.821	33.241	36.390	39.170	41.722	44.119	46.407
	U	1227.2	2587.2	2587.9	2623.0	2654.7	2724.2	2784.4	2839.1	2890.3	2939.2
	H	1235.7	2780.6	2781.6	2826.6	2867.5	2957.1	3035.1	3106.2	3172.7	3236.2
	S	3.0664	5.8601	5.8619	5.9425	6.0144	6.1675	6.2955	6.4072	6.5079	6.6006
6500 (280.82)	V	1.335	29.719	....	31.146	32.567	35.704	38.465	40.994	43.366	45.629
	U	1232.5	2586.3	....	2619.0	2651.2	2721.6	2782.4	2837.5	2888.9	2938.0
	H	1241.1	2779.5	....	2821.4	2862.9	2953.7	3032.4	3103.9	3170.8	3234.5
	S	3.0759	5.8527	....	5.9277	6.0008	6.1558	6.2849	6.3974	6.4986	6.5917
6600 (281.84)	V	1.338	29.223	....	30.490	31.911	35.038	37.781	40.287	42.636	44.874
	U	1237.6	2585.5	....	2614.9	2647.7	2719.0	2780.4	2835.8	2887.5	2936.7
	H	1246.5	2778.3	....	2816.1	2858.4	2950.2	3029.7	3101.7	3168.9	3232.9
	S	3.0853	5.8452	....	5.9129	5.9872	6.1442	6.2744	6.3877	6.4894	6.5828
6700 (282.84)	V	1.342	28.741	....	29.850	31.273	34.391	37.116	39.601	41.927	44.141
	U	1242.8	2584.6	....	2610.8	2644.2	2716.4	2778.3	2834.1	2886.1	2935.5
	H	1251.8	2777.1	....	2810.8	2853.7	2946.8	3027.0	3099.5	3167.0	3231.3
	S	3.0946	5.8379	....	5.8980	5.9736	6.1326	6.2640	6.3781	6.4803	6.5741
6800 (283.84)	V	1.345	28.272	....	29.226	30.652	33.762	36.470	38.935	41.239	43.430
	U	1247.9	2583.7	....	2606.6	2640.6	2713.7	2776.2	2832.4	2884.7	2934.3
	H	1257.0	2775.9	....	2805.3	2849.0	2943.3	3024.2	3097.2	3165.1	3229.6
	S	3.1038	5.8306	....	5.8830	5.9599	6.1211	6.2537	6.3686	6.4713	6.5655
7000 (285.79)	V	1.351	27.373	....	28.024	29.457	32.556	35.233	37.660	39.922	42.068
	U	1258.0	2581.8	....	2597.9	2633.2	2708.4	2772.1	2829.0	2881.8	2931.8
	H	1267.4	2773.5	....	2794.1	2839.4	2936.3	3018.7	3092.7	3161.2	3226.3
	S	3.1219	5.8162	....	5.8530	5.9327	6.0982	6.2333	6.3497	6.4536	6.5485
7200 (287.70)	V	1.358	26.522	....	26.878	28.321	31.413	34.063	36.454	38.676	40.781
	U	1267.9	2579.9	....	2589.0	2625.6	2702.9	2767.8	2825.6	2878.9	2929.4
	H	1277.6	2770.9	....	2782.5	2829.5	2929.1	3013.1	3088.1	3157.4	3223.0
	S	3.1397	5.8020	....	5.8226	5.9054	6.0755	6.2132	6.3312	6.4362	6.5319
7400 (289.57)	V	1.364	25.715	....	25.781	27.238	30.328	32.954	35.312	37.497	39.564
	U	1277.6	2578.0	....	2579.7	2617.8	2697.3	2763.5	2822.1	2876.0	2926.9
	H	1287.7	2768.3	....	2770.6	2819.3	2921.8	3007.4	3083.4	3153.5	3219.6
	S	3.1571	5.7880	....	5.7919	5.8779	6.0530	6.1933	6.3130	6.4190	6.5156
7600 (291.41)	V	1.371	24.949	....	....	26.204	29.297	31.901	34.229	36.380	38.409
	U	1287.2	2575.9	....	....	2609.7	2691.7	2759.2	2818.6	2873.1	2924.3
	H	1297.6	2765.5	....	....	2808.8	2914.3	3001.6	3078.7	3149.6	3216.3
	S	3.1742	5.7742	....	....	5.8503	6.0306	6.1737	6.2950	6.4022	6.4996
7800 (293.21)	V	1.378	24.220	....	....	25.214	28.315	30.900	33.200	35.319	37.314
	U	1296.7	2573.8	....	....	2601.3	2685.9	2754.8	2815.1	2870.1	2921.8
	H	1307.4	2762.8	....	....	2798.0	2906.7	2995.8	3074.0	3145.6	3212.9
	S	3.1911	5.7605	....	....	5.8224	6.0082	6.1542	6.2773	6.3857	6.4839
8000 (294.97)	V	1.384	23.525	....	....	24.264	27.378	29.948	32.222	34.310	36.273
	U	1306.0	2571.7	....	....	2592.7	2679.9	2750.3	2811.5	2867.1	2919.3
	H	1317.1	2759.9	....	....	2786.8	2899.0	2989.9	3069.2	3141.6	3209.5
	S	3.2076	5.7471	....	....	5.7942	5.9860	6.1349	6.2599	6.3694	6.4684

ABS. PRESS., kPa (SAT. TEMP., °C)		SAT. WATER	SAT. STEAM	TEMPERATURE, °C (TEMPERATURE, K)							
				450 (723.15)	475 (748.15)	500 (773.15)	525 (798.15)	550 (823.15)	575 (848.15)	600 (873.15)	650 (923.15)
5800 (273.35)	V	1.312	33.651	54.026	56.357	58.644	60.896	63.120	65.320	67.500	71.807
	U	1194.7	2591.9	2992.9	3038.8	3084.4	3129.8	3175.2	3220.7	3266.4	3358.5
	H	1202.3	2787.0	3306.3	3365.7	3424.3	3483.0	3541.2	3599.5	3657.9	3775.0
	S	3.0071	5.9066	6.7416	6.8223	6.8996	6.9740	7.0460	7.1157	7.1835	7.3139
5900 (274.46)	V	1.315	33.034	53.048	55.346	57.600	59.819	62.010	64.176	66.322	70.563
	U	1200.3	2591.1	2991.9	3037.9	3083.5	3129.0	3174.4	3220.0	3265.7	3357.9
	H	1208.0	2786.0	3304.9	3364.4	3423.3	3481.9	3540.3	3598.6	3657.0	3774.3
	S	3.0172	5.8986	6.7322	6.8132	6.8906	6.9652	7.0372	7.1070	7.1749	7.3054
6000 (275.55)	V	1.319	32.438	52.103	54.369	56.592	58.778	60.937	63.071	65.184	69.359
	U	1205.8	2590.4	2990.8	3036.9	3082.6	3128.2	3173.7	3219.3	3265.1	3357.4
	H	1213.7	2785.0	3303.5	3363.2	3422.2	3480.8	3539.3	3597.7	3656.2	3773.5
	S	3.0273	5.8908	6.7230	6.8041	6.8818	6.9564	7.0285	7.0985	7.1664	7.2971
6100 (276.63)	V	1.322	31.860	51.189	53.424	55.616	57.771	59.898	62.001	64.083	68.196
	U	1211.2	2589.6	2989.8	3036.0	3081.8	3127.4	3173.0	3218.6	3264.5	3356.8
	H	1219.3	2783.9	3302.0	3361.9	3421.0	3479.8	3538.3	3596.8	3655.4	3772.8
	S	3.0372	5.8830	6.7139	6.7952	6.8730	6.9478	7.0200	7.0900	7.1581	7.2889
6200 (277.70)	V	1.325	31.300	50.304	52.510	54.671	56.797	58.894	60.966	63.018	67.069
	U	1216.6	2588.8	2988.7	3035.0	3080.9	3126.6	3172.2	3218.0	3263.8	3356.3
	H	1224.8	2782.9	3300.6	3360.6	3419.9	3478.7	3537.4	3595.9	3654.5	3772.1
	S	3.0471	5.8753	6.7049	6.7864	6.8644	6.9393	7.0116	7.0817	7.1498	7.2808
6300 (278.75)	V	1.328	30.757	49.447	51.624	53.757	55.853	57.921	59.964	61.986	65.979
	U	1221.9	2588.0	2987.7	3034.1	3080.1	3125.8	3171.5	3217.3	3263.2	3355.7
	H	1230.3	2781.8	3299.2	3359.3	3418.7	3477.7	3536.4	3595.0	3653.7	3771.4
	S	3.0568	5.8677	6.6960	6.7778	6.8559	6.9309	7.0034	7.0735	7.1417	7.2728
6400 (279.79)	V	1.332	30.230	48.617	50.767	52.871	54.939	56.978	58.993	60.987	64.922
	U	1227.2	2587.2	2986.6	3033.1	3079.2	3125.0	3170.8	3216.6	3262.6	3355.2
	H	1235.7	2780.6	3297.7	3358.0	3417.6	3476.6	3535.4	3594.1	3652.9	3770.7
	S	3.0664	5.8601	6.6872	6.7692	6.8475	6.9226	6.9952	7.0655	7.1337	7.2649
6500 (280.82)	V	1.335	29.719	47.812	49.935	52.012	54.053	56.065	58.052	60.018	63.898
	U	1232.5	2586.3	2985.5	3032.2	3078.3	3124.2	3170.0	3215.9	3261.9	3354.6
	H	1241.1	2779.5	3296.3	3356.8	3416.4	3475.6	3534.4	3593.2	3652.1	3770.0
	S	3.0759	5.8527	6.6786	6.7608	6.8392	6.9145	6.9871	7.0575	7.1258	7.2572
6600 (281.84)	V	1.338	29.223	47.031	49.129	51.180	53.194	55.179	57.139	59.079	62.905
	U	1237.6	2585.5	2984.5	3031.2	3077.4	3123.4	3169.3	3215.2	3261.3	3354.1
	H	1246.5	2778.3	3294.9	3355.5	3415.2	3474.5	3533.5	3592.3	3651.2	3769.2
	S	3.0853	5.8452	6.6700	6.7524	6.8310	6.9064	6.9792	7.0497	7.1181	7.2495
6700 (282.84)	V	1.342	28.741	46.274	48.346	50.372	52.361	54.320	56.254	58.168	61.942
	U	1242.8	2584.6	2983.4	3030.3	3076.6	3122.6	3168.6	3214.5	3260.7	3353.5
	H	1251.8	2777.1	3293.4	3354.2	3414.1	3473.4	3532.5	3591.4	3650.4	3768.5
	S	3.0946	5.8379	6.6616	6.7442	6.8229	6.8985	6.9714	7.0419	7.1104	7.2420
6800 (283.84)	V	1.345	28.272	45.539	47.587	49.588	51.552	53.486	55.395	57.283	61.007
	U	1247.9	2583.7	2982.3	3029.3	3075.7	3121.8	3167.8	3213.9	3260.0	3353.0
	H	1257.0	2775.9	3292.0	3352.9	3412.9	3472.4	3531.5	3590.5	3649.6	3767.8
	S	3.1038	5.8306	6.6532	6.7361	6.8150	6.8907	6.9636	7.0343	7.1028	7.2345
7000 (285.79)	V	1.351	27.373	44.131	46.133	48.086	50.003	51.889	53.750	55.590	59.217
	U	1258.0	2581.8	2980.1	3027.4	3074.0	3120.2	3166.3	3212.5	3258.8	3351.9
	H	1267.4	2773.5	3289.1	3350.3	3410.6	3470.2	3529.6	3588.7	3647.9	3766.4
	S	3.1219	5.8162	6.6368	6.7201	6.7993	6.8753	6.9485	7.0193	7.0880	7.2200
7200 (287.70)	V	1.358	26.522	42.802	44.759	46.668	48.540	50.381	52.197	53.991	57.527
	U	1267.9	2579.9	2978.0	3025.4	3072.2	3118.6	3164.9	3211.1	3257.5	3350.7
	H	1277.6	2770.9	3286.1	3347.7	3408.2	3468.1	3527.6	3586.9	3646.2	3764.9
	S	3.1397	5.8020	6.6208	6.7044	6.7840	6.8602	6.9337	7.0047	7.0735	7.2058
7400 (289.57)	V	1.364	25.715	41.544	43.460	45.327	47.156	48.954	50.727	52.478	55.928
	U	1277.6	2578.0	2975.8	3023.5	3070.4	3117.0	3163.4	3209.8	3256.2	3349.6
	H	1287.7	2768.3	3283.2	3345.1	3405.9	3466.0	3525.7	3585.1	3644.5	3763.5
	S	3.1571	5.7880	6.6050	6.6892	6.7691	6.8456	6.9192	6.9904	7.0594	7.1919
7600 (291.41)	V	1.371	24.949	40.351	42.228	44.056	45.845	47.603	49.335	51.045	54.413
	U	1287.2	2575.9	2973.6	3021.5	3068.7	3115.4	3161.9	3208.4	3254.9	3348.5
	H	1297.6	2765.5	3280.3	3342.5	3403.5	3463.8	3523.7	3583.3	3642.9	3762.1
	S	3.1742	5.7742	6.5896	6.6742	6.7545	6.8312	6.9051	6.9765	7.0457	7.1784
7800 (293.21)	V	1.378	24.220	39.220	41.060	42.850	44.601	46.320	48.014	49.686	52.976
	U	1296.7	2573.8	2971.4	3019.6	3066.9	3113.8	3160.4	3207.0	3253.7	3347.4
	H	1307.4	2762.8	3277.3	3339.8	3401.1	3461.7	3521.7	3581.5	3641.2	3760.6
	S	3.1911	5.7605	6.5745	6.6596	6.7402	6.8172	6.8913	6.9629	7.0322	7.1652
8000 (294.97)	V	1.384	23.525	38.145	39.950	41.704	43.419	45.102	46.759	48.394	51.611
	U	1306.0	2571.7	2969.2	3017.6	3065.1	3112.2	3158.9	3205.6	3252.4	3346.3
	H	1317.1	2759.9	3274.3	3337.2	3398.8	3459.5	3519.7	3579.7	3639.5	3759.2
	S	3.2076	5.7471	6.5597	6.6452	6.7262	6.8035	6.8778	6.9496	7.0191	7.1523

ABS. PRESS., kPa (SAT. TEMP., °C)		SAT. WATER	SAT. STEAM	TEMPERATURE, °C (TEMPERATURE, K)							
				300 (573.15)	320 (593.15)	340 (613.15)	360 (633.15)	380 (653.15)	400 (673.15)	425 (698.15)	450 (723.15)
8200 (296.70)	V	1.391	22.863	23.350	25.916	28.064	29.968	31.715	33.350	35.282	37.121
	U	1315.2	2569.5	2583.7	2657.7	2718.5	2771.5	2819.5	2864.1	2916.7	2966.9
	H	1326.6	2757.0	2775.2	2870.2	2948.6	3017.2	3079.5	3137.6	3206.0	3271.3
	S	3.2239	5.7338	5.7656	5.9288	6.0588	6.1689	6.2659	6.3534	6.4532	6.5452
8400 (298.39)	V	1.398	22.231	22.469	25.058	27.203	29.094	30.821	32.435	34.337	36.147
	U	1324.3	2567.2	2574.4	2651.1	2713.4	2767.3	2816.0	2861.1	2914.1	2964.7
	H	1336.1	2754.0	2763.1	2861.6	2941.9	3011.7	3074.8	3133.5	3202.6	3268.3
	S	3.2399	5.7207	5.7366	5.9056	6.0388	6.1509	6.2491	6.3376	6.4383	6.5309
8600 (300.06)	V	1.404	21.627	....	24.236	26.380	28.258	29.968	31.561	33.437	35.217
	U	1333.3	2564.9	....	2644.3	2708.1	2763.1	2812.4	2858.0	2911.5	2962.4
	H	1345.4	2750.9	....	2852.7	2935.0	3006.1	3070.1	3129.4	3199.1	3265.3
	S	3.2557	5.7076	....	5.8823	6.0189	6.1330	6.2326	6.3220	6.4236	6.5168
8800 (301.70)	V	1.411	21.049	....	23.446	25.592	27.459	29.153	30.727	32.576	34.329
	U	1342.2	2562.6	....	2637.3	2702.8	2758.8	2808.8	2854.9	2908.9	2960.1
	H	1354.6	2747.8	....	2843.6	2928.0	3000.4	3065.3	3125.3	3195.6	3262.2
	S	3.2713	5.6948	....	5.8590	5.9990	6.1152	6.2162	6.3067	6.4092	6.5030
9000 (303.31)	V	1.418	20.495	....	22.685	24.836	26.694	28.372	29.929	31.754	33.480
	U	1351.0	2560.1	....	2630.1	2697.4	2754.4	2805.2	2851.8	2906.3	2957.8
	H	1363.7	2744.6	....	2834.3	2920.9	2994.7	3060.5	3121.2	3192.0	3259.2
	S	3.2867	5.6820	....	5.8355	5.9792	6.0976	6.2000	6.2915	6.3949	6.4894
9200 (304.89)	V	1.425	19.964	....	21.952	24.110	25.961	27.625	29.165	30.966	32.668
	U	1359.7	2557.7	....	2622.7	2691.9	2750.0	2801.5	2848.7	2903.6	2955.5
	H	1372.8	2741.3	....	2824.7	2913.7	2988.9	3055.7	3117.0	3188.5	3256.1
	S	3.3018	5.6694	....	5.8118	5.9594	6.0801	6.1840	6.2765	6.3808	6.4760
9400 (306.44)	V	1.432	19.455	....	21.245	23.412	25.257	26.909	28.433	30.212	31.891
	U	1368.2	2555.2	....	2615.1	2686.3	2745.6	2797.8	2845.5	2900.9	2953.2
	H	1381.7	2738.0	....	2814.8	2906.3	2983.0	3050.7	3112.8	3184.9	3253.0
	S	3.3168	5.6568	....	5.7879	5.9397	6.0627	6.1681	6.2617	6.3669	6.4628
9600 (307.97)	V	1.439	18.965	....	20.561	22.740	24.581	26.221	27.731	29.489	31.145
	U	1376.7	2552.6	....	2607.3	2680.5	2741.0	2794.1	2842.3	2898.2	2950.9
	H	1390.6	2734.7	....	2804.7	2898.8	2977.0	3045.8	3108.5	3181.3	3249.9
	S	3.3315	5.6444	....	5.7637	5.9199	6.0454	6.1524	6.2470	6.3532	6.4498
9800 (309.48)	V	1.446	18.494	....	19.899	22.093	23.931	25.561	27.056	28.795	30.429
	U	1385.2	2550.0	....	2599.2	2674.7	2736.4	2790.3	2839.1	2895.5	2948.6
	H	1399.3	2731.2	....	2794.3	2891.2	2971.0	3040.8	3104.2	3177.7	3246.8
	S	3.3461	5.6321	....	5.7393	5.9001	6.0282	6.1368	6.2325	6.3397	6.4369
10000 (310.96)	V	1.453	18.041	....	19.256	21.468	23.305	24.926	26.408	28.128	29.742
	U	1393.5	2547.3	....	2590.9	2668.7	2731.8	2786.4	2835.8	2892.8	2946.2
	H	1408.0	2727.7	....	2783.5	2883.4	2964.8	3035.7	3099.9	3174.1	3243.6
	S	3.3605	5.6198	....	5.7145	5.8803	6.0110	6.1213	6.2182	6.3264	6.4243
10200 (312.42)	V	1.460	17.605	....	18.632	20.865	22.702	24.315	25.785	27.487	29.081
	U	1401.8	2544.6	....	2582.3	2662.6	2727.0	2782.6	2832.6	2890.0	2943.9
	H	1416.7	2724.2	....	2772.3	2875.4	2958.6	3030.6	3095.6	3170.4	3240.5
	S	3.3748	5.6076	....	5.6894	5.8604	5.9940	6.1059	6.2040	6.3131	6.4118
10400 (313.86)	V	1.467	17.184	....	18.024	20.282	22.121	23.726	25.185	26.870	28.446
	U	1410.0	2541.8	....	2573.4	2656.3	2722.2	2778.7	2829.3	2887.3	2941.5
	H	1425.2	2720.6	....	2760.8	2867.2	2952.3	3025.4	3091.2	3166.7	3237.3
	S	3.3889	5.5955	....	5.6638	5.8404	5.9769	6.0907	6.1899	6.3001	6.3994
10600 (315.27)	V	1.474	16.778	....	17.432	19.717	21.560	23.159	24.607	26.276	27.834
	U	1418.1	2539.0	....	2564.1	2649.9	2717.4	2774.7	2825.9	2884.5	2939.1
	H	1433.7	2716.9	....	2748.9	2858.9	2945.9	3020.2	3086.8	3163.0	3234.1
	S	3.4029	5.5835	....	5.6376	5.8203	5.9599	6.0755	6.1759	6.2872	6.3872
10800 (316.67)	V	1.481	16.385	....	16.852	19.170	21.018	22.612	24.050	25.703	27.245
	U	1426.2	2536.2	....	2554.5	2643.4	2712.4	2770.7	2822.6	2881.7	2936.7
	H	1442.2	2713.1	....	2736.5	2850.4	2939.4	3014.9	3082.3	3159.3	3230.9
	S	3.4167	5.5715	....	5.6109	5.8000	5.9429	6.0604	6.1621	6.2744	6.3752
11000 (318.05)	V	1.489	16.006	....	16.285	18.639	20.494	22.083	23.512	25.151	26.676
	U	1434.2	2533.2	....	2544.4	2636.7	2707.4	2766.7	2819.2	2878.9	2934.3
	H	1450.6	2709.3	....	2723.5	2841.7	2932.8	3009.6	3077.8	3155.5	3227.7
	S	3.4304	5.5595	....	5.5835	5.7797	5.9259	6.0454	6.1483	6.2617	6.3633
11200 (319.40)	V	1.496	15.639	....	15.726	18.124	19.987	21.573	22.993	24.619	26.128
	U	1442.1	2530.3	....	2533.8	2629.8	2702.2	2762.6	2815.8	2876.0	2931.8
	H	1458.9	2705.4	....	2710.0	2832.8	2926.1	3004.2	3073.3	3151.7	3224.5
	S	3.4440	5.5476	....	5.5553	5.7591	5.9090	6.0305	6.1347	6.2491	6.3515
11400 (320.74)	V	1.504	15.284	....	....	17.622	19.495	21.079	22.492	24.104	25.599
	U	1450.0	2527.2	....	....	2622.7	2697.0	2758.4	2812.3	2873.1	2929.4
	H	1467.2	2701.5	....	....	2823.6	2919.3	2998.7	3068.7	3147.9	3221.2
	S	3.4575	5.5357	....	....	5.7383	5.8920	6.0156	6.1211	6.2367	6.3399

		SAT. WATER	SAT. STEAM	TEMPERATURE, °C (TEMPERATURE, K)							
ABS. PRESS., kPa (SAT. TEMP., °C)				475 (748.15)	500 (773.15)	525 (798.15)	550 (823.15)	575 (848.15)	600 (873.15)	625 (898.15)	650 (923.15)
8200 (296.70)	V	1.391	22.863	38.893	40.614	42.295	43.943	45.566	47.166	48.747	50.313
	U	1315.2	2569.5	3015.6	3063.3	3110.5	3157.4	3204.3	3251.1	3298.1	3345.2
	H	1326.6	2757.0	3334.3	3396.4	3457.3	3517.8	3577.9	3637.9	3697.8	3757.7
	S	3.2239	5.7338	6.6311	6.7124	6.7900	6.8646	6.9365	7.0062	7.0739	7.1397
8400 (298.39)	V	1.398	22.231	37.887	39.576	41.224	42.839	44.429	45.996	47.544	49.076
	U	1324.3	2567.2	3013.6	3061.6	3108.9	3155.9	3202.9	3249.8	3296.9	3344.1
	H	1336.1	2754.0	3331.9	3394.0	3455.2	3515.8	3576.1	3636.2	3696.2	3756.3
	S	3.2399	5.7207	6.6173	6.6990	6.7769	6.8516	6.9238	6.9936	7.0614	7.1274
8600 (300.06)	V	1.404	21.627	36.928	38.586	40.202	41.787	43.345	44.880	46.397	47.897
	U	1333.3	2564.9	3011.6	3059.8	3107.3	3154.4	3201.5	3248.5	3295.7	3342.9
	H	1345.4	2750.9	3329.2	3391.6	3453.0	3513.8	3574.3	3634.5	3694.7	3754.9
	S	3.2557	5.7076	6.6037	6.6858	6.7639	6.8390	6.9113	6.9813	7.0492	7.1153
8800 (301.70)	V	1.411	21.049	36.011	37.640	39.228	40.782	42.310	43.815	45.301	46.771
	U	1342.2	2562.6	3009.6	3058.0	3105.6	3152.9	3200.1	3247.2	3294.5	3341.8
	H	1354.6	2747.8	3326.5	3389.2	3450.8	3511.8	3572.4	3632.8	3693.1	3753.4
	S	3.2713	5.6948	6.5904	6.6728	6.7513	6.8265	6.8990	6.9692	7.0373	7.1035
9000 (303.31)	V	1.418	20.495	35.136	36.737	38.296	39.822	41.321	42.798	44.255	45.695
	U	1351.0	2560.1	3007.6	3056.1	3104.0	3151.4	3198.7	3246.0	3293.3	3340.7
	H	1363.7	2744.6	3323.8	3386.8	3448.7	3509.8	3570.6	3631.1	3691.6	3752.0
	S	3.2867	5.6820	6.5773	6.6600	6.7388	6.8143	6.8870	6.9574	7.0256	7.0919
9200 (304.89)	V	1.425	19.964	34.298	35.872	37.405	38.904	40.375	41.824	43.254	44.667
	U	1359.7	2557.7	3006.6	3054.3	3102.3	3149.9	3197.3	3244.7	3292.1	3339.6
	H	1372.8	2741.3	3321.1	3384.4	3446.5	3507.8	3568.8	3629.5	3690.0	3750.5
	S	3.3018	5.6694	6.5644	6.6475	6.7266	6.8023	6.8752	6.9457	7.0141	7.0806
9400 (306.44)	V	1.432	19.455	33.495	35.045	36.552	38.024	39.470	40.892	42.295	43.682
	U	1368.2	2555.2	3003.5	3052.5	3100.7	3148.4	3195.9	3243.4	3290.9	3338.5
	H	1381.7	2738.0	3318.4	3381.9	3444.3	3505.9	3566.9	3627.8	3688.4	3749.1
	S	3.3168	5.6568	6.5517	6.6352	6.7146	6.7906	6.8637	6.9343	7.0029	7.0695
9600 (307.97)	V	1.439	18.965	32.726	34.252	35.734	37.182	38.602	39.999	41.377	42.738
	U	1376.7	2552.6	3001.5	3050.7	3099.0	3146.9	3194.5	3242.1	3289.7	3337.4
	H	1390.6	2734.7	3315.6	3379.5	3442.1	3503.9	3565.1	3626.1	3686.9	3747.6
	S	3.3315	5.6444	6.5392	6.6231	6.7028	6.7790	6.8523	6.9231	6.9918	7.0585
9800 (309.48)	V	1.446	18.494	31.988	33.491	34.949	36.373	37.769	39.142	40.496	41.832
	U	1385.2	2550.0	2999.4	3048.8	3097.4	3145.4	3193.1	3240.8	3288.5	3336.2
	H	1399.3	2731.2	3312.9	3377.0	3439.9	3501.9	3563.3	3624.4	3685.3	3746.2
	S	3.3461	5.6321	6.5268	6.6112	6.6912	6.7676	6.8411	6.9121	6.9810	7.0478
10000 (310.96)	V	1.453	18.041	31.280	32.760	34.196	35.597	36.970	38.320	39.650	40.963
	U	1393.5	2547.3	2997.4	3047.0	3095.7	3143.9	3191.7	3239.5	3287.3	3335.1
	H	1408.0	2727.7	3310.1	3374.6	3437.7	3499.8	3561.4	3622.7	3683.8	3744.7
	S	3.3605	5.6198	6.5147	6.5994	6.6797	6.7564	6.8302	6.9013	6.9703	7.0373
10200 (312.42)	V	1.460	17.605	30.599	32.058	33.472	34.851	36.202	37.530	38.837	40.128
	U	1401.8	2544.6	2995.3	3045.2	3094.0	3142.3	3190.3	3238.2	3286.1	3334.0
	H	1416.7	2724.2	3307.4	3372.1	3435.5	3497.8	3559.6	3621.0	3682.2	3743.3
	S	3.3748	5.6076	6.5027	6.5879	6.6685	6.7454	6.8194	6.8907	6.9598	7.0269
10400 (313.86)	V	1.467	17.184	29.943	31.382	32.776	34.134	35.464	36.770	38.056	39.325
	U	1410.0	2541.8	2993.2	3043.3	3092.4	3140.8	3188.9	3236.9	3284.8	3332.9
	H	1425.2	2720.6	3304.6	3369.7	3433.2	3495.8	3557.8	3619.3	3680.6	3741.8
	S	3.3889	5.5955	6.4909	6.5765	6.6574	6.7346	6.8087	6.8803	6.9495	7.0167
10600 (315.27)	V	1.474	16.778	29.313	30.732	32.106	33.444	34.753	36.039	37.304	38.552
	U	1418.1	2539.0	2991.1	3041.4	3090.7	3139.3	3187.5	3235.6	3283.6	3331.7
	H	1433.7	2716.9	3301.8	3367.2	3431.0	3493.8	3555.9	3617.6	3679.1	3740.4
	S	3.4029	5.5835	6.4793	6.5652	6.6465	6.7239	6.7983	6.8700	6.9394	7.0067
10800 (316.67)	V	1.481	16.385	28.706	30.106	31.461	32.779	34.069	35.335	36.580	37.808
	U	1426.2	2536.2	2989.0	3039.6	3089.0	3137.8	3186.1	3234.3	3282.4	3330.6
	H	1442.2	2713.1	3299.0	3364.7	3428.8	3491.8	3554.1	3615.9	3677.5	3738.9
	S	3.4167	5.5715	6.4678	6.5542	6.6357	6.7134	6.7880	6.8599	6.9294	6.9969
11000 (318.05)	V	1.489	16.006	28.120	29.503	30.839	32.139	33.410	34.656	35.882	37.091
	U	1434.2	2533.2	2986.9	3037.7	3087.3	3136.2	3184.7	3233.0	3281.2	3329.5
	H	1450.6	2709.3	3296.2	3362.2	3426.5	3489.7	3552.2	3614.2	3675.9	3737.5
	S	3.4304	5.5595	6.4564	6.5432	6.6251	6.7031	6.7779	6.8499	6.9196	6.9872
11200 (319.40)	V	1.496	15.639	27.555	28.921	30.240	31.521	32.774	34.002	35.210	36.400
	U	1442.1	2530.3	2984.8	3035.8	3085.6	3134.7	3183.3	3231.7	3280.0	3328.4
	H	1458.9	2705.4	3293.4	3359.7	3424.3	3487.7	3550.4	3612.5	3674.4	3736.0
	S	3.4440	5.5476	6.4452	6.5324	6.6147	6.6929	6.7679	6.8401	6.9099	6.9777
11400 (320.74)	V	1.504	15.284	27.010	28.359	29.661	30.925	32.160	33.370	34.560	35.733
	U	1450.0	2527.2	2982.6	3033.9	3083.9	3133.1	3181.9	3230.4	3278.8	3327.2
	H	1467.2	2701.5	3290.5	3357.2	3422.1	3485.7	3548.5	3610.8	3672.8	3734.6
	S	3.4575	5.5357	6.4341	6.5218	6.6043	6.6828	6.7580	6.8304	6.9004	6.9683

TEMPERATURE, °C
(TEMPERATURE, K)

ABS. PRESS., kPa (SAT. TEMP., °C)		SAT. WATER	SAT. STEAM	340 (613.15)	360 (633.15)	380 (653.15)	400 (673.15)	420 (693.15)	440 (713.15)	460 (733.15)	480 (753.15)
11600 (322.06)	V	1.511	14.940	17.134	19.019	20.601	22.007	23.298	24.507	25.655	26.755
	U	1457.9	2524.1	2615.5	2691.7	2754.2	2808.8	2858.4	2904.7	2948.7	2990.9
	H	1475.4	2697.4	2814.2	2912.3	2993.2	3064.1	3128.7	3189.0	3246.3	3301.3
	S	3.4708	5.5239	5.7174	5.8749	6.0007	6.1077	6.2022	6.2880	6.3672	6.4413
11800 (323.36)	V	1.519	14.607	16.658	18.557	20.139	21.538	22.820	24.019	25.155	26.243
	U	1465.7	2521.0	2608.0	2686.3	2750.0	2805.3	2855.4	2902.1	2946.3	2988.8
	H	1483.6	2693.3	2804.6	2905.3	2987.6	3059.5	3124.7	3185.5	3243.1	3298.5
	S	3.4840	5.5121	5.6962	5.8579	5.9860	6.0943	6.1898	6.2763	6.3561	6.4305
12000 (324.65)	V	1.527	14.283	16.193	18.108	19.691	21.084	22.357	23.546	24.672	25.748
	U	1473.4	2517.8	2600.4	2680.8	2745.7	2801.8	2852.4	2899.4	2944.0	2986.7
	H	1491.8	2689.2	2794.7	2898.1	2982.0	3054.8	3120.7	3182.0	3240.0	3295.7
	S	3.4972	5.5002	5.6747	5.8408	5.9712	6.0810	6.1775	6.2647	6.3450	6.4199
12200 (325.91)	V	1.535	13.969	15.740	17.672	19.256	20.645	21.910	23.089	24.204	25.269
	U	1481.2	2514.5	2592.5	2675.2	2741.4	2798.2	2849.3	2896.8	2941.6	2984.6
	H	1499.9	2684.9	2784.5	2890.8	2976.3	3050.0	3116.6	3178.5	3236.9	3292.9
	S	3.5102	5.4884	5.6528	5.8236	5.9565	6.0678	6.1653	6.2532	6.3341	6.4094
12400 (327.17)	V	1.543	13.664	15.296	17.248	18.835	20.219	21.476	22.646	23.751	24.806
	U	1488.8	2511.1	2584.4	2669.5	2737.0	2794.6	2846.3	2894.1	2939.2	2982.5
	H	1508.0	2680.6	2774.0	2883.4	2970.5	3045.3	3112.6	3174.9	3233.8	3290.1
	S	3.5232	5.4765	5.6307	5.8063	5.9418	6.0546	6.1532	6.2418	6.3232	6.3990
12600 (328.40)	V	1.551	13.367	14.861	16.836	18.426	19.805	21.055	22.217	23.312	24.357
	U	1496.5	2507.7	2576.0	2663.7	2732.5	2790.9	2843.2	2891.4	2936.9	2980.4
	H	1516.0	2676.1	2763.2	2875.8	2964.6	3040.5	3108.5	3171.3	3230.6	3287.3
	S	3.5361	5.4646	5.6082	5.7889	5.9272	6.0416	6.1411	6.2305	6.3125	6.3888
12800 (329.62)	V	1.559	13.078	14.434	16.433	18.028	19.404	20.648	21.801	22.887	23.922
	U	1504.1	2504.2	2567.3	2657.7	2728.0	2787.2	2840.1	2888.7	2934.5	2978.2
	H	1524.0	2671.6	2752.1	2868.1	2958.7	3035.6	3104.3	3167.7	3227.4	3284.4
	S	3.5488	5.4527	5.5852	5.7715	5.9125	6.0285	6.1291	6.2193	6.3019	6.3786
13000 (330.83)	V	1.567	12.797	14.015	16.041	17.642	19.015	20.252	21.397	22.474	23.500
	U	1511.6	2500.6	2558.4	2651.6	2723.4	2783.5	2836.9	2886.0	2932.1	2976.1
	H	1532.0	2667.0	2740.6	2860.2	2952.7	3030.7	3100.2	3164.1	3224.2	3281.6
	S	3.5616	5.4408	5.5618	5.7539	5.8979	6.0155	6.1173	6.2082	6.2913	6.3685
13200 (332.02)	V	1.576	12.523	13.602	15.659	17.266	18.637	19.868	21.006	22.074	23.091
	U	1519.2	2497.0	2549.1	2645.4	2718.7	2779.8	2833.7	2883.2	2929.6	2973.9
	H	1540.0	2662.3	2728.7	2852.1	2946.6	3025.8	3096.0	3160.5	3221.0	3278.7
	S	3.5742	5.4288	5.5378	5.7362	5.8832	6.0026	6.1054	6.1972	6.2809	6.3585
13400 (333.19)	V	1.584	12.256	13.195	15.285	16.900	18.270	19.495	20.625	21.686	22.694
	U	1526.7	2493.2	2539.5	2639.1	2714.0	2776.0	2830.6	2880.5	2927.2	2971.7
	H	1547.9	2657.4	2716.3	2843.9	2940.5	3020.8	3091.8	3156.8	3217.8	3275.8
	S	3.5868	5.4168	5.5133	5.7183	5.8685	5.9897	6.0937	6.1862	6.2705	6.3486
13600 (334.36)	V	1.593	11.996	12.793	14.920	16.544	17.912	19.133	20.256	21.309	22.308
	U	1534.2	2489.4	2529.4	2632.6	2709.2	2772.2	2827.3	2877.7	2924.8	2969.5
	H	1555.8	2652.5	2703.4	2835.5	2934.2	3015.8	3087.6	3153.2	3214.6	3272.9
	S	3.5993	5.4047	5.4880	5.7002	5.8538	5.9769	6.0820	6.1753	6.2603	6.3388
13800 (335.51)	V	1.602	11.743	12.394	14.563	16.196	17.565	18.781	19.897	20.942	21.933
	U	1541.6	2485.5	2518.9	2625.9	2704.3	2768.3	2824.1	2874.9	2922.3	2967.4
	H	1563.7	2647.5	2689.9	2826.9	2927.9	3010.7	3083.3	3149.5	3211.3	3270.0
	S	3.6118	5.3925	5.4619	5.6820	5.8391	5.9641	6.0704	6.1645	6.2501	6.3291
14000 (336.64)	V	1.611	11.495	11.997	14.213	15.858	17.227	18.438	19.549	20.586	21.569
	U	1549.1	2481.4	2507.7	2619.1	2699.4	2764.4	2820.8	2872.1	2919.8	2965.2
	H	1571.6	2642.4	2675.7	2818.1	2921.4	3005.6	3079.0	3145.8	3208.1	3267.1
	S	3.6242	5.3803	5.4348	5.6636	5.8243	5.9513	6.0588	6.1538	6.2399	6.3194
14200 (337.76)	V	1.620	11.253	11.600	13.871	15.528	16.897	18.105	19.209	20.240	21.215
	U	1556.5	2477.3	2495.9	2612.1	2694.4	2760.4	2817.6	2869.3	2917.4	2962.9
	H	1579.5	2637.1	2660.6	2809.1	2914.9	3000.4	3074.6	3142.0	3204.8	3264.2
	S	3.6366	5.3679	5.4064	5.6449	5.8095	5.9385	6.0473	6.1431	6.2299	6.3099
14400 (338.87)	V	1.629	11.017	11.199	13.535	15.206	16.576	17.781	18.879	19.903	20.871
	U	1563.9	2473.1	2483.2	2605.0	2689.3	2756.4	2814.2	2866.4	2914.9	2960.7
	H	1587.4	2631.8	2644.9	2799.9	2908.3	2995.1	3070.3	3138.3	3201.5	3261.3
	S	3.6490	5.3555	5.3762	5.6260	5.7947	5.9258	6.0358	6.1325	6.2199	6.3004
14600 (339.97)	V	1.638	10.786	10.791	13.205	14.891	16.264	17.465	18.558	19.576	20.536
	U	1571.3	2468.8	2469.1	2597.6	2684.1	2752.4	2810.9	2863.6	2912.4	2958.5
	H	1595.3	2626.3	2626.6	2790.6	2901.5	2989.9	3065.9	3134.5	3198.2	3258.3
	S	3.6613	5.3431	5.3436	5.6069	5.7797	5.9130	6.0244	6.1220	6.2100	6.2910
14800 (341.06)	V	1.648	10.561	....	12.881	14.583	15.959	17.157	18.245	19.256	20.210
	U	1578.7	2464.4	....	2590.1	2678.9	2748.3	2807.5	2860.7	2909.8	2956.3
	H	1603.1	2620.7	....	2780.7	2894.7	2984.5	3061.5	3130.7	3194.8	3255.4
	S	3.6736	5.3305	....	5.5874	5.7648	5.9003	6.0130	6.1115	6.2002	6.2817

ABS. PRESS., kPa (SAT. TEMP., °C)		SAT. WATER	SAT. STEAM	TEMPERATURE, °C (TEMPERATURE, K)							
				500 (773.15)	520 (793.15)	540 (813.15)	560 (833.15)	580 (853.15)	600 (873.15)	625 (898.15)	650 (923.15)
11600 (322.06)	V	1.511	14.940	27.817	28.848	29.855	30.840	31.808	32.761	33.934	35.089
	U	1457.9	2524.1	3032.0	3072.3	3111.9	3151.2	3190.2	3229.1	3277.6	3326.1
	H	1475.4	2697.4	3354.7	3406.9	3458.2	3508.9	3559.2	3609.1	3671.2	3733.1
	S	3.4708	5.5239	6.5113	6.5779	6.6419	6.7034	6.7630	6.8209	6.8910	6.9590
11800 (323.36)	V	1.519	14.607	27.293	28.312	29.305	30.278	31.232	32.172	33.328	34.467
	U	1465.7	2521.0	3030.1	3070.5	3110.3	3149.7	3188.8	3227.8	3276.4	3325.0
	H	1483.6	2693.3	3352.2	3404.6	3456.1	3507.0	3557.3	3607.4	3669.6	3731.7
	S	3.4840	5.5121	6.5009	6.5678	6.6320	6.6938	6.7535	6.8115	6.8818	6.9499
12000 (324.65)	V	1.527	14.283	26.786	27.793	28.774	29.734	30.676	31.603	32.743	33.865
	U	1473.4	2517.8	3028.2	3068.8	3108.7	3148.2	3187.4	3226.4	3275.2	3323.8
	H	1491.8	2689.2	3349.6	3402.3	3454.0	3505.0	3555.5	3605.7	3668.1	3730.2
	S	3.4972	5.5002	6.4906	6.5578	6.6222	6.6842	6.7441	6.8022	6.8727	6.9409
12200 (325.91)	V	1.535	13.969	26.296	27.291	28.260	29.208	30.138	31.052	32.177	33.283
	U	1481.2	2514.5	3026.3	3067.0	3107.1	3146.7	3186.0	3225.1	3273.9	3322.7
	H	1499.9	2684.9	3347.1	3400.0	3451.8	3503.0	3553.7	3604.0	3666.5	3728.8
	S	3.5102	5.4884	6.4805	6.5480	6.6126	6.6747	6.7348	6.7931	6.8637	6.9321
12400 (327.17)	V	1.543	13.664	25.821	26.805	27.763	28.699	29.617	30.519	31.629	32.720
	U	1488.8	2511.1	3024.4	3065.2	3105.4	3145.2	3184.6	3223.8	3272.7	3321.6
	H	1508.0	2680.6	3344.5	3397.6	3449.7	3501.0	3551.8	3602.3	3664.9	3727.3
	S	3.5232	5.4765	6.4704	6.5382	6.6030	6.6654	6.7257	6.7841	6.8548	6.9234
12600 (328.40)	V	1.551	13.367	25.362	26.334	27.281	28.206	29.112	30.003	31.098	32.175
	U	1496.5	2507.7	3022.4	3063.5	3103.8	3143.7	3183.2	3222.5	3271.5	3320.4
	H	1516.0	2676.1	3342.0	3395.3	3447.6	3499.1	3550.0	3600.5	3663.3	3725.8
	S	3.5361	5.4646	6.4605	6.5286	6.5936	6.6562	6.7166	6.7752	6.8461	6.9148
12800 (329.62)	V	1.559	13.078	24.916	25.878	26.814	27.728	28.623	29.503	30.584	31.647
	U	1504.1	2504.2	3020.5	3061.7	3102.2	3142.2	3181.8	3221.2	3270.3	3319.3
	H	1524.0	2671.6	3339.4	3393.0	3445.4	3497.1	3548.2	3598.8	3661.7	3724.4
	S	3.5488	5.4527	6.4507	6.5190	6.5843	6.6471	6.7077	6.7664	6.8375	6.9063
13000 (330.83)	V	1.567	12.797	24.485	25.437	26.362	27.265	28.150	29.019	30.086	31.135
	U	1511.6	2500.6	3018.5	3059.9	3100.6	3140.6	3180.4	3219.9	3269.0	3318.2
	H	1532.0	2667.0	3336.8	3390.6	3443.3	3495.1	3546.3	3597.1	3660.2	3722.9
	S	3.5616	5.4408	6.4409	6.5096	6.5752	6.6381	6.6989	6.7577	6.8289	6.8979
13200 (332.02)	V	1.576	12.523	24.066	25.008	25.923	26.816	27.690	28.549	29.603	30.639
	U	1519.2	2497.0	3016.6	3058.1	3098.9	3139.1	3178.9	3218.5	3267.8	3317.0
	H	1540.0	2662.3	3334.3	3388.3	3441.1	3493.1	3544.5	3595.4	3658.6	3721.5
	S	3.5742	5.4288	6.4313	6.5003	6.5661	6.6292	6.6902	6.7492	6.8205	6.8896
13400 (333.19)	V	1.584	12.256	23.659	24.592	25.497	26.380	27.245	28.093	29.134	30.157
	U	1526.7	2493.2	3014.6	3056.4	3097.3	3137.6	3177.5	3217.2	3266.6	3315.9
	H	1547.9	2657.4	3331.7	3385.9	3438.9	3491.1	3542.6	3593.6	3657.0	3720.0
	S	3.5868	5.4168	6.4218	6.4910	6.5571	6.6205	6.6816	6.7407	6.8122	6.8814
13600 (334.36)	V	1.593	11.996	23.265	24.188	25.084	25.958	26.812	27.651	28.679	29.690
	U	1534.2	2489.4	3012.7	3054.6	3095.6	3136.1	3176.1	3215.9	3265.4	3314.8
	H	1555.8	2652.5	3329.1	3383.5	3436.8	3489.1	3540.7	3591.9	3655.4	3718.5
	S	3.5993	5.4047	6.4124	6.4819	6.5482	6.6118	6.6731	6.7323	6.8040	6.8734
13800 (335.51)	V	1.602	11.743	22.882	23.796	24.683	25.547	26.392	27.221	28.238	29.236
	U	1541.6	2485.5	3010.7	3052.8	3094.0	3134.5	3174.7	3214.5	3264.1	3313.6
	H	1563.7	2647.5	3326.4	3381.2	3434.6	3487.1	3538.9	3590.2	3653.8	3717.1
	S	3.6118	5.3925	6.4030	6.4729	6.5394	6.6032	6.6646	6.7241	6.7959	6.8654
14000 (336.64)	V	1.611	11.495	22.509	23.415	24.293	25.148	25.984	26.804	27.809	28.795
	U	1549.1	2481.4	3008.7	3051.0	3092.3	3133.0	3173.3	3213.2	3262.9	3312.5
	H	1571.6	2642.4	3323.8	3378.8	3432.4	3485.1	3537.0	3588.5	3652.2	3715.6
	S	3.6242	5.3803	6.3937	6.4639	6.5307	6.5947	6.6563	6.7159	6.7879	6.8575
14200 (337.76)	V	1.620	11.253	22.147	23.045	23.914	24.760	25.587	26.398	27.392	28.367
	U	1556.5	2477.3	3006.7	3049.1	3090.6	3131.5	3171.8	3211.9	3261.7	3311.3
	H	1579.5	2637.1	3321.2	3376.4	3430.2	3483.1	3535.2	3586.7	3650.6	3714.1
	S	3.6366	5.3679	6.3846	6.4550	6.5221	6.5863	6.6481	6.7078	6.7800	6.8497
14400 (338.87)	V	1.629	11.017	21.796	22.685	23.546	24.384	25.202	26.004	26.986	27.951
	U	1563.9	2473.1	3004.7	3047.3	3089.0	3129.9	3170.4	3210.5	3260.4	3310.2
	H	1587.4	2631.8	3318.6	3374.0	3428.0	3481.1	3533.3	3585.0	3649.0	3712.7
	S	3.6490	5.3555	6.3755	6.4463	6.5136	6.5780	6.6400	6.6998	6.7722	6.8420
14600 (339.97)	V	1.638	10.786	21.453	22.335	23.187	24.017	24.827	25.620	26.592	27.545
	U	1571.3	2468.8	3002.7	3045.5	3087.3	3128.4	3169.0	3209.2	3259.2	3309.0
	H	1595.3	2626.3	3315.9	3371.6	3425.8	3479.0	3531.4	3583.3	3647.5	3711.2
	S	3.6613	5.3431	6.3665	6.4376	6.5051	6.5697	6.6319	6.6919	6.7644	6.8344
14800 (341.06)	V	1.648	10.561	21.120	21.994	22.839	23.660	24.462	25.247	26.209	27.151
	U	1578.7	2464.4	3000.7	3043.7	3085.6	3126.8	3167.5	3207.9	3258.0	3307.9
	H	1603.1	2620.7	3313.3	3369.2	3423.6	3477.0	3529.6	3581.5	3645.9	3709.7
	S	3.6736	5.3305	6.3575	6.4289	6.4968	6.5616	6.6239	6.6841	6.7568	6.8269

ABS. PRESS., kPa (SAT. TEMP., °C)		SAT. WATER	SAT. STEAM	TEMPERATURE, °C (TEMPERATURE, K)							
				360 (633.15)	370 (643.15)	380 (653.15)	400 (673.15)	420 (693.15)	440 (713.15)	460 (733.15)	480 (753.15)
15000 (342.13)	V	1.658	10.340	12.562	13.481	14.282	15.661	16.857	17.940	18.946	19.893
	U	1586.1	2459.9	2582.3	2631.4	2673.5	2744.2	2804.2	2857.8	2907.3	2954.0
	H	1611.0	2615.0	2770.8	2833.6	2887.7	2979.1	3057.0	3126.9	3191.5	3252.4
	S	3.6859	5.3178	5.5677	5.6662	5.7497	5.8876	6.0016	6.1010	6.1904	6.2724
15200 (343.19)	V	1.668	10.1246	12.248	13.181	13.988	15.371	16.565	17.643	18.643	19.584
	U	1593.5	2455.3	2574.4	2624.9	2668.0	2740.0	2800.7	2854.9	2904.8	2951.7
	H	1618.9	2609.2	2760.5	2825.3	2880.7	2973.7	3052.5	3123.0	3188.1	3249.4
	S	3.6981	5.3051	5.5476	5.6491	5.7345	5.8749	5.9903	6.0907	6.1807	6.2632
15400 (344.24)	V	1.678	9.9136	11.939	12.885	13.700	15.087	16.279	17.354	18.348	19.283
	U	1600.9	2450.6	2566.2	2618.3	2662.5	2735.8	2797.3	2851.9	2902.2	2949.5
	H	1626.8	2603.3	2750.0	2816.7	2873.5	2968.2	3048.0	3119.2	3184.8	3246.4
	S	3.7103	5.2922	5.5272	5.6317	5.7193	5.8622	5.9791	6.0803	6.1711	6.2541
15600 (345.28)	V	1.689	9.7072	11.635	12.595	13.418	14.810	16.001	17.071	18.060	18.989
	U	1608.3	2445.8	2557.7	2611.5	2656.8	2731.6	2793.8	2849.0	2899.6	2947.2
	H	1634.7	2597.3	2739.2	2808.0	2866.2	2962.6	3043.4	3115.3	3181.4	3243.4
	S	3.7226	5.2793	5.5063	5.6142	5.7040	5.8495	5.9678	6.0701	6.1615	6.2450
15800 (346.31)	V	1.699	9.5052	11.334	12.311	13.142	14.539	15.729	16.796	17.780	18.703
	U	1615.7	2440.9	2549.0	2604.6	2651.1	2727.2	2790.3	2846.0	2897.0	2944.9
	H	1642.6	2591.1	2728.0	2799.1	2858.7	2957.0	3038.8	3111.4	3178.0	3240.4
	S	3.7348	5.2663	5.4851	5.5964	5.6885	5.8367	5.9566	6.0598	6.1519	6.2360
16000 (347.33)	V	1.710	9.3075	11.036	12.030	12.871	14.275	15.464	16.527	17.506	18.423
	U	1623.2	2436.0	2539.9	2597.4	2645.2	2722.9	2786.8	2843.0	2894.5	2942.6
	H	1650.5	2584.9	2716.5	2789.9	2851.1	2951.3	3034.2	3107.5	3174.5	3237.4
	S	3.7471	5.2531	5.4634	5.5784	5.6729	5.8240	5.9455	6.0497	6.1425	6.2270
16200 (348.34)	V	1.722	9.1141	10.742	11.755	12.605	14.016	15.205	16.265	17.239	18.151
	U	1630.6	2430.9	2530.6	2590.1	2639.2	2718.4	2783.2	2840.0	2891.8	2940.3
	H	1658.5	2578.5	2704.6	2780.6	2843.4	2945.5	3029.5	3103.5	3171.1	3234.4
	S	3.7594	5.2399	5.4411	5.5602	5.6572	5.8112	5.9343	6.0395	6.1330	6.2181
16400 (349.33)	V	1.733	8.9247	10.450	11.483	12.344	13.763	14.952	16.008	16.978	17.885
	U	1638.1	2425.7	2520.9	2582.6	2633.1	2714.0	2779.6	2837.0	2889.2	2938.0
	H	1666.5	2572.1	2692.3	2771.0	2835.5	2939.7	3024.8	3099.5	3167.7	3231.3
	S	3.7717	5.2267	5.4183	5.5417	5.6413	5.7984	5.9232	6.0294	6.1237	6.2093
16600 (350.32)	V	1.745	8.7385	10.160	11.216	12.087	13.515	14.704	15.758	16.724	17.625
	U	1645.5	2420.4	2510.9	2574.9	2626.9	2709.4	2776.0	2833.9	2886.6	2935.7
	H	1674.5	2565.5	2679.5	2761.1	2827.5	2933.8	3020.1	3095.5	3164.2	3228.2
	S	3.7842	5.2132	5.3949	5.5228	5.6253	5.7856	5.9120	6.0194	6.1143	6.2005
16800 (351.30)	V	1.757	8.5535	9.8712	10.952	11.835	13.272	14.462	15.514	16.475	17.372
	U	1653.5	2414.9	2500.4	2567.0	2620.5	2704.8	2772.4	2830.9	2883.9	2933.3
	H	1683.0	2558.6	2666.2	2751.0	2819.4	2927.8	3015.3	3091.5	3160.7	3225.2
	S	3.7974	5.1994	5.3708	5.5037	5.6091	5.7728	5.9009	6.0093	6.1050	6.1918
17000 (352.26)	V	1.770	8.3710	9.5837	10.691	11.588	13.034	14.225	15.274	16.232	17.124
	U	1661.6	2409.3	2489.5	2558.9	2614.0	2700.2	2768.7	2827.8	2881.3	2931.0
	H	1691.7	2551.6	2652.4	2740.7	2811.0	2921.7	3010.5	3087.5	3157.2	3222.1
	S	3.8107	5.1855	5.3458	5.4842	5.5928	5.7599	5.8899	5.9993	6.0958	6.1831
17200 (353.22)	V	1.783	8.1912	9.2964	10.434	11.344	12.801	13.994	15.041	15.994	16.882
	U	1669.7	2403.5	2478.0	2550.6	2607.4	2695.4	2764.9	2824.7	2878.6	2928.6
	H	1700.4	2544.4	2637.9	2730.0	2802.5	2915.6	3005.6	3083.4	3153.7	3219.0
	S	3.8240	5.1713	5.3200	5.4644	5.5762	5.7469	5.8788	5.9894	6.0866	6.1745
17400 (354.17)	V	1.796	8.0140	9.0084	10.179	11.104	12.573	13.767	14.812	15.762	16.645
	U	1677.7	2397.6	2466.0	2542.0	2600.6	2690.7	2761.2	2821.6	2875.9	2926.3
	H	1709.0	2537.1	2622.7	2719.1	2793.8	2909.4	3000.7	3079.3	3150.2	3215.9
	S	3.8372	5.1570	5.2931	5.4442	5.5595	5.7340	5.8677	5.9794	6.0775	6.1659
17600 (355.11)	V	1.810	7.8394	8.7187	9.9272	10.868	12.348	13.545	14.588	15.535	16.414
	U	1685.8	2391.6	2453.2	2533.1	2593.7	2685.8	2757.4	2818.4	2873.2	2923.9
	H	1717.6	2529.5	2606.6	2707.8	2784.9	2903.1	2995.8	3075.2	3146.6	3212.8
	S	3.8504	5.1425	5.2649	5.4235	5.5425	5.7209	5.8566	5.9695	6.0684	6.1574
17800 (356.04)	V	1.825	7.6674	1.9676	9.6775	10.635	12.129	13.328	14.369	15.313	16.188
	U	1693.7	2385.3	1743.6	2523.9	2586.5	2680.9	2753.6	2815.2	2870.5	2921.5
	H	1726.2	2521.8	1778.6	2696.2	2775.8	2896.8	2990.8	3071.0	3143.0	3209.6
	S	3.8635	5.1278	3.9466	5.4024	5.5253	5.7078	5.8456	5.9597	6.0593	6.1489
18000 (356.96)	V	1.840	7.4977	1.9465	9.4298	10.405	11.913	13.115	14.155	15.096	15.966
	U	1701.7	2378.9	1737.4	2514.5	2579.3	2675.9	2749.7	2812.1	2867.7	2919.1
	H	1734.8	2513.9	1772.4	2684.2	2766.6	2890.3	2985.8	3066.9	3139.4	3206.5
	S	3.8765	5.1128	3.9362	5.3808	5.5079	5.6947	5.8345	5.9498	6.0502	6.1405
18200 (357.87)	V	1.856	7.3302	1.9285	9.1838	10.178	11.701	12.906	13.945	14.883	15.750
	U	1709.7	2372.3	1732.0	2504.7	2571.8	2670.8	2745.8	2808.8	2865.0	2916.7
	H	1743.4	2505.8	1767.1	2671.9	2757.1	2883.8	2980.7	3062.7	3135.8	3203.3
	S	3.8896	5.0975	3.9272	5.3587	5.4902	5.6815	5.8234	5.9400	6.0412	6.1321

TEMPERATURE, °C
(TEMPERATURE, K)

ABS. PRESS., kPa (SAT. TEMP., °C)		SAT. WATER	SAT. STEAM	500 (773.15)	520 (793.15)	540 (813.15)	560 (833.15)	580 (853.15)	600 (873.15)	625 (898.15)	650 (923.15)
15000 (342.13)	V	1.658	10.340	20.795	21.662	22.499	23.313	24.107	24.884	25.835	26.768
	U	1586.1	2459.9	2998.7	3041.8	3084.0	3125.3	3166.1	3206.5	3256.7	3306.7
	H	1611.0	2615.0	3310.6	3366.8	3421.4	3475.0	3527.7	3579.8	3644.3	3708.3
	S	3.6859	5.3178	6.3487	6.4204	6.4885	6.5535	6.6160	6.6764	6.7492	6.8195
15200 (343.19)	V	1.668	10.1246	20.479	21.339	22.169	22.975	23.761	24.530	25.472	26.394
	U	1593.5	2455.3	2996.6	3040.0	3082.3	3123.7	3164.6	3205.2	3255.5	3305.6
	H	1618.9	2609.2	3307.9	3364.4	3419.2	3473.0	3525.8	3578.0	3642.7	3706.8
	S	3.6981	5.3051	6.3399	6.4119	6.4803	6.5455	6.6082	6.6687	6.7417	6.8121
15400 (344.24)	V	1.678	9.9136	20.171	21.024	21.847	22.645	23.424	24.185	25.117	26.030
	U	1600.9	2450.6	2994.6	3038.2	3080.6	3122.2	3163.2	3203.8	3254.3	3304.5
	H	1626.8	2603.3	3305.2	3361.9	3417.0	3470.9	3523.9	3576.3	3641.1	3705.3
	S	3.7103	5.2922	6.3311	6.4035	6.4721	6.5376	6.6005	6.6612	6.7343	6.8049
15600 (345.28)	V	1.689	9.7072	19.871	20.717	21.533	22.324	23.095	23.850	24.772	25.676
	U	1608.3	2445.8	2992.6	3036.3	3078.9	3120.6	3161.8	3202.5	3253.0	3303.3
	H	1634.7	2597.3	3302.6	3359.5	3414.8	3468.9	3522.0	3574.5	3639.5	3703.8
	S	3.7226	5.2793	6.3225	6.3952	6.4641	6.5298	6.5928	6.6536	6.7270	6.7977
15800 (346.31)	V	1.699	9.5052	19.579	20.418	21.227	22.011	22.775	23.523	24.436	25.330
	U	1615.7	2440.9	2990.5	3034.5	3077.2	3119.1	3160.3	3201.1	3251.8	3302.1
	H	1642.6	2591.1	3299.9	3357.1	3412.6	3466.8	3520.2	3572.8	3637.9	3702.4
	S	3.7348	5.2663	6.3139	6.3869	6.4561	6.5220	6.5852	6.6462	6.7197	6.7905
16000 (347.33)	V	1.710	9.3075	19.293	20.126	20.928	21.706	22.463	23.204	24.108	24.994
	U	1623.2	2436.0	2988.4	3032.6	3075.5	3117.5	3158.9	3199.8	3250.5	3301.0
	H	1650.5	2584.9	3297.1	3354.6	3410.3	3464.8	3518.3	3571.0	3636.3	3700.9
	S	3.7471	5.2531	6.3054	6.3787	6.4481	6.5143	6.5777	6.6389	6.7125	6.7835
16200 (348.34)	V	1.722	9.1141	19.015	19.841	20.637	21.409	22.159	22.892	23.788	24.665
	U	1630.6	2430.9	2986.4	3030.7	3073.8	3115.9	3157.4	3198.4	3249.3	3299.8
	H	1658.5	2578.5	3294.4	3352.1	3408.1	3462.7	3516.4	3569.3	3634.6	3699.4
	S	3.7594	5.2399	6.2969	6.3706	6.4403	6.5067	6.5703	6.6316	6.7054	6.7765
16400 (349.33)	V	1.733	8.9247	18.743	19.564	20.354	21.118	21.862	22.589	23.476	24.344
	U	1638.1	2425.7	2984.3	3028.8	3072.1	3114.3	3155.9	3197.1	3248.0	3298.7
	H	1666.5	2572.1	3291.7	3349.7	3405.9	3460.7	3514.5	3567.5	3633.0	3697.9
	S	3.7717	5.2267	6.2885	6.3625	6.4325	6.4991	6.5629	6.6243	6.6983	6.7696
16600 (350.32)	V	1.745	8.7385	18.478	19.293	20.076	20.835	21.572	22.292	23.172	24.032
	U	1645.5	2420.4	2982.2	3027.0	3070.4	3112.8	3154.5	3195.7	3246.8	3297.5
	H	1674.5	2565.6	3289.0	3347.2	3403.6	3458.6	3512.6	3565.8	3631.4	3696.5
	S	3.7842	5.2132	6.2801	6.3545	6.4247	6.4916	6.5556	6.6172	6.6913	6.7628
16800 (351.30)	V	1.757	8.5535	18.219	19.028	19.806	20.558	21.289	22.003	22.875	23.726
	U	1653.5	2414.9	2980.1	3025.1	3068.6	3111.2	3153.0	3194.3	3245.5	3296.4
	H	1683.0	2558.6	3286.2	3344.7	3401.4	3456.6	3510.7	3564.0	3629.8	3695.0
	S	3.7974	5.1994	6.2718	6.3466	6.4171	6.4841	6.5483	6.6101	6.6844	6.7560
17000 (352.26)	V	1.770	8.3710	17.966	18.770	19.542	20.288	21.013	21.721	22.584	23.428
	U	1661.6	2409.3	2978.0	3023.2	3066.9	3109.6	3151.6	3193.0	3244.3	3295.2
	H	1691.7	2551.6	3283.5	3342.3	3399.1	3454.5	3508.8	3562.2	3628.2	3693.5
	S	3.8107	5.1855	6.2636	6.3387	6.4095	6.4768	6.5411	6.6031	6.6776	6.7493
17200 (353.22)	V	1.783	8.1912	17.719	18.517	19.283	20.024	20.743	21.445	22.301	23.137
	U	1669.7	2403.5	2975.9	3021.3	3065.2	3108.0	3150.1	3191.6	3243.0	3294.1
	H	1700.4	2544.4	3280.7	3339.8	3396.9	3452.4	3506.9	3560.5	3626.6	3692.0
	S	3.8240	5.1713	6.2554	6.3308	6.4019	6.4694	6.5340	6.5961	6.6708	6.7426
17400 (354.17)	V	1.796	8.0140	17.478	18.270	19.031	19.766	20.480	21.175	22.024	22.852
	U	1677.7	2397.6	2973.8	3019.4	3063.4	3106.4	3148.6	3190.3	3241.8	3292.9
	H	1709.0	2537.1	3277.9	3337.3	3394.6	3450.4	3505.0	3558.7	3625.0	3690.5
	S	3.8372	5.1570	6.2473	6.3231	6.3944	6.4622	6.5269	6.5892	6.6640	6.7360
17600 (355.11)	V	1.810	7.8394	17.241	18.029	18.785	19.514	20.222	20.912	21.753	22.574
	U	1685.8	2391.6	2971.7	3017.5	3061.7	3104.8	3147.1	3188.9	3240.5	3291.7
	H	1717.6	2529.5	3275.2	3334.8	3392.3	3448.3	3503.1	3556.9	3623.4	3689.0
	S	3.8504	5.1425	6.2392	6.3153	6.3870	6.4550	6.5199	6.5824	6.6574	6.7295
17800 (356.04)	V	1.825	7.6674	17.011	17.793	18.544	19.268	19.970	20.654	21.488	22.303
	U	1693.7	2385.3	2969.6	3015.6	3060.0	3103.2	3145.7	3187.5	3239.2	3290.6
	H	1726.2	2521.8	3272.4	3332.3	3390.0	3446.2	3501.1	3555.2	3621.7	3687.6
	S	3.8635	5.1278	6.2311	6.3076	6.3796	6.4478	6.5130	6.5756	6.6507	6.7230
18000 (356.96)	V	1.840	7.4977	16.785	17.563	18.308	19.027	19.724	20.403	21.230	22.037
	U	1701.7	2378.9	2967.4	3013.6	3058.2	3101.6	3144.2	3186.1	3238.0	3289.4
	H	1734.8	2513.9	3269.6	3329.8	3387.8	3444.1	3499.2	3553.4	3620.1	3686.1
	S	3.8765	5.1128	6.2232	6.3000	6.3722	6.4407	6.5061	6.5688	6.6442	6.7166
18200 (357.87)	V	1.856	7.3302	16.564	17.337	18.077	18.791	19.483	20.156	20.977	21.777
	U	1709.7	2372.4	2965.3	3011.7	3056.5	3100.0	3142.7	3184.8	3236.7	3288.2
	H	1743.4	2505.8	3266.8	3327.2	3385.5	3442.0	3497.3	3551.6	3618.5	3684.6
	S	3.8896	5.0975	6.2152	6.2924	6.3650	6.4337	6.4992	6.5622	6.6377	6.7103

ABS. PRESS., kPa (SAT. TEMP., °C)		SAT. WATER	SAT. STEAM	TEMPERATURE, °C (TEMPERATURE, K)							
				380 (653.15)	390 (663.15)	400 (673.15)	410 (683.15)	420 (693.15)	440 (713.15)	460 (733.15)	480 (753.15)
18400 (358.77)	V	1.872	7.1647	9.9544	10.782	11.493	12.125	12.701	13.740	14.675	15.538
	U	1717.7	2365.6	2564.2	2619.5	2665.7	2705.9	2741.9	2805.6	2862.2	2914.3
	H	1752.1	2497.4	2747.4	2817.8	2877.2	2929.0	2975.6	3058.4	3132.2	3200.2
	S	3.9028	5.0820	5.4722	5.5794	5.6682	5.7446	5.8124	5.9302	6.0323	6.1237
18600 (359.67)	V	1.889	7.0009	9.7333	10.572	11.288	11.923	12.501	13.539	14.471	15.330
	U	1725.7	2358.5	2556.4	2613.2	2660.5	2701.4	2737.9	2802.4	2859.4	2911.8
	H	1760.9	2488.8	2737.4	2809.8	2870.4	2923.2	2970.4	3054.2	3128.6	3197.0
	S	3.9160	5.0661	5.4540	5.5641	5.6548	5.7326	5.8013	5.9204	6.0233	6.1154
18800 (360.55)	V	1.907	6.8386	9.5146	10.365	11.087	11.725	12.304	13.341	14.271	15.127
	U	1733.8	2351.3	2548.3	2606.8	2655.2	2696.8	2733.9	2799.1	2856.6	2909.4
	H	1769.7	2479.8	2727.2	2801.7	2863.6	2917.3	2965.2	3049.9	3124.9	3193.8
	S	3.9294	5.0498	5.4354	5.5486	5.6413	5.7205	5.7902	5.9106	6.0144	6.1071
19000 (361.43)	V	1.926	6.6775	9.2983	10.160	10.889	11.531	12.111	13.148	14.075	14.928
	U	1742.1	2343.8	2540.1	2600.3	2649.8	2692.2	2729.9	2795.8	2853.8	2906.9
	H	1778.7	2470.6	2716.8	2793.4	2856.7	2911.3	2960.0	3045.6	3121.3	3190.6
	S	3.9429	5.0332	5.4166	5.5330	5.6278	5.7083	5.7791	5.9009	6.0056	6.0988
19200 (362.30)	V	1.946	6.5173	9.0841	9.9593	10.695	11.340	11.922	12.959	13.884	14.733
	U	1750.4	2335.9	2531.6	2593.7	2644.3	2687.6	2725.8	2792.5	2851.0	2904.5
	H	1787.8	2461.1	2706.0	2784.9	2849.7	2905.3	2954.7	3041.3	3117.6	3187.4
	S	3.9566	5.0160	5.3973	5.5172	5.6141	5.6962	5.7679	5.8912	5.9967	6.0906
19400 (363.16)	V	1.967	6.3575	8.8720	9.7608	10.504	11.152	11.736	12.773	13.696	14.542
	U	1758.9	2327.8	2522.9	2586.9	2638.8	2682.8	2721.6	2789.1	2848.2	2902.0
	H	1797.0	2451.1	2695.0	2776.3	2842.5	2899.2	2949.3	3036.9	3113.9	3184.1
	S	3.9706	4.9983	5.3777	5.5012	5.6004	5.6840	5.7568	5.8814	5.9879	6.0825
19600 (364.02)	V	1.989	6.1978	8.6617	9.5649	10.315	10.968	11.553	12.590	13.511	14.355
	U	1767.6	2319.2	2514.0	2580.0	2633.1	2678.0	2717.5	2785.8	2845.3	2899.5
	H	1806.5	2440.7	2683.7	2767.5	2835.3	2893.0	2943.9	3032.5	3110.2	3180.9
	S	3.9849	4.9801	5.3577	5.4850	5.5866	5.6717	5.7456	5.8717	5.9791	6.0743
19800 (364.86)	V	2.012	6.0377	8.4530	9.3715	10.130	10.786	11.374	12.412	13.331	14.171
	U	1776.5	2310.3	2504.7	2572.9	2627.4	2673.2	2713.2	2782.4	2842.5	2897.0
	H	1816.3	2429.8	2672.1	2758.5	2828.0	2886.8	2938.4	3028.1	3106.4	3177.6
	S	3.9996	4.9610	5.3373	5.4686	5.5726	5.6594	5.7345	5.8620	5.9704	6.0662
20000 (365.70)	V	2.037	5.8766	8.2458	9.1805	9.9470	10.608	11.197	12.236	13.154	13.991
	U	1785.7	2300.8	2495.2	2565.7	2621.6	2668.3	2709.0	2778.9	2839.6	2894.5
	H	1826.5	2418.4	2660.2	2749.3	2820.5	2880.4	2932.9	3023.7	3102.7	3174.4
	S	4.0149	4.9412	5.3165	5.4520	5.5585	5.6470	5.7232	5.8523	5.9616	6.0581
20200 (366.53)	V	2.064	5.7138	8.0399	8.9918	9.7669	10.432	11.024	12.064	12.980	13.815
	U	1795.3	2290.8	2485.4	2558.3	2615.6	2663.3	2704.7	2775.5	2836.7	2892.0
	H	1837.0	2406.2	2647.9	2740.0	2812.9	2874.0	2927.4	3019.2	3098.9	3171.1
	S	4.0308	4.9204	5.2951	5.4352	5.5444	5.6345	5.7120	5.8427	5.9529	6.0501
20400 (367.36)	V	2.093	5.5484	7.8352	8.8052	9.5894	10.260	10.854	11.895	12.809	13.642
	U	1805.4	2280.1	2475.3	2550.8	2609.6	2658.3	2700.3	2772.0	2833.8	2889.5
	H	1848.1	2393.3	2635.2	2730.4	2805.2	2867.6	2921.7	3014.7	3095.1	3167.8
	S	4.0474	4.8984	5.2733	5.4181	5.5300	5.6220	5.7007	5.8330	5.9442	6.0421
20600 (368.17)	V	2.125	5.3793	7.6269	8.6206	9.4144	10.090	10.687	11.729	12.642	13.472
	U	1816.1	2268.5	2464.7	2543.1	2603.4	2653.1	2695.9	2768.5	2830.8	2887.0
	H	1859.9	2379.3	2621.9	2720.7	2797.4	2861.0	2916.0	3010.1	3091.3	3164.5
	S	4.0651	4.8750	5.2505	5.4007	5.5156	5.6094	5.6894	5.8233	5.9355	6.0341
20800 (368.98)	V	2.161	5.2050	7.4179	8.4380	9.2417	9.9225	10.522	11.566	12.478	13.305
	U	1827.6	2256.0	2453.4	2535.2	2597.2	2648.0	2691.4	2765.0	2827.9	2884.4
	H	1872.5	2364.2	2607.7	2710.7	2789.4	2854.3	2910.3	3005.6	3087.4	3161.2
	S	4.0841	4.8498	5.2265	5.3831	5.5010	5.5967	5.6781	5.8136	5.9269	6.0261
21000 (369.78)	V	2.202	5.0234	7.2076	8.2572	9.0714	9.7577	10.360	11.405	12.316	13.142
	U	1840.0	2242.1	2441.5	2527.1	2590.8	2642.7	2686.9	2761.5	2824.9	2881.9
	H	1886.3	2347.6	2592.8	2700.5	2781.3	2847.6	2904.5	3001.0	3083.6	3157.8
	S	4.1048	4.8223	5.2015	5.3652	5.4863	5.5840	5.6667	5.8040	5.9182	6.0182
21200 (370.58)	V	2.249	4.8314	6.9956	8.0781	8.9032	9.5953	10.201	11.248	12.158	12.981
	U	1853.9	2226.5	2428.9	2518.8	2584.4	2637.4	2682.4	2757.9	2822.0	2879.3
	H	1901.5	2328.9	2577.2	2690.1	2773.1	2840.8	2898.6	2996.4	3079.7	3154.5
	S	4.1279	4.7917	5.1754	5.3471	5.4714	5.5712	5.6553	5.7943	5.9096	6.0103
21400 (371.37)	V	2.306	4.6239	6.7811	7.9007	8.7372	9.4354	10.044	11.093	12.003	12.824
	U	1869.7	2208.4	2415.6	2510.4	2577.8	2631.9	2677.8	2754.3	2819.0	2876.7
	H	1919.0	2307.4	2560.7	2679.4	2764.7	2833.9	2892.7	2991.7	3075.8	3151.1
	S	4.1543	4.7569	5.1481	5.3286	5.4563	5.5583	5.6438	5.7846	5.9010	6.0024
21600 (372.15)	V	2.379	4.3918	6.5629	7.7248	8.5732	9.2777	9.8898	10.941	11.850	12.669
	U	1888.6	2186.7	2401.6	2501.7	2571.1	2626.5	2673.1	2750.7	2815.9	2874.1
	H	1940.0	2281.5	2543.2	2668.5	2756.2	2826.9	2886.7	2987.0	3071.9	3147.8
	S	4.1861	4.7154	5.1192	5.3098	5.4411	5.5453	5.6323	5.7750	5.8924	5.9945

ABS. PRESS., kPa (SAT. TEMP., °C)		SAT. WATER	SAT. STEAM	TEMPERATURE, °C (TEMPERATURE, K)							
				500 (773.15)	520 (793.15)	540 (813.15)	560 (833.15)	580 (853.15)	600 (873.15)	625 (898.15)	650 (923.15)
18400 (358.77)	V	1.872	7.1647	16.348	17.116	17.852	18.560	19.247	19.915	20.729	21.523
	U	1717.7	2365.6	2963.2	3009.8	3054.7	3098.4	3141.2	3183.4	3235.5	3287.1
	H	1752.1	2497.4	3263.9	3324.7	3383.2	3439.9	3495.4	3549.8	3616.9	3683.1
	S	3.9028	5.0820	6.2073	6.2849	6.3577	6.4267	6.4924	6.5555	6.6312	6.7040
18600 (359.67)	V	1.889	7.0009	16.136	16.900	17.631	18.335	19.017	19.680	20.487	21.274
	U	1725.7	2358.5	2961.0	3007.8	3052.9	3096.8	3139.7	3182.0	3234.2	3285.9
	H	1760.9	2488.8	3261.1	3322.2	3380.9	3437.8	3493.4	3548.1	3615.3	3681.6
	S	3.9160	5.0661	6.1995	6.2774	6.3505	6.4197	6.4857	6.5490	6.6248	6.6977
18800 (360.55)	V	1.907	6.8386	15.929	16.689	17.415	18.114	18.791	19.449	20.250	21.030
	U	1733.8	2351.3	2958.8	3005.9	3051.2	3095.2	3138.3	3180.6	3232.9	3284.7
	H	1769.7	2479.8	3258.3	3319.6	3378.6	3435.7	3491.5	3546.3	3613.6	3680.1
	S	3.9294	5.0498	6.1916	6.2700	6.3434	6.4128	6.4790	6.5424	6.6185	6.6915
19000 (361.43)	V	1.926	6.6775	15.726	16.481	17.203	17.898	18.570	19.223	20.018	20.792
	U	1742.1	2343.8	2956.7	3003.9	3049.4	3093.6	3136.8	3179.2	3231.7	3283.6
	H	1778.7	2470.6	3255.4	3317.1	3376.3	3433.6	3489.6	3544.5	3612.0	3678.6
	S	3.9429	5.0332	6.1839	6.2626	6.3363	6.4060	6.4723	6.5360	6.6122	6.6854
19200 (362.30)	V	1.946	6.5173	15.527	16.279	16.996	17.686	18.353	19.002	19.791	20.559
	U	1750.4	2335.9	2954.5	3002.0	3047.6	3091.9	3135.3	3177.9	3230.4	3282.4
	H	1787.8	2461.1	3252.6	3314.5	3374.0	3431.5	3487.6	3542.7	3610.4	3677.1
	S	3.9566	5.0160	6.1761	6.2552	6.3292	6.3992	6.4657	6.5295	6.6059	6.6793
19400 (363.16)	V	1.967	6.3575	15.333	16.080	16.793	17.479	18.141	18.785	19.568	20.330
	U	1758.9	2327.8	2952.3	3000.0	3045.9	3090.3	3133.8	3176.5	3229.1	3281.2
	H	1797.0	2451.1	3249.7	3312.0	3371.6	3429.4	3485.7	3540.9	3608.7	3675.6
	S	3.9706	4.9983	6.1684	6.2479	6.3222	6.3924	6.4592	6.5231	6.5997	6.6732
19600 (364.02)	V	1.989	6.1978	15.142	15.885	16.594	17.275	17.934	18.573	19.350	20.106
	U	1767.6	2319.2	2950.1	2998.0	3044.1	3088.7	3132.3	3175.1	3227.8	3280.1
	H	1806.5	2440.7	3246.9	3309.4	3369.3	3427.3	3483.8	3539.1	3607.1	3674.1
	S	3.9849	4.9801	6.1608	6.2406	6.3153	6.3857	6.4527	6.5168	6.5936	6.6672
19800 (364.86)	V	2.012	6.0377	14.955	15.694	16.399	17.076	17.730	18.365	19.136	19.887
	U	1776.5	2310.3	2947.9	2996.1	3042.3	3087.1	3130.8	3173.7	3226.6	3278.9
	H	1816.3	2429.8	3244.0	3306.8	3367.0	3425.2	3481.8	3537.3	3605.5	3672.6
	S	3.9996	4.9610	6.1532	6.2334	6.3084	6.3790	6.4462	6.5105	6.5875	6.6613
20000 (365.70)	V	2.037	5.8766	14.771	15.507	16.208	16.881	17.531	18.161	18.927	19.672
	U	1785.7	2300.8	2945.7	2994.1	3040.5	3085.4	3129.3	3172.3	3225.3	3277.7
	H	1826.5	2418.4	3241.1	3304.2	3364.7	3423.0	3479.9	3535.5	3603.8	3671.1
	S	4.0149	4.9412	6.1456	6.2262	6.3015	6.3724	6.4398	6.5043	6.5814	6.6554
20200 (366.53)	V	2.064	5.7138	14.592	15.324	16.021	16.690	17.335	17.962	18.722	19.461
	U	1795.3	2290.8	2943.4	2992.1	3038.7	3083.8	3127.7	3170.9	3224.0	3276.5
	H	1837.0	2406.2	3238.2	3301.6	3362.3	3420.9	3477.9	3533.7	3602.2	3669.6
	S	4.0308	4.9204	6.1381	6.2191	6.2946	6.3658	6.4334	6.4981	6.5754	6.6495
20400 (367.36)	V	2.093	5.5484	14.415	15.144	15.838	16.502	17.144	17.766	18.521	19.254
	U	1805.4	2280.1	2941.1	2990.1	3036.9	3082.1	3126.2	3169.5	3222.7	3275.4
	H	1848.1	2393.3	3235.3	3299.0	3360.0	3418.8	3476.0	3531.9	3600.6	3668.1
	S	4.0474	4.8984	6.1305	6.2120	6.2878	6.3593	6.4271	6.4919	6.5694	6.6437
20600 (368.17)	V	2.125	5.3793	14.243	14.968	15.658	16.318	16.956	17.574	18.323	19.052
	U	1816.1	2268.5	2939.0	2988.1	3035.1	3080.5	3124.7	3168.1	3221.5	3274.2
	H	1859.9	2379.3	3232.4	3296.4	3357.6	3416.6	3474.0	3530.1	3598.9	3666.6
	S	4.0651	4.8750	6.1231	6.2049	6.2811	6.3528	6.4208	6.4858	6.5635	6.6379
20800 (368.98)	V	2.161	5.2050	14.073	14.795	15.481	16.138	16.771	17.385	18.130	18.853
	U	1827.6	2256.0	2936.7	2986.1	3033.3	3078.8	3123.2	3166.7	3220.2	3273.0
	H	1872.5	2364.2	3229.5	3293.8	3355.3	3414.5	3472.1	3528.3	3597.3	3665.1
	S	4.0841	4.8498	6.1156	6.1978	6.2744	6.3463	6.4146	6.4798	6.5576	6.6322
21000 (369.78)	V	2.202	5.0234	13.907	14.625	15.308	15.961	16.591	17.201	17.940	18.658
	U	1840.0	2242.1	2934.5	2984.1	3031.5	3077.2	3121.7	3165.3	3218.9	3271.8
	H	1886.3	2347.6	3226.5	3291.2	3352.9	3412.4	3470.1	3526.5	3595.6	3663.6
	S	4.1048	4.8223	6.1082	6.1908	6.2677	6.3399	6.4084	6.4737	6.5518	6.6265
21200 (370.58)	V	2.249	4.8314	13.743	14.459	15.138	15.787	16.413	17.019	17.754	18.467
	U	1853.9	2226.5	2932.2	2982.1	3029.6	3075.5	3120.2	3163.9	3217.6	3270.6
	H	1901.5	2328.9	3223.6	3288.6	3350.6	3410.2	3468.1	3524.7	3594.0	3662.1
	S	4.1279	4.7917	6.1008	6.1838	6.2610	6.3335	6.4022	6.4677	6.5460	6.6208
21400 (371.37)	V	2.306	4.6239	13.583	14.295	14.971	15.617	16.239	16.841	17.571	18.279
	U	1869.7	2208.4	2930.0	2980.0	3027.8	3073.9	3118.6	3162.5	3216.3	3269.5
	H	1919.0	2307.4	3220.6	3286.0	3348.2	3408.1	3466.2	3522.9	3592.3	3660.6
	S	4.1543	4.7569	6.0935	6.1769	6.2544	6.3271	6.3961	6.4618	6.5402	6.6152
21600 (372.15)	V	2.379	4.3918	13.425	14.135	14.807	15.450	16.068	16.667	17.392	18.095
	U	1888.6	2186.7	2927.7	2978.0	3026.0	3072.2	3117.1	3161.1	3215.0	3268.3
	H	1940.0	2281.5	3217.7	3283.3	3345.8	3405.9	3464.2	3521.1	3590.7	3659.1
	S	4.1861	4.7154	6.0862	6.1700	6.2478	6.3208	6.3900	6.4559	6.5345	6.6096

ABS. PRESS., kPa (SAT. TEMP., °C)		SAT. WATER	SAT. STEAM	TEMPERATURE, °C (TEMPERATURE, K)							
				380 (653.15)	390 (663.15)	400 (673.15)	410 (683.15)	420 (693.15)	440 (713.15)	460 (733.15)	480 (753.15)
21800 (372.92)	V	2.483	4.1151	6.3399	7.5503	8.4112	9.1222	9.7379	10.792	11.700	12.517
	U	1913.1	2158.3	2386.3	2492.8	2564.2	2620.9	2668.4	2747.0	2812.9	2871.5
	H	1967.2	2248.0	2524.5	2657.4	2747.6	2829.8	2880.7	2982.3	3068.0	3144.4
	S	4.2276	4.6622	5.0886	5.2906	5.4257	5.5322	5.6207	5.7653	5.8839	5.9867
22000 (373.69)	V	2.671	3.7278	6.1105	7.3771	8.2510	8.9689	9.5883	10.645	11.552	12.368
	U	1952.4	2113.6	2370.0	2483.6	2557.2	2615.2	2663.6	2743.3	2809.9	2868.9
	H	2011.1	2195.6	2504.4	2645.9	2738.8	2812.6	2874.6	2977.5	3064.0	3141.0
	S	4.2947	4.5799	5.0559	5.2711	5.4102	5.5190	5.6091	5.7556	5.8753	5.9789
22120 (374.15)	V	3.170	3.1700	5.9689	7.2738	8.1558	8.8779	9.4995	10.558	11.465	12.279
	U	2037.3	2037.3	2359.4	2478.0	2553.0	2611.8	2660.7	2741.1	2808.0	2867.3
	H	2107.4	2107.4	2491.5	2638.9	2733.4	2808.2	2870.9	2974.6	3061.6	3139.0
	S	4.4429	4.4429	5.0350	5.2593	5.4007	5.5110	5.6021	5.7498	5.8702	5.9742

ABS. PRESS., kPa (SAT. TEMP., °C)		SAT. WATER	SAT. STEAM	TEMPERATURE, °C (TEMPERATURE, K)							
				500 (773.15)	520 (793.15)	540 (813.15)	560 (833.15)	580 (853.15)	600 (873.15)	625 (898.15)	650 (923.15)
21800 (372.92)	V	2.483	4.1151	13.271	13.977	14.646	15.285	15.900	16.495	17.216	17.915
	U	1913.1	2158.3	2925.4	2976.0	3024.1	3070.5	3115.6	3159.6	3213.7	3267.1
	H	1967.2	2248.0	3214.7	3280.7	3343.4	3403.7	3462.2	3519.2	3589.0	3657.6
	S	4.2276	4.6622	6.0789	6.1631	6.2413	6.3146	6.3839	6.4500	6.5288	6.6041
22000 (373.69)	V	2.671	3.7278	13.119	13.822	14.488	15.124	15.736	16.327	17.043	17.737
	U	1952.4	2113.6	2923.1	2973.9	3022.3	3068.9	3114.0	3158.2	3212.4	3265.9
	H	2011.1	2195.6	3211.7	3278.0	3341.0	3401.6	3460.2	3517.4	3587.4	3656.1
	S	4.2947	4.5799	6.0716	6.1563	6.2347	6.3083	6.3779	6.4441	6.5231	6.5986
22120 (374.15)	V	3.170	3.1700	13.029	13.731	14.395	15.029	15.638	16.228	16.941	17.632
	U	2037.3	2037.3	2921.7	2972.7	3021.2	3067.8	3113.1	3157.4	3211.7	3265.2
	H	2107.4	2107.4	3209.9	3276.4	3339.6	3400.3	3459.0	3516.3	3586.4	3655.2
	S	4.4429	4.4429	6.0672	6.1522	6.2309	6.3046	6.3743	6.4406	6.5198	6.5953

Index

The letter *e* following a page number indicates that the entry refers to an Example, and similarly the letter *p* refers to a Problem.